T0140426

Lecture Notes in Networks and Systems

Volume 84

The series "Lecture Notes in Networks and Systems" publishes the latest developments in Networks and Systems—quickly, informally and with high quality. Original research reported in proceedings and post-proceedings represents the core of LNNS.

Volumes published in LNNS embrace all aspects and subfields of, as well as new challenges in, Networks and Systems.

The series contains proceedings and edited volumes in systems and networks, spanning the areas of Cyber-Physical Systems, Autonomous Systems, Sensor Networks, Control Systems, Energy Systems, Automotive Systems, Biological Systems, Vehicular Networking and Connected Vehicles, Aerospace Systems, Automation, Manufacturing, Smart Grids, Nonlinear Systems, Power Systems, Robotics, Social Systems, Economic Systems and other. Of particular value to both the contributors and the readership are the short publication timeframe and the world-wide distribution and exposure which enable both a wide and rapid dissemination of research output.

The series covers the theory, applications, and perspectives on the state of the art and future developments relevant to systems and networks, decision making, control, complex processes and related areas, as embedded in the fields of interdisciplinary and applied sciences, engineering, computer science, physics, economics, social, and life sciences, as well as the paradigms and methodologies behind them.

** **Indexing: The books of this series are submitted to ISI Proceedings, SCOPUS, Google Scholar and Springerlink** **

More information about this series at http://www.springer.com/series/15179

Svetlana Igorevna Ashmarina ·
Marek Vochozka ·
Valentina Vyacheslavovna Mantulenko
Editors

Digital Age: Chances, Challenges and Future

Editors
Svetlana Igorevna Ashmarina
Applied Management Department
Samara State University of Economics
Samara, Russia

Marek Vochozka
Institute of Technology and Business
Ceske Budejovice, Czech Republic

Valentina Vyacheslavovna Mantulenko
Department of Applied Management
Samara State University of Economics
Samara, Russia

ISSN 2367-3370 ISSN 2367-3389 (electronic)
Lecture Notes in Networks and Systems
ISBN 978-3-030-27014-8 ISBN 978-3-030-27015-5 (eBook)
https://doi.org/10.1007/978-3-030-27015-5

This Springer imprint is published by the registered company Springer Nature Switzerland AG
The registered company address is: Gewerbestrasse 11, 6330 Cham, Switzerland

Contents

Evolution of the World in the Era of Digitalization

Past in the Future vs. Future Without Past: Challenges of the Economic Education . . . 3
N. F. Tagirova, E. I. Sumburova, and Yu. A. Zherdeva

Transformation of the Banking System as a Way to Minimize Information Asymmetry . . . 12
O. Y. Kuzmina, M. E. Konovalova, and E. S. Chulova

Circular and Sharing Economy Practices and Their Implementation in Russian Universities . . . 19
B. A. Nikitina

Relationship Between the Economy Digitalization and the “Knowledge” Production Factor . . . 27
A. M. Mikhailov and A. A. Kopylova

Priority Directions of Digital Economy Development and Effectiveness of State Policy in the Informatization Field . . . 39
O. A. Bulavko, N. N. Belanova, and L. R. Tuktarova

Transformation of Worldview Orientations in the Digital Era: Humanism vs. Anti-, Post- and Trans-Humanism . . . 47
A. V. Guryanova and I. V. Smotrova

Digital Transformation of Education, Science and Innovations . . . 54
E. V. Pogorelova and T. B. Efimova

Personal Brand of University Teachers in the Digital Age . . . 62
V. V. Mantulenko, E. Z. Yashina, and S. I. Ashmarina

Key Priorities of Business Activities Under Economy Digitalization . . . 71
E. V. Volkodavova, A. P. Zhabin, G. I. Yakovlev, and R. I. Khansevyarov

Research of Efficiency of Tax Stimulation of Innovative Entrepreneurship . . . 80
D. V. Aleshkova, M. V. Greshnova, E. S. Smolina, and L. E. Popok

Technological Development 1.0. of the Russian Federation . . . 85
A. V. Krivtsov and L. S. Valinurova

Digital Era and Consumer Behavior on the Internet . . . 92
P. Martiskova and R. Svec

The Role of Planetary Property as a Basis for Sustainable Harmonious Economic Development . . . 101
O. E. Ryazanova and V. P. Zolotareva

Sustainable Economic Development in the Context of Digitalization: Threats and Opportunities

Stages of Innovative Production in the Traditional Factory . . . 109
R. Sh. Bikmetov, N. V. Starun, and D. V. Aleshkova

Digital Economy as a Way to Ensure Economic Growth . . . 116
E. B. Razuvaeva, N. V. Starun, and L. G. Elkina

Strategies for Obtaining Added Value in Developing Technological Innovations . . . 128
M. V. Simonova and N. V. Kozhuhova

Digitalization of Labor Regulation Management: New Forms and Content . . . 137
V. A. Schekoldin, I. V. Bogatyreva, and L. A. Ilyukhina

Digital Transformation of Tax Administration . . . 144
M. A. Nazarov, O. L. Mikhaleva, and K. S. Chernousova

Comparative Analysis of Automatized Systems for Management Processes Information Support . . . 150
Yu. A. Pertulisov, E. S. Smolina, and L. A. Vodopyanova

The Impact of Digitalization on the Economic Security Index of GDP . . . 159
O. A. Naumova, I. A. Svetkina, and T. A. Korneeva

Digital Transformation of Municipal Management Under Sustainable Development . . . 165
A. A. Sidorov, N. V. Lazareva, and N. V. Starun

Digital Technologies as a Tool for Solving Basic Industrial Problems in the Agro-Industrial Complex . . . 172
E. P. Gusakova, A. V. Shchutskaya, and E. P. Afanaseva

Digital Technologies as a Factor of Expanding the Investment Opportunities of Business Entities . . . 180
M. E. Konovalova, O. Y. Kuzmina, and S. A. Zhironkin

Participatory Budgeting in City of Prague: Boosting Citizens' Participation in Local Governance Through Digital Tools (Case Study) . . . 189
E. Velinov, S. I. Ashmarina, and A. S. Zotova

The Possibilities of a Paperless Company Concept . . . 198
P. Šuleř and V. Machová

Digitalization as a Driver of the New Economic Order and Regional Development

Regional Digital Maturity: Design and Strategies . . . 205
E. K. Chirkunova, G. A. Khmeleva, E. N. Koroleva, and M. V. Kurnikova

Information Society Development in Regions of the Russian Federation . . . 214
N. V. Kulikova, N. P. Persteneva, and T. V. Ruslanova

The Impact of the Digital Economy on the Development of the Stock Market in Russia . . . 225
A. V. Vaulin and E. V. Pogorelova

Trends in Optimizing the Formation of Consolidated Reporting in Holding Companies in the Context of Global Digitization . . . 233
V. P. Fomin and E. S. Potokina

Special Economic Zones as Instrument of Industry and Entrepreneurship Development . . . 243
A. V. Streltsov, G. I. Yakovlev, and N. V. Nikitina

The Transformation of the Customer Value of Retail Network Services Under Digitalization . . . 252
D. V. Chernova, N. S. Sharafutdinova, I. I. Nurtdinov, Y. S. Valeeva, and L. I. Kuzmina

Digitalization of the Public Sector of the Regional Economy . . . 261
T. M. Tarasova, L. V. Averina, and E. P. Pecherskaya

Assessment of Quality of Services of Public Transport in Digital Economy . . . 269
A. R. Rakhmatullina, A. N. Sivaks, and E. P. Pecherskaya

Analysis of Innovative Activity of Russian Oil and Gas Companies in the Context of World Experience . . . 281
E. S. Smolina, E. Yu. Kuzaeva, O. B. Kazakova, and N. A. Kuzminykh

Internet-Marketing in the Sphere of Higher Education 290
S. Ziyadin and A. Serikbek

Modelling of Software Producer and Customer Interaction: Nash Equilibrium 298
T. Czegledy, R. V. Fedorenko, and N. A. Zaichikova

The Readiness of the Economy for Digitalization: Basic Methodological Approaches 308
E. L. Sidorenko and Z. I. Khisamova

Influence of Digitalization on Motivation Techniques in Organizations 317
E. P. Troshina and V. V. Mantulenko

Digitalization of Business and Management: Realities and Prospects

Digital Talents: Realities and Prospects 327
E. P. Barinova, E. N. Sheremetyeva, and A. S. Zotova

Harmonization of Financial and Credit Resources of Commercial Organizations in the Digital Economy 335
E. A. Serper, O. A. Khvostenko, and M. A. Pershin

Improving Russian Agribusiness Competitiveness Within the Digital Transformation Framework 342
N. V. Molotkova, M. N. Makeeva, M. A. Blium, B. I. Gerasimov, and E. B. Gerasimova

The Process of Production Digital Transformation at the Industrial Enterprise 351
A. A. Chudaeva, I. A. Svetkina, and A. S. Zotova

Digitization of the Agricultural Sector of Economy as an Element of Innovative Development in Russia 359
O. V. Mamai, I. N. Mamai, and M. V. Kitaeva

Competitiveness of Project-Oriented Recreational Organizations in the Context of Marketing Technologies Transformation 366
E. N. Sheremetyeva, E. P. Barinova, and N. V. Mitropolskaya-Rodionova

XBRL Reporting in the Conditions of Digital Business Transformation 373
O. V. Astafeva, E. V. Astafyev, E. A. Khalikova, T. B. Leybert, and I. A. Osipova

Digital Transformation in the Management of Contemporary Organizations 382
O. V. Astafeva, E. P. Pecherskaya, T. M. Tarasova, and E. V. Korobejnikova

Digital Reality and Perspective of the Management of Educational Organizations 390
E. A. Mitrofanova, I. V. Bogatyreva, and V. V. Tarasenko

Strategic Purchasing Control of the Industrial Enterprise: Digitalization and Logistics Approach 398
I. A. Toymentseva, N. P. Karpova, and T. E. Evtodieva

Digital Transformation in Business 408
S. Ziyadin, S. Suieubayeva, and A. Utegenova

Advantages and Disadvantages of Automated Control Systems (ACS) 416
M. Vochozka, J. Horák, and T. Krulický

Using Artificial Intelligence in Company Management 422
J. Vrbka and Z. Rowland

Digital Infrastructure of the Economy: Tools, Platforms and Mechanisms

Model of Sustainable Development of Regional Ecological and Economic Systems 433
N. V. Lazareva, G. S. Rosenberg, and O. A. Sapova

Digitalization of Education as a Basis for the Competence Approach 441
E. G. Repina, O. V. Bakanach, and N. V. Proskurina

Optimization of Higher Education in Economy Digitalization 448
N. V. Speshilova, V. N. Shepel, and M. V. Kitaeva

Socio-Technical Approach to a Research of Information Economy 458
O. V. Bakanach, O. F. Chistik, M. Y. Karyshev, and N. V. Proskurina

Development of International Production Cooperative Relations in the Digital Economy 466
E. M. Pimenova, A. V. Streltsov, and G. I. Yakovlev

Development of Corporate Digital Training 473
M. V. Lovcheva, V. G. Konovalova, and M. V. Simonova

Staff Responsibility as Efficiency-Driven Factor of ERP-Systems 480
V. M. Svistunov, V. V. Lobachev, and M. V. Simonova

Evaluation of Event Marketing in IT Companies 487
D. V. Chernova, N. S. Sharafutdinova, E. N. Novikova, I. T. Nasretdinov, N. G. Xametova, and Y. S. Valeeva

Trends in the Effectiveness of Russian Logistics in the Digital Economy 494
L. A. Sosunova, S. V. Noskov, K. P. Syrova, and I. G. Bakanova

Modeling Enterprise Architecture Using Language ArchiMate 506
L. A. Opekunova, A. N. Opekunov, I. N. Kamardin, and N. V. Nikitina

Students' and Their Parents' Choice of Higher Education Institution in the Era of Digitalization 514
S. I. Ashmarina, L. G. Lebedeva, Yu. A. Tokarev, and A. M. Izmailov

Digitization, Digital Technology, and Importance of Digital Technology in Teaching 526
I. Kmecová

Institutional Transformation of the Legal Environment in the Context of Digitalization of the Economy

Subject Structure of the Offense in Artificial Intelligence (AI) and Robotics 541
A. V. Sidorova

Analysis of Legal and Economic Risks for Entrepreneurs in Digital Economy 548
F. F. Spanagel, O. A. Belozerova, and M. K. Kot

Personal Data and Digital Technologies: Problems of Legal Regulation 557
N. V. Deltsova

The State Sovereignty in Questions of Issue of Cryptocurrency 564
S. P. Bortnikov

Digitization: The Bar's Aspect 574
J. A. Dorofeeva

Using Source Code Escrow as a Way to Develop Information Technology Industry 581
M. A. Tokmakov

Responsibility in the Telemedicine Area 589
S. N. Revina

Digital Technologies in Counteracting to Extremism Among Young People 597
I. E. Milova, M. A. Yavorskiy, and N. I. Razzhivina

Cryptocurrency – Money of the Digital Economy 604
G. S. Panova

Prospects for the Legal Regulation of Central Bank Digital Currency 613
E. L. Sidorenko and A. A. Lykov

Approaches Determining the Applicable Law Using Internet Technologies in the Digital Economy 622
K. K. Taran

Stablecoin as a New Financial Instrument 630
E. L. Sidorenko

Author Index 639

Evolution of the World in the Era of Digitalization

Past in the Future vs. Future Without Past: Challenges of the Economic Education

N. F. Tagirova(✉), E. I. Sumburova, and Yu. A. Zherdeva

Samara State University of Economics, Samara, Russia
tag-nailya@yandex.ru

Abstract. The article is dedicated to the analysis of formation of the technical and economic paradigm of the industrial revolution in Russia, its influence on economic education and on creation of planned (economic) educational institutions. The theory and practice of planning is considered through the prism of changes in economic science and education. At the same time, the authors check an opportunity of application of the recurrent model of Large wave by C. Pérez and create its soviet case. The authors drew conclusions about continuity and innovations in deployment of the technical and economic paradigm of industrialization in Russia, phases of its distribution in realities of the Soviet Russia of the 1920–1930s.

Keywords: Conception of "large waves" · Recurrent model · Social project · Economic discussions of the 1920–1930s · Economic education · Planning institutions

1 Introduction

The society, which is presented by different social groups and institutes, thinks on the future, builds its prospects and forms new images of "bright future" at each new stage of historical development. It becomes particularly evident in conditions of large social and economic "shifts". Technological revolutions are among them. As a rule, science carries out functions of forecasting and scientific justification of the future, and the education system performs functions of scaling and distribution of new knowledge. State and political support and financial provision of studies are necessary conditions for these processes. Technological, knowledge and educational spheres with appropriate support determine and create new future [3]. The authors consider these processes through the prism of the concept of Carlota Pérez on the basis of materials of the Russian economic history of the 1890–1930s.

2 Methods

The key question of authors' interests is how the technical and economic paradigm of the second technological revolution in Russia was generated and how it became a new ideal, which was professed by millions of Soviet people.

S. I. Ashmarina et al. (Eds.): ISCDTE 2019, LNNS 84, pp. 3–11, 2020.
https://doi.org/10.1007/978-3-030-27015-5_1

The purpose of this research consists in determination of forms of interrelations of economic science and higher business education and conceptual features of the Soviet technical and economic paradigm of the 1920–1930s. We consider three main aspects.

1. Establishment of the Soviet economic science of the 1920–1930s in the context of the technical and economic paradigm of industrialization (or the second technological revolution).
2. Modernization of economic education as a respond to challenges of the industrial era and its technical and economic paradigm.
3. The Soviet version of this scientific concept as an option of the recurrent model of C. Pérez.

The genesis of the technological revolution is connected with the moment of formation of its scientific paradigm [13]. C. Pérez notes that it goes through two stages: genesis (20–30 years) (somewhere in the middle there comes the crucial point when resistance of the old idea is broken), and the second period - deployment of a new paradigm. Each stage in its turn has two "phases, which differ in the nature of assimilatory process": introduction – aggression – a turning point between them – synergy – maturity [12]. This process influences the micro- and macrolevel of the economy, affects the whole system of social and political regulation. The economy, technologies, and institutions are the driving forces of wavy development of the technological revolution [12]. This complex multilevel system is poorly formalized. C. Pérez presented it as a recurrent model, a returning sequence. We consider her working model check on the basis of Russian historical material (the imperial and Soviet period) in relation to the second technological revolution.

3 Discussion

The main issue of interest to the authors of the article is a place of higher business education in conditions of birth of a new technological revolution. The way in which a new technical and economic paradigm becomes common property. These processes are investigated on the basis of the Russian material of the second Large technological wave.

The second technological revolution in Russia, which started in the 1890s with S. Yu. Witte's program, moved to another phase only in two and a half decades, that is in the middle of the Large wave. Riding the crest of this Large wave, Russia was in conditions of the First World War (1914–1918), and the Great Russian Revolution (1917–1922). During this decade (1914–1924) economic, technological and institutional spheres "were developed" in the socialist plane. Economic changes were connected with strengthening of regulatory functions of the state, replacement of market orientation of the economy by planned management of the national economy. These processes took place against the background of the accruing industrial crisis [6]. Interaction of technologies, science and education determines success of the technological revolution [13]. At the same time, universities are the place of interaction of social capital and public benefit [7]. While studying the origins of the industrial theory, we rely on works, written by its contemporaries, generally in the 1890–1930s.

4 Results

4.1 Establishment of the Technical and Economic Paradigm of Industrialization in the 1920–1930s. Discussion About Economic Management

The main elements of the new economic concept stretched back to the 19th century. There can be conditionally allocated two tendencies: social and economic, technical and economic. Let us cursorily designate the key theses: the future lies in the machine industry, in concentration of the industrial capital, in the financial capital and in financial and industrial groups. The state regulation objectively increases in such conditions and gets the forms of control or coordination of certain parties of reproduction. The capitalist competition extends beyond the national boundaries and reaches the international level. The new era was called the "imperialism" of "the financial capital", but the leading place in the economy was given to the industry and industrial technologies.

The industrial future of the Soviet industry was accepted without any reservations after the Great Russian revolution of 1917–1922. The Lenin's thesis that "large-scale industry can be the only material basis for socialism" [11] became a guide for action for the Power and the Party in issues of planned development of the heavy industry. Lenin considered development of natural productive forces, electrification, scientific organization of labor and management as the most important objectives of the economic policy. The planned economy seemed to Lenin as a "uniform factory". At the same time all the functions of economic management concentrated in the government apparatus.

The main topic of the Soviet scientific economic thought of the 1920s was management methods. The most argumentative issues were about regulators of the socialist economy, the plan-market ratio, types of advance planning, economic laws, which should be taken into account (general economic, peculiar to both capitalism and socialism, or specifically socialist) [10].

The most important task of the economic practice – "how actually elimination of spontaneous market regulation will happen and how its place will be taken by the socialist plan" was solved differently [14]. And whether the political economy of socialism may exist? Evgenii Preobrazhenskiy (1886–1937) believed that production in the socialist economy is managed both by market (the value law) and planned (the law of primary accumulation) regulators. On the contrary, Aleksander Bogdanov (Malinovskii) (1873–1928) claimed that the regulator of the Soviet economy is the value law, which is modified in the law of proportional labor costs.

By the middle of the 1920s, there appeared two methodological approaches in understanding of the national economic planning in the USSR – "genetic" and "teleologic", which differed in understanding of the nature of spontaneity (genetic) or orientation (teleologic) of public processes. Both notions were introduced by one of the key workers of the State Planning Committee of the USSR - Vladimir Bazarov-Rudnev (1874–1939). In his understanding, the "genetic" approach is predictive and is based on extrapolation of the existing trends of economic development. As for the "teleologic" approach, it is directive and it leans on the priority of target prescriptions, planned tasks

[2]. In other words, whether the future is possible without the past, or it should rely on achievements of the past?

The theorists of the genetic approach – Nikolai Kondratiev (1892–1938), Vladimir Bazarov (1874–1939), Vladimir Groman (1874–1940) – defended the market mechanism of economic management, which is based on careful studying of spontaneous processes, account of conjuncture, analysis of the past experience and effectiveness of extrapolation of conclusions for the future. The followers of the teleologic approach – Gleb Krzhizhanovsky (1872–1959), Stanislav Strumilin (1877–1974), Nikolai Kowalewski (1892–1958) – supported management and reconstruction of the national production by means of target transformations. In the early thirties these scientific discussions were curtailed.

The special State commission GOELRO (The state electrification of Russia) which was created in February, 1920, determined the state industrial policy. By December, 1920 it elaborated the development plan for the power industry (GOELRO plan), and in February, 1921 it acquired the status of the State general planned commission for development of a unified nation-wide plan (Gosplan). The long-term development plan of the power industry for 10–15 years turned into a predictive tool for planning of the whole national economy of the country. By the end of the 1920s, the technology of economic planning became an enforceable directive.

K. Marx wrote about capitalism as the "steam era", Lenin pushed the idea of socialism as the "electricity era" [8]. The Lenin's conception of electrification of the country as a basis of reorganization of the whole industry relied on views of the professor of the Berlin University, the economist of the Weimar Republic, Karl Ballod (1864–1931) and the director of the Moscow higher technical college, the professor of heating engineering, the author of the project on reform in professional education, Vasiliy Grinevetsky (1871–1919). The Ballod's ideas, which were stated in his book "The State of the Future" (1898, 1919) formed the basis for "the scientific plan of the socialist reorganization of the whole national economy of Germany" and became a starting point for development of a unified economic plan of the GOELRO commission [16]. V. Grinevetsky described the nearest future of the Russian economy after the First World War in the book "Postwar Prospects of the Russian Industry" (1919). His statistical calculations, data and conclusions set the structure and the logic of "The plan for electrification of the RSFSR" [15].

It is important to note that both scientists considered economic transformations in a broad context of social changes. Accordingly, K. Ballod considered general distribution of the rationalization system of labor actions and business management (scientific management), which were developed by the American engineer Frederick W. Taylor, as means of achievement of high labor productivity. V. Grinevetsky was the first in Russia, who connected the prospects of the industry with solution of social problems. In his book he disproved the idea of technocrats that "mechanization of production destroys the need for skilled labor, it turns a worker into a supplement to an automatic machine". V. Grinevetsky believed that initially it is necessary to increase the cultural and professional level of everyone, who is engaged in production [15]. The same principles formed the basis for practice of "the cultural revolution" in the USSR.

Unification of equipment, economy and spiritual culture into one system received completed embodiment in works of Bogdanov. He saw the total "split of a man" in the human history - divergence of the highest "managerial" and the lowest "performing" labor forms, entrenched in technical and economic relations. Bogdanov developed the program of "a universal organizational science", tectology. This universal theory of organization of human knowledge was based on "coherent organizational thinking", "all-social regularity" and "comprehensive mobility of labor". It was designed to provide "integrated education" and to erase borders between knowledge of "managers" and "performers" [4]. Rejected by Lenin, the Bogdanov's theory anticipated a number of structural changes in the society of the 20th century, but influenced the Lenin's idea of "the scientific organization of labor and management" (NOT).

Consequently, the ideas of industrialization received scientific justification in works starting from the 1880–1890s. Approximately in 20–30 years they took the form of scientific concepts with accurate identification of industrialization tendencies, forms of its realization (planning, production rationalization) and social orientation. The concept was formed in conditions of serious scientific discussions. We consider social orientation of industrial transformations on the example of higher education.

4.2 Modernization of Higher Economic Education as a Respond to Challenges of the Second Technical and Economic Paradigm

Reorganization of higher business education in the USSR in the 1920–1930s was carried out in the framework of the second technological revolution concurrently with change of state, political and economic targets. The authors distinguish three transformation stages of the system of business higher education institutions of the Soviet country: 1917–1921, 1921–1929, 1930–1937.

At the first stage, the basis for the developing system was the former pre-revolutionary structure of economic educational institutions. The Soviet power kept all the commercial institutes, but submitted them to nationalization and changed their names. For example, the Petrograd Higher commercial courses of M.V. Pobedinsky and similar courses of E.V. Tarle were renamed into the Petrograd institute of national economy named for F. Engels. The Moscow commercial institute was changed to the Moscow institute of national economy of K. Marx, the Higher cooperative school attached to the Moscow peoples' university in the name of A.L. Shanyavsky became the All-Russian cooperative institute, the Kiev commercial institute – the Kiev institute of national economy, the Kharkiv commercial institute – the Kharkiv institute of national economy. New names of economic higher education institutions reflected the desirable prospect – training of specialists for the national economy in general.

At the second stage, in the period of the New Economic Policy (1921–1927), narrowly specialized higher education institutions - financial and economic, industrial and economic, national economic accounting, etc. were opened in various regions of the country. Their features were: randomness of creation, brevity of existence, unreasoned supply of teaching staff and material resources, which were often not available. At the same time, there was no clear understanding of what scientific ideas should be the basis for training.

At the third stage, search for the most appropriate model of higher economic school for the Soviet society was generally finished. New economic higher education institutions were created on the basis of economic faculties of universities. For example, in 1931 the Voronezh planned-economy institute was created on the basis of the planned-economy faculty of the Voronezh State University. However, in 1932 it was reorganized into the Voronezh institute of national economic accounting [9]. At the same time, the departmental affiliation of higher education institutions was unified, and disciplines on planning of the national economy and Marxism were introduced into curricula. The ideological control over pedagogical activity was tightened.

Over these years, the principles of massification, proletarization and free of charge education became firmly established at the Soviet higher school, including economic ones [1, 5]. In such a way, the Bolsheviks implemented political slogans about available and classless education. Receiving higher education became possible for all segments of the population (especially in the period of cancellation of educational qualification), but mostly the preference was given to representatives of the working class – workers and peasants.

In summary, the ideology of economic education, which was implemented in the Soviet Union in the 1920–1930s, was rooted in origins of the second technological revolution – the period of market industrialization in Russia (industrialization of S.Yu. Witte) in the 1890–1900s. At that time, economic qualification was an obligatory part of training experts of technical specialties (engineers), which were highly demanded by the second technological revolution. The new Bolshevik leadership of the country continued this streak, believing that experts, who plan development of the national economy, should also have good understanding of the principles of changes in physical processes. However, in the second half of the 1930s the most part of technical departments was eliminated in economic higher education institutions, because their significant amount did not correspond to the objectives of planned education. The authors believe that the essential role in reduction in the share of technical disciplines was played by inclusion of the disciplines connected with studying Marxism and the political economy of socialism in training programs.

4.3 The Soviet Version of the Recurrent Model of C. Pérez

We schematically displayed the changes in management technology, economic science and the system of economic education, which took place over the life cycle of two generations of Russians, in Table 1. As well as in the model of C. Pérez, the duration of formation of the paradigm in Russia was approximately half a century. In the Russian version, the tipping point from one phase to another can be dated back to the Great Russian revolution (1917–1922) and political changes, or to 1927–1928 when the practice of five-year planning for the whole national economy was approved in the USSR. In contrast to the recurrent model of C. Pérez, the feature of the Soviet version consists in the fact that it is almost impossible to divide the periods of aggression and synergy at the deployment stage of the industrial paradigm in the Soviet Russia. Disputes and the scale of economic discussions in the late twenties were so deep that they require special study. In reality many participants of these discussions were repressed.

Table 1. The recurrent model of the evolution of technical and economic paradigm of C. Pérez: the Soviet version of the industrial scientific paradigm

	The deployment period of first industrial theories	The formation period of the industrial paradigm		The turning point	The deployment period the industrial paradigm		The formation period of the postindustrial paradigm
phase	maturity	introduction	agression		agression - synergy	maturity	introduction
years		1890-1900s	1910s	1917-1920	1920-1930s	1940-1950s	1960-1980s
Management technologies		Intracompany management of factory machine production. Private and state sector in the industry. Market regulation of the economy	Intracompany, sectoral management (military-industrial complex), inter-sectoral state coordination (special meetings) of the First World War	Nationalization of the industry. GOELRO plan	Five-year planning for the whole national economy (from 1927/28) Attempts at implementation of the NOT system at certain enterprises	Total planning as a public project of production management, Social sphere and science	Five-year planning for the national economy
Economic science		Concepts of the national economy in the S.Yu. Witte's program	Concepts of Ballod and Grinevetsky in the Lenin's program		Economic scientific discussions (till 1930-1932). Repressions and "purge" of scientific personnel	The Soviet economic science, justification for directive planning	Trashing of the industrial model, the beginning of economic discussions 1960s
Economic education		Private commercial and technical education	Private commercial and technical education	Nationalization of higher education institutions	Strengthening of social and political orientation in training of economists	Separate training of economic and technical experts.	

5 Conclusion

Russian industrialization became the time of ideological, state and political, technical and economic reconsideration. At the same time there was the change of public ideals. The Soviet industrial policy was continuation of the idea about industrialization, which appeared at the end of the 19th century. Its features were made in the form of the concept of socialist industrialization in the course of scientific discussions, which were artificially interrupted in the first half of the 1930s. It did not give a chance "to couch seeds" of the past era (scientific and technical achievements of the 1900–1910s) in the Soviet future. Prevention of a possibility of their studying by the Soviet economic science had long-term effect and created a thinking "track" (scientific and practical). This "track" was at the beginning of the new Large technological wave in the second half of the 20th century.

The attempts of the Soviet leadership to build industrial economy without reliance on the past were successful regarding the social project. The paradigm of the socialist industrialization was actively implemented and expanded through the system of higher business education. This project can be considered successful as there was industrial growth in the country in the 1930s, despite considerable social costs.

Acknowledgments. The authors express their gratitude for support given by Russian Found for Basic Research (RFBR) and the Government of Samara Region to the project № 18-49-630008 "Experience of social design in the sphere of the higher economic education of the Samara region in the first half of the 20th century".

The main ideas of this article were discussed at the student scientific seminar on economic history. The authors are especially grateful to its participants – Vladislav Eliseev and Pavel Ivliev.

References

1. Act of Sovnarkom: Act of Sovnarkom and Central Committee of the Communist Party "About work of higher educational institutions and about the management of the higher school" 23.06.1936. In: The Higher School: Main Resolutions, Orders and Instructions. Sov. Nauka., Moscow (1957). (in Russian)
2. Bazarov, V.A.: To the issue of an economic plan. Econ. Surv. **6**, 10–15 (1924). (in Russian)
3. Bodrunov, S.D.: The Future. New Industrial Society: Reboot. Cultural Revolution, Moscow (2016). (in Russian)
4. Bogdanov, A.A.: General Labor Organization (Tectology). Knigoizdatel'stvo pisatelej v Moskve, Moscow (1917). (in Russian)
5. Decree of Sovnarkom: Decree of Sovnarkom "About some changes in composition and structure of the state educational and higher educational institutions of the Russian Republic" 01.10.1918. In: Code of Justice and Government Decrees of 1917–1918, pp. 999–1000. Administrative Office of Sovnarkom of the USSR, Moscow (1942). (in Russian)
6. Gruzinov, A.S.: Economic collapse in 1917: the consequences of a long war or the result of the revolution? In: Economic History. Annals 2016/17, pp. 201–240. Institute of Russian History, Moscow (2017). (in Russian)
7. Fuller, S.: What makes universities unique? Updating the ideal for an entrepreneurial age. Educ. Stud. Moscow **2**, 50–76 (2005). (in Russian)
8. Gvozdeckij, V.L.: GOELRO plan. Myths and reality. Sci. Life **5**, 102–109 (2001). (in Russian)
9. Karpachev, M.D.: Forgotten higher education institutions: from history of the Voronezh higher school of the 1930s. In: Proceedings of Voronezh State University, Series: Problems of Higher Education, vol. 1, pp. 208–213 (2012). (in Russian)
10. Koritskiy, E.B. (ed.): What a Plan Would Be: Discussions of the 20s: Articles and Modern Comment. Lenizdat., Leningrad (1989). (in Russian)
11. Lenin, V.I.: Report theses on tactics of RCP. In: Lenin, V.I. (ed.) Complete Set of Works, 5th edn, vol. 44, pp. 3–12. Gosudarstvennoe izdatel'stvo politicheskoj literatury, Moscow (1921). (in Russian)

12. Pérez, C.: Technological Revolutions and Financial Capital: The Dynamics of Bubbles and Golden Ages. Delo, Moscow (2011). (in Russian)
13. Pérez, C.: Technological revolutions and techno-economic paradigms. Camb. J. Econ. **34**(1), 185–202 (2009)
14. Preobrazhenskiy, E.A.: New Economics (Theory and Practice): 1922–1928 (Edited by Gorinov, M.M., Tsakunov, S.V.) Izdatel'stvo Glavarhiva Moskvy, Moscow (2008). (in Russian)
15. Shcherbakova, O.M.: VI Grinevetsky about the ways of developing the country in 1918 (on the occasion of the centenary of the book «Post-War Perspectives of Russian Industry»). Bull. Moscow Region State Univ. Ser. Hist. Polit. Sci. **2**, 115–125 (2018). (in Russian)
16. Sneps-Sneppe, M.: On the tasks of the digital economy, looking back at the works of Professor Karl Ballod. Int. J. Open Inf. Technol. **6**(4), 80–91 (2018). (in Russian)

Transformation of the Banking System as a Way to Minimize Information Asymmetry

O. Y. Kuzmina(✉), M. E. Konovalova, and E. S. Chulova

Samara State University of Economics, Samara, Russia
pisakina83@yandex.ru

Abstract. The banking market is not an exception among other markets that are subjects to cases of inefficient operation of the market mechanism. In this regard, this article is focused on addressing the issues of prevention and finding ways to deal with failures in the banking segment, most of which are related to the digitalization process. The system approach was the methodological research tool that allowed studying the problem of combating imperfections in the functioning of the banking market in a comprehensive manner, affecting not only individual problems of modernizing the banking system due to the introduction of individual IT technologies in the business processes of these structures, but also the transformation of the banking system in an ever-increasing competition with information investment platforms. The article examines the genesis of market fiasco, which made it possible to identify not only the factors contributing to the reproduction of failures, but also to see the tools with which these failures can be eliminated; The analysis of theoretical approaches to the classification of market failures was carried out, which made it possible to identify the most characteristic fiasco; A number of recommendations have been proposed for eliminating and minimizing the socioeconomic consequences of failures of the Russian banking market, taking into account the process of transformation of the banking system in modern conditions. The materials of the article are of practical value for the banking business community, offering a number of tools to improve the competitiveness of this financial segment.

Keywords: Banking system · Market failures · Information asymmetry · Transformation of the banking system · Fiasco of the banking market · Digitalization

1 Introduction

It is necessary to understand that the transformation of the banking system is a dynamic, multifactorial and multilateral process which is interrelated with the processes of other systems. Therefore, the products of the transformation of the banking system can and should be used to combat the information failures of the banking market [1, 8].

The purpose of the study is to assess the possibilities for leveling or minimizing the negative social and economic consequences of failures of the banking market in the context of the process of transformation of the banking system of Russia. To achieve

S. I. Ashmarina et al. (Eds.): ISCDTE 2019, LNNS 84, pp. 12–18, 2020.
https://doi.org/10.1007/978-3-030-27015-5_2

the goal, the study analyzed theoretical approaches to the study of market fiasco and their typologization, on the basis of which the failures most characteristic of the Russian banking market were identified and a number of tools were proposed to combat them. The analysis of the imperfections in the functioning of the banking market in Russia was carried out taking into account the ongoing transformation of the financial sector, largely due to the transition to a new informational level of economic development.

The economic literature suggests various classifications of market failures [2]. The most standard classification divides failures into three types, depending on the nature of the underlying failure phenomenon: lack of information asymmetry; imperfect competition; externalities (external effects) [3, 10]. We also add that for our study it was important to study the typology of "external effects", in particular, the technological external effect, the monetary external effect, the network effect and the overflow effect were considered by the example of the banking market [10]. A detailed analysis of various types of market failures and the factors determining their education made it possible to see the most characteristic fiasco of the banking market for the Russian economy and suggest some ways to eliminate them.

2 Methodology

In the theoretical part of the genesis of market fiasco and their classification study, we have widely used not only formal logic tools, in particular, analysis and synthesis, induction and deduction, but also the causal method, evolutionary and genetic methods. Measures to eliminate the failures characteristic of the Russian banking market are presented taking into account the use of tools of the systems approach, which makes it possible to approach the study of the problem of combating imperfections in the functioning of the banking market in an integrated way, affecting not only individual issues of modernization of the banking sector, but the transformation of the banking system as a whole.

3 Results

The first block of problems, which is the source of failures of the banking market, is associated with the necessity of owning and using objective and complete information. Let us give two examples of the asymmetry of information when making transactions in the markets of bank sales, and also consider in detail the problem of the "principal-agent" that is relevant to supervisory measures.

The asymmetry of information in the banking market for transactions of various nature (credit, investment, trust, etc.) has a bilateral nature: distorted or incomplete information can be provided by both a bank representative and a client. Asymmetry is expressed in its uneven distribution between the parties to the transaction [1, 14]. To protect themselves, banks use various methods of checking the client (scoring tests, studying credit histories, checking by the security service, underwriting, etc.). These measures in most cases help to optimize the risk of banking organizations. But the risks of the client are often expressed in ignorance of the specifics of banking operations, the bankruptcy of a commercial bank, hidden fees, poor-quality and unfair contracts. In the matter of security for the client, there are some difficulties regarding the detection of information asymmetry on the part of banking organizations. The level of completeness of information is different for ordinary citizen, specialist in the banking industry, small and large enterprises. Often, a potential client has to receive information additionally, with the help of specialists, rating agencies, specialized websites, discussion on forums, etc.

It is necessary to involve the state, in order to increase the effectiveness of the fight against information failures in the banking segment, or, in other words, solving the "bank-client" problem. To ensure the security and stability of the banking market, the state adopts laws, regulations, decrees, etc. Examples of such actions by the state are the creation of a deposit insurance system, the licensing of banking detail, and the conduct of supervisory measures.

The digital transformation of the banking system is also able to offset some of the market failures. At the present, the banking organizations' tools for obtaining "fast" and high-quality customer information are expanding, in turn, the accessibility of banking information for customers is also increasing, and potential users of banking services are able to obtain objective information about the banking market with a significantly smaller amount of transaction costs [6]. During the implementation of these measures the reverse side of the "clean market" should be taken into account, which may adversely affect competition and the development of the banking segment of the economy.

In the course of supervisory measures within the banking system of Russia, the problem of partial insolvency of the regional divisions of the Bank of Russia was revealed in terms of evaluating the activities of the territorial branches of commercial banks under their jurisdiction. At the end of 2018, a reform of centralized banking supervision began to be implemented, which is designed to solve the problem of the "principal-agent." The Central Bank of the Russian Federation acts as a principal, and its territorial divisions play the role of an agent. Agents (territorial divisions) endowed with supervisory functions and could purposefully transmit incorrect information about the results of supervisory measures to the principal (Central Bank of the Russian Federation), in particular, this was revealed after bankruptcy and large-scale inspections of commercial banks. These incidents indicate the desire of organizational units to justify their effectiveness and the necessity to have a sufficient number of commercial banks under their jurisdiction, therefore they turned a blind eye to their precarious financial situation, and this also indicates that their duties were being carried out illegally and dishonestly. In turn, the problem of the "principal-agent" prevented the implementation of effective measures for the widespread development of the country's banking system [4]. As we can see, this failure of the banking market was partially smoothed over by upgrading the functionality of the Russian banking system.

The second reason for the market failures is marked by imperfect competition, and, in our opinion, its causes are simple and understandable. First of all, larger banks can easily absorb smaller regional banks that do not stand up to competition. In such a situation, one can observe either a merger or a voluntary takeover, or another option - the yielding bank goes bankrupt, its private reorganization takes place by a large bank.

The number of alternative banks for clients decreases, the regional banking market becomes monopolized or disappears altogether.

Secondly, the reason that adversely affects the competition of the banking market is the existence of collusion between bank managers regarding the policies pursued by banks, which in practice means a sharply decreasing number of players in the market. Examples of collusion can be the pre-divided market and its segments, the target audience, the quota sales of various types of banking services. The so-called market failure in the form of an external monetary effect has a very negative effect on the domestic banking market [7]. Larger and more influential banks in their pricing policies are putting pressure on small regional commercial banks, since their cash flow and the wide range of banking services provided, the accumulated customer base can reduce the cost of services provided [11].

Thirdly, there is an "objective" bankruptcy of commercial banks, both under the influence of external factors, a targeted reduction in the number of banks during competition, and under the influence of internal causes. It is not a secret to anyone that the presence of insolvent organizations is characteristic of the banking market. Another point is the proportion of failed banks in the market and how sensitive their bankruptcy is for the economy as a whole.

Fourthly, a large-scale audit of the activities of commercial banks in Russia has led to the understanding that even large enough banks are at risk and, therefore, it is necessary to take measures to keep the country's banking market from failing.

The reorganization of the banking segment is one of the most frequently used tools in the fight against the failures of the banking market in Russia, facilitating its purification from insolvent and unscrupulous players, as well as the recovery of key participants in the banking market.

As practice has shown, at this stage new measures were required in the field of antitrust policy and regulation of competition in the market. The first changes in the banking system affected the procedure for licensing the activities of banking organizations. The application of new requirements for banking organizations allowed us to divide commercial banks into two categories: with a basic and universal license, thereby limiting the possible risks of customers and the banks themselves.

It is worth noting that by solving the problem of the failure of the banking market, expressed through the problems of imperfect competition, by licensing commercial banks and their activities, the state thereby creates a legal monopoly [5].

And we must understand that any monopoly, one way or another, leads to inevitable negative consequences. Competition in the banking market suffers because of the decline in the number of banking organizations. There is one more negative point that you should pay attention to. In modern conditions, access to the banking services market by "new banks" is quite problematic. This controversial situation leads to reflections not only about failures in the banking market, but also about the fiasco of the state itself due to the "effect of special interests".

Let us consider in more detail the mechanism for the recovery of commercial banks proposed by the Central Bank of the Russian Federation for the country's banking system.

The establishment of the Banking Sector Consolidation Fund is a tool to prevent the bankruptcy of large banking organizations. The creation of the fund has become possible for several reasons, including the growing threat of a financial crisis and an increase in the number of bank structures in bankruptcy, as well as a sanctions ban on domestic commercial banks from resorting to international lending. The sanctions mainly included large banks whose assets are largely owned by the Central Bank of the Russian Federation: PJSC Bank Otkritie Financial Corporation, PJSC Binbank, Promsvzbank PJSC, etc. The country's banking system, by proposing an effective way to rehabilitate the banking market in the form of the Fund for the Consolidation of the Banking Sector, simultaneously increased the level of risk of ineffective returns from the measures implemented. The concentration of a significant number of banking assets in the hands of the Bank of Russia and the state, which later, according to experts, will only grow, makes the Russian banking market less flexible and increasingly dependent on government decisions and actions.

The Russian banking market is under the influence of technological and network external effects. Their feature is duality, they simultaneously act as both causes and consequences of market failures, and they can also be considered as tools while minimizing the risk of failure. The negative impact of the technological external effect is manifested in the reduction in the number of participants in financial transactions. So, smart contracts, digital platforms and applications allow us to exclude a bank as an intermediary from a number of financial operations. Consequently, the income of banks, especially those based on commission operations, currency transfers, cross-border payments, etc., decreases, in the context of individual banking services, competition with representatives of the non-banking sector increases, increasing bank expenses lead to an increase in the cost of banking products. A reasonable solution is the intensive use of digital technologies in banking, which will definitely affect both the functionality of the bank and its organizational structure. The process of digitalization of the Russian banking segment has been launched. In modern conditions on the Russian banking market, the blockchain technology and Big Data have attracted great interest. Many large banks offer pilot versions of blockchain solutions on the Masterchain platform. These solutions include the messaging system (Central Bank of the Russian Federation), the exchange of powers of attorney and factoring (PJSC Sberbank of Russia), the exchange of digital letters of credit (Alfa Bank JSC), the distributed register of digital bank guarantees (PJSC VTB), the service on accounting fraudsters (Bank Otkritie FC), fast exchange of payments (Payment system "QiWi"). In our opinion, the digital transformation of the banking system is an effective measure to prevent failures of the banking market.

Do not forget that one of the features of the banking market is its close relationship with other economic segments. Given that mutual relations are bilateral in nature, the effect of the flow of capital cannot be ignored. For the banking market, this phenomenon is widespread and requires more careful study in order to strengthen its regulation and control as a set of measures to prevent the occurrence of market failures. It is well known that changes in the foreign exchange market will certainly lead to changes in indicators on the banking market, this situation is aggravated by economic and political events, which at a certain stage of development of the Russian banking system required the creation of a national payment system.

4 Discussion

As it can be seen from the examples in this study, the prevention of the occurrence and elimination of failures of the Russian banking market comes down mainly to three mechanisms that find application in modern realities: the first is strengthening government regulation in the banking services market; the second is an objective transformation of the banking system in the conditions of transition to a new stage of development – the information economy; the third is the growth of control over the banking segment by buyers of financial services [13]. Launching these mechanisms as tools to effectively combat market imperfections and their negative consequences requires much more attention to studying issues such as what interests the central bank of the Russian Federation should be guided in when pursuing its policy, how objective the economic interests and needs of the banking system are, how they relate to state interests.

It should be understood that the dynamic development of the economy, as well as other spheres of society, is accompanied by the constant emergence of market failures, the occurrence of which is determined by various reasons. The most universal and effective way of fighting the fiasco is to strengthen the role of the institution of state regulation of this segment while taking into account the objective factors of the transformation of the banking system in the context of the development of the digitalization process of the economy [9, 12].

Changing the functionality of banks in modern conditions, their search for a new profile of activity will contribute to successfully combating failures in the banking market. The process of banking transformation must be diversified, and the instruments of influence on the market mechanism of the banking system must have a wide range of applications. The main criteria for effective work to eliminate market failures in the banking segment should be long-term forecasting of processes at both micro and macro levels, quick response to external economic challenges, intensive use of modern digital solutions as tools for preventing the occurrence of failures and leveling negative manifestations in the banking market.

5 Conclusion

The study examined a detailed classification of market failures. In order to successfully achieve the goal, the most frequent fiasco of the Russian banking market was studied. Having identified and identified the main reasons leading to the insolvency of the market mechanism for the functioning of the economy, we selected those that are more relevant for Russia, that is, they are the causes of the fiasco in the domestic banking market. We also note that some phenomena and processes, initially classified as "causes of failures" and rated as negative, became less explicit when studied in detail, some could be interpreted as factors of growth and development of the banking market.

Digital transformation of the banking system of the country, undoubtedly, can act as conditions for the implementation of measures to prevent and eliminate the failures of the banking market. The work analyzed and assessed the changes observed in the domestic banking segment, including the centralization of banking supervision, the use of digital technologies, two-level licensing, reorganization with the help of FCBS, the creation of the NPS MIR, etc. But it should be understood that all these measures, transforming the banking system of Russia, are not perfect; they can both contribute to its development and, at a certain stage, become an inhibiting factor. Applying any measures, it is necessary to withstand the "golden mean", which will help to establish the effective work of the market mechanism, without prejudice of the interests of some market participants in favor of others.

References

1. Asongu, S.A., Odhiambo, N.M.: Information asymmetry, financialization, and financial access. Int. Finance **21**(3), 297–315 (2018). https://doi.org/10.1111/infi.12136
2. Borochin, P., Ghosh, C., Huang, D.: Target information asymmetry and takeover strategy: insights from a new perspective. Eur. Financ. Manag. **25**(1), 38–79 (2019). https://doi.org/10.1111/eufm.12199
3. Cook, S.: Selfie banking: IIS it a reality? Biometric Technol. Today **3**, 9–11 (2017). https://doi.org/10.1016/S0969-4765(17)30056-5
4. Danilov, Y.A.: Financial markets reform programs: success factors. World Econ. Int. Relat. **60**(10), 52–61 (2016). https://doi.org/10.20542/0131-2227-2016-60-10-52-61
5. Guo, J.: Study on situation, problem and countermeasure in the development of banking industry. Paper presented at the 2016 International Conference on Industrial Economics System and Industrial Security Engineering, IEIS 2016 – Proceeding (2016). https://doi.org/10.1109/ieis.2016.7551897
6. Guo, Y., Liang, C.: Blockchain application and outlook in the banking industry. Financ. Innov. **2**(1) (2016). https://doi.org/10.1186/s40854-016-0034-9. https://jfin-swufe.springeropen.com/articles/10.1186/s40854-016-0034-9
7. Khudyakova, L.S.: Reform of global finance in the context of sustainable development. World Econ. Int. Relat. **62**(7), 38–47 (2018). https://doi.org/10.20542/0131-2227-2018-62-7-38-47
8. Landes, X., Néron, P.-Y.: Morality and market failures: asymmetry of information. J. Soc. Philos. **49**(4), 564–588 (2018). https://doi.org/10.1111/josp.12260
9. Lee, B.S., Mauck, N.: Informed repurchases, information asymmetry and the market response to open market share repurchases. Rev. Pac. Basin Financ. Mark. Polic. **21**(3) (2018). https://doi.org/10.1142/s0219091518500212
10. Machan, T.R.: Has capitalism been invalidated? Glob. Bus. Econ. Rev. **11**(3–4), 225–233 (2009). https://doi.org/10.1504/GBER.2009.031170
11. Prates, D.M., Farhi, M.: The shadow banking system and the new phase of the money manager capitalism. J. Post Keynesian Econ. **37**(4), 568–589 (2015). https://doi.org/10.1080/01603477.2015.1049925
12. Sissoko, C.: The plight of modern markets: How universal banking undermines. Capital Mark. Econ. Notes **46**(1), 53–104 (2017). https://doi.org/10.1111/ecno.12071
13. Soloviev, V.: Fintech ecosystem in Russia. In: Proceedings of 2018 11th International Conference "Management of Large-Scale System Development", MLSD 2018 (2018). https://doi.org/10.1109/mlsd.2018.8551808
14. Zvonova, E.A., Kuznetsov, A.V.: Scenarios of world monetary and financial system development: opportunities and risks for Russia. World Econ. Int. Relat. **62**(2), 5–16 (2018). https://doi.org/10.20542/0131-2227-2018-62-2-5-16

Circular and Sharing Economy Practices and Their Implementation in Russian Universities

B. A. Nikitina(✉)

Samara State University of Economics, Samara, Russia
belanik@yandex.ru

Abstract. This article is devoted to the consideration of the impact of the digital age on modern Russian society in the environmental aspect, the basis of the study were the materials describing the environmental initiatives carried out today by students at universities across the Russia. It should be noted that the most important impact of digitalization is the access of the population, especially young people, to information on examples of best practices to resolve environmental problems abroad. It is this access that made it possible for young people to initiate projects that are more advanced than what the Russian state offers today in the field of waste management. The second important aspect of the impact of digitalization on the solution of environmental problems is the possibility of using virtual communication for the organization of circular economy and sharing economy practices that contribute to the reduction of waste. Despite the high potential of virtual platforms in the optimization of reverse logistics, modern Russian society has not yet matured to make full use of it. This becomes especially clear when comparing the best practices available in Russia, implemented by student teams, with what is proposed by different institutions of the EU. The article points out that transformation of organizational culture of universities would be more important than optimization of waste management, and it may have more significant environmental, economic and social benefits. However, to reach this level, it is necessary to develop a specific sustainable systems thinking.

Keywords: Circular/sharing economy · Waste management

1 Introduction

The level of development of environmental education in the Russian Federation has never been high, but specific scientific and educational activity was significantly higher in the 90s of the 20th century. Later the concept of Education for Sustainable Development appeared, but it had not any serious positive impact on educational landscape in Russia, where environmental aspects of life studying with minimal attention. At the same time, the development of Internet communications has brought to life the younger generation of Russians a lot of information coming from developed countries, many of them personally visited these countries and noticed serious differences in the state of the environment in general and in the field of waste management in

S. I. Ashmarina et al. (Eds.): ISCDTE 2019, LNNS 84, pp. 19–26, 2020.
https://doi.org/10.1007/978-3-030-27015-5_3

particular. The ideas of the Sharing and Circular economy, which play a very important role in reducing the volume of household waste, are little known and innovative for Russians. At the same time, it is the increase in the amount of household waste in recent years, as well as the lack of development of management of this process has led to an aggravation of the environmental crisis, which particularly affected the residents of cities.

The waste management system, mal-organized in Russia, finally came to the critical point with wide public protests. While EU develop its new Waste Framework Directive [4], working on extended producer responsibility and waste prevention activity, Russia making new system of pseudo taxation without changing of core waste treatment institutions. While developed countries create waste prevention programs using power of crowdsourcing, Russian population was put aside of the discussion about planned transformations. Conceptions of Circular and Sharing Economy hardly discussed in Russia as a way of waste prevention. Russians have a lot of spontaneous practices of resources savings, but no legislative support or state investments in such kind of projects. Civil society again is ahead of state in that field, especially an environmental movement, which started to develop and implement ideas of Circular Economy in a casual life of population and first of all – among young people, especially students.

A Circular Economy is an economic and industrial system where resources are kept in use for as long as possible and it has several layers of its execution. At first glance, the circular economy is the recycling of products and their elements, as well as the extension of the life cycle of various things that can be considered as the responsibility of industrial corporations arising from the environmental policy of the state. In this case, consumers, who are in fact the main engines of economic development, turn into passive observers, and models of environmentally friendly and socially responsible behavior will be designed and imposed on the population by corporations and the state. However, it is important to see not only business models and strategies for the transition to a circular economy, but also the development of society, social prerequisites for the transition to sustainable consequences and prerequisites, the perception of the population and the impact on society as a whole.

Even the earliest Circular Economy concept of 3R actions (recycle, reuse, reduce), which is in large measure embodied in the developed countries, still not sufficiently implemented in Russia, therefore it needs the disclosure of social mechanism of its organization. That mechanism includes material infrastructure, laws, social institutions emerged in this context and different kinds of relationships between people and those sophisticated environment. The role of legislation, local community, private household and individual culture in this process has its matter [8, 9].

After the implementation of the concept of 3-R there are concepts with a much larger number of “R”: 6-R, 8-R, 10-R with following demands: recycle, reduce, recover, reuse, rebuild, redesign, remanufacture, refurbish, repair, rethink, refuse [4]. These ideas were invented for industrial production, but not only. Some of circular economy levels demand including more and more social compliance and solidarity and human creativity in the same time [11]. These ideas are implemented in the urban design, in the way the infrastructure for everyday life of people is formed, how local communities are involved in innovative projects [13]. The development of a variety of virtual sharing platforms and the provision of access to the Internet EP are also important factors for the development of this sphere [2]. All those conditions are important for implementation of following principles of Circular Economy:

(1) Slowing resource loops: Through the design or consumption of long-life goods and product-life extension (i.e. service loops to extend a product's life, for instance through repair, remanufacturing), the utilization period of products is extended and/or intensified, resulting in a slowdown of the flow of resources.
(2) Closing resource loops: Through recycling, the loop between post-use and production is closed, resulting in a circular flow of resources.
(3) Resource efficiency or Narrowing resource flows, aimed at using fewer resources per product [7].

All these principles can be realized being both a consumer and a producer, which mean not only formation of clear understanding of a new consumer ethics, but turning citizens into designers of their own life on the scale of their own household or organization of their own community [12]. Thus, it is not enough to study consumer preferences that stimulate the circulation economy, and how they are formed in the population. It's necessary to understand how people can implement their own paths that contribute to "Slowing resource loops", "Closing resource loops", or "Narrowing resource flows". The study of job creation opportunities for the implementation of the principles of Sharing and Circular Economy have to be a special area. That's why it's important to consider in more detail what prosumerism is and how innovative human qualities can contribute to the development of the Circular Economy in all its aspects taking in consideration some of modern theories [15]. But the most important thing is to move to the part of the Circular Economy that concerns the change of goal-setting, which connected very closely with theories of moral economy, the economy of merit, their existence today at various levels (global, national, corporate, municipal, individual) and the prospects for their development.

An important area of research is the analysis of the impact of digital technologies on the individual possibilities of environmental optimization of their own life – in relation to the organization of movement, food, sanitary and hygienic standards of living conditions, their own health, etc. and accumulation of the cases of best Circular and Sharing economy practices. Sharing economy which often also referred to as the Collaborative economy usually defined as a business models meeting following criteria: interaction includes three parties – the service provider, the online platform and the customer; goods, services or resources provided on a temporary basis, they are otherwise unused and they can be offered with or without compensation (i.e. for profit or non-profit/sharing) [10].

2 Methodology

The study is based on a comparative analytical methodology, within which a double comparison is carried out. In this article we rely on the information collected as a result of the project "Green universities of Russia" [6]. First of all, the best environmentally friendly practices carried out in modern Russian universities that have joined the Association of Green Universities were compared. Secondly, these practices are compared with the some case studies from the sphere of circular economics, given on the website-catalog of the environmental friendly inventions [5]. Such a comparison makes it possible to determine the stage of comprehension, even by the most progressive representatives of modern Russian universities, of the principles of circular and Sharing economy, as well as to reveal the barriers to their development [10].

3 Results

The Association of "Green universities" of Russia, which is an All-Russian youth ecological Association of representatives of student teams of Russian universities, was established in 2017 by the all-Russian green Movement "ECA" and the Foundation for support of youth initiatives "ERA" (Association of "green" universities of Russia) [6]. Due to the lack of state incentives to implement the principles of environmental management in educational institutions, the Association "Green universities" actually took over the function of ideological, theoretical and methodological support for education for sustainable development in this area.

The Association began to gain strength by 2019, covering 33 regions and 50,000 students from 288 Russian universities as participants of the action "Ecological quest". Some of these universities participate only in the actions proposed by the Association, while others are more independent, working in several areas of environmental optimization. Thus, 58 universities that are members of the Association do not just hold one-time actions, but have permanent waste separation systems.

The movement "Green universities of Russia" has absorbed the most active in the environmental aspect of the Russian students and faculty. In this regard, the experience of universities, collected in the book "Environmental initiatives of modern Russian universities", can be considered representative of the most competent and effective actions taken by social institutions of modern Russia. Using this collection of case studies for analysis, we consider the experience of 14 universities, 9 of which implementing separate waste collection nowadays and 5 realizing alternative environmentally friendly practices.

Among the 9 universities under consideration, in four cities - Arkhangelsk, St. Petersburg, Rostov-on-Don and Kaliningrad waste separation was organized by student organizations independently, and in five cities (Astrakhan, Kazan, Moscow, Rubtsovsk, Nizhny Novgorod) it was the result of joining the initiative of the "ECA" movement. In all cases separately collected were plastic, paper, aluminum and glass waste and in case of St. Petersburg 8 more other types of waste are gathering separately. As for small universities the collection of glass and metal waste separately is unprofitable and inconvenient, since it is too slow. In 4 cases, the collection of accumulators is also carried out.

In 4 cases, University teams received small grants from the ECA Movement to organize a separate collection, but in most cases the allocation of separate fractions and their sale allows to pay for the cost of installing containers. Participants pointing out that in cases of serious legal study of the issue, it is possible to significantly reduce the payment for the waste. Despite the fact that the projects are mostly non-commercial in nature, in three cases, volunteers involved in the control of the process are paid, and in one case (in Rostov-on-Don University) municipal waste separation is carried out in the form of a commercial project that brought profit to its initiator. In all the described cases, the organization of separate waste collection became a trigger of the students environmental culture growth, stimulated other environmental actions – tree planting, bookcrossing and participation in other environmental actions.

In most cases, the organizers face problems of insufficient understanding of separate waste collection environmental importance not only by the University administrations, but by the students themselves. That problem could be overcame quickly enough, but requires perseverance and moral stability of the organizers and volunteers. The question of fire safety immediately was raised in large universities, but in presented cases mostly it was solved quickly after negotiations with the administrative stuff. Project participants often point to administrative and regulatory barriers to the implementation of their activities, as well as the importance of cooperation with recycling organizations. In some cases, activists are quite satisfied with the non-commercial nature of the interaction, but in half of the cases, the initiators seek to obtain economic benefits from the delivery of recyclables.

Another important conclusion is that great importance for the success of the projects belongs to by the presence of a cohesive team, ready to persistently implement the plan and take responsibility for voluntary duties.

As for non-standard environmentally friendly practices, these include the project on water saving (St. Petersburg), the project on book-crossing in combination with the practices of collecting waste paper (Moscow), the project on collecting clothes (Moscow), as well as projects on the exchange of folders for papers (Kazan), the production of pens and notebooks from used posters (Moscow). Of all these projects only the last one uses the terminology from the sphere of Circular Economy, rightly naming activities upcycling, means a secondary use of materials with an increase in their cost, but without processing as raw materials. At the same time, the projects of office supplies exchange and charity gathering of clothes can be attributed to practices of Collaborative/Sharing Economy.

4 Discussion

The given experience of implementing environmental innovations in universities should be considered as the diffusion of socio-cultural patterns in the field of environmental management from developed countries to Russian society. These innovations meet both support and resistance from different social subsystems. In some cases, a negative factor is the low level of understanding of the meaning of separate waste collection, which is combined with the antisocial behavior of some individual actors, as a result, the waste is poorly separated and requires additional sorting. Due to the fact that environmental education works effectively as a tool for changing behavior and ensures compliance with specific requirements for waste collection, the initiators of student projects are ready to involve wider segments of the population in their projects, bringing "eco-points" outside the University campuses. It is interesting that the technological tools are seen by the organizers of the projects less important, so the press for PET bottles they consider as an extra purchase.

Generalizing, we emphasize that in most of the above cases the principles of circular economy have been developed, but at the lowest level of this approach, when the loop is closing by involving resources in recycling. Higher levels of circular economy are less represented, only in three cases there is a method of extending the life cycle of products – for example, the collection of clothes, as well as the collection of folders for papers. Extends the life cycle of products and the use of books for their transfer to the public in the framework of bookcrossing.

In each of the cases of extension of the product life cycle, we can talk about the implementation of the principles of non-commercial type of sharing economy. As for the separate collection of waste, among the cases in a single case, an example of commercialization is presented.

It should be noted that according to some research [1], students are more likely to discuss the possibility of optimizing those aspects of the circular economy that are closer to them as a result of their professional education. It is not surprising that most students turn to the level of recycling, i.e. separate waste collection for their subsequent processing. At the same time, the organizers of the project pay great attention to interconnections with the waste management subsystems, because they have realized from their own experience how important it is to be able to deliver the collected recyclables to the place of processing. As the research emphasizes, it is "reverse logistics" that is crucial for the successful development of closed cycles [3]. At the same time, it should be recognized that the approaches of social design and consumer selectivity are poorly developed, which, as studies show, requires more active information impact on the population (so called Persuasive Communication) for the formation of Pro-Circular Values [14].

At the same time, we note the lack of examples of the design of social processes in order to change the very organization of the economy, preventing the production of waste. Also at this stage, there is no special attention to the environmental characteristics of goods consumed by internal institutions; there is no understanding of the need for greater emphasis on environmental criteria in the procurement of certain types of goods by the University.

In this regard, we point out that the hierarchy of waste management directly indicates that the primary and most effective activity is not sorting waste for their further processing, but changing the ways of organizing life, based on the analysis of alternative ways to achieve terminal goals [7, 16].

In the field of wildlife protection, there are so-called "umbrella species", which include such animals as tigers or bears. In their protection, there is a side effect when both safety and improvement of living conditions are provided to all species of animals and plants that are part of the ecosystem of the umbrella species.

A similar situation exists in the development of the circular economy. Reducing the amount of plastic entering the landfill can be achieved through its transfer to recycling, but the organizational culture of the University can lead to minimizing the appearance of plastic debris or a complete ban on its use, which will not only bring more environmental benefits, but also can create new jobs and change the understanding of the meaning of using this invention. However, at this stage, the initiators of environmental projects in Russian universities not only do not swing at the level of rethinking the organizational culture, but also probably do not think.

5 Conclusion

As a result of the analysis, the following conclusions can be drawn:

- the activity of modern youth in the field of waste management is growing, in the University communities there is a variety of forms of environmentally friendly projects implemented by student teams;
- students' attention is focused mainly on the least effective forms of optimization, such as separate waste collection, poorly changing their way of thinking
- there is a clear lack of implementation of innovative models of resource management, allowing to prevent the occurrence of the problem, and not to stop it after the fact;
- share projects implemented by students have the character of charity, which does not include the transformation of their own way of life, its greening;
- the revealed trends indicate a pronounced backwardness of modern Russian household practices from the organization of daily life in developed countries, one of the reasons for which can be considered a low level of environmental education in Russia, the lack of awareness of young people about such concepts as circular and sharing economy, and their practical socio-environmental potential.
- during the implementation of projects, digital platforms play a minor role and their function is mainly to inform students mostly about environmental activity in Russia, and much less about circular and sharing economy innovations and mode of thinking.

References

1. Andrews, D.: The circular economy, design thinking and education for sustainability. Local Econ. **30**(3), 305–315 (2015). https://doi.org/10.1177/0269094215578226
2. Taranic, I., Behrens, A., Topi, C.: Understanding the circular economy in Europe, from resource efficiency to sharing platforms: the CEPS framework (2016). https://www.ceps.eu/publications/understanding-circular-economy-europe-resource-efficiency-sharing-platforms-ceps. Accessed 7 Mar 2019
3. Cobo, S., Dominguez-Ramos, A., Irabien, A.: From linear to circular integrated waste management systems: a review of methodological approaches. Resour. Conserv. Recycl. **135**, 279–295 (2018). https://doi.org/10.1016/j.resconrec.2017.08.003
4. Directive (EU) 2018/851 of the European Parliament and of the Council of 30 May 2018 amending Directive 2008/98/EC on Waste. https://eur-lex.europa.eu/legal-content/EN/TXT/PDF/?uri=CELEX:32018L0851&rid=1. Accessed 2 Mar 2019
5. European Economic and Social Committee: Sustainable development is not a zero-sum game (2019). https://publications.europa.eu/s/lnsg. Accessed 15 Apr 2019
6. Environmental Initiatives in Russian Universities: Successful practice and a guide to action, Moscow (2019). http://erafoundation.ru/assets/Sbornik-eco-practik.pdf. Accessed 12 Apr 2019. (in Russian)
7. European Commission: Closing the Loop – An EU Action Plan for the Circular Economy; COM 614 Final; European Commission, Brussels (2015). https://ec.europa.eu/transparency/regdoc/rep/1/2015/EN/1-2015-614-EN-F1-1.PDF. Accessed 5 June 2019

8. European Commission: Towards a circular economy: a zero waste programme for Europe (2014). http://ec.europa.eu/environment/circular-economy/pdf/circular-economy-communication.pdf. Accessed 7 Apr 2019
9. European Commission: Study to monitor the economic development of the collaborative economy in the EU. Final Report. Part A: Final Report. 23 February 2018 (2018). https://publications.europa.eu/en/publication-detail/-/publication/0cc9aab6-7501-11e8-9483-01aa75ed71a1, Accessed 12 Apr 2019
10. Gullstrand Edbring, E., Lehner, M., Mont, O.: Exploring consumer attitudes to alternative models of consumption: motivations and barriers. J. Clean. Prod. **123**, 5–15 (2016). https://doi.org/10.1016/j.jclepro.2015.10.107
11. Kalmykova, Y., Sadagopan, M., Rosado, L.: Circular economy – from review of theories and practices to development of implementation tools. Resour. Conserv. Recycl. **135**, 190–201 (2017). https://doi.org/10.1016/j.resconrec.2017.10.034
12. Kalmykova, Y., Rosado, L., Patrício, J.: Resource consumption drivers and pathways to reduction: economy, policy and lifestyle impact on material flows at the national and urban scale. J. Clean. Prod. **132**, 70–80 (2016). https://doi.org/10.1016/j.jclepro.2015.02.027
13. Levchenko, N.V.: Ecological education in Russian universities: civil activism in formal education. Human. South of Russia **6**(5), 276–286 (2017). https://doi.org/10.23683/2227-8656.2017.5.24. (in Russian)
14. Muranko, Z., Andrews, D., Newton, E.J., Chaer, I., Proudman, Ph: The pro-circular change model (P-CCM): proposing a framework facilitating behavioural change towards a circular economy. Resour. Conserv. Recycl. **135**, 132–140 (2018). https://doi.org/10.1016/j.resconrec.2017.12.017
15. Ritzer, G., Jurgenson, N.: Production, consumption, prosumption: the nature of capitalism in the age of the digital 'prosumer'. J. Consum. Cult. **10**(1), 13–36 (2010). https://doi.org/10.1177/1469540509354673
16. Whalen, K.A., Berlin, C., Ekberg, J., Barletta, I., Hammersberg, P.: All they do is win': lessons learned from use of a serious game for circular economy education. Resour. Conserv. Recycl. **135**, 335–345 (2018). https://doi.org/10.1016/j.resconrec.2017.06.021

Relationship Between the Economy Digitalization and the "Knowledge" Production Factor

A. M. Mikhailov(✉) and A. A. Kopylova

Samara State University of Economics, Samara, Russia
2427994@mail.ru

Abstract. Economies of the leading countries of the world have been evolving towards digitalization, as a process of expansion and penetration of digital technologies into entire spheres of the economy of the post-industrial society during the past two decades. The article is a scientific research focused on solving the problem of grounding the digitalization of the modern economy as an objective process subject to creating the information society, knowledge-based economy and innovative economy in the post-industrial era. The purpose of the work is to study the relationship between the economy digitalization and the transformation of knowledge into a leading factor of the post-industrial production. Processes of digitalization of the economy in the post-industrial society are discussed in the article; transformations in the interaction of the post-industrial production factors are highlighted, and the separation of knowledge into an independent fifth production factor is substantiated here. The transformation of knowledge into a leading production factor is associated with the separation of intellectual activity in addition to labor and entrepreneurial activities. A relationship between functioning of the knowledge-based economy and its digitalization appears. Additional relevance of the study is to consider a status of digitalization of the Russian economy, to identify the strengths and weaknesses of the state program of the country digitalization, which should contribute to the process of the Russian economy digitalization.

Keywords: Knowledge · Information · Post-industrial society · Production factors · Digital economy · Digitalization · Knowledge economy

1 Introduction

The digitalization process is thought of as using "digitized data in organizational and public processes (including economic activities)" [3, 4]. The digitalization process is increased in the modern economy every year, as evidenced, first of all, by the development of information and communication technologies. The most developed countries of the world entered the stage of the post-industrial society, which economy can be described as an information economy, a knowledge-based economy, an innovative economy and an increasingly more digitalized economy. A modern developed economic system is impossible to be imagined without any of these elements.

S. I. Ashmarina et al. (Eds.): ISCDTE 2019, LNNS 84, pp. 27–38, 2020.
https://doi.org/10.1007/978-3-030-27015-5_4

The term “digital economy” was introduced by the Organization for Economic Co-operation and Development (OECD) in 2014. The digital revolution, often called as the 4th industrial revolution, is a transition from analog, mechanical and electronic technology to digital one. Technological and digital revolutions do not create new societies, but they change conditions in which social, political and economic relations are implemented.

The digital economy becomes a driving force of growth in recent years. There are the following indicators of the developed digital economy: a level of human capital, researches and innovations; a status of information infrastructure; and using digital technologies in public administration, business and the production process. The digital economy creates opportunities for the development of innovations, and the adoption of new technical and technological solutions. Digital transformation as integration of digital technologies into the economy results into essential changes in the strategy of doing business, communication and development at the national and international levels.

Many things depend on the state position in the process of digitalization of the economy. It should be noted that the program “Digital economy in the Russian Federation” was approved in Russia in July 2017. The program certainly defines a number of priorities, but it does not pay enough attention to e-commerce, economic mobility, the sharing economy and online platforms – those spheres that provide the basis of the country’s digital economy. The problems of the digital economy and the digitalization of the modern economy are one of the most discussed in the world economic literature in recent years [4, 5, 8, 21, 29–31].

At the same time, in our opinion, the problems of the development of the digital economy and the digitalization of the economy of the post-industrial society are not associated with transformation in the interaction of factors of the post-industrial production with the separation of knowledge into an independent, fifth and leading production factor in most of the studies. Namely, the latter determined the purpose of the study, which is to substantiate the relationship of the economy digitalization with the transformation of knowledge into a leading factor of the post-industrial production.

To achieve this purpose, we should perform the following tasks:

- to consider the processes of digitalization of the economy in the post-industrial society;
- to reveal transformations in the interaction of factors of the post-industrial production;
- to substantiate the separation of knowledge in an independent leading production factor;
- to explore the relationship between the knowledge-based economy and its digitalization;
- to show an objective background for developing the digital economy in Russia.

2 Methodology

The authors use a combination of general scientific and special methods of studying economic phenomena. Such an integrated approach includes methods of analysis and synthesis, concrete historical approach, abstract logical approach, method of terminological analysis, as well as positive and normative analysis. The use of these methods to study the relationship of digitalization of the economy with the factor of "knowledge" production made it possible to analyze the constituent elements of economy of the post-industrial society, to consider transformations in the interaction of factors of the post-industrial production due to the transformation of knowledge into a leading factor, substantiate objective reasons and the need to digitalize the economy processes.

The following materials were used for the study:

- Reports, reviews, studies and publications of the World Bank;
- OECD Digital Economy Outlook 2017 – a biennial review prepared by the OECD (Organization for Economic Cooperation and Development), which reflects new opportunities and challenges in the digital economy;
- Russia 2025. Joint research by the Boston Consulting Group (BCG), Sberbank PJSC, the Sberbank Contribution to the Future Charitable Foundation, the WorldSkills Russia Union and the Global Education Futures.

The study includes more than 90 interviews with senior management of the largest Russian employer organizations from 22 industries collectively providing jobs for more than 3.5 million people: with representatives of boards of directors and shareholders, managers and their deputies on strategic and personnel issues, HR directors, as well as with representatives of state administration bodies, the education system, small and medium businesses, start-ups, business associations, Russian and international experts in the field of human capital.

In addition to the interview, an online survey of Russian employers was conducted, which aimed to collect opinions on the company's priorities and objectives up to 2025, plans and barriers to development, changes in staff numbers and categories of employees, expectations from employees and requirements for them. In addition, the survey made it possible to form a vision of the influence of global trends and current realities on the image of the future labor market in Russia.

- The Global Innovation Index is an annual study of the level of development of innovations in countries around the world. The research was organized by INSEAD, an International Business School; Cornell University; World Intellectual Property Organization (WIPO).
- Digital Evolution Index 2017 reflects the state and rate of digital evolution and identifies the implications for investment, innovation and policy priorities.

3 Results

The formation of the economy of the post-industrial society includes a number of processes that characterize the post-industrial economy:

– firstly, this is a process of informatization, which allowed talking about the transition to the information society where information is a main economic resource and the economy becomes informational;

- secondly, the economy based on knowledge appears; knowledge becomes a main factor of the production; an economy of knowledge appears;
- thirdly, the economy becomes innovative; it is based on innovations, not only knowledge, but new knowledge in economic activity is used;
- fourthly, the digitalization process of the economy is underway, and the digital economy is being formed now.

These processes took place in the world economy in stages as the economy of the post-industrial society developed, but they can occur simultaneously. The latter is most typical for countries which are in transition from the industrial to post-industrial stage. The process of the economy digitalization is objectively due to the transformation of information into a main economic resource and knowledge into a leading factor of the production. Transformation in the interaction of factors in the post-industrial production is a consequence of this. Fundamental changes in the factors of the production take place; the process of their interaction is being transformed. And the technological method of the production of the post-industrial society, behind the development of which driving forces are information and knowledge, replaces the technological mode of the production of the industrial society.

These transformations did not occur spontaneously. People began considering information as a full-fledged economic resource in the middle of the last century, and economic development was always carried out at the expense of new knowledge and the development of new ideas, which were subsequently used in the production process. Knowledge was applied in the production process from the moment of their appearance, as far as their knowledge grew, so did their role. If we consider the genesis of economic history, knowledge has always been a part of the production process. Over time, knowledge and experience of each generation was accumulated, systematized and underwent changes. A life style, features, rules and production experience were formed based on knowledge. Empirical knowledge acquired a formal nature and became a basis for creating a regulatory and legal framework necessary for the functioning of society. Inventions and achievements of the scientific and technological revolution – all this is knowledge, thanks to which new products were created or labor productivity increased.

The transformation of the interaction of factors of the production and the deepening division of labor in the industrial society led to the separation of entrepreneurial activity as independent one, and entrepreneurship to an independent factor of the production. The division of labor in the post-industrial production leads to the separation as an independent production activity, along with labor and entrepreneurial activity, and that one for the knowledge application, related to innovations, innovations, as well as the introduction of scientific and technological revolution in the production process. In our opinion, the term "intellectual" is most suitable for defining this activity. This is an activity of applying and using knowledge, i.e. acquired and processed information. It was not separated as an independent one in the industrial society, but was carried out in the process of labor and business activity [19]. The modern production process becomes impossible without intellectual activity. An earthshaker in the interaction of factors of the production is connected with the fact that the activity on using and applying knowledge becomes determinative in the modern production process. As a result, technological relations and interrelations between factors of the production fundamentally change; the process of their interaction is transformed, and a technological method of the post-industrial production appears.

The activity on applying knowledge in the post-industrial society is not directed like in the industrial society to the improvement of other factors of the production, the improvement of their organization and interaction. It is associated with the development and increase of knowledge itself and the production of an intellectual product [17]. The modern system of social reproduction can be described as innovative reproduction [18], which is based on new scientific knowledge [16], information technologies, services and products. Innovative processes fundamentally change the role of information and knowledge in the system of the production factors. They begin to play an independent and crucial role in it [19].

A significant number of concepts characterizing knowledge as an economic category, on the one side, is undoubtedly conditioned by the importance and relevance of this resource, but on the other side, they do not allow revealing its meaning and significance. No doubt that any product of the production includes a certain amount of knowledge having a certain value. Capitalization of knowledge in the production takes place through the transfer of their value to the value of the production product. And knowledge in the modern economy is not just equivalent, but also the prevailing factor in creating value.

Understanding the fact that the production process requires the combination of all factors of the production is one of the fundamental in economic theory. Modern economic science recognizes knowledge as a limited, valuable and unique production resource, but it is still not always ready to consider it as a full-fledged production factor. Knowledge acts as a factor that integrates and creates innovative content of other production factors by enriching them with fundamentally new content. If we consider the properties of knowledge as a production factor, then the following things can undoubtedly be included:

- the ability to create added value of a product;
- limitations;
- a rarity, in connection with which the process of obtaining knowledge requires capital investments, and the cost of knowledge is high;
- knowledge is the first principle of skill; it is impossible to form the necessary professional competence of an employee to create a product without knowledge;
- the inability to be replaced by any other resource;
- the ability to generate income (as intellectual income is payment for knowledge).

This is only a part of those properties that knowledge possesses. Production factors affect its quality at the stage of manufacturing a product. Knowledge is no exception. Unlike other production factors, there is no direct relationship between investment in knowledge and its effectiveness. It is level of intelligence, education, skills and experience of a carrier of knowledge that is a pledge to the quality of knowledge. Based on the aforesaid, it follows the conclusion that knowledge is undoubtedly a production factor.

Knowledge is a result of using human and intellectual capital [12, 15, 20]. Summing up the results of 2017 year, the World Bank notes that the human capital is two-thirds of the world's wealth. Investing in people results in greater wealth and faster economic growth than investing in material production factors. Human capital, such as investing in skills, experience, knowledge and efforts of the population, is the biggest asset in the world. It accounts for more than 65% of global wealth. However, in low-income countries, human capital makes up no more than 41% of wealth – it is human capital. As the country's economy develops, the share of human capital becomes more and more significant. Among other problems, the acceleration of technological development requires that countries should urgently invest in their populations if they hope to compete in the economy of the future [1].

Using new digital technologies, the digitalization of the economy of the post-industrial society and the formation of the digital economy is the inevitable consequence of the processes of the economy intellectualization, the transformation of knowledge into a leading production factor and the development of the human capital.

According to rough estimate, a share of the digital economy in the world economy is about 5% of world GDP and it covers 3% of the world labor market. The digital economy is unevenly distributed on a worldwide scale – most of the digital economy is concentrated in the countries of the global North, but the countries of the global South demonstrate the most significant growth rates. However, potential growth may be even higher, so there is a need for further research on the current limitations and long-term impact of the digital economy [4]. The rate of the transformation processes that occur in the global economy is very high: many of urgent questions in 2016 are currently implemented and fully applied in business processes.

The value terms of the high-tech sector, which is especially in demand in the context of ubiquitous digitalization, exceeds the cost of traditional industrial production. The price of companies is influenced by the value of owning intangible assets – intellectual property and intellectual capital. Currently, the cost of raw materials, other materials and labor costs affect the economic result to a lesser extent. The development and creation of a beneficial environment for accumulating the experience and knowledge of employees, attracting highly qualified specialists, as well as maintaining and encouraging innovative activity are key factors for the development of a modern enterprise. This allows not only achieving an increase in production efficiency, but to create new knowledge and embody it into unique products. Companies with a high level of organizational and human capital that intensively use knowledge are more competitive now.

At the same time, the process of understanding the patterns of functioning of the digital economy is just beginning. It is possible that the countries with developed economies are still at the "installation stage" by concentrating on finding new ways of doing business in violation of the established practice. Successful digitalization requires government programs to manage the transition to the digital economy, which should maximally adapt existing institutions to this transition.

Understanding the need to move to the innovation-based development has been existing in Russia for a long time. But, despite the adoption of various programs, including at the state level and the implementation of large-scale projects, the production in the country remains at the industrial stage even at the level of the largest corporations in the fuel and energy complex. In our opinion, the lack of sufficient interest in the development of an innovative economy is one of the reasons for this.

A significant leap can occur if interests of main owners of production factors and, first of all, the owners of knowledge, are implemented in the process of innovation development [16].

The digital economy plays a key role in the new economic realities of the knowledge economy. At the same time, "Digital Economy in the Russian Federation", a Russian program, is focused, first of all, not on the acquisition of knowledge, but on the acquisition of skills and groups of skills or competences. The development of digital skills of all members of society provided for by the state program is certainly necessary, but rather as a measure of social support of the population than as a priority direction of the development. The knowledge society is based on unique products of intellectual labor – namely, these intangible assets must be highlighted as a priority-oriented. The main incomes of the developed countries have been already formed due to the formation of intellectual capital.

Russia needs to focus on foreign experience in development and knowledge management: the creation of national hubs, free economic zones to attract specialists and the intellectual elites. Digital platforms, new technologies, etc. – all of these are products created by knowledgeable professionals. Now, namely digital flows, but not trade in traditional goods, determine global GDP growth.

Turning to assessing the objective development level of information and communication technologies in the Russian Federation, let us turn to the Global Innovation Index. In 2018, Russia ranked 46th out of 126 countries participating in the study. The index analyzes such indicators as human capital, research and developments, infrastructure and market potential. It should be noted that the result of innovation activity is evaluated not only from the point of view of specific technological innovations, but also takes into account the products of creative activity.

The next analytical indicator, Digital Evolution Index 2017, reflects both the status and speed of digital evolution and determines the implications for investment, innovation, and policy priorities. In 2017, Russia ranked 39th out of 60 countries analyzed. Components of the index refer the country to the category of "promising", while China is a leader. The Chinese experience in building the digital economy is very important for Russia. China competes with Western mobile technologies to a large extent due to a combination of economic growth, huge investments in 4G infrastructure and a competitive mobile phone market formed by such companies as Xiaomi, Oppo, Huawei and Vivo. Having overcome the rest of the country in the field of mobile payments, online dating and lending through mobile services, China will soon become a leader in the export of mobile technologies. The economy of the PRC, the second largest after the USA, will be the largest by many estimates by mid-century [7, 9, 10, 28].

China's unique growing progress in the digital economy is fueled by such Internet giants as Alibaba, Baidu and Tencent, which form fundamentally new business models in the international market. These three companies together have 500–900 million active monthly users in their respective sectors.

In terms of the attractiveness of the labor market for talents, Russia lags behind not only developed, but also many developing countries continuing to lose talents. This is largely due to the fact that the Russian economy continues to be primarily orientated on raw materials and focused on the export of natural resources. The demand for labor in general remains primitive, and the state dominates in the structure of employers. The model of "social employment" is encouraged, i.e. inefficient jobs are preserved even in conditions of GDP reducing. A share of small and medium-sized businesses is growing very slowly (16%) in the country; the digital economy is stagnating (2–2.5%), and the venture capital market is represented in its infancy (it is hundreds of times smaller than in the USA, 12 times smaller than in Israel and 6 times smaller than in Japan) [26].

Such an approach to the formation of the labor market threatens in the near future that with onset of the "second era of machines", more than a million professions can be automated in connection with using innovative technologies. As noted earlier, there is a problem of preserving inefficient jobs to maintain a relatively low unemployment rate in Russia; therefore, the Russian economy will have to face the dilemma of prioritization in the near future: either develop and implement digital technologies in the production, or maintain the current number of jobs. Automation of production, artificial intelligence, etc. – all this will help develop knowledge and intellectual potential of the economy, and eliminate routine work, but it can lead to increased social inequality.

Plans to digitize Russian economy are impossible to be implemented without a systematic approach to the development of human capital including both attracting and retaining the best minds and providing conditions for the growth of progressive employing companies. To create new highly skilled jobs that are inherent in a technological, diversified and creative economy – knowledge economy – is one of the most important tasks of these companies [26].

Several innovative companies, such as Yandex, Kaspersky Lab, M2M Telematics, Optogan, NPO Saturn, and even state-owned Sberbank, were successfully created in Russia. In addition, individual entrepreneurs have considerable innovative potential. However, these are only islands of innovation. Creating an innovative potential for productivity growth requires investing in capital to conduct researches, fully disclose creative human resources of Russia and strengthen the institutional environment in support of innovations including the protection of intellectual property. Only in this case, a path to innovation and increasing productivity will begin.

The relevance of the formation of the Russian digital economy is confirmed by the presence of global trends in the digitalization of international society and the need to preserve the digital sovereignty of the country. The Russian economic model, with difficulty moving from raw material to post-industrial ones, turned out to be at the crossroads. While the entire world turned to the development of the non-production sphere and the transition to digital markets, Russia cannot orient itself in its priorities. On the one side, attention is paid to the development of digital services and markets, but on the other side, there is virtually no development in the field of knowledge and no reception of real GDP growth from this. Production and innovation infrastructure do not work together to create and use innovations. Since infrastructure is still being developed, Adam Smith's invisible hand will not work because of the lack of many markets [21].

The Russian economy is historically difficult to change itself. And changes are inevitable in the modern world. Over-dependence on oil issues challenge to the long-term sustainability of the economy. To create a competitive advantage, the main task of the Russian economy is to focus on the intellectualization of social and economic processes, active innovation, as well as the intensive development of intellectual human potential and digital technologies

4 Discussion

The global community monitors the state of the digital economy every year. Digital technologies potentially may increase the effectiveness in all sectors, while providing companies the possibility to increase their income and market share, as well as promoting continuous innovations. The Organization for Economic Cooperation and Development (OECD) publishes the official report that considers and documents the evolution, new features and problems of the digital economy. By the opinion of OECD specialists, the innovations managed by data, new business models and digital applications influence the current state of the science, governments, cities, and such industries and healthcare and agriculture. Political measures for support of digital innovations are mostly focused on innovative networks, access to finances and using data, but pay less attention to investments in information-communication technologies (ICT), the capital based on knowledge, and the analysis of data [23].

Recently, ICT draw the attention of economists as one of the sources of economic growth. ICT favorably influence the economic growth through two mechanisms: on the one hand, they operate as a production technology that directly promotes the growth of production due to increase in labor efficiency; on the other hand, they have specific features of knowledge and indirectly increase the productivity and the economic growth [27].

The digital economy includes a wide variety of types of economic activity, where using digital information and knowledge is the key factor. Contemporary information networks become an important sphere of activity, and the effective use of ICT acts as an important motive force in increasing the efficiency and optimizing the structure of the economy. Owing to digitalized, network and intellectual ICT, the today's economic activity becomes even more flexible, dynamic and elaborate [8].

Meanwhile, the professor Van Ark (2016) from the Groningen University in his work "The paradox of the efficiency of the new digital economy" notes that only a limited number of companies in the USA, the Great Britain and Germany have completely passed on to the digital economy [31].

As Poloz, Bank of Canada Governor, notes, "In terms of economic models, it is worth considering whether the relationship between inflation and economic growth could change as the economy evolves [24]. Certainly, the concept of an output gap is gradually changing, as services capacity depends mainly on people and skills rather than industrial capacity, while some parts of our old industrial capacity could become redundant in the face of major structural changes. The concept of investment is shifting away from plants and machinery toward human capital.

The Professor of Economy of the Stanford University Bloom, the representative of the Harvard Business School Sadun, the Professor of Management and Economy of the Massachusetts Institute of Technology Van Reenen found that companies that use high-quality managerial and organizational practice and a qualified human force, or have the access to it ("talented human resources"), usually obtain more profit from their investments in ICT [2].

In the "Post-Capitalist Society" Drucker states: the knowledge is the only valuable resource as of today. Traditional "production factors" (i.e. natural resources), the labor and the capital had not disappeared, but became secondary [6]. They can be easily obtained if one has the knowledge. In this new sense the knowledge is the utility, and the means to obtaining social and economic results.

Turning the knowledge into the main production factor alters the structure and transforms the interaction between the factors of production [17]. Working in the sphere of knowledge, which includes a complex identification of a problem, solving problems or a high-technological design, and which results in emerging of new products and services or creation of new means of using markets, had soon become the center of economic growth, individual and organizational prosperity [22].

While the digital economy is being formed, the need arises to effectively manage knowledge that is characterized by the highest level of development of the creative potential of the human person as the bearer and the generator of the knowledge. Within the economy of knowledge, the human capital is the most precious resource. The share of intellectual labor increases, and the labor content obtains the intellectual and mental significance [12].

Economical and political realities infuse with technological innovations and lead to a rapid development of the digital economy; meanwhile, such increase is most seen in the developing countries. The strategy for development of the digital economy should be designed by the private businesses; the role of the government is to forward this increase; the role of the civil and academic society is to analyze the growth of the digital economy [4].

Discussing innovations, the digital economy, and the increase of labor effectiveness and competitiveness on the basis of digitalization are paid top-level attention in the Russian economical literature. Changes historically never came easy to the Russian economy; nevertheless, changes are inevitable in the today's world. An excessive dependence on the oil poses a challenge for long-term stability of the economy. To create a competitive advantage, the main goal for the Russian economy is to fix on intellectualization of social-economic processes, the active innovative activity, as well as the intensive development of the intellectual human potential and digital technologies [11, 13, 14, 25].

5 Conclusion

The new global economy includes an information economy, a knowledge economy, an innovation economy and a digital economy. The knowledge economy is a fundamental basis of the digital economy that appeared in the last decade as a separate industry.

Knowledge becomes an independent fifth factor of the post-industrial production because the modern production process becomes impossible without intellectual activity. The earthshaker in the interaction of production factors is connected with the fact that the activity on the use and application of knowledge – intellectual activity – is essential to the production process. As a result, technological relations and interrelations between production factors fundamentally change, and the technological way of their interaction changes as well.

Information becomes a driving force of the production, and intellectual activity becomes a leading type of the production activity in the economy of the post-industrial society. These processes require overall digitalization of the economy, and they are impossible without the formation of digital economy.

Russia has sufficient resources and advanced IT industry to achieve a competitive advantage when using digital platforms for processing big data masses using block-chain technologies. At the same time, the Russian innovation system is disorganized and unbalanced; its key elements exist separately from each other. The industrial economy of Russia needs a transition to the economy, in which knowledge and information are key factors. The program for the development of digital economy in Russia is based on the basic principles of the global development of digitalization, but does not create objective prerequisites for the general digitalization of the Russian economy.

The growth of the digital economy in Russia can be considered as another step for change that will play a leading role in subsequent events, such as creating an information society, a network society and a knowledge-based society. Russia needs not just a transition to a digital economy, but a transition to a convergence of economic innovations, in which knowledge will become competitive advantage of the country in the international market.

References

1. Barne, D., Khokhar, T.: Year in Review: 2017 in 12 Charts. World Bank (2017). https://www.worldbank.org/en/news/feature/2017/12/15/year-in-review-2017-in-12-charts
2. Bloom, N., Sadun, R., Van Reenen, J.: Americans do IT better: US multinationals and the productivity miracle. Am. Econ. Rev. **102**(1), 167–201 (2012). https://doi.org/10.1257/aer.102.1.167
3. Brennen, S., Kreiss, D.: Digitalization and digitization. Culture Digitally (2014). http://culturedigitally.org/2014/09/digitalization-and-digitization/
4. Bukht, R., Heeks, R.: Defining, conceptualising and measuring the digital economy. Int. Org. Res. J. **13**(2), 143–172 (2018). https://doi.org/10.17323/1996-7845-2018-02-07
5. Dahlman, C., Mealy, S., Wermelinger, M.: Harnessing the digital economy for developing countries. In: OECD Development Centre Working Papers, vol. 334, OECD Publishing (2016)
6. Drucker, P.F.: Post Capitalist Society. Harper Business, New York (1993)
7. Franklin, D., Andrews, J. (eds.): MegaChange: The World in 2050 (2012). https://www.theguardian.com/books/2012/mar/21/megachange-2050-franklin-andrews-review
8. Gnezdova, J.V.: Development of digital economy in Russia as a factor of global competitiveness increase. In: Intelligence. Innovations. Investments, vol. 5, pp. 16–19. Orenburg State University, Orenburg (2017)
9. Hawksworth, J., Chan, D.: World in 2050: The BRICs and Beyond – Prospects, Challenges, and Opportunities (2013). http://globaltrends.thedialogue.org/publication/world-in-2050-the-brics-and-beyond-prospects-challenges-and-opportunities/
10. Hu, A.: China in 2020: A New Type of Superpower. Brookings Institution Press, Washington D.C. (2011)
11. Kapranova, L.D.: The digital economy in Russia: its state and prospects of development. Econ. Taxes Law **11**(2), 58–69 (2018)
12. Kaufman, N.Yu., Shirinkina, E.V.: Features of formation of managerial innovations in the development of human capital. Fundam. Res. **1**, 169–172 (2017)

13. Komarov, A.V., Borisova, E.S., Kuzbenova, E.R.: Innovations as a tool for effective development of the digital economy. In: Digital Economy and Modern Society, vol. 1, pp. 71–75. Science Center "Science Plus", Moscow (2018)
14. Konovalova, M.E.: Institutional and Technological Conditions of Structural Balance of Social Reproduction in the Transformational Economy of Russia (Monograph). Center for the Development of Scientific Cooperation, Novosibirsk (2009)
15. Mikhailov, A.M.: Role of human capital in realization of economic and institutional interests of factors of production owners. Bull. Saratov State Agrarian Univ. named after N.I.Vavilov **11**, 72–76 (2009). Saratov State Agrarian University in honor of N.I. Vavilov, Saratov
16. Mikhailov, A.M.: The realization of economic and institutional interests in the process of innovational economy formation in Russia. Econ. Sci. **8**(105), 39–43 (2013). Economic Sciences, Moscow
17. Mikhailov, A.M.: The evolution of production factors interaction under the shifts of technological modes of production. Econ. Sci. **2**(135), 19–22 (2016). Economic Sciences, Moscow
18. Mikhailov, A.M., Karova, E.A.: The key transformation aspects of the industrial production mode into the innovational. Vestnik Samara State Univ. Econ. (3), 54–60 (2014). Samara State University of Economics, Samara
19. Mikhailov, A.M., Mikhailov, M.V.: Information and knowledge in the system of the postindustrial production factors. Econ. Sci. **68**(7), 49–56 (2010). Economic Sciences, Moscow
20. Mikhailov, A.M., Pronina, E.U.: Economic nature of intellectual capital and its interconnection with human capital assets. Bull. Samara State Univ. Econ. **5**, 85–89 (2013). Samara State University of Economics, Samara
21. Nalebuff, B.J., Brandenburger, A.M.: Co-opetition: competitive and cooperative business strategies for the digital economy. Strateg. Leadersh. **25**(6), 28–33 (1997)
22. Neef, D. (ed.): The Knowledge Economy. Butterworth-Heinemann, Boston (1998)
23. OECD iLibrary: OECD Digital Economy Outlook 2017 (2017). https://doi.org/10.1787/9789264276284-en
24. Poloz, S.S.: From hewers of wood to hewers of code: Canada's expanding service economy (2016). https://www.bis.org/review/r161129c.pdf
25. Prazdnov, G.S.: Innovation as a factor of acceleration and increase of efficiency of industrial enterprises functioning. Bus. Educ. Law **4**, 60–64 (2017). Volgograd Institute of Business, Volgograd
26. Russia 2025: Resetting the Talent Balance. Joint Research by the Boston Consulting Group (BCG), Sberbank PJSC, the Sberbank Contribution to the Future Charitable Foundation, the WorldSkills Russia Union and the Global Education Futures (2017). https://worldskills.ru/media-czentr/dokladyi-i-issledovaniya.html
27. Shahabadi, A., Kimiaei, F., Afzali, M.A.: The evaluation of impacts of knowledge-based economy factors on the improvement of total factor productivity (a comparative study of emerging and G7 economies). J. Knowl. Econ. **9**(3), 896–907 (2018)
28. Stephens, P.: China has thrown down a gauntlet to America (2013). http://collapsechina.blogspot.com/2013/11/china-has-thrown-down-gauntlet-to.html
29. Teece, D.J.: Profiting from innovation in the digital economy: enabling technologies, standards, and licensing models in the wireless world. Res. Policy **47**(8), 1367–1387 (2017)
30. Yudina, T.N., Tushkanov, I.M.: Digital economy through the prism of philosophy of economy and political economy. Philos. Econ. **1**, 193–200 (2017)
31. Van Ark, B.: The productivity paradox of the new digital economy. Int. Prod. Monit. **3**(31), 1–15 (2016)

Priority Directions of Digital Economy Development and Effectiveness of State Policy in the Informatization Field

O. A. Bulavko, N. N. Belanova(✉), and L. R. Tuktarova

Samara State University of Economics, Samara, Russia
Belanova.nata@yandex.ru

Abstract. In the given research the priority directions of the development of digital economy are considered as an integratively distributed technology and a digital platform for the transformation and development of structures aimed at the effectiveness of state policy in the field of informatization. As part of this study, the priority directions of the development of digital economy are considered as an integratively distributed technology and a digital platform for the transformation and development of structures with the aim of the effectiveness of government policy in the field of informatization. A critical assessment is made of the features of new technologies and the consequences of their introduction for organizations and enterprises of the real sector of the economy are considered. Based on the assessment, the authors substantiate the approach to the study of actual problems in accordance with the needs and goals of society within the framework of the transformational institutional paradigm caused by the transition to digital economy.

Keywords: Target indicators · National state program · Costs of digital economy

1 Introduction

According to some authors, the result of the socio-economic revolution taking place in the modern world is the construction of a post-industrial society in which information technologies, computerized systems, high production and innovative technologies play an important role [10]. According to Semenov "The spreading of digital technologies gives a reason to talk about the formation of new socio-economic relations, about the digital economy" [14]. Currently global economy does not expect a global breakthrough of technological innovations. However, it is impossible to deny the perspective transformations associated with the industrial revolution "Industry 4.0" which is based on the principles of digitizing of vertical processes within an enterprise and horizontal links between companies - manufacturers, customers, intermediaries, partners and other counterparties. Nowadays you can find a lot of research of Russian and foreign authors on technological breakthrough, solving of technological problems proposed for the reorganization of transaction processes and the exchange of information for private purposes. In this regard issues related to digital economy are studied. Many authors such as

S. I. Ashmarina et al. (Eds.): ISCDTE 2019, LNNS 84, pp. 39–46, 2020.
https://doi.org/10.1007/978-3-030-27015-5_5

Brynjolfsson and Kahin [3], Bukht and Heeks [4], and others [1, 2, 17] study the term "digitalization" and the patterns of development of digital economy. Haltiwanger and Jarmin [9], Moulton [11] and Sheehy [15] pay more attention to the questions of its measurement. At present there are practically no studies devoted to issues of state management of digitalization processes and identifying of strategic development directions. There are some studies of management that cover only certain areas or are limited by public regulation [5]. However, digital technologies are developing rapidly; digitalization covers all aspects of modern life. The level of development of society and the rate of economic growth are directly dependent on the nature and directions of development of digital economy. Therefore, planning and determining of the priority directions of development in the field of digitalization are the key tasks of the state. The development of programs and the assessment of the effectiveness of their implementation are the most important state decisions. This article is devoted to the questions mentioned above.

2 Methodology

The methodological basis of the research is the system approach, which allows considering digital economy as a holistic object, including a multitude of elements. The following research methods were used in the work: formal-logical (deduction, induction, justification, argumentation); abstract-logical (when setting goals, research tasks); empirical (observation and experimentation); economic-statistical, economic and mathematical. Data processing was performed using the Microsoft Office software package (Excel, Word).

3 Results

Let's have a look at the proposed sources of financing and their structure for the national program for the development of digital economy. Extra-budgetary funds will make a predominant share in financial sources in 2020. In other years the federal budget will become the key source of funding for the national program. Analysis of the distribution of financial resources for projects shows that the largest expenditures are planned for the projects "Information Infrastructure" and "Digital Technologies". This national project will be implemented within the framework of several state programs, including the State Program "Information Society". It is going to be implemented during the period from 2011 to 2020. The program has target development indicators. You can view them, comparing planned and actual indicators.

From Table 1 it can be seen that the actual values of most indicators do not reach the planned ones. Only one indicator value exceeds the planned value - the share of citizens using the mechanism of receiving state and municipal services in electronic form. The development of information society, information and telecommunication technologies is a key task in a digital economy. Let us gave a look at the sub-programs of the State Program "Information Society". The sub-programs "Information and Telecommunication Infrastructure of the Society and Services Provided on its Basis", "Information Environment", "Security in the Information Society" and "The

Table 1. Target development indicators of the state program "Information Economy" in 2014–2017

	2014		2015		2016		2017	
	Plan	In fact	Plan	In fact	Plan	In fact	Plan	In fact
The place of the Russian Federation in the international ranking of information technology development index	40	45	20	–	10	43	42	45
The share of citizens using the mechanism of receiving state and municipal services in electronic form, %	35	35,3	40	39.6	39,6	–	60	64,3
The share of the population that does not use the Internet for security reasons, %	–	–	7	0,4	5	0,5	–	–
The degree of differentiation of the subjects of the Russian Federation on the integral indicators of information development, units	2,3	2,3	2	–	2	–	1,9	–
The share of households with access to the Internet, %	–	–	75	66,7	90	70,7	83	76,3
The number of high-performance jobs engaged in the sphere of communications, thousand units	–	–	–	–	401,5	–	290,2	–

Source: compiled by the authors on the base of the Passport of the National Program "The Digital Economy of the Russian Federation", 2018) [12]

Information State" are directly related to the development of the digital economy of the country. Let us evaluate the degree of achievement of target indicators for these sub-programs. The implementation of the planned development indicators for sub-program 1 "Information and Telecommunication Infrastructure of the Society and Services Provided on its Basis" can be called satisfactory. In 2014, 2 of 7 indicators were not achieved, in 2015 - 2 of 8, in 2017 - 3 of 7, respectively. In 2016, none of the planned indicators for this sub-program was achieved. This can be viewed as a consequence of the crisis development of the economy in face of external challenges. The fulfillment of the planned development targets for sub-program 2 "Information Environment" is successful, because during the period of research only in 2015 one development target was not achieved. The fulfillment of development targets for sub-program 3 "Security in the Information Society" is extremely important as the pace of development of information, communication and computer technologies, and hence the development of the digital economy in the country will depend on the degree of user confidence. However, the level of development targets fulfillment is rather low. In 2015, 1 of 4 indicators was not fulfilled; in 2017, the only planned development target was not achieved. The analysis of the dynamics of the fulfillment of planned development targets for the sub-program "The Information State" shows an improvement in 2017 [6]. Only one planned target was not achieved (compared with 5 in 2014 and 11 in 2015). Let us assess the effectiveness and efficiency of the state program by calculating several indicators. The coefficient of implementation of the state program (K) activities is defined as the number of completed activities to the total number of planned activities:

$$K = \frac{Mr}{Mn} \quad (1)$$

Mr – the number of completed activities,
Mn – the number of planned activities.
The results of the calculation of the indicator for the sub-programs are presented in Table 2.

Table 2. The degree of fulfillment of activities for the sub-programs of the state program "The Information Economy" in 2015–2017

Sub-programs	2015	2016	2017
Sub-program 1	0,71	0,75	0,57
Sub-program 2	1	0,92	1
Sub-program 3	1	0,75	0
Sub-program 4	0,8	0,69	0,9

Source: calculated by the authors on the basis of cumulative implementation of control figures [13]

The coefficient of fulfillment of the sub-programs activities shows the share of activities implemented in the total number of planned activities. As it can be seen from the table, during the analyzed period the average value of the indicator was 0.76. It means that about 76% of the planned activities were carried out. The integral evaluation of the program performance is estimated on the basis of the comparison of planned and factual development indicators and is calculated by the formula:

$$R = \frac{1}{N} \times \sum\nolimits_{n=1}^{N} \frac{Xnf}{Xnp} \quad (2)$$

N – the number of indicators,
Xnf – factual value of the indicator,
Xnp – planned value of the indicator.
This value was 1.02 in 2014, 0.65 in 2015, 0.63 in 2016, and 1.13 in 2017. In 2014 and 2017 the factual values of the indicators on average slightly exceeded the planned ones. In 2015–2016 the level of program performance was significantly lower: on average, the factual values of the indicators amounted to 61% of the level of the planned indicators. The level of financial support of the program is calculated by the formula:

$$F = \frac{Ff}{Fp} \quad (3)$$

Ff – factual costs aimed at the implementation of the state program
Fp – planned costs aimed at the implementation of the state program.

The calculation of the level of financial support shows relatively low results. In 2014–2015, only 54.3% of the planned indicators were allocated for the implementation of the state program. In 2016 F = 0.067, that is, in fact, only 6.7% of financial resources were allocated from the planned level of expenditures. Only in 2017 the factual costs are the same as the planned ones (F = 1.13). Visually these ratios are presented in Fig. 1.

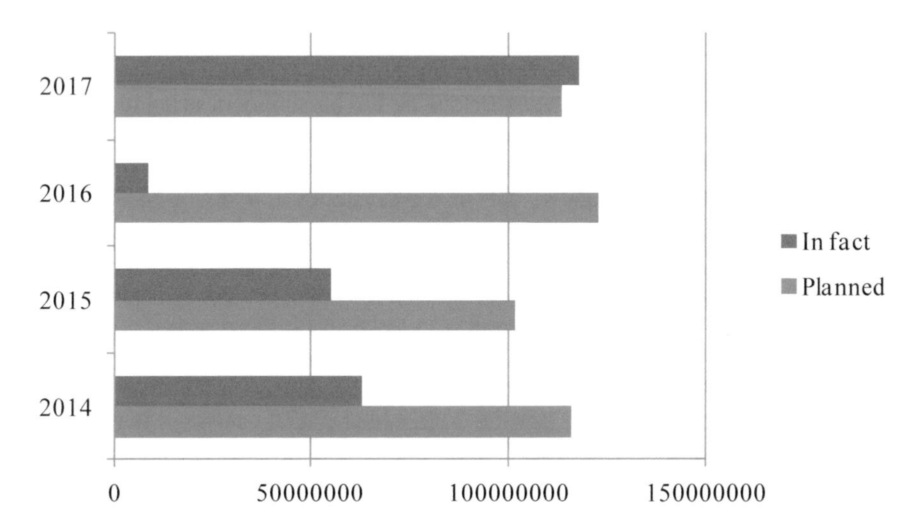

Fig. 1. Planned and factual budget allocations in 2014–2017. (Source: compiled by the authors on the basis of [13])

Evaluation of the effectiveness of the use of budget funds is calculated as the ratio of the coefficient of activities implementation to the level of financial support for the program

$$E = K/F \tag{4}$$

The calculation of this indicator reflects the high efficiency of the use of the budget allocations. In 2014, this indicator was 1.55, in 2015 - 1.29, in 2017 - 0.74. In 2017, the factual costs exceeded the planned value, but 5 out of 30 indicators were not achieved. The evaluation of the effectiveness of the implementation of the program is calculated by the formula:

$$Э = R \times E \tag{5}$$

This indicator can be estimated on the following scale:

$Э \geq 0.9$ – the program implementation efficiency is high,

$0.75 \leq Э < 0.9$ – the efficiency is medium,

$0.6 \leq Э < 0.75$ – the efficiency is satisfactory,

$Э < 0.6$ – the efficiency is unsatisfactory.

On the Basis of the calculation of this indicator we can make a conclusion that in 2014 the efficiency of the program was high and in 2015–2017 it was medium.

4 Discussion

Most of the research devoted to new information technologies, as a rule, focuses on great opportunities on the one hand, and technological problems on the other hand, but ignores features arising between these extreme points, such as implementation, compromises, limitations, materiality and aspects of control that may limit opportunities. As part of this study, we will consider the boundaries expanding the possibility of using digital technologies in order to increase the level of professional training in the field of digital technologies, as well as in the field of educational programs. In the first half of 2019 it is planned to adopt about 50 regulatory acts on the development of digital economy. The key focus is the development of the national program "The Digital Economy of the Russian Federation" [12]. The main objective of this study is to consider the directions of the development of digital economy as a system multifactorial process, reflecting, among other things, the effectiveness of state policy in the field of informatization. The main criterion is going to be the development of a mechanism based on the example of the so-called target indicators for the implementation of digital economy: the share of households and socially important infrastructure objects that can be connected to broadband Internet access. According to the World Bank report [16], information technologies cover all areas of human activity and the state and are becoming increasingly important in the development of the economy, improving of the national welfare and receiving of the so-called "digital dividends". The report defines digital economy as a paradigm of accelerating economic development with the help of digital technologies [16]. The existing approaches to understanding the phrase "digital economy" can be divided into two directions. If we consider the "classic" approach to "digital economy", then this is an economy based on digital technologies and representing the field of electronic goods and services. An "advanced" approach says that "digital economy" is economic production with the use of "digital technologies" [5]. Six projects will be implemented within the framework of the national program "Digital Economy of the Russian Federation": normative regulation of the digital environment, information structure, personnel for digital economy, information security, digital technologies and digital state administration. The implementation of each project is aimed at achieving a specific strategic task. These tasks include:

1. Creating a normative and legal environment for the functioning of digital economy: legal conditions for creating a digital environment of trust, the formation of electronic documents management and other measures aimed at legislative regulation of the digital economy.
2. Creating a global sustainable competitive infrastructure based on mainly domestic developments: creating advanced data transmission infrastructure between households, state administrative bodies and educational institutions; developing a new generation of mobile and satellite communications infrastructure and medicine, introducing digital technologies and platform solutions, etc.

3. Training and retraining of personnel, provision of competent human resources necessary for the development of the digital economy.
4. Providing information security in transmission, processing and storage of data by way of the preferential use of domestic developments.
5. The development of perspective digital technologies and the creation of digital platforms.
6. The use of digital technologies and platform solutions in the sphere of state management and state services provision.

5 Conclusion

The results of the research confirmed that the global modern economy is an information, net, intellectual and psychological economy with its inherent hypercompetitive technologies and methods of information, psychological, programmable and controlled influence on the consciousness, mind and will of people (producers and consumers) [7]. Within the frame of this research work it has been proved that the level of financial support shows relatively low results. Nevertheless, the implementation of the state program is an innovative fundamental technology that offers new ways to organize the recording of transactions, events, certificates and access rights. Based on the analysis of the emerging patterns of digital transformation of the world and national economies, it can be concluded that according to the results of the research of some authors [8] the recent large-scale processes of digital transformation, convergence and integration of information spaces, as well as the widespread introduction of the block chain technology in all spheres launch the process of "creative destruction" of the old world financial and economic system and the formation of the new global neural network hypercompetitive economy and its neuron network regulatory institutions.

References

1. Balenko, E.: The industry is greedy as far as digitalization is concerned. Experts have evaluated the willingness of enterprises to introduce modern technologies. RBK Newspaper **117**(2841) (2018). (in Russian). https://www.rbc.ru/newspaper/2018/07/03/5b3a26a89a794785abc9f304. Accessed 8 May 2019
2. Brennen, S., Kreiss, D.: Digitalization and digitization (2014). http://culturedigitally.org/2014/09/digitalization-and-digitization/. Accessed 8 May 2019
3. Brynjolfsson, E., Kahin, B. (eds.): Understanding the Digital Economy - Data, Tools, and Research. MIT Press, Cambridge (2000)
4. Bukht, R., Heeks, R.: Defining, conceptualizing and measuring of digital economy. Int. Organ. Res. J. **13**(2), 143–172 (2018). https://doi.org/10.17323/1996-7845-2018-02-07
5. Chirkunova, E.K., Kireeva, E.E., Kornilova, A.D., Belanova, N.N.: Innovative development of the building complex on the basis of environmental and energy-efficient technologies. In: International Science Conference SPbWOSCE-2016 "SMART City", MATEC Web of Conferences, vol. 106, p. 08002 (2017). https://doi.org/10.1051/matecconf/20171060
6. Digital Economy: How experts understand this term. In: RIA Novosti–2017 (2017). https://ria.ru/20170616/1496663946.html. Accessed 8 May 2019

7. Dyatlov, S.A., Bulavko, O.A., Balanovskaya, A.V., Nikitina, N.V., Chudaeva, A.A.: Principles of the organization of the global economic system. Int. J. Environ. Sci. Educ. **11** (10), 3783–3790 (2016)
8. Dyatlov, S.A., Lobanov, O.S., Selischeva, T.A.: Information space convergence as a new stage of e-governance development in eurasian economic space. Paper presented at the ACM International Conference Proceeding Series, pp. 99–106. Part F13028299-106 (2017). https://doi.org/10.1145/3129757.3129775
9. Haltiwanger, J., Jarmin, R.S.: Measuring the Digital Economy (2000). http://citeseerx.ist.psu.edu/viewdoc/download?doi=10.1.1.38.4433&rep=rep1&type=pdf. Accessed 8 May 2019
10. Ioda, E.V., Bulavko, O.A., Khmeleva, G.A., Ioda, Yu.V.: Modernization Mechanisms for the Formation of a New Technological Structure (Monograph). Samara Academy of State and Municipal Administration, Samara (2013). (in Russian)
11. Moulton, B.R.: GDP and the digital economy. In: Brynjolfsson, E., Kahin, B. (eds.) Understanding the Digital Economy, pp. 34–48. MIT Press, Cambridge (2000)
12. Passport of the National Program “The Digital Economy of the Russian Federation” (2018). (in Russian). http://government.ru/info/35568/. Accessed 8 May 2019
13. Regulatory framework. State programs. Official site of the State Program of the Russian Federation. Quarterly monitoring. (in Russian). https://programs.gov.ru/Portal/programs/. Aaccessed 8 May 2019
14. Semenov, Yu.A.: IT-economy in 2016 and 10 years later. Econ. Strateg. **1**(143), 126–135 (2017). (in Russian)
15. Sheehy, A.: GDP cannot explain the digital economy. Forbes (2016). https://www.forbes.com/sites/andrewsheehy/2016/06/06/gdp-cannot-explain-the-digital-economy/#7e1afa2f18db. Accessed 8 May 2019
16. The World Bank: Developing the Digital Economy in Russia (2016). (in Russian). http://www.worldbank.org/en/events/2016/12/20/developing-the-digital-economy-in-russia-international-seminar-1#4 (accessed 8 May 2019)
17. What is Digital Economy? Unicorns, Transformation and the Internet of Things. https://www2.deloitte.com/mt/en/pages/technology/articles/mt-what-is-digital-economy.html. Accessed 8 May 2019

Transformation of Worldview Orientations in the Digital Era: Humanism vs. Anti-, Post- and Trans-Humanism

A. V. Guryanova(✉) and I. V. Smotrova

Samara State University of Economics, Samara, Russia
annaguryanov@yandex.ru

Abstract. The article deals with the problems of the worldview transformations of the human and the society in the era of digitalization. The phenomenon of «humanism» and its modern worldview alternatives – «antihumanism», «posthumanism» and «transhumanism» are analyzed. The concepts of «human» and «posthuman» are compared as the subjects of biological and technological nature. The final conclusion of the article comes from a critical view on the transhumanist initiatives of human transformation. Transhumanism improves technically the human being but destroys its ethical dimension. The authors are sure, it's unacceptable in the modern conditions of the digital era.

Keywords: Antihumanism · Digital world · Humanism · Immortalism · Posthumanism · Transhumanism

1 Introduction

Under an influence of the new digital technologies the world around us is greatly changing. As it is well-known in the field of philosophy, being determines consciousness (Karl Marx). So, the human living in the transforming world can't stay unchanged. He's changing much. He's inevitably forced to changes. This process touches his worldview, self-consciousness, the system of values and ethical norms that determine his way of life and social behavior [9].

Modern human is closely surrounded by high technologies – nanotechnology, biotechnology, information and cognitive technologies, etc. He's living in the world of neural networks, computer diagnostics and Big Data using in various fields of human activity. He's applying online services, using actively the Internet of things, communicating virtually and spending his life in Smart cities. This list can be added by an increasing of social development, an impact of information flows and information overload, by pressure from the fields of mass culture and mass consciousness. All this leads to the feeling of uncertainty as a main character of the modern human existence.

It's clear that such an «uncertain» human faces a problem of self-understanding. His own essence suddenly becomes very important for him. Who is he? What place does he occupy in the digital world? What role does he play in it? [8] Is human an instrument for technology's prosperity, or technology must serve human interests? To answer all these existential questions and to help humans to adapt the changes around

S. I. Ashmarina et al. (Eds.): ISCDTE 2019, LNNS 84, pp. 47–53, 2020.
https://doi.org/10.1007/978-3-030-27015-5_6

them, it's necessary to understand their main way. What is the direction (positive or negative) of the changes taking place nowadays? Do they lead to regression, degradation of human mind and domination of artificial intelligence? Does their last stage correspond with the progress, prosperity and real perfection of the humanity itself.

In the context of all these actual questions an idea of humanism becomes really important. It has been developed by the best minds of European culture for centuries. Humanism was always considered as an indisputable civilization value. It seemed to be something essential in the very understanding of the human. But not very long ago humanism was faced with the first serious challenges.

Today we can often hear an opinion that traditional understanding of humanism is outdated, as if it has lost its meaning and significance for the society development. So, the problem of humanism (its essence and actuality) becomes an important subject for discussions. This problem is becoming much more serious due to the presence of alternative worldview orientations such as «posthumanism», «transhumanism» and «immoralism». This paper is an attempt to analyze and to compare them all from the philosophical point of view.

2 Theoretical Background

2.1 Humanism as a Worldview Value of Humanity

The concept «humanism» has a traditional meaning. It's an attempt to reach humanness, to create conditions for a decent human life. Humanism proclaims the value of human as a person, his rights to freedom, happiness, development and realization of all his abilities.

We can see humanism partially manifesting itself already in the Ancient East and in the era of Antiquity (humanistic ideas pass through the philosophical constructions of Confucius, Socrates, Aristotle and Epicurus). But it was fully declared during the Renaissance epoch. At that time humanism has become a basis for the further development of science, art, social thought, politics, economics. Francesco Petrarch, Pico Mirandola and many other cultural figures of Renaissance have laid foundations of the future humanistic worldview in their works.

The greatest humanist without any doubt was a founder of German classical philosophy Immanuel Kant. In his famous «Critique of practical reason» (1788) he formulated his famous imperatives of moral behavior – categorical and practical [10]. The first of them calls us to act in such a way that can become a universal law at any time. The second one claims to estimate every human (yourself or everyone else) as a purpose but not as an instrument for its realization. It's clear that in the works of Kant (it was the second half of the XVIII century) a new line of the modern humanist worldview was born for the first time.

2.2 Antihumanism as an Opposite Worldview Orientation

In the XIX century the principle opposite to humanism – the so called «antihumanism» – was widely spread in philosophy. It is correctly associated with a name of Friedrich Nietzsche [14]. He is an author of the idea of «superhuman» appearance which is inhumane by its essence. Nietzsche's antihumanism is closely connected with his ideas of the «death of man» and the «death of God». The «death of God» (in other words, criticism of religion) is a necessary procedure because God is standing in the way of human self-improvement. So, it's time to «kill God» and to make a free way for human development. The «death of human» according to Nietzsche means his disappearance and future transformation into the form of superhuman.

It's clear that Nietzsche abandons the human and calls for a superhuman. Thus, he's a real antihumanist. But on the other hand (despite all his shocking statements), he still hasn't break with the human completely. His superhuman simply develops and purposefully cultivates the most significant qualities of the human. These are such qualities as the strongest creative impulse (such as Goethe or Beethoven have had), the courage of the spirit, the willpower, the ability to overcome suffering, etc.

It's interesting, but love, compassion and even happiness aren't the characters of Nietzsche's superhuman. These qualities are for the weak, but not for the strong people. They are for the slaves and not for masters. They belong to the human and they aren't needed by the posthuman.

2.3 Posthumanism as an Alternative Worldview Orientation

In the middle of the XX century a new «posthuman condition» [15] began to form. An idea of «superhuman» developed by Nietzsche was replaced by an idea of «posthuman» (or as it was sometimes called «transhuman»). These ideas were presented in the new worldview currents of «posthumanism» and «transhumanism». These two concepts are usually used as synonyms [6].

The concepts of «posthuman» and «transhuman» were used for the first time by the famous Iranian futurist and science fiction writer Esfandiary [5]. He is one of the founders of the transhumanist current. Esfandiary has used an idea of Nietzsche about the «human» as an intermediate link between the «animal» and the «superhuman». This idea has become a theoretical basis of his own conception, but he has also made a few changes in it. Esfandiary puts the «human» on the lowest level. «Transhuman» for him is an «intermediate link» between the «human» and the «posthuman». And what is for the «posthuman», he is modified in such a great way that we can't call him no longer a real «human».

The posthuman has a body improved by different implants. He is a gender-neutral creature. He can reproduce himself in the artificial manner only. His consciousness and personality are distributed in several bodies (biological and technological). In epy later versions of transhumanism they all can be transferred into the computer or some other device. In fact the posthuman is a cyborg (a gender-neutral hybrid of the human and the machine). This is an image of the future posthuman created by the first transhumanists.

We can compare Nietzsche's idea of the superhuman and Esfandiary's idea of the posthuman. At first they seem similar, but then significant differences between them can be found. For example, Nietzsche's superhuman is an aristocrat by his worldview.

He respects only those who are equal to him, those who could become superhumans, who were ready to overcome the «human, all too human» [13] in themselves. The superhuman looks with contempt at all those who stand below him, who could not pass the way from the animal and the human to the superhuman. In contrast, the principle of aristocracy and the superhuman's advantage don't mean something important for the posthuman. Anyone who wants can become a posthuman.

Therefore, it may seem that posthumanism is much more humane than antihumanism. But it isn't so. In the case of the real existence the posthuman society will be absolutely inhumane by its qualities.

2.4 Transhumanism as a Worldview Orientation of the Digital Era

Today transhumanism is a well-known philosophical conception, a large-scale worldview current of thought. It's also a wide international project of the digital era. Transhumanism's first forerunner is F. Nietzsche with his ideas of the «death of human» and the «superhuman», which have already been analyzed above. The second ones are the theoretical constructions of Russian cosmism, especially conceptions of Fedorov and Tsiolkovsky. Transhumanists have borrowed from cosmism an idea of technical improvement of the human. The cosmists themselves associate it with space exploration [7] while transhumanists believe such an improvement is quite possible on the Earth, for example as a result of the artificial life forms distribution.

Transhumanists are sure, with a help of the achievements of modern science and advanced technology we can change fundamentally the human nature and the humanity itself. And we must be able to do it soon because of the modern human – a very imperfect biological creature. He thinks slowly than the modern computers, he can't move fast, he needs a long rest and sleep, he eats too much food, his behavior is determined by a number of social values such as freedom, love, courage, creativity, etc. The last ones aren't necessary for life from the point of view of transhumanists.

To make a human more perfect it's necessary to overcome his biological limits, to correct his physical defects, to eliminate suffering and diseases, to strengthen significantly his mental and psychological characters, to conquer against aging and to defeat death (the last point is a main subject of interest of immortalists). As a result a human can transform himself into a posthuman. Transhumanists proclaim disappearance of the human as a biological type. And even more, he must completely lose his biological, social, spiritual and even sexual specificity on his way to the posthuman existence.

Unlike traditional humanists, transhumanists don't believe that the human is a last step or a top of evolution development. They think the human is continuously evolving and changing. This means he can improve himself indefinitely. It's also necessary to mention that the evolutionary formation of the posthuman isn't natural and biological, but artificial and technological in its character. Posthuman finally moves away from the biosphere and becomes an essential part of the technosphere.

One of the most famous transhumanists of our time is a futurist and inventor Raymond Kurzweil [11]. He made a detailed list of forecasts for the future development of the humanity. According to Kurzweil's forecasts in 2045 the era of «technological singularity» will start. By this time the computers will become a billion times more intelligent than an individual human mind. They'll think and communicate so quickly that the humans won't be able to comprehend what's really happening. Then the computers will begin to develop themselves, faster and faster with each new generation of artificial intelligence. And the Earth will turn into a giant supercomputer, a great computer system. And the human will be deeply integrated into it, with the numerous implants and nanorobots prolonging his life and giving him additional opportunities. And finally, after 1945 the process of «technological singularity» will «wake up» the Universe.

3 Results

The problem of correlation between humanism and its worldview alternatives is often considered by classic and modern philosophers. For example, antihumanist idea of «human's death» was discussed not only by F. Nietzsche but by the famous structuralists of the XX century R. Barthes and M. Foucault. Then it was developed by neomarxist L. Althusser who has made on its basis his program of «theoretical antihumanism». After all, the idea of «human's death» was accepted by postmodernists. Therefore this idea seems to be familiar and even obvious in the modern philosophical literature.

The problems of transhumanism and posthumanism are analyzed in detail in the research works of contemporary philosophers and scientists – Esfandiary [5] and Kurzweil [11], Circovic, Sandberg and Bostrom [3], Eckersley and Sandberg [4]. All these authors are transhumanists. The authors of this article, conversely, consider transhumanism from the critical point of view. We agree at this point with the opinion of the famous Russian academician Lektorskiy [12]. So, this problem is debatable, especially, from the ethic and cultural points of view.

Ethical problems of the digital epoch are well analyzed in the scientific works of Bounfour [1] and Capurro [2]. This problematic is very actual nowadays. So, it requires more attention and additional reflection in conditions of increasing digitalization.

4 Conclusion

Today transhumanism is a utopia, but not a social one (as a most part of the famous utopias of the XX century). It's a biotechnological utopia. Social utopias frightened us with excesses of totalitarism. But biotechnological utopias are much more terrible in our mind. It may seem transhumanists pursue the good goals: they fight diseases; they try to make humans immortal and much more perfect in all other aspects. But as it's said in Russia «good intentions pave the way to hell». In fact, transhumanists offer us to go beyond the «human» as we know him in our days. Both trans- and posthumanism impress us by their fantastic technological prospects and opportunities. But at the same time they frighten us much by their hypertrophied technicism and antihumanism in the true sense of this word.

What can we answer to the arguments of post- and transhumanism? Any interference (especially thoughtless) into the complex genetic and neural human structures is very dangerous! It could have disastrous consequences for the human future. Humanity has once intervened in the natural course of events. The people have tried to transform nature with a help of technology. And what came out of this? Environmental disaster! And if we put experiments to change the human mind and body, it can get even worse. In exchange for more or less healthy creature (physical and mental) we can create even not a cyborg, but a real monster.

Let's assume that modern scientists will find out all the necessary consequences and understand all the complex gene and nerve human structures (although modern science is still very far from it). But this doesn't give us an absolutely guarantee that the new «posthuman» won't destroy the world of culture. But it's a culture that makes him being a real human, gives him morality, ethical values and ideals. The human life is empty and insignificant without them!

We think the main weakness of transhumanism is the same: it improves technically the human being, but destroys its ethical dimension. Even today the total domination of technology begins to transform the basic values, makes a human an egoist unable to love, sacrifice and compassion. But at the same time the human still continues to be a human. And he has both physical and spiritual nature. Transhumanists shouldn't forget about it!

Discussions about humanism, posthumanism and transhumanism will be continued because of the great scientific and technological achievements and the fact of technosphere's expansion. However, rejection of humanism in favor of its post- and trans-varieties is unacceptable – neither theoretically nor practically. Rejection of human is impossible. And rejection of humanism is doubly impossible. Posthumanism leads to the collapse of both the human and the humanity. Nowadays it's more obvious than ever before. So, the main task of social development in the digital world is to save a human as a human and to revive humanism which has been a main value of human culture. It must keep the same value level in future too.

References

1. Bounfour, A.: Digital Futures, Digital Transformation. Springer, Cham (2016)
2. Capurro, R.: Who are we in the digital age? Bibliotecas. Anales de Investigation **13**(2), 113–115 (2017)
3. Cirkovic, M.M., Sandberg, A., Bostrom, N.: Anthropic shadow: observation selection effects and human extinction risks. Risk Anal. **30**(10), 1495–1506 (2010). https://doi.org/10.1111/j.1539-6924.2010.01460.x
4. Eckersley, P., Sandberg, A.: Is brain emulation dangerous? J. Artif. Gen. Intell. **4**(3), 170–194 (2013)
5. Esfandiary, F.M.: Are you a Transhuman? Monitoring and Stimulating your Personal Rate of Growth in a Rapidly Changing World. Warner Books, New York (1989)
6. Ferrando, F.: Posthumanism, transhumanism, metahumanism and new materialism: differences and relations. Existenz **8**(2), 26–32 (2013)
7. Guryanova, A., Astafeva, N., Filatova, N., Khafiyatullina, N., Guryanov, N.: Global crisis: overcoming the uncertainty of the concept in the philosophical paradigm of globalization. In: Popkova, E.G. (ed.) The Future of the Global Financial System: Downfall or Harmony. Lecture Notes in Networks and Systems, vol. 57, pp. 836–843 (2019). https://doi.org/10.1007/978-3-030-00102-5_9
8. Guryanova, A., Khafiyatullina, E., Kolibanov, A., Makhovikov, A., Frolov, V.: Philosophical view on human existence in the world of technic and information. In: Popkova, E.G. (ed.) Advances in Intelligent Systems and Computing, vol. 622, pp. 97–104 (2018)

9. Guryanova, A.V., Krasnov, S.V., Frolov, V.A.: Human transformation under an influence of the digital economy development. In: Ashmarina, S., Mesquita, A., Vochozka, M., (eds.) Digital Transformation of the Economy: Challenges, Trends and New Opportunities. Advances in Intelligent Systems and Computing, vol. 908, pp. 140–149 (2020)
10. Kant, I.: Critique of Practical Reason. Cambridge University Press, Cambridge (2015)
11. Kurzweil, R.: The Age of Spiritual Machines: When Computers Exceed Human Intelligence. Penguin Books, New York (1999)
12. Lektorskiy, V.A.: Human and Culture. SPBGUP, Saint-Petersburg (2018). (in Russian)
13. Nietzsche, F.: Human, All Too Human. Cambridge University Press, Cambridge (1996)
14. Nietzsche, F.: Thus spoke Zarathustra. Cambridge University Press, Cambridge (2006)
15. Pepperell, R.: The Posthuman Condition: Consciousness Beyond the Brain. Intellect Books, Portland (2003)

Digital Transformation of Education, Science and Innovations

E. V. Pogorelova(✉) and T. B. Efimova

Samara State Economic University, Samara, Russia
jour.ru@gmail.com

Abstract. The article is devoted to the study of digital transformation of the «knowledge triangle» - science, education and innovations. Problems facing the Russian society with respect to interaction of these components are considered, available applied solutions are analyzed. Importance of the platform approach is substantiated: it is shown how the project "Creation and launching of the digital platform of knowledge exchange and copyrights management" will be implemented on the basis of the digital platforms available in the Russian Federation, the result of which will be the IPUniversity platform.

Keywords: Digital economics · «Triangle of knowledge» · Digital transformation · Knowledge exchange · Copyrights management

1 Introduction

In January 2019 in the vicinity of Perm the interregional meeting "Leaders of Digital Development" took place. Following it passports of the national programme "Digital Economy" and the federal projects included into it which had long been awaited by the Russian community were published [10]. In conformity with the passport of the federal project "Staff for Digital Economics" a network of centres of development of models "Digital University" including the Fund of new forms of education development, Federal State Autonomous Educational Institution of Higher Education "National Research University "Higher School of Economics", Autonomous Non-commercial Organization «University of National Technological Initiative 20.35», Autonomous Non-commercial Organization "Institute of Internet Development", Autonomous Non-commercial Organization "Digital education and technologies", federal State Budgetary Educational Institution of Higher Education "The Russian Presidential Academy of National Economy and Public Administration", Companies of digital economy, was set up in 2019 and has been functioning in RF since that time.

Adoption of the normative act of the Government of RF regulating provision of grants in the form of a subsidy to higher educational institutions for the purpose of realization of models "Digital University" is declared as a control measure. Establishing the network of international scientific-methodological centres designed for popularization of the best international practices of training, retraining and on-the-job training of advanced personnel in digital economics is also mentioned in the project passport [1]. Methods of forecasting the demand of economy sectors in personnel for

S. I. Ashmarina et al. (Eds.): ISCDTE 2019, LNNS 84, pp. 54–61, 2020.
https://doi.org/10.1007/978-3-030-27015-5_7

digital economy are suggested to be realized; on the basis of this forecast the list of training programmes (specialities) is to be compiled, as well as control figures of enrollment of applicants to educational institutions are to be calculated. A complex accelerated educational environment, where all the materials are to be interactive, will be functioning; students should have certain channels for contacts with the teaching medium (video conferences, computer conferences, electronic mail, telephone, fax, postal communication lines and personal meetings) [3]. Owing to this environment the necessary knowledge, competences and experience of business activity, as well as advanced "cross-cutting" technologies of digital economics will be accessible. The environment is supposed to be formed by all types of educational institutions, business communities, regions and active territorial institutions [1].

Judging by these measures the significance of realization of interaction of the spheres of education, science and innovations is clear. Projects encompassing a small number of participators and carried out within technological parks and clusters by individual companies are becoming history: on the modern stage of development innovative ecological systems and open innovations on the basis of platforms attracting a wide range of users are becoming a priority [11]. Such platforms create an innovative space of interaction of participators- representatives of scientific and business communities.

2 Theoretical Background

At all times scientific achievements and innovations have been the engine of economic development of society facilitating creation of new professions, technologies, competitiveness of enterprises.

Coefficient of efficiency of knowledge transfer with business partners may be represented as follows:

$$E = \frac{U}{T}$$

Where T is the total number of knowledge units transferred to business partners during the given period; U is the number of knowledge units transferred and used by business partners during the given period. A knowledge unit is considered to be a method, an algorithm, a software tool ready for use.

The symbolic notion of the knowledge triangle denotes interaction between education, scientific research and innovations which jointly are the main motive power of modern economy. The key problem of the university scientific and innovative activity is the transfer of fundamental scientific knowledge and innovations to the level of applied technological developments, as well as transfer of knowledge with business partners.

Therefore under modern economic conditions partnership of scientists, higher schools and external service providers is becoming of vital importance: knowledge data base for provision of services oriented towards the services user on the basis of open innovations is expanding [11]. Having left the limitations of the "live laboratory" and "experimental medium", priority is being passed over to service and management aspects in the context of economics of digital platforms [2]. The practice of concluding partnerships on the initial stage of innovations creation is becoming history: now innovative services are based on the systematic ecological approach and are realized on the last stages, when the product is entering the market. As the result of these services has a commercial form, they may be classified as market mechanisms, not mechanisms of innovations stimulation [6]. At present of interest is integration of platforms of open innovations with the "knowledge triangle".

Open access to platforms stimulates innovative activity and provides mutual advantages of parties in creation of the new cost.

At the fifth International congress "Production. Science. Education in Russia: technological revolutions and social and economic transformations" (PSE-2018) problems facing the Russian community within the scope of present-day interaction of science, education and innovations were mentioned; at this congress representatives of China, USA, Italy, Greece, the Commonwealth of Independent States and constituent territories of RF took part [9]:

1. S.D. Bodrunov (doctor of economic sciences, professor, president of the Free economic society of Russia, president of the International Union of Economists, director of the Institute of new industrial development named after S. U. Vitte) underlined, that at the present moment knowledge constituting the essence of Industry 4.0. Comes to the foreground in materials production. In his report he mentions that growth of knowledge capacity and of the creative constituent part of labour makes new challenges to society. The continuously accelerating evolution transformation is leading to the new economic pattern – noonomics which is characterized by the man's leaving the production sphere limits.
2. R.S. Grinberg (doctor of economic sciences, professor, corresponding member of the Russian Academy of Sciences, scientific supervisor of the Economics Institute of the Russian Academy of Sciences, of the Institute of new industrial development named after S.U. Vitte, noted that the coming fourth industrial revolution will cause many professions disappear, it will also lead to social tensity; this process will be accompanied by stagnation in the development of scientific institutes of RF, demand decline for research conducted by them. However RF and post-Soviet countries possess a great integration potential.
3. R.I. Nigmatullin (doctor of physical and mathematical sciences, professor, academician of the Russian Academy of Sciences, scientific supervisor of the Institute of oceanology of the Russian Academy of Sciences named after P.P. Shirshov) stressed insufficiency of initiatives of the RF present scientific elite, scarce investments into the cross-cutting high-technology trends.
4. O.N. Smolin (doctor of philosophy, professor, academician of the Russian Academy of Sciences, chairman of the State Duma Committee for Education and Science) underlined, that at present human potential is being economized on, and

science (and education) lack funding (which is lower than that in the Western countries). As a result the per cent of human capital emigration remains increased. His report dwelt upon the theme of electronic education as well: the latter should be carefully introduced saving the key role of the teacher and personality at the same time.

5. B.S. Kashin (doctor of physical and mathematical sciences, professor, academician of the Russian Academy of Sciences) noted that there is no complex initiative of national science development "from below", scientific activity is not attractive for society, talented researchers are leaving the country, scientific potential is being restricted.
6. Lack of qualified personnel in the sphere of material production was also mentioned: disproportion between training of specialists and demand of the real sector is observed. It was mentioned, that numerous problems of the Russian economics, science and education are connected with the subjective factor, unwillingness to understand the scientific society, lack of close interaction between production, scientific research and educational spheres. The necessity of uniting the new economic programme connected with the issues of technological development and landmarks of the educational process has been pointed out.

Studying the world experience of interaction of science, education and innovations, one may make a note of the following: possibility of provision of scientific equipment and materials with simultaneous technical and consultative support, provision of knowledge obtained by a particular scientific organization which is used by the consumer, performance of research & development or of a project with transfer of the results of these activities to the consumer, other methods [15]. Elements of organizational and informational support include the process of handling applications of services consumers, the process of concluding agreements, control of their fulfillment, submission of reports. This became possible only due to the use of digital platforms. While providing a single access point to informational resources of the project participators, national systems-aggregators allowed solution of one of the most important tasks – popularization of the idea of open access to scientific publications, as well as introduction of scientific publications into international scientific search systems and services [5]. In this connection social networks are of great importance, scientific blogs and forums are actively functioning, previews of articles and researches are constantly appearing; on account of the above mentioned the role of science in society is increasing, volumes of financing are rising. Transition to electronic publications has made publication of articles cheaper and their popularization easier, thus expanding the number of people interested in a particular scientific problem. On-line libraries, repositories, informational resources of scientific and technical information are actively developing. In order to simplify access, identifiers of a digital object (DOI) have been introduced; they allow solving the problem of changing the site of physical placement of the resource in the network. The meta-data set (Dublin Core) describing a digital object has been standardized; the system of public licenses (Creative Commons) has been introduced.

Consequently, realization of the platforms approach allowing transformation of behaviour of economic subjects and formation of the «sharing economy» is essential for solution of problems facing the scientific and business community of RF [11]. In this way new forms of production and innovations cooperation arise: innovations are created as a result of integration of various forms of knowledge obtained from the spheres of science, production and business (from different bases) while attracting social capital. Owing to digital platforms, symbiosis of participators-representatives of business and scientific communities is organized. The innovations process is transformed from the initial stage – knowledge initiation to entering the market, at the same time qualitative orientation towards the consumer's needs takes place, as he takes an active part in the process of these transformations correcting these needs («live laboratories», prototypes, etc.) at the same time [11]. Business- and innovative ecological systems facilitate open innovations development [4]. Thus, digital platforms provide new services promoting the network effect, scaling of production and reduction of costs; the appearance of the former is stipulated by the necessity of services stimulating development of «knowledge triangle» [2].

3 Results

Analyzing the situation in RF, one should underline, that digital infrastructure of science and education refers to the initial stage of development. There are some fragmentary solutions, but they solve a problem only partially. Not only the corresponding infrastructure is necessary, but also reasonable structuring of heterogenous information, support of large amounts of data on the basis of which it is possible to assess which spheres of science are developing in the country and in what way they are being financed. At present only assessments regarding results of scientific publications are obtained, however, as practice shows, a publication may appear much later than an interesting scientific project is completed. It is also important to track the history of this project realization, whether patents have been obtained, or it continued being further developed or remained on the level of a report in an organization.

In order to solve the task set by the company Digital Science the analytical platform of scientific and technological information Dimensions has been developed; it is the most highly developed platform which is widely used by the community at the present moment (among its partners there are the leading universities and scientific organizations of RF). It combines data about grants, patents, publications, industry standards. While simplifying the process of applications choice and not permitting refinancing of the same project, the platform facilitates the process of making decisions in financing research, in this way providing a map of scientific research worldwide. Combining of information which was searched in various services earlier is an obvious advantage of Dimensions. On the whole the use of the platform is free, an insignificant analytic customization (for a reasonable payment) allows making large complex reports and finding advanced trends of scientific research, as well as interests conflicts.

Another successful experience is the platform Altmetric which allows defining the types of scientific research attracting public interest at present, as well as the new trends; it also attracts business for financing on the early stages, thus developing the ecological system of innovations.

In RF individual solutions allowing simplification of work in formatting of scientific articles are being implemented; it is planned to develop the programme of start-ups support jointly with universities, funds and business representatives. Modernization and combining of university repositories, storage systems and data processing systems are being planned as well. Scientific communication is already conducted in the digital format, it requires a modern single combining digital platform.

In November 2019 it is planned to complete the project «Creation and launching of the digital platform of knowledge exchange and copyrights management». The result will be introduction of the platform IPUniversity. Introduction of the module of gamification is also planned for the purpose of creation of a new virtual ecological system for scientists which will stimulate generation of knowledge, research, services, mutual reviews, etc. [14].

Thus, IPUniversity will allow tracking of complex chains of use, processing and creation of new objects simplifying their initiating directly by authors, meanwhile employers may see the volume of consumption and define the value of legal objects. Thus, a principally new advanced culture of work with the intellectual property necessary for solution of tasks of digital economics development in RF will be formed [12].

4 Discussion

According to the opinion of V.R. Kovalev [7] digital economics cannot replace intellectual and cultural activity of a man in various industries of the national economic complex. Growth of intellectual capital of a man and an enterprise as global components of society will be the driver of formation of the new economy, the intellectual one, where the leading role belongs to the intellectual capital of innovative enterprises. Academician of the Russian Academy of Sciences I.A. Sokolov [15] thinks that now it is necessary to unite efforts of scientists, data and instruments by means of new technologies. Kuznetsova [8] analyzes the role of the technology blockchain with respect to actualization of scientific publications and objects of intellectual property.

5 Conclusion

As a result of the conducted problem analysis it should be underlined that creation of a single digital space for production, reproduction and exchange of results of intellectual activity will allow widening of the horizon of cooperation of scientific and business communities and activate generation of new knowledge establishing intellectual rights. The realized platform will be able to carry out the necessary control by means of integration of block-chain technologies, technologies of mutual citation, knowledge exchange and remote expertise of objects innovation. Further creation of a single system of digital regulations integrated into the platform will provide possibilities of creating a separate virtual ecological system enabling increase of a scientist's (collective's, collaboration's) transparency and performance, as well as introduce elements of natural creativity inspiration and support and objects for both free and commercial use [12].

References

1. APPENDIX № 3 to the minutes of the meeting of the Government commission in digital development, use of digital technologies for improvement the quality of life and conditions of management of business activity, of December, 25th, 2018. № 1 PASSPORT of the federal project "Personnel for digital economics" of the national programme "Digital economy of the Russian Federation" (materials www.tadviser.ru)
2. Brynjolfsson, E., McAfee, A.: Machine, Platform, Crowd: Harnessing Our Digital Future. W.W. Norton and Company, New York (2017)
3. Efimova, T.B., Filatov, D.A.: Clever education in Southern Korea. In: Khasaev, G.R., Ashmarina, S.I. (eds.) Science of XXI Century: Actual Development Trends. Collection of Scientific Articles of VII International Scientific and Practical Conference, Samara, Russia, 5 February 2019, pp. 12–18. Samara State University of Economics, Samara (2018). (in Russian)
4. Huhtamaki, J., Rubens, N.: Exploring innovation ecosystems as networks: four European cases. In: Bui, T.X., Sprague Jr., R.H. (eds.) Proceedings of the Annual Hawaii International Conference on System Sciences, Kauai, Hawaii, 5–8 January 2016, Los Alamitos, California, Washington, pp. 4505–4514. IEEE Computer Society, Tokyo (2016). https://doi.org/10.1109/hicss.2016.560
5. Kachan, D.A., Bogatko, A.V., Bogatko, I.N., Enin, S.V., Kulazhanko, V.G., Lazarev, V.S., Lis, P.A., Skalaban, A.V., Yuryk, I.V.: Integration of information resources of open access for provision of the scientific and educational process in higher educational institutions. Open Educ. **22**(4), 53–63 (2018). https://doi.org/10.21686/1818-4243-2018-4-53-63. (in Russian)
6. Katzy, B., Turgut, E., Holzmann, T., Sailer, K.: Innovation intermediaries: a process view on open innovation coordination. Technol. Anal. Strateg. Manag. **25**(3), 295–309 (2013). https://doi.org/10.1080/09537325.2013.764982
7. Kovalev, V.R., Lukin, G.I., Tarasov, S.V.: Concept of interaction of investment, innovation and integration in regional professional education system (3I concept) under technological modernization of production and transition to digital economy. J. Legal Econ. Stud. **3**, 187–194 (2017). (in Russian)
8. Kuznetsova, V.P., Bondarenko, I.A.: Blockchain as an instrument of digital economics in education. J. Econ. Regul. **9**(1), 102–109 (2018). https://doi.org/10.17835/2078-5429.2018.9.1.102-109. (in Russian)
9. Maslov, G.A., Yakovleva, N.G.: Production, science, education in Russia: technological revolutions and social and economic transformations. Econ. Revival Russia **1**(59), 58–64 (2019). (in Russian)
10. National Program "Digital economy of the Russian Federation" (2019). http://government.ru/info/35568/. Accessed 6 June 2019. (in Russian)
11. Raunio, M., Nordling, N., Kautonen, M., Rasanen, P.: Open innovation platforms as a knowledge triangle policy tool – evidence from Finland. Foresight STI Gov. **12**(2), 62–76 (2018). https://doi.org/10.17323/2500-2597.2018.2.62.76

12. Russia creates a digital platform for knowledge sharing and copyright management. Integral-Russia, 9 October 2017. http://integral-russia.ru/2017/10/09/rossiya-sozdaet-tsifrovuyu-platformu-obmena-znaniyami-i-upravleniya-avtorskimi-pravami. Accessed 15 Apr 2019
13. Sokolov, I.A., Kupriyanovsky, V.P., Talashkin, G.N., Dunaev, O.N., Zazhigalkin, A.V., Raspopov, V.V., Namiot, D.E., Pokusaev, O.N.: Digital joint economics: technologies, platforms and libraries in industry, construction, transport and logistics. Int. J. Open Inf. Technol. **5**(6), 56–75 (2017). (in Russian)
14. The Russian Ministry of education and science discussed the interim results of the creation of a digital knowledge exchange platform. News of Siberian science, 7 February 2019. http://www.sib-science.info/ru/fano/minobrnauki-rossii-obsudili-06022019. Accessed 15 Apr 2019. (in Russian)
15. Zatsarinnyy, A.A., Shabanov, A.P.: System aspects of management technology for scientific and educational services. Open Educ. **2**, 88–96 (2017). https://doi.org/10.21686/1818-4243-2017-2-88-96. (in Russian)

Personal Brand of University Teachers in the Digital Age

V. V. Mantulenko(✉), E. Z. Yashina, and S. I. Ashmarina

Samara State University of Economics, Samara, Russia
mantoulenko@mail.ru

Abstract. Personal brand is becoming extremely important in any field of human activity nowadays: politics, show business, sports, business and science. On the one hand, the educational sphere of developing countries of the former Soviet Union is characterized by private conflicts, including the "teacher-student" field, the low prestige of the teacher's profession because of the weak financing of this sphere and devaluing the teachers' work during many years. On the other hand, the orientation of economies to innovative markets, the digital transformation of socio-economic systems necessitate the maintenance of strong, competitive education and science. This is impossible without top universities, high-class research centers, and therefore without unique teachers and researchers. An important task in this context is the formation of a new generation of highly qualified specialists, unique pedagogues and scientists, competitive in the world market. A special role here belongs to personal branding of such specialists. The purpose of this article is to study how a teacher's personal brand can be built, what opportunities it opens, what are its value and specific features. In addition, the authors explore what factors contribute to the creation of a personal brand in the context of education and science. To solve these research tasks, the authors use methods of analysis, comparison and synthesis of information, survey methods (questioning, interviewing, narrative technique), modeling. One of the most significant results of the study is the model of the teacher's personal brand developed by the authors, which shows what attributes of the image are of great importance for the personal brand of a teacher and scientist.

Keywords: Digital age · Higher education · Personal brand · Branding

1 Introduction

A personal brand is a name, values, a unique way of development and special worldview, which is enriched over the years, forming a valuable experience. We live in the time of great changes. If earlier it was enough to get one education, and this knowledge was enough for the whole working life of a person, now progress occurs every 2–3 years. We see that the younger generation is developing together with the progress faster than adults, mastering more and more new technologies. On the one hand, humanity today has many ways to learn, develop and promote itself, on the other hand, trying to be aware of the modern digital world, we can easily get lost in the existing diversity and variety of information resources which are available to us both in

S. I. Ashmarina et al. (Eds.): ISCDTE 2019, LNNS 84, pp. 62–70, 2020.
https://doi.org/10.1007/978-3-030-27015-5_8

educational and social terms. A teacher is a figure, a person who plays a central role in the formation of the younger generation, starting with pre-school education and ending with professional training and retraining courses. A modern teacher is not just a translator of knowledge, his/her role is becoming more complex, and the importance is increasing in the modern digital world.

In the innovative economy, built on knowledge and information, we are trying to get not so only into certain educational organisations and courses, but also to specific teachers, coaches with unique competencies in their professional field, who have proven themselves as excellent psychologists, mentors, etc. Therefore, today we are increasingly talking about personal brands, including the field of education.

Personal pedagogical brand is particularly relevant for the post-Soviet space. The educational systems of many developing countries after the collapse of the Soviet Union were in a very difficult situation, for decades without receiving the necessary funding. Difficult economic and political processes have led to the fact that the teaching profession has ceased to be prestigious and respected. As a result, the new generation is taught by "burnt out people" who need to survive between salaries. More and more often in the news, we learn about situations when teachers behave unprofessionally, and the pedagogical space is filled with mutual disrespect and incorrect behavior, both from the part of students and teachers. In our opinion, only teachers themselves can increase the prestige of pedagogical work and maintain respect for their profession, transmitting their value, importance and uniqueness into society, promoting themselves as specialists.

The main difficulty and specificity of the formation, positioning, promotion of a personal brand, that distinguish it from the corporate or product one, is that its creation is a joint work of the "I-brand" and the brand manager. The issue of turning a person into a brand is also a question of personal readiness and his/her desire. Managing your own brand is a complex work carried out at all stages of the brand life cycle, both at the external and internal levels. This work involves the constant strengthening of the brand, taking into account current trends in the market and the needs of target groups through the consolidation of the personal image in the minds of the target audience.

The main stages of branding should be the analysis of the market situation and the target audience. It is necessary to choose a strategy for creating a brand and planning the content of the brand, which includes a unique set of values, a brand mission, a style and foresight.

2 Methodology

2.1 Theoretical Background

The concept of personal branding was introduced by Peter in 1997, stating a very new idea: "Brand YOU, everything you do – and everything you choose not to do – communicate the value and character of your brand . Therefore brand yourself is a way to differentiate yourself from the "crowd" and a concept that fosters individual's creativity [5]. Different issues of personal branding were studied and presented in the works of modern scientists, for example [1–3, 8].

Let's analyze a bit more its definition (not univocal), derivation and its functions. Personal branding is a phenomenon characterized in the field of marketing, defined as sets of traits crucial for achieving success - that distinguish one person from others - and it is used to generate values for themselves and others. In other words, it is a positioning strategy. As arising from the research 'Personal branding: A systematic review of the research and design strategies used reported in journal articles relating to critical elements of personal branding': "A personal brand is normally based on an individual's personal or professional reputation, which could result in personal or organisation revenue from a career management perspective if maintained and sustained according to the market needs. As with any business brand, this could be seen as a critical element of life and/or career success".

For the variety of application it can have, personal branding is a technique applied to various categories and so, to ordinary people as well. Not only to actors and politicians and big communicators. One category, for example, is the one of entrepreneurs, who are concerned about the future uncertainty of their position. In their cases, the consequences of the process of personal branding are: enhancing the number of customers, promoting their reputation, the progresses of an entrepreneur are facilitated, and so to leverage brand for new fields (having a higher chance of success), gaining inner satisfaction. Last but not least, the moral social consequences of becoming a brand. The increased visibility in society makes them under more attention and judgment so that they are more responsible for their actions.

Social networks fostered the branding process of each person, giving the possibility of self managing it. Therefore, to the person depends its creation, promotion and communication and the future self-surveillance. The managing of this branding process consists mainly of two phases: (a) keeping the brand up to date given changing circumstances and (b) measuring its effectiveness. Time passing, the informations' updating phase is faster and faster and the measurement of the effectiveness is given by how quick is the response to these changes or how quickly you pass from one job to another. Rapidity and adaptability are characteristics of personal branding.

Whether a person is working for a company, or he is a solo entrepreneur, personal branding should be supported. The reason is, if personal branding is considered as your personal identity, as a consequence, "your personal identity is the full identity of your business". So, a well built personal brand, boosts your business - increasing recognition and credibility. Eventually, success or fails of the business can be signals of how the person should work on their brand/personal identity. Supposedly, this reasoning can be spread to all the sectors, not only when working in a firm, or being an entrepreneur, but also in the educational field, teaching and learning side. For example, students, future employed, consider their online presence as a normalized reflection of themselves. The everyday increasing importance of personal branding - especially in the form of online personal branding - suggests that personal branding is arising academic interest and researches are in demand.

2.2 Research Stages

The study included 3 stages:

- at the first stage, the social status of the teacher in Russia for recent years was analyzed;
- at the second stage, we conducted our own study of the attitude of the Russian society to branding in the education system (questionnaire);
- at the third stage, possibilities of a personal brand for teachers were considered.

In the framework of the study, we used methods of analysis, synthesis and synthesis of information, the method of survey (questioning). The survey was attended by 214 people, residents of the Russian Federation, aged 18 to 60 years. The survey was conducted through an electronic questionnaire and personal interviews from February to April 2019.

3 Results

3.1 Teacher's Social Status in Russia in Recent Years

Let us consider here the social status of teachers in Russia. Previously, it was largely determined by public opinion, which gave representatives of this profession worthy for imitation certain moral and ethical qualities. The profession was considered a respected and prestigious. What is happening with the status of a teacher today? According to many experts, it has heavily fallen. This can be judged by two main parameters characterizing the social status of teachers: the level of professional competence and the economic situation.

Thus, according to a survey conducted by the Public Opinion Foundation (FOM) in August 2004, the Russians already believed that the number of good teachers in educational institutions had decreased. The survey on the quality of education conducted by the same fund in July 2015 showed that only a third of respondents recognized it as good and believed that it is improving [6]. It means that a lot of people are not satisfied with the quality of education, and therefore they estimate the work of the teachers involved in it with low scores.

In the early 2000s, the status of a teacher in the minds of Russians has significantly decreased, primarily because of the low, inadequate payment of pedagogical labor. A detailed study on this topic was conducted by the Center for Sociological Research of the Moscow State University in 2005. The results of this study were presented in the article by Parabuchev [4].

In this regard, we find interesting the schedule of changes in the public attitude to the teachers' profession for the period from 2001 to 2014 presented on the website of the FOM [7]. According to the surveys conducted during these years, the status of a teacher has increased significantly by 2014 compared with 2001. Opinions about the prestige of the profession today are not clear, moreover, teachers themselves still do not feel the growth of their social status.

3.2 Research on the Attitude of the Russian Society to Branding in the Educational System

Our own survey has shown the following results. Almost 70% of respondents gave affirmative answers to the question whether a University teacher should have a brand. Highly likely, this is due to the fact that these people met teachers who they consider as "branded" (66.7%) during their education.

Answering the question about the aim of the teacher's brand, the survey participants pointed out the following aspects:

1. Brand is necessary for a successful career. A "brand-teacher" is recognizable and popular. He is trusted by students, the target audience has understanding of his/her "depth" as a specialist, expert.
2. The brand is associated by the interviewees with authority, with what ultimately attracts students to courses of a particular teacher and even to university (students go to a certain "name", assuming that the "legend teacher"/'brand-teacher' can really teach, help to get unique knowledge. That means that the interest in learning increases.
3. Brand teachers are important for any educational institution. His/Her figure and name increase the prestige of training, create a competitive advantage for the organisation where the teacher gives courses.

Those respondents, who believe that the brand is not necessary for a modern teacher, justify their position by the fact that they do not understand the aim of pedagogical brand, its tasks and functions. They also note that "the brand should be at the university, and teachers should correspond to this brand."

We also asked respondents to give a few key characteristics of a "brand teacher" (maximum 3 characteristics) to form a portrait of this figure and understand how he/she differs from a "non-brand" teacher, in their opinion. A list of qualities is headed by "being in demand", "wide fame as an expert, fame in scientific community, outside the educational institution", and "professionalism, competence, experience, high qualification". Among the key features of a brand-teacher, the respondents also named creativity, uniqueness (unique style/format of lectures), recognition, authority, readiness and willingness to development, versatility, focus on results, charisma. His/her image is like a business card not only of the university or a specific training course, but also the knowledge that students receive from him/her.

One third of respondents believe that the percentage of so-called "brand-teachers" occupies 10% of the total teaching staff of an educational organisation. The remaining two thirds are even less optimistic in their estimates (from 1 to 8%).

3.3 Opportunities of a Personal Brand for Teachers

Education as a service sector, the sphere of acquisition and development of unique competencies is an essential element of the digital economy. The modern labor market determines the need to create competitive advantages of the highest order and individual characteristics that form a personal brand. The function of the brand is reflected in the possibility of capitalization in the market, increasing its market value.

The concept of a personal is based on the idea of oneself as a trademark, and the conceptual construction of a personal brand means a way to build not only a successful professional, but also a personal life [9]. Personal brand has many functions. Let's consider some of them, in relation to what advantages they can give for a teacher.

First, personal brand is a way to increase the own capitalization in the relevant market of specialists. Moreover, capitalization can be carried out both in monetary terms and in the form of intangible social bonuses. If you know how to attract and hold public attention, then you will find a way to convert it into relevant benefits for yourself. So, a bright personal brand allows a teacher to get respect of the target audience (students and their parents), authority among colleagues and increase self-evaluation.

Secondly, a person with a strong personal brand becomes less vulnerable during the period of economic difficulties and various crisis phenomena. A competent teacher knows his/her worth and is not afraid of losing the job during economic crises and mass layoffs. Brand-teacher feels more confident in changing circumstances and in a situation of high uncertainty. In such periods it is especially important and necessary to present yourself, to show what you are capable of.

Personal brand allows you to "become visible" in the eyes of a potential target audience. Within the formation and promotion of the brand, fame/publicity is drawing attention to e teacher's work and public coverage of its results, and accordingly personally to the employee as a person of interest. Personal brand is focused on the formation of unique personality, increasing awareness in the labor market.

Thirdly, the work on creating a personal brand helps to establish a balance between the internal and external world, professional and personal life (work-life balance). Awareness of the need to develop in different directions, on the one hand, and the importance of focusing on a certain range of strategically important tasks, effectively forming the personal and professional trajectory, on the other hand, stimulates to find the most optimal balance of the own presence in different areas, find the own rhythm of life. Branding contributes to the effective structuring of activities. This is a practical tool that helps to realize the professional growth and systematize knowledge and skills.

A strong personal brand frees you from the need to climb the corporate ladder, allows to gain a greater degree of social freedom, to expand the scope of activities, while carrying less expenses, because firstly the teacher works for his/her own name, and then the name works for him/her. Thus, the brand allows to save personal energy.

Fourthly, branding allows to look at many aspects of activities like through a magnifying glass, from a new perspective, which enables to see new opportunities there. Increase in choice opportunities implies that the owner of a strong personal brand has a competitive advantage over other players in the labor market with similar qualifications and always takes a strong position by the selection of projects, in matters of cooperation with one or another employer, has the opportunity to participate in interesting projects due to the reason – "I-brand".

In the fifth, personal brand clearly defines the core of personality. A brand, that reflects true values of a teacher, creates a clear communication space between the pedagogue and the target audience. As a rule, the stronger and more realistic the brand is, the less conflicts, tensions and disagreements arise in the communication. Brand-teacher more effectively promotes his/her so-called "ideal student". As a rule, a talented teacher has successful students (winners of various competitions). The more truthful the teacher's personal brand is, the more likely it is that the target audience will be with the teacher.

Improvement, more effective organization of business communications is also the capital of the brand, confirming the professional status. During the period of employment, there is building a wide network of professional connections, which contributes to a significant leveling of the risk of getting lost in the information field. Business communications provide fame at the expert level, which is valued no less than the mass popularity. In the professional environment, a business reputation is formed, which facilitates social interaction and allows flexibility in the labor market, especially in a crisis that results in mass layoffs.

3.4 Teacher's Personal Brand Model

Based on the analysis of the available works on the subject of personal branding and the results of our own research, we offer the following model of the teacher's personal brand (Fig. 1).

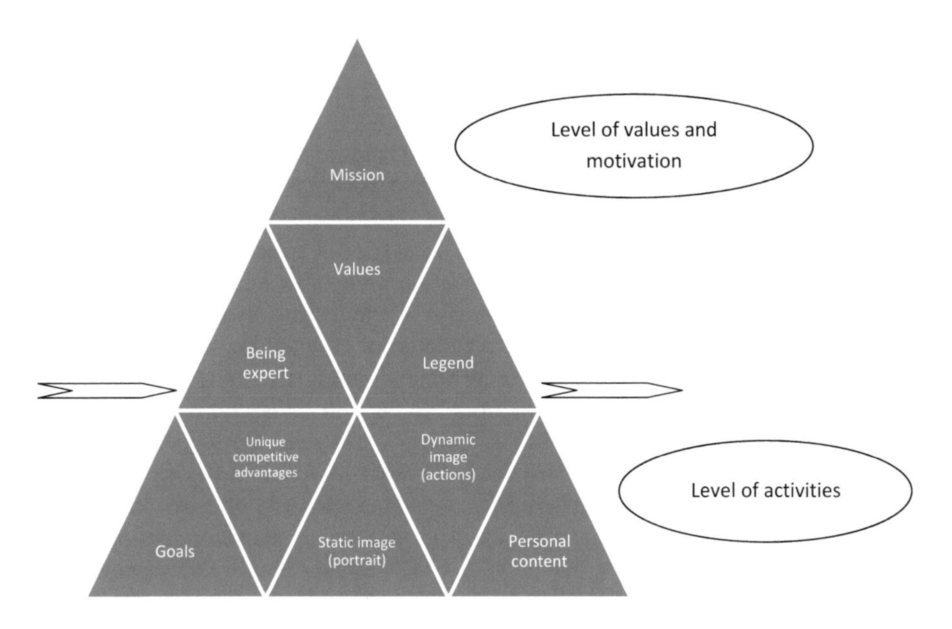

Fig. 1. The model of a personal brand of a teacher (Source: compiled by the authors).

The level of values and motivation is presented by the mission, values themselves, the expert status and the legend. The mission is the most important global, fundamental idea of your activity. Values reflect the life philosophy, basic personal and professional principles, orientations, personal worldview. Being an expert means to be highly qualified, be a real professional in your area(s). The legend includes the best personal qualities and unique life experience, individual history of achievements and success.

The level of activities is built by goals, unique competitive advantages, static and dynamic image and personal content. Goals are our aspirations, socially significant aims. Unique competitive advantages are qualities that distinguishes a particular expert, teacher from others, his/her strengths. The static image is a portrait in various sources of information (official webpage of the educational institution, social networks, etc.), visual image, style, appearance, manners. The dynamic image reflects our actions, projects, manner of speech, channels and style of communication. Personal content is individual content, key ideas that are promoted.

In the digital age, this model of a personal brand can be implemented through social networks, where the teacher can promote it through expert opinions, professional recommendations, reviews, video lectures, photos (family, colleagues, photos from various professionally significant events), expression of opinion, personal position in relation to important social events, storytelling (stories about things that inspire, motivate, or, on the contrary, stories about something "behind the scenes" but that can be interesting to the target audience, etc.) and so on.

4 Conclusion

During this research work, we have studied how the personal brand of a teacher can be built, what opportunities it opens for the educational sphere, and what are its value and features. The authors investigated some factors that contribute to the creation of a personal brand in the context of education and science. We considered the theoretical background of the research issues, analyzed the teacher's social status in Russia in the recent 19 years, conducted our own research on the attitude of the Russian society to branding in the educational system, determined opportunities of a personal brand for teachers and developed the model of the teacher's personal brand. This model shows what attributes of the image are of great importance for the personal brand of a teacher and scientist and in what direction a person can form and promote the personal brand in the conditions of the information society and actively developing digital economy.

References

1. Ekhlasi, A., Talebi, K., Alipour, S.: Identifying the process of personal branding for entrepreneurs. Asian J. Res. Mark. **4**(1), 100–111 (2015)
2. Gorbatov, S., Khapova, S.N., Lysova, E.I.: Personal branding: interdisciplinary systematic review and research agenda. Front. Psychol. 9, article 2238 (2018). https://doi.org/10.3389/fpsyg.2018.02238
3. Mohammed, E., Steyn, R.: 'Personal branding: a systematic review of the research and design strategies used reported in journal articles relating to critical elements of personal branding. In: Twum-Darko, M. (Ed.), Proceedings of the International Conference on Business and Management Dynamics 2016: Sustainable Economies in the Information Economy, pp. 27–35. Cape Town: AOSIS (2016). https://doi.org/10.4102/aosis.2016. icbmd10.04

4. Parabuchev, A.I.: The teacher in the era of public transformations – to the self-portrait of the profession. Issues of Education **4,** 246–259 (2005). https://cyberleninka.ru/article/n/uchitel-v-epohu-obschestvennyh-transformatsiy-k-avtoportretu-professii. Accessed 14 June 2019. (in Russian)
5. Peters, T.: The Brand Called You (1997). https://www.fastcompany.com/28905/brand-called-you. Accessed 14 June 2019
6. Public Opinion Foundation (FOM) Parents of schoolchildren about education and the role of teachers (2015). https://fom.ru/Nauka-i-obrazovanie/12243. Accessed 14 June 2019. (in Russian)
7. Public Opinion Foundation (FOM) Teacher: The prestige of the profession and the necessary qualities (2014). https://fom.ru/Nauka-i-obrazovanie/11773. Accessed 14 June 2019. (in Russian)
8. Rangarajan, D., Gelb, B.D., Vandaveer, A.: Strategic personal branding - and how it pays off. Keller Cent. Res. Rep. **11**(3) (2018). https://www.baylor.edu/business/kellercenter/doc.php/316076.pdf. Accessed 14 June 2019
9. Zabojnik, R.: Personal branding and marketing strategies. Eur. J. Sci. Theol. **14**(6), 159–169 (2018)

Key Priorities of Business Activities Under Economy Digitalization

E. V. Volkodavova(✉), A. P. Zhabin, G. I. Yakovlev, and R. I. Khansevyarov

Samara State University of Economics, Samara, Russia
vev.sseu@gmail.com

Abstract. The goal of the study is to identify and describe the key features of the all-penetrating new production paradigm, which assumes a fundamental change in the role of industrial entrepreneurship in the digital economy. On the one hand, it acquires obvious features of service, and on the other hand, it acquires the features of the horizontal infrastructure technological platform for all socio-economic processes of society. The objective of the study is to find and identify large organizational and economic reserves to increase the efficiency of industrial entrepreneurial activities of products and increase the competitiveness through the possibilities of innovative digital solutions, new methods of working with suppliers and consumers of different levels, value streams that can be used at enterprises of different forms of ownership.

Keywords: Competitiveness · Digital technologies · Entrepreneurship · Industry · Priorities

1 Introduction

The current profound transformation of socio-economic and industrial relations is seen as unprecedented challenges for the competitiveness of industrial enterprises and their systems throughout the world. The impact of destruction technologies that have arisen as a result of the fourth industrial revolution makes industrialists and entrepreneurs establish priorities for further development and their survival, since it is necessary to develop priority areas for modernization of technical, organizational and economic components of industrial entrepreneurial activity. New entrepreneurial ideas arising on old production and marketing decisions in accordance with the concept of "creative destruction" can increase the productivity of economic factors several times in case of successful implementation. The modern economic development of society has become universally linked with digital technology. It is defined as "big challenges" in relation to the further development of productive forces and entrepreneurship. In order to remain competitive in the post-industrial era, national economies must become really involved in the achievements of destruction technologies [7]. Under rapid changes in technologies and methods of organizing production, the unpredictability of the environment significantly increases the role of the correctly established priority as the initial stage of planning.

S. I. Ashmarina et al. (Eds.): ISCDTE 2019, LNNS 84, pp. 71–79, 2020.
https://doi.org/10.1007/978-3-030-27015-5_9

It is no coincidence that the business environment constantly maintains an increased interest in such formalized procedures of planning and forecasting work as the Eisenhower matrix, Covey, SWOT analysis, and similar tools of strategic analysis. Successful entrepreneurial choice in the natural risk environment is based on the correct priorities: what organizational and managerial task to undertake in the first place, how to reveal your obvious and hidden reserves, competitive advantages, true goals and values of the company, reveal all essential environmental factors. In the context of hypercompetition, it is vital for an entrepreneur to have time to solve as many problems as possible in a unit of time. For this, digital technologies make it possible to solve definitely more tasks than it is possible now, taking into account the fact that procrastination is a postponement of problems "for later", which can be fatal to the success of a business. The business is based on economic efficiency, profit, therefore the introduction of digital technologies should be considered as a direct way to reduce costs and increase profitability.

2 Methodology

The national transition from "small scale" economic growth, limited by the interval of annual GDP growth of the country in the range of 1–1.5%, to dynamic growth rates of three to five percent, capable of doubling the country's GDP by 2025, can be solved at the expense of correct priorities of intensive renewal of industrial technologies, advanced production of high processing in the framework of new production and technological systems and prioritization of entrepreneurial activity. It becomes clear, in accordance with the innovative concept of "creative destruction" put forward by Schumpeter [7], that the latest digital technologies, nature-like solutions, robotics, artificial intelligence and machine learning technologies, cyber-physical systems, virtual and augmented reality, etc. are the real possibility to produce new types of machinery and equipment using fundamentally new physical principles, both capable of destroying reproduction models that were previously implemented, and creating conditions for super efficient entrepreneurial style of actions that can generate, improve and develop them.

The penetration of digital technologies into everyday life transforms social and production relations, through the immediate dual unity of extremely contradictory trends, such as: the possibility of mass production and individualized service provision; the combination of "real", virtual and augmented reality; protection of identity, personal data of a person and transparency of his life and behavior; gradual alienation of the inner world from its carrier - a man, which leads to a person's loss of freedom and ability to form his own self; principles of democracy and strengthening the control of human masses through the all-pervasive technologies of social engineering; further polarization of human communities, strengthening the enrichment of the wealthy, and further impoverishment of the poor.

Under digitalization, the whole life of citizens turns out to be in full view – movements, social connections, communication, preferences, payments. The use of digital technologies involves actors in a new n-dimensional interaction at the interface of the physical, digital, social and biological worlds (citizens, government, entrepreneurs, society, technology, property, inanimate world – "smart" factories, objects and tools, real estate), which decrease transaction and organizational costs.

The model of strategic decision making, international comparisons, technical and economic foresight, as well as meaningful economic interpretation of the results of applying digital technologies in modern enterprises, their production capacities and innovative developments in accordance with the concept of their digital transformation, were used to uncover the phenomenon of digital transformation of society. The solution to priority problems of digital transformation and the use of drivers in moving towards the goal of optimizing business processes are methodologically new and they are identified in the works of digitalization pioneer Negroponte (1995) [4].

It turned out that the new competitiveness of enterprises depends largely on society's readiness to "digest" and master the existing digital technologies and participate in digital processes, using mainly the Internet. The business needs a flexible approach to the introduction of technologies and a corresponding change in business models that will enable it to take advantage of emerging opportunities for digitalization. There is a clear link between high rates of the country and the level of labor productivity – this is most strongly reflected in the "readiness for the future" factor, that is, the ability to manage the breakthrough changes carried by digital technologies. Enterprise competitiveness and entrepreneurship are depressed by high investment risk, localized approaches to globalization, and poor use of big data technology and analytics.

According to Womack [12], the priorities of entrepreneurial activity are to carry out successful transformation of enterprises in the face of destruction technologies, digitalization, to find and identify leaders and "agents of change", who have the necessary competencies to transform the value stream, decisively destroying the routine production procedure and reformatting it into perfect business processes, penetrating into all the functional areas of the enterprise's activity. Upon completion of digital transformation of the value stream, this process is expanded vertically, including suppliers and consumers, and horizontally, forming network structures and reliable social connections.

3 Results

It is known that the competitiveness of enterprise products is based on low production costs with high quality, the possibility of earning a profit above the industry average in the growing market, which is ensured by the transparency and controllability of production at all stages of this process, which today are effectively solved by digitalization and automation of business processes. It is assumed that the controlling role of a person will change, whose functions in a single information space will be taken over by automation systems in a constant mode. But with the introduction of digitalization in the workplace, new threats are emerging, associated with the changing role of living labor, completely new requirements for people's competences. Over the next twenty years, it is expected that almost half of the jobs in the modern world will be automated, leaving millions of people unemployed. On the other hand, a significant number of digital professionals will be required. According to various estimates, in the next five to seven years Russia may experience a shortage of up to ten million analysts in the field of computing technologies and information management, which is typical in proportion to the scale of production for other developed economies.

It is obvious that digitalization brings fundamentally new solutions in all spheres of life, bringing productivity to a new level, forcing to diversify its activities, changing the paradigm of routine production. As a result, it is expected to optimize the costs of enterprises, to increase the profitability of existing assets, to expand geographically, since it is caused by the accelerated product delivery to consumers and the workforce with necessary qualifications. Under hypercompetition, each progressive organization must undergo its genetic development in order to define modern business goals and basic tools for their achievement, which is the main condition for survival in the rapidly changing environment. The enterprise should extend communication lines through its current state by taking "readiness for flexibility" initiatives with the desired, targeted state of high competitiveness, characterized by the maturity of the enterprise in key areas of operations. The maturity of digital transformation of the enterprise is characterized by the effective state of key elements creating value added, such as:

- Development and management of the product life cycle.
- Methods of production management.
- Audit and business analytics, controlling.
- Convergence and integration into data management.
- Information security and the amount of funds to maintain it.
- Innovative organizational culture and staff ready for change.

When digital transformation is carried out, a "digital maturity model" is better formed in virtual reality, fundamentally necessary for the enterprise to determine the compliance of the current state of the production system with the target, complementing management decisions by analyzing the state of the main functional areas that determine the sustainability of the business [5]. At the same time, the cornerstone is the positive interaction with customers and consumers, who have a high interest in the supplier's performance, his products and services. Positive customer experience allows us to justify the presence of the company in the market, and to increase sales, thereby attracting financial resources to modernize production, improve operational efficiency in the course of digital transformation.

Studies of a number of well-known consulting companies, based on a comparison of investment in technology initiatives with the depth of transformation in organizational and management activities in a single matrix, revealed a clear dependence of financial indicators on new digital technologies and management methods. It turned out that firms whose technology and new management methods are organized on a digital platform, have a profitability of 26% higher than their competitors. At the same time, the profitability of organizations that are interested in digital technologies to the detriment of management is 11% less than the former. Traditional firms, that improve only the organizational and managerial mechanism without digitalization, get plus 9% of profits losing the opportunity to have three times more using digital technology [11].

The practice in Russia shows that having thousands of digital solutions in production, it is necessary to improve the efficiency of use. The overwhelming majority of enterprise managers venture steps to introduce automation of production processes for fashion reasons, often "provoked" by generous promises of sales department specialists, results of scientific conferences, articles in popular print, successful examples from foreign experience and wonderful dreams to manage all production operations only with a mouse in your hand. At the same time, the actual production tasks are often missed, since the virtual digital twin is created specifically for an actually working enterprise. Table 1 shows the most popular and sought-after options for automation systems of enterprise management in matters of ensuring high productivity and high sales margins.

Table 1. Priority areas for digital solutions in Russian enterprises

№	Digital transformation	Event content	Expected event effectiveness
1.	Comprehensive solutions for forecasting and modeling the customer behavior	Ensure the production of such a volume of products that consumers will buy for sure	Reduce the project execution time and increases its efficiency. This is the key for most small and medium-sized companies to increase efficiency, to become flexible and adaptable
2.	Systems for monitoring the work of own equipment	Ensure 100% reliability of the equipment and the quality of products	Implementing a project on time and within budget, the client is satisfied, and then he will reapply and bring partners
3.	Automation of decisions for target production planning, creation of virtual production systems	A number of factors and criteria are taken into account: availability of production capacities, order lines, enterprise environment parameters, risks, requirements of stakeholders and influence groups	Visual implementation of the learning effect. When implementing a project, even agreeing on a strategy, potential risks may arise and disrupt plans. They need to be warned. Provide sustainable development of the company
4.	Integrated automated systems for monitoring fixed assets of production in the given parameters	Quickly mark areas with deviations in equipment reliability, compliance with technical processes of production	Provide the best competitive position. The concomitant effect improves the company's reputation. Prevent the rise in price of products due to the prevention of equipment malfunction or reducing the number of scrap in the early stages of assembly
5.	Advanced AR/VR design visualization technologies and 3D modeling	A virtual counterpart of the real work of assembly lines is being created	This advantage is important not only within the company, but also outside it. It is associated with increased efficiency

Source: compiled by the authors.

For Russian businessmen, four areas of innovative development provided by digital technical methods are now the most relevant:

- Restructuring of the existing infrastructure of software and hardware enterprise management;
- Using technologies for managing data about the product and the life cycle of products, respectively PDM and PLM in production;
- Implementing optimizing analytical systems for making more calculated decisions online;
- Adapting employees to innovations and new positions of hierarchy in a digital enterprise.

Many Russian enterprises need to start not with digitalization, but with primary automation and informatization of traditional production.

4 Discussion

It is necessary to agree with Voronin, that creating a digital enterprise in Russia is a time-consuming task that requires careful calculation and placement of production lines, information systems and security tools [10]. In this case, Korotovskaya and Smolov emphasizes the need to identify promising areas for applying achievements of the Fourth Industrial Revolution, including developing strategies and explaining the benefits of digitalization at all levels of management, developing an adequate culture, the spirit of innovation and implementing new technologies [2].

The key problems of the Russian industry when implementing Industry 4.0 are weak innovation, low involvement in international trade, inaccessibility of investments and infrastructure, and weaknesses in government regulation. Based on the works of some experts [1, 3, 6], we can note the following barriers in digital transformation of Russian enterprises:

- Routine and unwillingness of enterprises to produce finished products and provide services through network technologies;
- Lack of the necessary number of specialists in the field of digital technologies, typical for all developed economies;
- Threat of mass unemployment due to the widespread introduction of automated production lines, smart plants and factories.
- Particular challenges in ensuring information and cyber security, especially in the context of US unprecedented hacker attacks against other states and various organized structures in accordance with the Presidential Policy Directive 20, or PPD-20, which came into force in 2012, allowing information networks cause damage to their opponents.
- Resource constraints, as the national economy, like any complex mechanism gaining speed, has a certain optimal level of power when all available resources and effective ways to use them are involved, and it is difficult to mobilize new ones, especially in the short term.

- Availability of the market and effective sufficient demand, which has its limits, since incomes of the population rarely experience noticeable leaps.
- Passive thinking and expectations of people, obligations to previous strategies of enterprises. Private capital begins to invest and work with the prospects of making a profit above the refinancing rate of the Central Bank of the country. The modern speed of technology development is significantly ahead of the ability of most people's consciousness to adapt to new realities, causing a kind of internal conflict, resistance to everything new, even where the effectiveness of changes is obvious.
- Undeveloped information infrastructure, critical for the development of BigData and IoT technologies, the need to introduce 5G standard data networks for enterprises in many industries – metallurgy and coal mining, agribusiness, industry, energy, construction, proceeding with digital transformation.

Although we live in the era of the digital economy, officials, representatives of the public sector, especially at the regional and municipal level need the target digitalization program more. It will help them incorporate into the processes of realizing the future, as private enterprises, even Russian ones, have been living in the digital environment for over ten years [9].

The reasons for such concerns are the following:

- The existing education system of Russia orients specialists to solve only standard rather than searchable, creative tasks — use modern gadgets, rather than reinvent them.
- A distorted wage system, which practically equalizes wages of highly skilled labor with the routine skills and skills of low-skilled workers. A teacher, doctor, engineer in the Russian Federation often receives less than a driver or a janitor.
- A low initial level of automation and digitalization of technological processes, a low level of statistical data analysis in stable production conditions that could be digitized.
- Lack of skilled digital technology professionals in the industry;
- Outdated technical regulation of enterprises, complicating the introduction of new technologies.
- A low level of digital culture of enterprise management, lack of motivation to use digital methods and gain effect, the withdrawal of capital abroad.

At the first stage, it is necessary to conduct primary industrialization, restore the lost industrial sectors which are the basis for the application of high-performance digital technologies that can actually increase labor productivity by 1.8 times (or even significantly more) and create 25 million high-performance jobs that are planned by the RF Government. Weak infrastructure, disorder in Russian cities and settlements, which require deep modernization of housing and public utilities, garbage disposal, normal and storm sewers and thousands of other "little measures" should be done even under declared deep digitalization.

Under exhaustion of the former development paradigms, the impoverishment of large-scale investment and innovative drivers in the domestic economy, a new wave of growth is required, which representatives of the elite instinctively represent in the form of digitalization. Although it was promptly overruled and the corresponding state program was approved, it was not supported by any kind of financing and bypassed actual industrial production in its target priorities. Despite all the obvious bias of this topic, society should be soberly aware that digitalization is a procedure that provides. It is no coincidence that the authors of the Expert magazine note that this is a pseudo-solution, meaning: how to do it, but not what to do. How to do – is the secondary question. Today, no production, no process can do without information and communication tools and technologies. At this stage, much depends on the government, and it is necessary to properly organize digital mobilization, infrastructure, finance, and direct it [13].

Digitalization can dramatically change the position of a person in his environment. It can affect the spiritual world, consciousness, social relationships, involvement in reproduction processes, property relations, etc. Shevchenko, Korsukova note that blockchain is an unprecedented efficient way of storing data in digital form, which is a register of transactions, information which can be stored on any of the computers on the World Wide Web, and in real time it provides access to any Internet user [8]. However, there are obvious significant contradictions in the use of new technology, since risks and vulnerabilities to security increase with information transparency of human participation in exchange relationships that can be exploited by attackers, depriving citizens of their property, savings and reputation.

5 Conclusion

Domestic entrepreneurs, business leaders need to identify the most promising areas of digital technology. Horizons of prospective development should be based on the implementation and development of digital culture and actively attract and develop their own and invited specialists in digital technologies, and provide them with conditions for high-quality creativity. If you want to be ahead, you should not wait, you have to act now, implementing the strategy of "overtaking, not catching up," i.e. to switch over to new technological high-performance industries without outdated technological structures.

Before moving to the achievements of the digital economy, domestic industrial entrepreneurs need to learn and produce technically complex things again: airplanes, machine tools, electronic component base, and then digitally digitize production processes since manufacturing industry faces the most profound decline. With proper prioritization, primary industrialization and secondary digitization of existing assets, it is possible to mobilize all your strengths and available resources in order to successfully crush a competitor and become world leaders in economic development without suffering enormous casualties. It is obvious that the challenge of digitalization is one of those big tasks that will require mobilization of the forces and resources of the entire Russian business community in the name of the country's prosperity.

Digital enterprises under digital transformation require priority restoration of the industrial economic sector on the new technological basis in order to reduce organizational and transaction costs, accelerate and cheapen business processes, through the implementation of new, ultra-efficient technologies. To develop a competitive model of industrial entrepreneurship, it is necessary to take priority decisions regarding the priority achievements of the fourth industrial revolution, including extensive robotization of mass production, cloud technologies, the Internet of things, big data, taking into account sociality, creativity and high standards of the newly emerging man-machine morality.

In further research, you need to identify new production concepts as "points of growth" for the new digital economy, since "breakthrough" technologies offer fundamentally new, super-efficient methods for organizing industrial research and development, production and marketing of knowledge-intensive consumer value. A typification of industrial entrepreneurial activity is required, depending on the place taken in the global chain of a new use value and the degree of complexity, the universality of technologies and equipment used and the concepts of well-balanced and lean manufacturing.

References

1. Buklemishev, O.: The alternative to reforms - the inertial continuation of stagnation. Kommersant Sci. **5**, 20 (2017). https://www.kommersant.ru/doc/3343851. Accessed 22 Apr 2019
2. Korotkovskaya, E.S., Smolov, F.M.: Technology "Industry 4.0" as a tool for digital transformation of industry. In: Yu, O., Chelnokova (Ed.) Economics in Saratov University: Past and Present: Proceedings of the International Conference, in the framework of the International Scientific Symposium on the 100th anniversary of liberal arts education at SSU "The Century of Humanitarian Education in Saratov State University: the dialogue of times – past, present and future": Collection of scientific articles, pp. 40–43. Saratov: Publishing House "Saratovskiy Istichnik" (2017)
3. Makhalin, V.N., Makhalina, O.M.: Management of calls and threats in digital economy of Russia. Management **2**(20), 57–60 (2018)
4. Negroponte, N.: Being Digital. New York: Alfred A. Knopf (1995). http://web.stanford.edu/class/sts175/NewFiles/Negroponte.%20Being%20Digital.pdf. Accessed 02 Jan 2018
5. Neligan, A.: Digitalisation as enabler towards a sustainable circular economy in Germany. Intereconomics **53**(2), 101–106 (2018). https://doi.org/10.1007/s10272-018-0729-4
6. Petrov G (2018) America began to apply the Third Compensation Strategy. Expert Online 2018. Available at: http://expert.ru/2018/08/17/amerika-nachala-primenyat-tretyu-kompensatsionnuyu-strategiyu/. Accessed 22 Apr 2019
7. Schumpeter, J.: Theory of economic development. Progress, Moscow (1982)
8. Shevchenko, K.V., Korsukova, N.D.: Blockchain, as a new stage in the development of the digital economy. Actual Probl. Aviat. Cosmonautics **3**, 524–526 (2017)
9. Shikarin, A., Klimova, M.: Digital distances. Expert Siberia 34(518) (2018). https://expert.ru/siberia/2018/30/tsifrovyie-dali/. Accessed 22 Apr 2019
10. Voronin, S.A.: Digital industrial enterprise of knowledge economy within the framework of the concept "Industry 4.0". In: Shirokova, S.V. (Ed.) Collection: SPbPU Science Week. Proceedings of the International Scientific Conference, pp. 96–99. St. Petersburg: Peter the great St. Petersburg Polytechnic University (2016)
11. What is "Digital enterprise" and how to become it?. http://www.docflow.ru/news/analytics/detail.php?ID=32175. Accessed 22 Apr 2019
12. Womack, J.P.: Lean manufacturing: How to get rid of losses and achieve the prosperity of your company. Alpina Publisher, Moscow. Translated from English-11th ed (2004)
13. Yakovlev, G.I.: Development of the competitiveness of enterprises on the basis of digital technologies. Electron. Sci. J. Synergy **3**, 29–34 (2018). https://orel.vepi.ru/wp-content/uploads/sites/10/2019/01/Sinergiya-2018.-3.pdf. Accessed 8 Mar 2019

Research of Efficiency of Tax Stimulation of Innovative Entrepreneurship

D. V. Aleshkova[1(✉)], M. V. Greshnova[1], E. S. Smolina[1], and L. E. Popok[2]

[1] Samara State University of Economics, Samara, Russia
dashajuly343@gmail.com
[2] Kuban State Agrarian University, Krasnodar, Russia

Abstract. The transition to digital economy is impossible without innovations. Therefore, an important task for the development of digital society is to stimulate innovative activities of entrepreneurship. Stimulation of innovative activity of enterprises is currently one of the most promising areas for the economic development of different countries. This issue will remain relevant for a long time due to the demand for innovations in our society, where innovations are often considered as one of the best way to meet the human needs. The aim of the study was to analyze the innovative activity of Russian enterprises and methods of its stimulation. The authors also studied the foreign experience of tax stimulation of innovative entrepreneurship in detail. On the basis of the obtained results, conclusions were made about possible measures of tax stimulation of innovative activity of enterprises.

Keywords: Digitalization · Digital economy · Innovation activity · Innovations · Entrepreneurship · Tax stimulation

1 Introduction

At the current stage of the society development, issues related to the reorientation of the economy to innovative paradigm are becoming increasingly relevant. Innovative entrepreneurship is able to bring the country economy to a new level, increasing the competitiveness of products and the ability to meet changing social needs. This topic is relevant for both Russian and foreign researchers: different aspects were considered by Hhan [2], Metlyuk [3], Pucihar et al. [5], Revina, Paulov, Sidorova [6], Ryabova [7], Ryumina [8], Spescha and Woerter [9].

Nowadays, a lot of countries (among them the Russian Federation), are moving from a commodity economy to an innovative and digital economy. The transformation process affects various industries. In cases of analysis of innovation activity in a particular sector of the economy, it is possible to clearly show its status, prospects and possible development directions in the current situation. However, there are still some questions left concerning the causes of fluctuations of the innovative activity of enterprises and possible methods of its stimulation.

The purpose of this study is to analyze the dynamics of innovative activity of organizations and develop recommendations for the development of measures to

S. I. Ashmarina et al. (Eds.): ISCDTE 2019, LNNS 84, pp. 80–84, 2020.
https://doi.org/10.1007/978-3-030-27015-5_10

stimulate the innovative entrepreneurship. Having answered the above questions, it is possible to achieve this goal.

2 Methodology

Data on the level of innovation activity of Russian organizations, as well as indicators of innovation activity of the other countries were a subject to comparison. The results were analyzed and synthesized. Based on the obtained results, the answers to the main research questions were formulated.

3 Results

In the world practice, "innovation" is interpreted as a replacement of potential scientific and technological progress with the real one in the form of new technologies or products. Usually, the process of development, implementation and commercial use of technical and technological innovations is called innovative entrepreneurship. It is believed that the basis of entrepreneurship is innovative products or services that contribute to the creation of new markets, as well as meet the needs of modern society. In real business, innovations are a specific tool of an organized search for opportunities to increase the competitiveness of enterprises.

Innovative activities can be implemented by companies in various formats:

- at existing enterprises with divisions engaged in research and innovation activity (as a rule, these are enterprises of medium and large business operating on the general tax regime);
- small innovative enterprises, just starting their activities and applying special tax regimes (it should be noted that many innovative enterprises are startups that belong to small business, this fact determines the necessity to take into account their features in the development of stimulation methods.

Based on the format of innovative entrepreneurial activity of companies, they can use different types of preferences. Large enterprises which pay income tax can use the depreciation bonus, the investment tax credit or reduced taxable profit by size of expenses on R&D in the amount of up to 75%.

Small innovative enterprises, that use the simplified taxation system (STS), can pay insurance premiums at a reduced rate only in the implementation of some activities. Thus, this measure does not cover all activities where innovative breakthroughs are possible. STS itself is aimed at stimulating development of small business, not innovative entrepreneurship.

While analyzing Article 427 of the Tax Code, that concerns reduced rates of insurance premiums, we can say that this measure is aimed at stimulating heterogeneous industries. Innovation activity is not determined as a separate position in the All-Russian classifier of economic activities (OKVED). So, the question arises: what types of activities can relate to innovation? According to Popova, innovative activity includes the scientific and technical activity relating in OKVED to classes 71–74: Section M. Professional, scientific and technical activity [4].

In our opinion, can be innovative activities from Section F. Construction (classes 41 "Construction of buildings" and 42 "Construction of engineering structures") and Section J. Activities in the field of information and communication (classes 61–63) can be also innovative.

Many other industries, which also have high innovation potential, remain uncovered. Innovative products are in all sectors: from the production of mineral fertilizers to the development of alternative energy sources, in agriculture, in the mining industry, finance. Thus, binding mechanisms of tax stimulation of innovation to OKVED can be inappropriate.

One of the factors hindering the development of innovative entrepreneurship in Russia is an inefficient system of tax stimulation. It is believed that nowadays there is a wide range of measures of tax stimulation in Russia, including the use of reduced tax rates and insurance premiums, tax base reduction, simplified tax system, investment tax credit and accelerated depreciation, as well as other tools that can be associated only with innovations. Most methods of tax stimulation are used throughout the Russian Federation, but some function only in certain areas, for example, in the innovation center "SKOLKOVO".

Meanwhile, speaking about startups and small business, it can be noted that they are the main "suppliers" of innovations in various industries. To assess the effectiveness of the current set of tax stimulation measures, let us consider the dynamics of innovative entrepreneurship development in Russia (Fig. 1, Table 1).

Table 1. Innovative activity of organizations in Russia, (% of total number of organizations)

2007	2008	2009	2010	2011	2012	2013	2014	2015	2016	2017	2018
9,7	8,6	8,5	8	7,7	7,9	8,9	9,1	10,1	10,9	10,6	10,1

Source: compiled by the authors according to the presentation of the State statistics service "Innovation Statistics in Russia 2018" [1]

As it can be seen from Fig. 1, it is impossible to identify periods of innovation activity increase. For the past 15 years, it remains at about the same level, the fluctuation rate is less than 5%, moreover, there has been a tendency for reduction of innovation activity in recent years.

Considering the foreign experience of tax stimulation for companies that generate and use innovative products and technologies, we can note the most promising techniques. For example, in the United States since 1981, one of the main methods has been tax credit for R&D (deduction from the tax base of R&D expenditures). In the Netherlands, the tax credit is calculated on the basis of the salary costs of employees engaged in R&D.

The experience of Japan is also quite interesting. This country uses different approaches to stimulate innovations: for example, tax rebates on development costs, tax breaks on research costs (a preferential rate is applied which is 20% of the amount of increase in R&D costs compared with the highest amount spent for these purposes in any year after 1966).

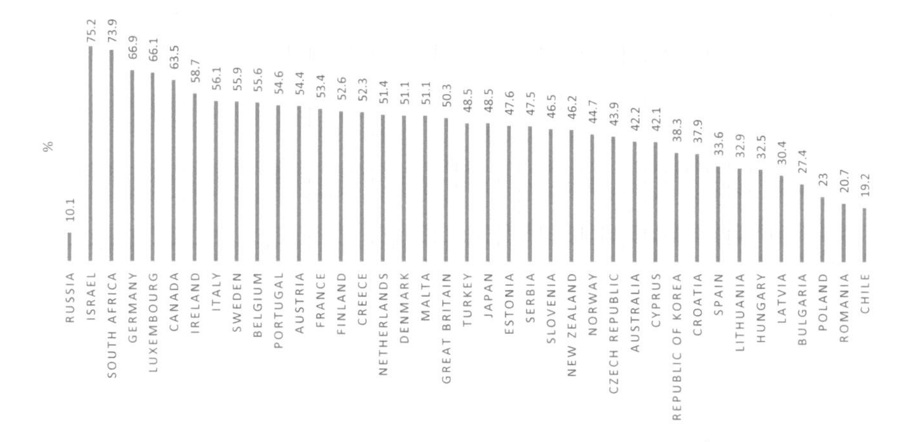

Fig. 1. Comparison of innovative activity of organizations in the world (Source: compiled by the authors according to the presentation of the State statistics service "Innovation Statistics in Russia 2018" [1])

Tax benefits can be divided into volume ones (proportional to the amount of R&D costs) and incremental ones (calculated depending on the increase in costs in the current period compared with the base or average level for a certain period). The highest level of a volume benefit is applied in Singapore (200%) and Australia (150%). In these countries, they deduct the taxable income by the amount which exceeds 1, 5–2 times the expenditures on R&D.

A key element of foreign models of tax stimulation is the reduction of corporate income tax. Basically, this decline is linked to the level of innovation activity achieved by companies. Therefore, the maximum benefit is obtained by companies which realize a successful commercialization of R&D results [8]. In 2001 in France, as well as in a number of other European countries, a law "Patent box" was adopted in the 2000s. The essence of this phenomenon consists in the decrease in the rate of taxation for income received as a result of the implementation of the patented products. The difference of the "Patent box" regime from others is determined by the fact that tax incentives are aimed at successful commercialization of scientific developments. If other methods of stimulation can be used during the period of product creation (reducing the research costs), in this case, the stimulation of innovation is carried out at the stage of generating income from the created product.

In most countries where the "Patent box" has been implemented (to include an object of intellectual property into the list of objects giving the right to apply this regime), a necessary condition is that R&D is carried out by a taxpayer, while R&D can be carried out abroad.

4 Conclusion

There are no standard methods of tax stimulation of innovative activity that can be taken for application in Russia, but it is possible to identify the most promising ones that are not used yet, but they may find their place in the tax system:

- application of the "Patent box" regime (the application of this regime in Russia, in such a way that it is used in foreign countries, is irrational, but it is possible to consider some elements that can be used in the Russian practice; by its acceptance, first of all, it is necessary to define and make the list of results of innovative activity which will allow to apply privileges);
- revision of the procedure for recognition of R&D expenditures for innovative enterprises, not only at the final stage of works, but also in the process, as well as providing the possibility of their recognition even if a negative result is obtained, including the usage of special tax regimes;
- use of fixed tax benefits for specified periods of time (for example, during development and implementation);
- tax holidays for enterprises with R&D expenditures exceeding 50%.

References

1. Analytical report Collection of the State statistics service "Russia in Figures 2018" (2018). http://www.gks.ru/free_doc/doc_2018/rusfig/rus18.pdf. Accessed 11 Apr 2019. (in Russian)
2. Hahn, K.: Innovation in times of financialization: do future-oriented innovation strategies suffer? Examples from German industry. Res. Policy **48**(4), 923–935 (2019)
3. Metlyuk, N.V.: The role and place of innovative entrepreneurship in the modern economy. Probl. Mod. Econ. **4**(44), 178–181 (2016)
4. Popova, O.P.: Innovative activity in the field of science-intensive technologies as an integral part of the Russian economy. Innov. Manag. Theory Methodol. Pract. **12**, 155–158 (2015)
5. Pucihar, A., Lenart, G., Borštnar, M.K., Vidmar, D., Marolt, M.: Drivers and outcomes of business model innovation-micro, small and medium-sized enterprises perspective. Sustainability (Switzerland) **11**(2), 344 (2019). https://doi.org/10.3390/su11020344
6. Revina, S.N., Paulov, P.A., Sidorova, A.V.: Regulation of tax havens in the age of globalization and digitalization. In: Ashmarina, S., Mesquita, A., Vochozka, M. (eds.) Digital Transformation of the Economy: Challenges, Trends and New Opportunities. Advances in Intelligent Systems and Computing, vol 908. Springer, Cham (2020). https://doi.org/10.1007/978-3-030-11367-4_8
7. Ryabova, N.Yu.: The role of innovative entrepreneurship in the development of economic systems. Probl. Mod. Econ. **4**(32), 31–33 (2009)
8. Ryumina, Yu.A: Foreign experience of tax stimulation of innovative activities. Bull. TSU **3** (19), 80–85 (2014)
9. Spescha, A., Woerter, M.: Innovation and firm growth over the business cycle. Ind. Innov. **26** (3), 321–347 (2019)

Technological Development 1.0. of the Russian Federation

A. V. Krivtsov[1(✉)] and L. S. Valinurova[2]

[1] Samara State University of Economics, Samara, Russia
2030202@gmail.com
[2] Bashkir State University, Ufa, Bashkortostan Republic, Russia

Abstract. The study considers the competitiveness of Russia in the global innovation market. A series of articles studied the directions of the state's activity in seven spheres: human resources, business environment, infrastructure, institutions, knowledge and technology and creative activity. The research is based on studies of the country's innovation index by domestic and foreign experts. This indicator is composed of 82 different variables that describe in detail the innovative development of countries of the world at different levels of economic development. The authors of the study believe that the success of the economy depends on the innovation potential and the conditions for its implementation. The authors suggest ways to improve the state-implemented strategy in the medium term, developed by means of SWOT analysis.

Keywords: Digital economy · Digitalization · Global innovation development index · Human development index · Knowledge economy · Science and technology development

1 Introduction

The development of any country cannot take place chaotically and it is necessary to determine benchmarks and indicators of this development. In particular, in the Russian Federation such "beacons" are designated by V.V. Putin in the Address to the Federal Assembly. During this period, the president identified them as issues of social and economic development. National projects, established by the last May Decree, are the "road map movements" of these landmarks.

National projects are built around a person to achieve a new quality of life for citizens, which can only be achieved with the dynamic development of Russia, expressed in the following indicators:

1. The development of new technologies and digitalization, the formation of competitive industries;
2. Improving the business climate and quality of national jurisdiction. Growth in investment in 2020 should increase by 6–7%. Achieving this level will be one of the key criteria for evaluating the work of the Government.
3. Removing infrastructure constraints for economic development, for unlocking the potential of our regions.

S. I. Ashmarina et al. (Eds.): ISCDTE 2019, LNNS 84, pp. 85–91, 2020.
https://doi.org/10.1007/978-3-030-27015-5_11

4. Training of modern personnel, the creation of a powerful scientific and technological base.

These four directions are currently implemented in the following programs:

- National project SCIENCE (includes Federal projects "Development of scientific and scientific-production cooperation"; "Development of advanced infrastructure for research and development in the Russian Federation"; "Development of human resources in the field of research and development");
- National project EDUCATION (includes Federal projects "Young professionals", "New opportunities for everyone", "Export of education");
- National project DIGITAL ECONOMY (includes the Federal project "Personnel for the digital economy").

The numerical characteristics of these programs are provided for in the Federal Budget, and Federal Law No. 459 provides for measures to implement the programs and criteria for their success.

The field of science, technology and innovation (hereinafter NT-sphere) is a complex, dynamically changing object, the state of which depends on political, economic, social and other external factors. Many of the elements of this system are directly immeasurable. Therefore for their study and formation of the effective policy, we need a comprehensive assessment, which can be based, in particular, on the index method of analysis.

Studies on the basis of indices make it possible to assess the success, the level of development of quite complex objects (the country as a whole, innovation ecosystems, etc.), to discover their advantages and disadvantages, as well as factors that contribute to or hinder progress. In addition, these indicators make it possible to correlate the growth rate of the global innovation market with the growth rate of this market in the Russian Federation. Such studies are provided by the Joint Publication of WIPO, the World Intellectual Property Organization, Cornell University, INSEAD Business School and Information Analytical Partners GII-2018 - GII Rating (Global Innovation Index) 2018, which was devoted to "Innovation as an Energy Source for the World".

The introduction presents a brief overview of areas to be developed in Russia and identifies the country's national development projects in the area under consideration. In order to determine the need and relevance of issues that the development of the Russian Federation is aimed to solve, it is necessary to analyze GII ratings and make a list of "pain points" that need to be addressed. Methods, which can solve these issues, will be discussed in a series of articles after the SWOT analysis.

Research questions are posed as follows:

What spheres of human activity have a special impact on the development and competitiveness of our country?

What areas are priority in the development of the Russian Federation to increase the rating of GII and what can be done for their development?

The study examined the studies of domestic and foreign analysts on the issue. The main part of the work and discussion concerns various spheres of human activity, society and the state - these are human resources, business environment, infrastructure, institutions, knowledge and technology results, as well as creative activity in the Russian Federation and its competitiveness in the global market. In conclusion, brief conclusions are given on the issue and further areas of work are outlined to develop the innovative potential of Russia.

2 Methods

The theoretical and methodological basis of the study is theoretical fundamental concepts of innovation and economic analysis. The study used general scientific methods of cognition and special data processing techniques, in particular, methods of economic analysis, scientific abstraction, logical, induction and deduction methods, which allow considering the problem under study most fully. The authors based their research on resources [1–4].

3 Results

In general, to determine the place of Russia in the global scientific space, it is worth noting the following positive changes. Russia rose by 19 lines - from the 62th to the 43th places only for the period from 2013 to 2016 in the GII global rating. But, despite the dynamic results, it is difficult for Russia to keep pace with growth rates in other countries. Its position is unsustainable in the ranking, between the most innovative and developing economies. The validity of this statement is proved by the fact that the Russian Federation was only the 46th in the ranking in 2018.

In accordance with the rating, countries are divided into 4 large groups: countries with developed (high-income) economy, with above-average income economy, with below-average income economy, and countries with low income economy. The countries grouped by the economic development and the development of NT:

Russia has above-average income economy. According to GII analysts, our country should be more developed in the field of NT. One of development indicators of scientific thought and NT as a whole is, in our opinion, the number of publications and registered patent applications in the international environment. During the study of these indicators, unconditional leaders were identified.

As for cities rating by the number of publications and patents in the considered areas (calculated per 100 sq. Km), Moscow occupies only the 30th place (18th and 49th place on publications and patents, respectively). The study indicates that for the Russian Federation, the leading specialization of innovation is physics, when in other countries such sciences as chemistry, engineering, oncology are in the lead. Comparing the number of publications in the USA (5,339,705 and patents - 803,058), in Spain (668,199, patents - 26,791) and in Russia, it is clear that the Russian Federation is far behind indicators with the number of publications - 279,909, and patents – 15 347.

Russia is included in the list of those countries that were affected by the 2008 financial crisis (in terms of innovation). Thus, studies show that the budget for scientific research (R & D) in the Russian Federation did not decline during the 2008 crisis, and even had a positive trend. It is worth noting that the majority of developed countries reduced financial investments in NT in 2009 (Israel, Germany, Italy, Greece, Great Britain), and many countries, such as Spain, Portugal, Iceland, Finland showed a negative dynamics of the investment rate in NT until 2016.

GII conducts a comparative analysis of NT of different countries at different stages of their economic development. Such an analysis, correlated with other countries with the same level of income economy, can illustrate important competitive advantages of the country and help interested parties to learn lessons to improve the efficiency of the state.

According to the results of the GII 2019 study, Russia belongs to countries with "above-average income economy". In is should be mentioned that only a few countries in the "above-average income economy" group have separate indicators above the average for the countries with developed economy. This happens only in four areas: for example, Croatia and the Russian Federation show the best results in the field of infrastructure; Thailand, South Africa, Colombia, Peru, Kazakhstan, Mauritius, Azerbaijan and Albania show the best results in terms of market diversity; Russia, Colombia and Brazil - in business development; Croatia, Thailand, Romania and the Islamic Republic of Iran - in the field of knowledge and technology.

This analysis allows us to trace in which areas of NT development Russia has achieved significant success and has reached the level of countries from the "developed economy" group. These areas need all the forces to support and develop. Other areas where the Russian Federation is lagging behind or is on the level with countries from the "above-average income economy" group should be developed and pay more attention in the strategic planning of the country's technical development.

It is worth noting that at the moment the strategic planning of technological development of the Russian Federation is aimed at the "lagging behind" areas of NT development. This means that the programs implemented in the Russian Federation are particularly relevant. Thus, the country's priority is to develop the potential of young scientists, increase the level of secondary and higher education, invest in research centers and, of course, improve the results of the creative work of science workers. This is evidenced by grants and presidential prizes, for example, the State Prize of the Russian Federation in the Field of Science and Technology and the Prize of the President of the Russian Federation in the Field of Science and Innovations for Young Scientists, which are important support and incentives for young scientists. Of course, we must not forget about the business environment. In recent years, this particular area - representatives of small/medium/large businesses of the Russian Federation closely cooperate with the state. According to the results of the rating compiled by the Competence Center of the Federal Project "Digital Technologies", the level of development of the Russian business environment is slightly behind some developed countries. Nevertheless, among the countries included in the rating, Russia is ahead of only Bulgaria, Greece, Brazil and Argentina.

As you know, the necessary factor in creating and running an innovative business is the availability of debt financing. Moreover, the business is in dire need of venture capital. According to the Russian Venture Capital Association, the venture capital market in our country is at the stage of intensive development, rapidly increasing from 2015. But at the moment Russia is still in the very "tail" of the countries in terms of the share of venture financing in GDP (0.004%). This indicator is two times higher among leaders, about 0.36%. This explains the growth of the "business development" indicator according to the points awarded by the GII 2018, and the drop in the cross-country rating by 4 lines.

Summing up the assessment of the business environment in the Russian Federation, it is necessary to mention the opinion of analysts that the main problem in the Russian Federation is not the opening of a commercial enterprise, but its further development and prosperity. Together with the low availability of borrowed capital, the underdevelopment of the financial market, the desire of the population to open their own business is constantly at a low level. As for the benefits, these include a high level of tax regime stability, as experts note. Nevertheless, the level of business development in NT in Russia, although it remains at a low level, is quite consistent with the average for countries with highly developed economy, and in 2015 and 2018 it reached the European average (according to GII). The dynamics of the environment is positive in comparison with other countries. Thus, in the cross-country rating in the "above-average income economy" group, the RF demonstrates an increase in the rating line from the 52th place to the 33th; this is 19 lines in 5 years. Some other indicators pose a significant threat to the country. For example, a clearly low score in the Russian Federation is observed according to the criteria of "institutions", "infrastructure" and "development of creative activity". In addition, the dynamics of these criteria in the ranking of countries is unstable and negative, which means that the rates of development of these functional areas in the Russian Federation are inferior to groups with above-average income economy in other countries. The criterion "infrastructure" is of higher concern.

In January, the World Bank Group published a report on the problem of infrastructure development "Beyond the Gap. How Countries Can Afford the Infrastructure They Need while Protecting the Planet" [3]. The study revealed how much money the countries should invest in the infrastructure development in order to achieve a permanent and significant development of NT, while remaining "eco-friendly". The report explores many possible ways to invest in infrastructure from 2019 to 2030 in order to achieve a certain level of development in the country. It turned out that the primary element is the goal that the states set for themselves, as well as the means to achieve it, allowing to convey the real value of investment to citizens (Table 1).

So, if a country adheres to the first scenario with limited ambitions, and plans to provide water supply and sanitary conditions at a basic level by 2030, the amount of electricity satisfying only the daily expenses of the urban population, as well as access to high-level personal transport, then these countries need to spend 2% of GDP on the infrastructure development.

If countries set more ambitious goals (scenario 3) and want to provide high-quality electricity everywhere, high-quality water supply and high standards of sanitary standards, to get a developed transport system - highways and an extensive system of metro and high-speed trains by 2030 - such countries need to invest 8% of GDP for the infrastructure development.

Table 1. Country and infrastructure investment scenarios

	Electricity	Transport	Water supply and sewage	Flood protection	Irrigation	Σ
Scenario 1. Low cost: less ambitious plans, high efficiency						
Measures	Strongly reduce energy demand through energy efficiency measures Invest in renewable energy and energy efficiency Gradually increase access to electricity in the poorest areas	Increase the use of public rail transport "Group" (i.e., make less spacious) cities	Provide only basic water supply and sanitary standards	Maintain the risk of coastal flooding being relatively constant Accept increased river flood risks based on cost-benefit analysis	Subsidize only the irrigation infrastructure Promote low-meat diets	2% of GDP
Cost	0.9% of GDP	0.53% of GDP	0.32% of GDP	0.06% of GDP	0.12% of GDP	
Scenario 2: Preferable: ambitious plans, high efficiency						
Measures	Invest in renewable energy and energy efficiency Gradually increase access to electricity in the poorest areas	Increase the use of public rail transport "Group" (i.e. make it less spacious) cities Promote the use of electric vehicles	Ensure safe water supply and sanitation using expensive technologies in cities and low-cost technologies in rural areas	Adopt Dutch flood protection standards in coastal areas Accept increased risks from river floods based on cost-benefit analysis	Subsidize only the irrigation infrastructure	4.5% of GDP
Cost	2.2% of GDP	1.3% of GDP	0.55% of GDP	0.32% of GDP	0.13% of GDP	
Scenario 3. Maximum cost: ambitious plans, low efficiency						
Measures	Do not invest in energy efficiency or demand management Provide high access to electricity using fossil energy for 10 years and give up on the possibility of switching to low-carbon raw materials in advance	Let cities grow Promote rail investment without a concomitant policy to increase rail utilization	Ensure safe water supply and sanitation using expensive technologies in all regions	Adopt Dutch flood protection standards in coastal areas Control the risk of river flood in absolute terms	Subsidize the irrigation infrastructure and electricity to extract water	8.2% of GDP
Cost	3% of GDP	3.3% of GDP	0.65% of GDP	1% of GDP	0.2% of GDP	

Source: compiled by the authors.

But ambitions and goals of countries are not the only lever of development. So, sound strategic planning can help reduce the above costs by up to 50%. Electric solar panels can smooth out electricity demand surges through smart meters and storage, while the government invests in renewable energy - solar panels. With properly built land use policies and "compacted" cities, transport planning will make public transport more reliable and affordable. Planning of water supply and sewage systems will make water available at much lower costs if innovative water treatment systems are used that are supplied with electricity from solar panels in urban areas. The number of leaks and pipe ruptures can be reduced by new control technologies in cities. Using such technologies and correctly exploiting them, countries can spend only 4.5% of GDP on infrastructure, while obtaining results similar to the third scenario discussed above.

Thus, development paths using low-carbon technologies may actually cost less than paths with higher levels of pollution, and they also provide better services to the public. If Russia adheres to the priority path in its development and investment in infrastructure, it will be able to rise in the ranking to higher positions in the coming years.

4 Conclusion

So, the study reviewed the position of Russia in the global innovation market. In particular, attention was paid to the development of scientific activity of the population, business environment and infrastructure. The study considered the directions of the state's activities in the development of these areas and proposed the ways to improve the strategy being implemented.

Some of the indicators that were not considered in this study are a threat to the country and therefore require separate consideration. In addition, it is impossible to consider 3 separate indicators in isolation from the others (7 indicators, which are based on 80 sub-indicators). An analysis of the strengths and weaknesses of all indicators of Russia will be conducted in the forthcoming studies.

References

1. Di Bella, G., Dynnikova, O., Slavov, S.T.: The Russian state's size and its footprint: Have they increased? (2019). https://www.imf.org/en/Publications/WP/Issues/2019/03/09/The-Russian-States-Size-and-its-Footprint-Have-They-Increased-46662. Accessed 10 June 2019
2. European bank for reconstruction and development (2018) Transition Report. People in transition. https://www.ebrd.com/transition-report-201819. Accessed 10 June 2019
3. Rozenberg, J., Fay, M.: Beyond the Gap: How Countries Can Afford the Infrastructure They Need while Protecting the Planet. Sustainable Infrastructure. Washington, DC: World Bank (2019). https://openknowledge.worldbank.org/handle/10986/31291. Accessed 10 June 2019
4. Summary of the 2016 Human Development Report (2015) Moscow: Ves' Mir (2015). https://www.vesmirbooks.ru/books/catalog/state/%203338%20/. Accessed 10 June 2019

Digital Era and Consumer Behavior on the Internet

P. Martiskova and R. Svec(✉)

Institute of Technology and Business in České Budějovice,
Ceske Budejovice, Czech Republic
svec.roman78@gmail.com

Abstract. Living in the digital era is connected with changes that logically influence also entrepreneurs who operate on the internet and run shops in the digitized form (i.e. e-shops). In general, e-shops are very popular with final customers; however, consumers are aware of leaving digital traces on the internet so they are more and more interested in privacy concerns. The aim of this paper is to answer the following research question: "Do men and women behave equally in selected situations?" Necessary data was collected by means of a printed questionnaire (n = 431) and subsequent data analysis was carried out by Chi-square test of independence and descriptive statistics. The overall results show that there are almost as big groups of those who have done it and those who have not. As for purchase refusal due to e-shop's requirement for too much personal data, there is no difference between men and women when shopping online. The majority of respondents has ever refused the purchase.

Keywords: Digital era · Internet consumer behavior · Data privacy concerns

1 Introduction

Changes in communication have been visible from the second half of the twentieth century. These changes, based on movement from analog to digital, have been so considerable that terms like "communication revolution" [6] and "information revolution" [15] have appeared. Another term is "digital explosion" which tries to capture the essence of the today's world full of various mediums we can use for mutual communication [3]. These terms are unambiguously connected with digital era/digital age which is explained as "the present time, when most information is in a digital form, especially when compared to the time when computers were not used" [4].

These changes logically influence also entrepreneurs who operate on the internet and run shops in the digitized form (i.e. e-shops). In general, e-shops are more and more popular with final customers (consumers); according to Lin, Miao and Liu [12], it has led to transformation of shopping habits. This grow in e-shops popularity is mentioned also by Zhang, Wang, Cao and Wang [18] who emphasize the role of advanced internet technologies (such as virtual try-on tool) which are able to enhance consumer experience. The importance of customer experience management is described by Gubiniova and Bartakova [9] who state that outstanding experience for customer is able to tie them emotionally and as a result, it brings profit for a company.

S. I. Ashmarina et al. (Eds.): ISCDTE 2019, LNNS 84, pp. 92–100, 2020.
https://doi.org/10.1007/978-3-030-27015-5_12

However it is necessary to keep in mind that enhancing consumer experience requires data, especially when the content of a particular web page is changed with regard to users' online behavior.

Leaving digital traces on the internet is an inevitable part of the today's internet activities. Cookies are probably the most known method how to get data about users. Bielova [1] focuses in her research on web tracking technologies and points out that every mouse move and button click may be useful for data brokers. Consumers are aware of this fact so they are more and more interested in privacy concerns [10].

Understanding consumers' behavior on the internet and their exact motives leading to buying decisions are a subject of many research studies (e.g. by Mokryn, Bogina and Kuflik [14]). Contemporary consumers are described as impatient, digital-savvy and cost-conscious shoppers [11].

2 Aims and Methods

This paper presents a part of own conducted research, focused on sensitivity perception of personal data. Necessary data for analysis was collected by means of a printed questionnaire. Totally 431 fully completed questionnaires were collected (from October until December 2017). The sample consisted of 217 men and 214 women from the Czech Republic and was balanced (this was tested by means of Chi-square goodness of fit test: Chi-square = 0.021, degrees of freedom = 1, p-value = 0.8851). The youngest respondent was 18 years old and the oldest one was 85 years old. This paper deals with the following research question: "Do men and women behave equally in selected situations?" For purposes of this paper, there is a defined set of selected situations which are related to online behavior:

#1 Have you ever deliberately removed cookies or your browsing history from a computer or tablet?
#2 Have you ever refused to provide personal information that was not relevant to the transaction (from your point of view)? (transaction = e.g. closing a deal, sending your order in an e-shop)
#3 Have you ever used a temporary name?
#4 Have you ever used a temporary e-mail address? (an e-mail address was different than you usually use)
#5 Have you ever given inaccurate or misleading information about yourself? (for instance: a fake phone number during the registration for a loyalty program)
#6 Have you ever decided not to buy a particular product because the e-shop required too much personal data? (for instance: necessity to register when purchasing on-line)

Respondents could answer the subquestions choosing one option: "Yes"/"No"/"I cannot remember, I do not know" (these options served for expression of the level of respondents' agreement). In order to answer the above stated research question, Chi-square test of independence and descriptive statistics were used.

3 Results and Discussion

At first, the Chi-square test of independence was used in order to find out whether men and women behave equally in selected situations. The null hypothesis (Ho) stated gender and level of agreement with a particular situation are not associated. The alternative hypothesis (Ha) stated gender and the level of agreement with a particular situation are associated, i.e. behavior is dependent on respondents' gender. Based on Table 1, it can be said there are three situations when behavior of men and women differ significantly: #1 (unambiguously), #2 and #4 (if the level of significance is 0.10).

Table 1. Statistic results – (in)dependence between the level of agreement and respondents' gender

#	Situations (described in the form of subquestions)	Chi-square statistic	P-value	Interpretation (if level of significance = .05)
1.	Have you ever deliberately removed cookies or your browsing history from a computer or tablet?	8.732	0.0127	Ha
2.	Have you ever refused to provide personal information that was not relevant to the transaction (from your point of view)?	5.933	0.0515	Ho (if level of significance = .10, then Ha)
3.	Have you ever used a temporary name?	2.543	0.2804	Ho
4.	Have you ever used a temporary e-mail address?	5.040	0.0805	Ho (if level of significance = .10, then Ha)
5.	Have you ever given inaccurate or misleading information about yourself?	2.110	0.3482	Ho
6.	Have you ever decided not to buy a particular product because the e-shop required too much personal data?	1.388	0.4996	Ho

Source: compiled by the authors.
Note: df = 2; level of significance = .05 (than chi-square statistic has to be higher than 5.991 for supporting Ha); p-value is rounded to 4 decimal places.

3.1 Deliberate Removal of Cookies or Browsing History from a Computer/Tablet (#1)

In this case, there is a clear difference between men and women – Chi-square test of independence has confirmed it (df = 2; chi-square statistics = 8.732; p-value = .0127). Figure 1 shows the difference between men and women in a visual form. Men tend more to remove cookies or their browsing history from a computer or a tablet (69% men versus 57% women). Tendency to remove cookies is also obvious from the amount of results given by Google search engine – if the key words "clear cookies" are entered, then 1 970 000 000 are found within 0.34 s (found out on May 27, 2019). There are many descriptions on the internet how to remove cookies and these descriptions have also a form of videos (e.g. Google Help [8] etc.).

It is interesting to point out that only a quarter of respondents has never removed cookies or browsing history (22% men and 26% women) – in other words it can be said they have not need to hide anything (it can be a situation when they do not share a computer/tablet so they do not care about cookies or their browsing history). Women more often were not able to remember whether they have or have not removed it (17% women versus 9% men).

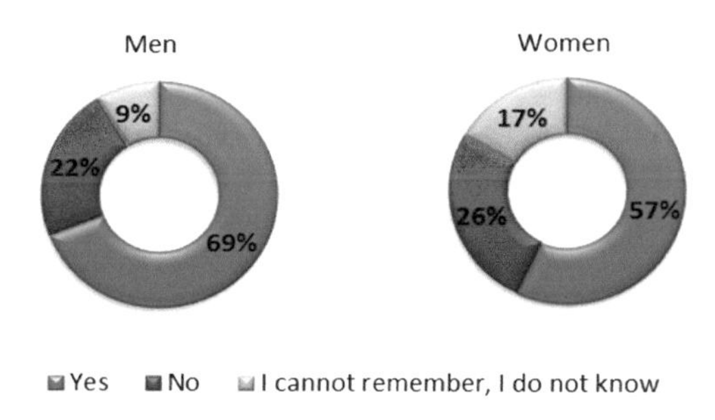

Fig. 1. Deliberate removal of cookies or browsing history from a computer/tablet (#1) (Source: compiled by the authors).

3.2 Refusal to Provide Personal Information (#2)

As for refusal to provide personal information which was considered not to be relevant to the transaction, men and women differ a bit: if the level of significance is 0.10, there is a difference; however, if the level of significance is 0.05, the difference is not provable (df = 2; chi-square statistics = 5.933; p-value = .0515). Still it is useful to show the results in a visual form – see Fig. 2.

Men seem to be more cautious when providing personal information (72% men versus 63% women refuse to provide personal information). Willingness to provide personal information is connected with the perceived degree of sensitivity of such information – the higher the degree is, the higher probability of refusal can be expected [13]. As for potential differences between age categories, behavior of adolescent + young adult and adult individuals was studied in a paper by Bietz, Cheung, Rubanovich, Schairer and Bloss [2]: their results show that there are similar levels of general privacy concern. Again, women are more forgetful (or they do not attach importance to this) because they claim twice more they are not able to remember (14% women versus 7% men do not know or cannot remember), Fig. 2.

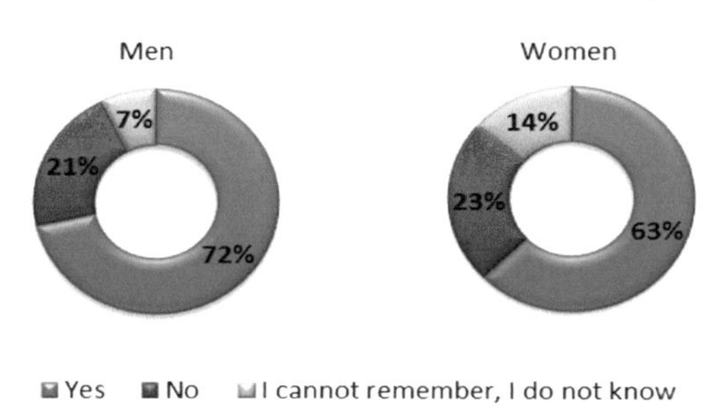

Fig. 2. Refusal to provide personal information (#2) (Source: compiled by the authors).

For business practice it will be useful to find out what kinds of information respondents consider as information which is not relevant to the transaction (typically sending an order in an e-shop etc.) – that can be a part of future research activities.

3.3 Use of a Temporary Name (#3)

Chi-square test of independence indicates there is no difference between men and women in case of using a temporary name (df = 2; chi-square statistics = 2.543; p-value = .2804). For this reason, Fig. 3 presents the results regardless the respondents' gender. As for use of a temporary name, only a quarter of respondents claims they have used it (27%). More than a majority of respondents has not used any temporary name (64%) – it indicates using a temporary name is not used very often. It indicates people do not have a need to hide themselves and create another identity. It can be solved by other solutions how to protect own online identity (some pieces of advice how to do it is described e.g. by Fisher [7].

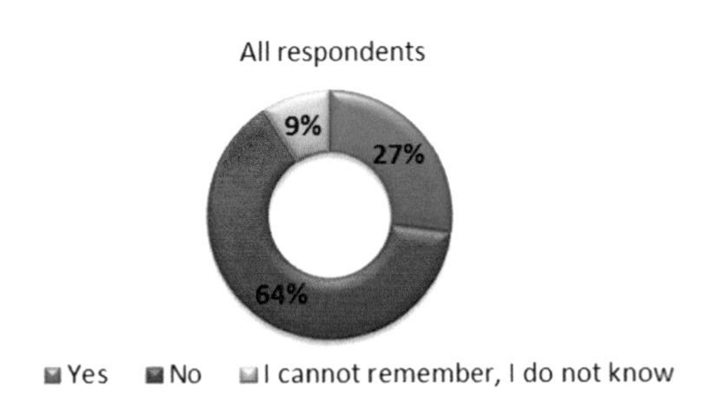

Fig. 3. Use of a temporary name (#3) (Source: compiled by the authors).

3.4 Use of a Temporary E-Mail Address (#4)

As for using a temporary e-mail address, men and women differ a bit: if the level of significance is 0.10, there is a difference; however, if the level of significance is 0.05, the difference is not provable (df = 2; chi-square statistics = 5.040; p-value = .0805). Still it is useful to show the results in a visual form – see Fig. 4. A temporary/disposable e-mail address is different than a respondent usually uses.

The majority of respondents has never used a temporary e-mail address (52% men versus 60% women). However, there is still a considerable number of those who has already done it – 41% men and 31% women. In some situations (e.g. signing up for a contest or buying online) it can be useful to use a disposable e-mail address in order to avoid spam in the private e-mail box because receiving spam (= junk e-mails) is considered to be very annoying [11, 17].

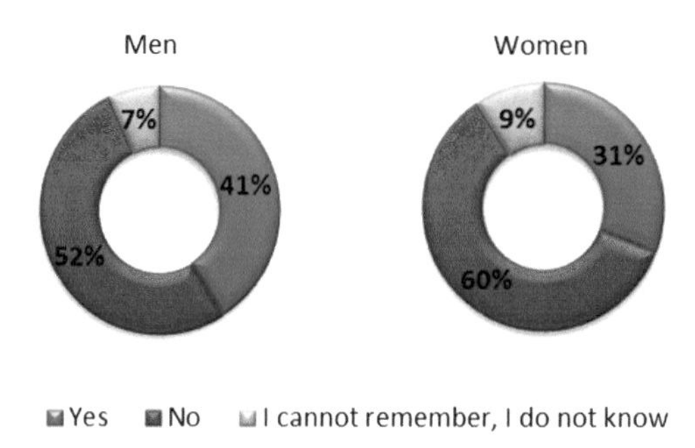

Fig. 4. Use of a temporary e-mail address (#4) (Source: compiled by the authors).

3.5 Giving Inaccurate or Misleading Information About Yourself (#5)

Inaccurate or misleading information can be for instance a fake phone number which is provided during the registration for a loyalty program. Misleading information is also connected with the term obfuscation [5]. Chi-square test of independence indicates there is no difference between men and women in case of giving inaccurate or misleading information about yourself (df = 2; chi-square statistics = 2.110; p-value = .3482) so Fig. 5 presents the results regardless the respondents' gender. The results show that there are almost as big groups of those who have done it and those who have not (yes = 40%, no = 49%).

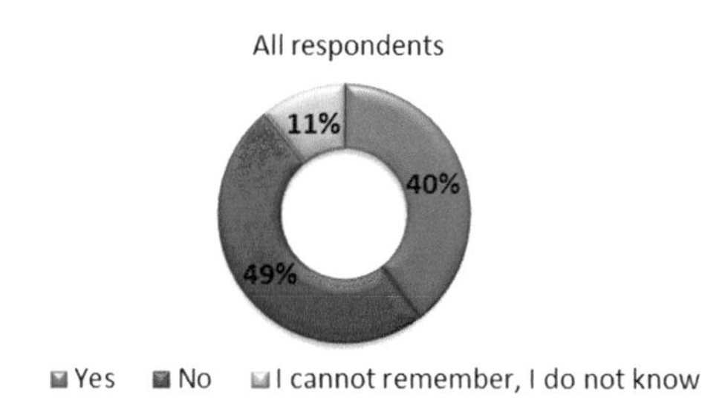

Fig. 5. Giving inaccurate or misleading information about yourself (#5) (Source: compiled by the authors).

3.6 E-Shop's Purchase Refusal Due to Its Requirement for Too Much Personal Data (#6)

E-shop's purchase refusal due to its requirement for too much personal data can occur for example in a situation when there is a necessity to register in order to complete the order. Also in this case, Chi-square test of independence indicates there is no difference between men and women when shopping online (df = 2; chi-square statistics = 1.388; p–value = .4996). For this reason, Fig. 6 shows the results regardless the respondents' gender. The majority of respondents has ever refused the purchase (52%). This is quite alarming for e-shops and it indicates that requirements for too much personal data can be a very serious obstacle for customers shopping on-line. When comparing this finding with usually published reasons why customers leave e-shops, requirements for personal data can be a part of mandatory creating a user/customer account (e.g. compared by Serrano [16]). Forcing to create a new user account is the second most common reason why customers abandon their carts or even abandon during checkouts [16].

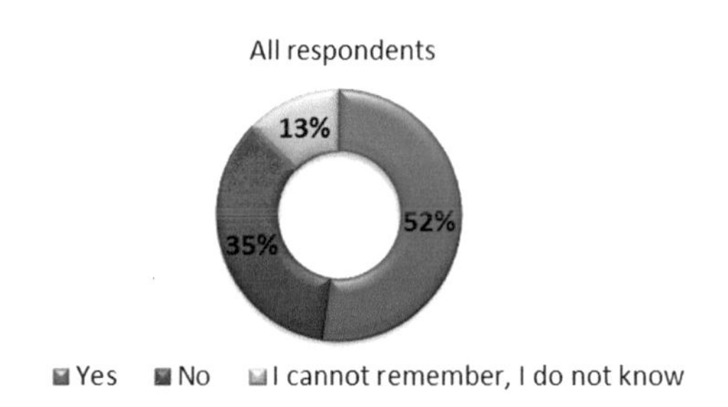

Fig. 6. E-shop's purchase refusal due to its requirement for too much personal data (#6) (Source: compiled by the authors).

4 Conclusion

The aim of this paper was to answer the following research question: "Do men and women be-have equally in selected situations?" In order to find answers, there was a defined set of selected situations which are related to online behavior. There are presented findings in brief.

The first situation was about deliberate removal of cookies or browsing history from a computer/tablet. It was found out men tend more to remove cookies or their browsing history (69% men versus 57% women). It is interesting to point out that only a quarter of respondents has never removed cookies or browsing history (22% men and 26% women) – in other words it can be said they have not need to hide anything (it can be a situation when they do not share a computer/tablet so they do not care about cookies or their browsing history).

The second situation was focused on refusal to provide personal information which was considered not to be relevant to the transaction. In this case, men and women differ a bit because men seem to be more cautious when providing personal information (72% men versus 63% women refuse to provide personal information).

The third situation dealt with using a temporary name. There is no difference between man and women: only a quarter of respondents claims they have used it (27%) and more than a majority of respondents has not used any temporary name (64%). It indicates people do not have a need to hide themselves and create another identity.

The fourth situation was about use of a temporary e-mail address. As for using a temporary e-mail address (a temporary/disposable e-mail address is different than a respondent usually uses), men and women differ only a bit. The majority of respondents has never used a temporary e-mail address (52% men versus 60% women).

However, there is still a considerable number of those who has already done it – 41% men and 31% women.

As for the fifth situation, inaccurate or misleading information can be for instance a fake phone number which is provided during the registration for a loyalty program. Chi-square test of independence indicates there is no difference between men and women in case of giving inaccurate or misleading information about yourself. The overall results show that there are almost as big groups of those who have done it and those who have not (yes = 40%, no = 49%).

Last but not least situation was about e-shop's purchase refusal due to its requirement for too much personal data which can occur for example in a situation when there is a necessity to register in order to complete the order. Also in this case, Chi-square test of independence indicates there is no difference between men and women when shopping online. The majority of respondents has ever refused the purchase (52%). This is quite alarming for e-shops and it indicates that requirements for too much personal data can be a very serious obstacle for customers shopping on-line.

References

1. Bielova, N.: Web tracking technologies and protection mechanisms. In: Proceedings of the ACM Conference on Computer and Communications Security, Dallas, United States, 30 October 2017, pp. 2607–2609 (2017)
2. Bietz, M.J., Cheung, C., Rubanovich, C.K., Schairer, C., Bloss, C.S.: Privacy perceptions and norms in youth and adults. Clin. Pract. Pediatr. Psychol. **7**(1), 93–103 (2019)
3. Burger, I.: Are we over-communicating in today's digital era? (2018). https://www.dvt.co.za/news-insights/insights/item/316-are-we-over-communicating-in-today-s-digital-era. Accessed 29 May 2019
4. Cambridge University Press: Cambridge Dictionary – "Digital age" (2019). https://dictionary.cambridge.org/dictionary/english/digital-age. Accessed 30 May 2019
5. Doyle, T.: Privacy, obfuscation, and propertization. IFLA J. **44**(3), 229–239 (2018)
6. Elliott, D., Spence, E.H.: Ethics for a Digital Era. Wiley, Hoboken (2018)
7. Fisher, T.: Ten Ways You Can Hide Your Online Identity (2019). http://www.lifewire.com/hide-online-identity-3483172. Accessed 28 May 2019
8. Google Help: Clear cache and cookies in Google Chrome (2019). http://www.youtube.com/watch?v=-kJvSmlLC28. Accessed 27 May 2019
9. Gubiniova, K., Bartakova, G.P.: Customer experience management as a new source of competitive advantage for companies. In: Kubickova, V., Brevenikova, D. (eds.) Proceedings of the 5th International Scientific Conference on Trade, International Business and Tourism: Application of Knowledge in Process of Business Dynamization in Central Europe, Mojmirovce, Slovakia, pp. 154–160 (2014)
10. Jiang, Y., Wang, C., Wang, Y., Gao, L.: A privacy-preserving e-commerce system based on the blockchain technology. In: IWBOSE 2019 - 2019 IEEE 2nd International Workshop on Blockchain Oriented Software Engineering, article number 8666470, Hangzhou, China, pp. 50–55 (2019)
11. Kalms, B., Freestone, O. (eds.): Re-Thinking Retail in the Digital Era. LID Publishing Ltd., London (2014)

12. Lin, G., Miao, Y., Liu, S.: Decision-behavior based online shopping. In: 15th International Conference on Control, Automation, Robotics and Vision, ICARCV 2018, Singapore, article number 8581336, pp. 1321–1326 (2018)
13. Martiskova, P., Svec, R., Slaba, M.: Perceived sensitivity of personal data in the globalised world: common personal data and health-related data. In: Kliestik, T. (ed.) 18th International Scientific Conference Globalization and Its Socio-Economic Consequences, University of Zilina, Faculty of Operation and Economics of Transport and Communications, Department of Economics, pp. 2229–2236. Rajecke Teplice, Slovak Republic (2018)
14. Mokryn, O., Bogina, V., Kuflik, T.: Will this session end with a purchase? Inferring current purchase intent of anonymous visitors. Electron. Commer. Res. Appl. **34**, 100836 (2019)
15. Pradhan, R.P., Arvin, M.B., Nair, M., Bennett, S.E., Hall, J.H.: The information revolution, innovation diffusion and economic growth: an examination of causal links in European countries. Qual. Quant. **53**(3), 1529–1563 (2019)
16. Serrano, S.: Top 10 Reasons (and solutions) for Shopping Cart Abandonment (2019). http://www.barilliance.com/10-reasons-shopping-cart-abandonment/. Accessed 28 May 2019
17. Stucken, A.: How can I stop getting spam emails? (2010). http://www.bbc.co.uk/webwise/guides/stopping-spam-emails. Accessed 29 May 2019
18. Zhang, T., Wang, W.Y.C., Cao, L., Wang, Y.: The role of virtual try-on technology in online purchase decision from consumers' aspect. Internet Research (2019, article in press)

The Role of Planetary Property as a Basis for Sustainable Harmonious Economic Development

O. E. Ryazanova[1(✉)] and V. P. Zolotareva[2]

[1] Moscow State Institute of International Relations (University) of the Ministry of Foreign Affairs of the Russian Federation, Moscow, Russia
afinna2011@yandex.ru
[2] Moscow Polytechnic University, Moscow, Russia

Abstract. The penetration of digital technologies into the main areas of activity does not always have a positive effect. The anthropogenic factor, intruding into natural processes, has a great influence on the quality and degree of harmlessness of vital sources of the human environment. In the digital era there is a threat of substitution of initial data, indicators, hiding any facts that contribute to the growth of negative processes in the economy. The purpose of this study is to understand the essence of planetary property as a fundamental institution of sustainable planetary development, its role in the realization of the right to common resources for all countries and peoples, ensuring the creation and appropriation of planetary rent and the creation of a biocompatible economy as a basis for solving global problems. In this regard, the most important task facing scientific community is to present economic factors of the development and functioning of society as basic and common for all states and peoples. Using a hypothetical method and a harmonization approach in stating the problems and contradictions that arise in the era of digitalization, the conclusion is drawn about the role of planetary property as the basis for sustainable economic development.

Keywords: Biocompatible economy · Digitalization · Economic development · Natural resources · Planetary property · Planetary rent

1 Introduction

The aggravation of the global systemic crisis, as the crisis of modern civilization as a whole, makes it necessary to search for new forms of the system of international economic relations and its sustainable development. The basis should be the harmony of all subsystems: economic, political, social, cultural with each other and with nature. Unlike the Sustainable Development Concept ("On the Concept of the Transition of the Russian Federation to Sustainable Development", 1996 [7]), the Harmony Concept [11] focuses on harmonizing world development, creating a noospheric civilization based on resource-saving and replacement production, a biocompatible economy, the subjects of which are humanity (concept of controlled harmony). A biocompatible economy is an economy capable of performing its functions, developing in harmony

S. I. Ashmarina et al. (Eds.): ISCDTE 2019, LNNS 84, pp. 101–106, 2020.
https://doi.org/10.1007/978-3-030-27015-5_13

with nature, providing economy and restoration of natural resources and without causing significant negative consequences for living organisms.

The synergistic approach that prevails today in the scientific community leads to underestimation of the harmonization approach, the importance of which is noted by Muratov [5]. He emphasizes that harmonization is the coordination of actions, interests and goals, and consistency implies compatibility (non-antagonistic) of the parts of a whole whose realization is manifested through efficiency, quality and effectiveness. It results in:

- Elimination of imbalances;
- Coordination of various contradictions in interests, actions, purposes;
- Continuous improvement of the features of the object, bringing them into line with changing requirements and conditions.
- Thus, the objectives of sustainable development of the modern world in terms of the harmonization approach are the following:
- Provide the material and financial basis for global reformation and modernization activities without prejudice to the existing economic systems;
- Conduct cumulative political, social, cultural, economic and historical research aimed at harmonizing world development, based on resource-saving and replacement production;
- Settle the ownership of planetary resources, form institutions and the mechanism for their assessment, seize planetary rent as a special income belonging to all mankind.

The basis of the normative base of civilization sustainable development can be the Planetary Project proposed by the Institute of Planetary Development [6]. The European Commission's plan for the transition to a closed-cycle economy model, adopted in December 2015, fits into the scope of this project, since this has the advantages of a global level. They reduce the negative environmental impact by reducing the use of resources in manufacturing, reducing production costs due to a decrease in the amount of primary resources used, the emergence of new markets and the creation of new jobs and an increase in overall well-being [8]. The hypothesis of the study: to provide arguments about the phenomenon and content of planetary property, which play an important role in the concept of sustainable harmonious economic development.

2 Methodology

The mechanism for the realization of planetary property is most intense manifested in the exploitation of natural resources. In the course of economic theory, the benefits are classified into economic and non-economic, and natural resources - preemptible and non-preemtible. Non-economic (free) goods are those that are available in unlimited quantities in relation to human needs (example: air, water). As Samuelson and Nordhaus believed, preemptible natural resources are when firms or consumers can get a benefit from their use (examples: vineyards or oil fields). Natural resources are called non-preemtible when the associated costs or benefits are not fully transferred to the owners (for example, air quality and climate, which are related to external effects generated by activities such as fossil fuel combustion) and external effects can occur when using these resources [9].

However, the practice shows that the relation to a property owned by someone is somewhat different than that of an ownerless resource. But here is a paradox - there are no nobody's resources. Obviously, as soon as a resource is directly or indirectly involved in the production process, it is automatically involved in the relations by a business entity, and thus property relations arise.

"Renewable resources, according to Akimova and Haskin, are substances and forces created on the Earth due to the flow of solar energy: heat, atmospheric moisture, water of sediments and fresh water bodies, river flow and hydropower, winds, waves, soil, some minerals, all living organisms, ecosystems, biosphere, man" [1]. In the man-made conditions of the XXI century and under the influence of human activity, the planetary balance of resources as a result of anthropogenic qualitative transformations of the environment changed dramatically, the amount of resources that seemed inexhaustible was significantly limited, and the law of limitation (exhaustion) of all natural resources began to apply. From the standpoint of the ecological-economic classification based on exhaustion and renewability, resources are divided into space and planetary [10].

In more detail the classification of planetary resources presented by Bezgodov [4] (Table 1).

Table 1. Planetary resources

№	Resource	Components	Valuable Items
1	Near-Earth space	Space bodies (the Sun, planets, meteorites, comets …), their orbits and energy (gravitational and magnetic fields, radiation, electromagnetic radiation, solar radiation …)	Sunlight and thermal energy
2	Atmosphere of the Earth	Troposphere, stratosphere, mesosphere, thermosphere, exosphere)	Air, oxygen and ozone
3	Solid and liquid substance of the Earth's surface	Continents and islands of the Earth's land, soil, geothermal energy, oceans, inland waters, hydrothermal energy	Fertile soils and fresh water
4	Climate as a set of climatic zones of the Earth	Air temperature, water temperature, atmospheric pressure, light regime	Climates suitable for the existence and reproduction of complex forms of protein life
5	Earth subsoil	Continental crust, oceanic crust, minerals, outer mantle, inner mantle, outer core, core	Irreplaceable exhaustible mineral resources
6	Biosphere as a complete system	Biota, including man and man-made systems	Man (due to his ability to influence the environment and be responsible for this)
7	Noosphere	Information, civilization-historical experience, languages and ideological paradigms, cultures, scientific knowledge, aesthetic and ethical values, the intellectual potential of humanity - the collective mind	Intellectual potential of humanity

The noosphere is singled out from this classification, since it is not a product of human activity, but an integral part of nature, a planetary asset. This is a unified system resulting from the interaction of nature and man, the natural and artificial worlds, living and indirect matter.

Thus, it is difficult to draw a clear distinction between "planetary resources" and "natural resources", but still the concept of "planetary resources" is somewhat broader than natural ones. Planetary resources have the following characteristics:

- These resources are fundamental and strategic for all other types of resources, i.e. without which sustainability and reproduction of the Earth as an integrated system are impossible;
- They are non-preemptible, because of their scale and system-forming nature, they exceed norms, claims and limitations of any national systems of law.

The efficiency of the distribution and use of any kind of resources is primarily determined by the form of ownership and only then by technology and the organization of production. Therefore, planetary property plays an important role in the concept of sustainable harmonious economic development.

3 Results

In the economic and legal literature there have been attempts to study the category of planetary property [3], but it has not become a full-fledged economic and legal institution with a special role in macroeconomic processes due to the unclear structuring of this definition. With a certain degree of conditionality, it is necessary to distinguish between planetary and planetary property. Planetary property includes state-owned property within state borders under the national jurisdiction of countries and planetary property that is outside state borders, outside national jurisdiction, managed and regulated by supranational structures. Planetary property is the highest form of public ownership of objects of joint consumption, which can become a real financial, economic and legal basis for universal integration in the digitalization era implementing the strategy of sustainable economic development of countries, regions and the world community as a whole.

The objective material basis of any activity is not only resources, but also the established form of ownership of them and its economic and legal implementation. As for the powers of planetary property, they can be implemented through the institutionalization of planetary corporations that manage the systemic use of space resources, the atmosphere, the oceans and the global information system.

The rights of ownership and realization of the rights of ownership of planetary resources, in principle, belong to all of humanity. However, humanity as a collective biosocial, planetary subject is partially presented situationally, in one or another aspect of law enforcement, in one or another concrete historical, socio-legal, and production-economic form. Therefore, all 11 powers of planetary property are shared by structures, institutions, organizations acting on behalf of all mankind and on behalf of individual groups.

The right to own planetary property due to difficulties in understanding and delimiting it from the national state ownership of countries is problematic, so the primary task for the scientific community is to find a consensus on a common understanding of the categorical apparatus.

The right to use planetary resources should belong to all the people of the Earth, but in reality, the modes of operation of the atmosphere, the World Ocean, and space are not defined with a due degree of consistency and coherence at the international level.

The right to income (rent) from the use of planetary resources is specific: the aggregate net income is distributed only between the entities participating in the use of planetary resources. The right of the sovereign belongs to humanity: it is impossible to violate the laws of nature, the natural processes of reproduction of the atmosphere, the oceans or the cosmos. The right of the sovereign is closely related to the right to security. The implementation of the right to transfer by inheritance is manifested in the preservation of the planet Earth for future generations. The right to indefinite is realized in the fact that the planet Earth indefinitely belongs to humanity and all species, elements of the biosphere.

A ban on use to the detriment of nature and society should be considered as natural, categorical, adopted at the level of the planetary environmental agreement and, most importantly, with absolute responsibility under all lease agreements for planetary resources.

4 Discussion

Most types and volumes of natural resources are part of national-state ownership systems and are governed by the current legislation of countries and their specific economic and legal institutions. At the same time, globalization processes introduced requirements to nature management for harmonizing national interests with common human rights on the management of the natural wealth of the Earth, which are the property of all its inhabitants. The natural reaction was decisions of the UN Conference on the Environment in 1992 in Rio de Janeiro, taken in accordance with the strategy of sustainable development, in the framework of which planetary property was recognized.

5 Conclusion

The conducted study allows drawing the following conclusions:

Globalization has introduced requirements in nature management to harmonize national interests with universal in terms of the use of natural and planetary resources and the private appropriation of the corresponding rent;

In the 21st century, the law of limited natural resources and consequences of its implementation is becoming more and more noticeable in new technospheric conditions, since man has greatly changed the planetary balance of resources as a result of anthropogenic impact on the environment.

The important role belongs to planetary property, the specification of the rights of which can be realized only under creation and functioning of the corresponding information and institutional infrastructure of the digital economy;

Supranational institutions (ECOSOC, TNCs), as subjects of planetary rent, should solve the issues of mandatory introduction of international and national special taxes, which will make it possible to efficiently use global property and ensure sustainable economic development [2];

The state, as the owner of natural resources, should create funds based on resource revenues, and a person engaged in economic activities that exploits planetary resources should be socially and legally responsible not only at the national but also at the international level for violating environmental standards and requirements according to international agreements adopted by the UN concerning planetary resources.

Acknowledgements. The research was conducted by the authors with the financial support of the Russian Foundation for Basic Research, Project no. 19-011-20091.

References

1. Akimova, T.A., Haskin, V.V.: Resources and dynamics of the biosphere. Energy Econ. Technol. Ecol. http://naukarus.com/resursy-i-dinamika-ekosfery. Accessed 19 June 2019. (in Russia)
2. Batrakova, L.G., Grigoriev, A.V.: Payments for the use of natural resources: economic and historical aspect. Yarosl. Pedagog. Gaz. **2**(1), 95–100 (2012). (in Russia)
3. Bazylev, N.I., Krivulya, D.S.: Planetary Ownership. Misanta, Moscow (2012). (in Russia)
4. Bezgodov, A.: Planetary Rent as a Tool for Solving Global Problems. Peter, St. Petersburg (2017). (in Russia)
5. Muratov, A.S.: Harmonization approach to enterprise management. Russ. Bus. **4**(2), 64–72 (2011). (in Russia)
6. Planetary Project: from Sustainable Development to Managed Harmony. http://www.ru.planetaryproject.com. Accessed 19 June 2019. (in Russia)
7. Presidential Decree of April 4, 1996 No. 440. On the Concept of the Transition of the Russian Federation to Sustainable Development. http://www.consultant.ru/cons/cgi/online.cgi?req=doc;base=EXP;n=233558#09599027536089231. Accessed 19 June 2019. (in Russia)
8. Ryazanova, O.E.: Small business in a closed-cycle economy. Econ. Horiz. **6**(39), 17–20 (2017). (in Russia)
9. Samuelson, P., Nordhaus, V.: Economics. Williams, Moscow (2018). (in Russia)
10. Sundahl Mark, J.: The Outer Space Convention. Martinus Nijhoff Publishers, Leiden (2013)
11. The Concept of Controlled Harmony. http://ru.planetaryproject.com

Sustainable Economic Development in the Context of Digitalization: Threats and Opportunities

Stages of Innovative Production in the Traditional Factory

R. Sh. Bikmetov[1], N. V. Starun[2], and D. V. Aleshkova[2](✉)

[1] Sterlitamak Branch of Bashkir State University, Sterlitamak, Russia
[2] Samara State University of Economics, Samara, Russia
dashajuly343@gmail.com

Abstract. The majority of enterprises are to introduce innovative production for sustainable development. According to the authors, modern Russian enterprises face the challenge of a radical update. Enterprises around the world need to learn how to produce high-demand, high-quality products, electric cars and electric scooters. The authors believe that 3D-printers of various types, various gadgets with more and more advanced characteristics, the latest medical devices are becoming more and more in demand. The purpose of the study is to develop an implementation strategy for innovative production in traditional factories. The study presents the strategy for the transition to innovative production that is in demand in various countries of the world today. The authors propose a basic formula that can calculate the necessary amount of investment required for the transition to innovative production.

Keywords: High-tech products · Innovation · Investments · Management · Strategy

1 Introduction

The modern world is a world of changing technologies, a world of breakthrough in various fields, the world of innovation. Production of innovative products is the opportunity to receive high profits, the ability to successfully compete with the best transnational companies of the world in various markets, the opportunity to attract the best highly skilled workers from around the world. For example, the Alphabet Corporation receives 1 million resumes every year and all creative people want to work here. Production of 3D-printers, electric cars, various gadgets, space rockets, medical equipment - all these goods are extremely in demand in various markets. The main problem is that factories producing unclaimed, rapidly aging product is to answer the question - how quickly and without devastating consequences to go to innovative production and therefore regain competitive positions. The purpose of the study is to introduce innovative production in a traditional factory with outdated equipment and producing unclaimed products. The authors will consider the introduction of innovative production in a traditional factory as a sequential process consisting of certain stages, as well as requiring certain investments, which will be calculated by the determined basic formula.

S. I. Ashmarina et al. (Eds.): ISCDTE 2019, LNNS 84, pp. 109–115, 2020.
https://doi.org/10.1007/978-3-030-27015-5_14

2 Methodology

Research methods are system analysis, structural-functional method, analogy method and expert assessment. At the first stage, a systematic analysis, a method of analogies, and expert assessment were used to form a consistent strategy for the transition to innovative production. At the second stage, a structural-functional method and expert assessment were used to determine a basic investment formula that is necessary for the introduction of innovative production.

3 Results

3.1 Theoretical Basis of the Considered Issues

Today Russia has signs of a rather serious socio-economic crisis. The reasons for this crisis are the inability of the former raw material model of the economy to work effectively and attract investment. At this stage, it becomes clear that the Russian economy needs deep-seated structural reforms, a new economic model should be based on high-tech products, like those of Japan or South Korea. The gross regional product of one state in California (at the expense of Silicon Valley, which is located in the territory of this state) is almost twice as large as Russia's GDP. In the modern world, organizations benefit from high-tech, high-quality products that have a stable demand [20]. As an example, look at the activities of Apple, the capitalization of this company as of September 2018 has already exceeded $ 1 trillion. The pre-sale presentation clips of this company collect millions of views, and the products of this company are in crazy demand. Huge queues for new products from Apple are in New York, Moscow, Berlin, London and Melbourne. It is also necessary to take into account that the products of this company are extremely expensive (at least for Russians), but their high-tech component attracts everyone's attention. Taking into account the experience of such successful projects is the most important task of any manager. The technological project is one of the main activities when getting ready for innovations, modernization or reconstruction of production. The technological project is the first model in the structure of future production, which allows modeling future production, including its economic consequences. The technological project is a system of activities related to the analysis, planning, design and visualization of new, extended, reconstructed and modernized production facilities or parts thereof [10].

The innovative development strategy for any traditional factory should begin with the definition of the basis for development. At the heart of this development is the fact that products manufactured before are no longer in demand and they do not meet the requirements of consumers. A new challenge is the development of an organization through innovative high-tech products [13]. As an example, we can mention the founder of SpaceX and Tesla companies, the American billionaire Ilona Mask. Today, his projects, which only recently seemed to be science fiction, are actively being carried out. The newest project - Hyperloop, a vacuum superfast train, has interested a variety of countries, including Russia. The construction options for the Hyperloop tunnel in the Far East, as well as between Moscow and Kazan, are being considered. Using the example of Tesla, you can see and explore Ilona Mask's step-by-step strategy. Its strategy has the following components:

1. Develop a plan, a project. Even if it is unrealizable from the standpoint of individual skeptics, who are always here when it comes to innovation, in fact it is "non-Luddite" [17]. Consider the possible demand for the product produced by the project. Study social and economic efficiency. For example, the social efficiency of Tesla electric vehicles is that the urban environment is improving due to the absence of emissions to the atmosphere.
2. Form a team of like-minded people with creative, courageous, extraordinary thinking. A cool team of highly qualified specialists is the basis for the project (startup) promotion. To implement Tesla Motors I. Mask gathered the best automotive industry specialists from the USA and Canada.
3. Search for the place of production. Ilona Mask's company bought an abandoned automobile factory on the edge of Silicon Valley and turned it into a super-technological and extremely robotic factory to produce electric vehicles.
4. Launch a test unit production of an innovative product with a high cost, then mass production. Finally, if it is successful, there is mass production of an innovative product with a relative low price for the consumer. The first Tesla electric cars were extremely expensive, about 50 thousand dollars. Subsequently, the company started mass production of relatively cheap electric vehicles worth $ 35 thousand. In the US, there are government subsidies for buyers of electric vehicles, so Tesla Model 3 will cost even less than 35 thousand dollars. New Tesla will cost even less - 21 thousand dollars.
5. Expand markets. Tesla has announced the construction of factories in Germany and China. The world is expanding the network of free gas stations equipped with solar panels. The number of consumers of electric vehicles is growing in the countries of the world. Today, there is a boom of electric cars in the United States, Norway, Canada, Japan, Thailand and South Korea.

A similar strategy of success in promoting an innovative product can also be used in developing the innovative development strategy for any traditional enterprise.

3.2 The Strategy of Creating Innovative Production at a Traditional Enterprise

At present, the developed countries of the world experience a real technological boom, when all new innovative developments in various fields of activity change the previous approaches. If earlier the most important factor in the development of the economy of many countries was mineral resources (oil, gas, ore, etc.), now creative individuals and teams capable of creating breakthrough innovations have come to the fore [7]. Until recently, many innovations seemed fiction but they are already firmly entering the everyday life of people in different countries of the world. These are electric cars and electric scooters, 3D printers, various gadgets with more and more advanced features, online trading, blockchain technology and newest medical products [3, 5]. These innovative developments are drivers of economic growth in their countries and regions. One of these major innovation technology centers is Silicon Valley in California. In a situation when the technological revolution is before our eyes, we cannot stand aside. It is necessary to ensure a technological boom, a breakthrough growth of the entire economy by promoting innovation on the territory of the entire Russian Federation.

We can insure ourselves against excessive risks of mineral depletion using the example of Bashkir Soda Company located in Sterlitamak. Today, the company experiences a shortage of minerals for its production. The problem is so serious that the management of the company is going to have massive reductions and even close its production. To ensure sustainable development of the company, it is proposed to create an additional division that will be focused on high-tech electrical products [4].

First, you need to understand the idea of the project. In our case, it would be extremely interesting to establish a very popular production of electric cars today. Today, the boom of electric vehicles is observed in the USA, Canada, Norway, China and the EU countries. The leader is I. Mask's successful project, which has become a successful multinational corporation Tesla. There is a study of the successful production of electric cars and electric buses in Shenzhen with the support of the state [12]. Another possible project is to launch a line producing 3d-construction printers that allow printing cheap houses and offices.

At the second stage, it is necessary to have a team that will implement the project. The project requires highly qualified specialists who are able to work in a team, ambitious and not conflict. The "head hunters" of enterprises need to pay attention to local universities where there are often many undervalued potential employees - students and teachers. Many companies, which are currently at the top of the capitalization rating, were created by ordinary students (Apple, Alphabet, Microsoft, etc.). Denote the key provisions of the strategy following the example of the Tesla strategy:

1. Define the innovative image of the traditional enterprise and the beginning of its positioning as an innovative company which is open to "smart" investments and to interesting developments [14]. Develop aggressive marketing aimed at promoting a new company image; targeted advertising of new products, aimed at young people. Use sports and movies stars in commercials.
2. Invest into the company's human capital - investments in training, medical services, assistance in purchasing housing and developing infrastructure. This will attract creatively thinking, well-educated and healthy people who will be interested in innovative breakthrough projects [9]. In 2018, the article described the impact of employee creativity on innovative solutions within the company. The described problem is an important issue in the field of production [19].
3. Create places for creative, intellectual work - technology parks, coworking on the territory of the company. Attract investment and create opportunities for real business angels.
4. Produce the first high-tech products, if successful, increase production volumes. It is great to have the support of the state, which is interested in high-tech products, especially against the background of sanctions. It is possible to start production of electric scooters, phones, tablets, laptops, 3D printers of various types, including construction ones.

5. Enter the market of other regions, and subsequently of other countries. Here you need to take advantage of online commerce. The most striking example is the Chinese one, and now the international platform Alibaba. Here you can buy anything from a flashlight to powerful laptops. Such a platform for the promotion of high-tech products will attract their customers in various countries around the world.

3.3 Formula for Calculating the Investment Required to Launch Innovative Production

The authors determine a formula that will allow calculating the amount of investments necessary to create a new unit. To create a unit, there are costs for the project team and additional staff training (costs on them are S1), as well as costs for attracting highly qualified employees from outside, who could act as consultants for launching new equipment (S2) and costs for management activities (S3). In addition, it is necessary to purchase equipment that will launch a new production line at the enterprise. This is the most significant part of costs, C1. Very important costs are for developing new high-tech products, T1, which include costs for creating a special open space environment (a kind of coworking city) that will encourage employees to think creatively and create innovative products. For a sample, you can take a coworking city Facebook or Alphabet: cozy parks, coworking, recreation, which would become the place of work of the project team. Refining urban spaces will be an additional social bonus for all residents of the city in which this city-forming enterprise is located. One should not forget about a variety of risks that will accompany updating, including the risks of increased inflation and deterioration of the macroeconomic situation due to external sanction pressure, which have been growing lately, R.

Add up all the costs for a new production department that produces high-tech products and the total cost (TC) is the following:

$$TC = S1 + S2 + S3 + C1 + T1 + R \tag{1}$$

The most important question is where you can get the funds to launch such a costly but extremely necessary project for sustainable development of the enterprise. The answer is to attract investors from all over the world, as well as lobbying the possibility of obtaining government subsidies in government circles. This requires good marketing, an open, interesting guide for investors and IPO.

4 Discussion

The scientific literature does not pay enough attention to innovative production at traditional enterprises. In general, the introduction of the digital economy is considered at small and medium-sized enterprises in the US. It is also argued that the digital revolution will ensure mass production of innovations [2]. There is a study that indicates the irreversible transition to the digital economy, as well as the fact that Russia is no exception in this regard [8]. The same study emphasizes the fact that information technologies are being actively implemented in all Russian sectors of the economy, and thanks to this process, enterprises need to change their organizational structure. In addition, it points out the various advantages of using information and communication technologies for sustainable development of enterprises and organizations in modern Russia [8]. In another study, it is indicated that the world is becoming increasingly digital. In fact, the era of "smart production" is coming when it should become extremely adaptive [1]. There is an interesting study concerning the assessment tool for innovative production; a method assessing innovative aspects in production development projects is developed [11]. Seyoum Eshetu Birkie writes that nowadays businessmen face changes. In order to remain competitive it is necessary to introduce innovations in business models, namely innovations provide a mechanism for strategic renewal. However, we have only a few studies that are devoted to such challenges in sustainable production systems [6]. There are articles that affect the need for innovation for entire regions or countries.

One study emphasizes the need for Russia's transition to a knowledge-based economy, high-tech products, high technology and intensive innovation, and this requires the transformation of the higher education system and the formation of entrepreneurial universities. An analysis of the higher education system in Russia has shown that the development of universities is contrary to the global educational trends and the social and economic priorities of Russia. High efficiency of cluster policy in the leading countries of the world has led to the promotion of initiatives in the field of innovative development of regional clusters.

The results of econometric analysis show that the presence of a strong university as an "anchor" increases the influence of innovative factors on the socio-economic development of regions. The formation of innovation clusters in the regions of Russia requires "strong" universities that are engaged in applied and basic research, work with industry and introduce knowledge into practice [16]. Jorge A. Heredia Pérez is developing a new framework for analyzing internal and external factors that influence the types of innovations and their relationship with business efficiency in the manufacturing sector. The proposed theoretical model is tested and used to assess the innovation process in countries (Peru and Chile) and companies by size, industry type, financial aspects and level of patenting. In Chile, drivers are technological innovations in processes, while in Peru they are non-technological innovations. Foreign investment is very important when developing innovations [15]. The article of 2017 states that innovative production allows for cleaner, environmentally friendly production that ensures sustainable development of the company [18].

5 Conclusion

The introduction of the additional division producing innovative high-tech products at the enterprise based on the digital economy is an important basis for a sustainable growth strategy of the traditional enterprise. In the future, the range of manufactured high-tech products expands, and the range of obsolete, unclaimed products decreases. Due to online sales, the enterprise will expand the market, enters the markets of foreign countries. This strategy will provide the former traditional enterprise, far from innovation, with high competitiveness in the battle against the giants - Apple, Alphabet, Tesla, Amazon, etc.

References

1. Bauer, W., Hämmerle, M., Schlund, S., Vocke, C.: Transforming to a hyper-connected society and economy – towards an "Industry 4.0". Procedia Manuf. **3**, 417–424 (2015)
2. Beckmann, B., Giani, A., Carbone, J., Koudal, P., Salvo, J., Barkley, J.: Developing the digital manufacturing commons. A national initiative for US manufacturing innovation. Procedia Manuf. **5**, 182–194 (2016)
3. Bikmetov, R.Sh.: Formation of the edhocratic structure of domestic organizations in the conditions of innovation development. Discussion **1**, 17–21 (2016). (in Russian)
4. Bikmetov, R.Sh.: Social aspects of strategic management of a mining enterprise in crisis conditions. Bull. Peoples' Friendsh. Univ. Russia. Ser. Sociol. **3**, 60–69 (2010). [in Rus.]
5. Bikmetov, R.Sh.: Strategy of social and economic development of city-forming enterprises in conditions of sanctions. Econ. Manag. Sci. Pract. J. **1**, 17–21 (2016). (in Russian)
6. Birkie, S.: Exploring business model innovation for sustainable production. Lessons from Swedish manufacturers. Procedia Manuf. **25**, 247–254 (2018)
7. Drucker, P.F.: Managing in the Next Society. Truman Talley Books/St. Martin's Press, New York (2002)
8. Garifova, L.: The economy of the digital epoch in Russia. Development tendencies an place in business. Procedia Econ. Financ. **15**, 1159–1164 (2014)
9. Hobikoğlu, E., Şanlı, B.: Comparative analysis in the frame of business establishment criteria and entrepreneurship education from the viewpoint of economy policies supported by innovative entrepreneurship. Procedia - Soc. Behav. Sci. **195**, 1156–1165 (2015)
10. Kováč, J., Rudy, V.: Innovation production structures of small engineering production. Procedia Eng. **96**, 252–256 (2014)
11. Larsson, L., Stahre, J., Warrol, C., Öhrwall Rönnbäck, A.: An assessment model for production innovation. The program Production2030. Procedia Manuf. **25**, 134–141 (2018)
12. Li, Y., Zhan, C., de Jong, M., Lukszo, Z.: Business innovation and government regulation for the promotion of electric vehicle use. Lessons from Shenzhen, China. J. Clean. Prod. **134** (A), 371–383 (2016)
13. Markatou, M.: Incentives to promote entrepreneurship in Greece: results based on the 'New Innovative Entrepreneurship' program. Procedia - Soc. Behav. Sci. **195**, 1113–1122 (2015)
14. Mas-Tur, A., Moya, V.: Young innovative companies (YICs) and entrepreneurship policy. J. Bus. Res. **68**(7), 1432–1435 (2015)
15. Pérez, J., Geldes, C., Kunc, M., Flores, A.: New approach to the innovation process in emerging economies. The manufacturing sector case in Chile and Peru. Technovation **79**, 35–55 (2018)
16. Pogodaeva, T., Zhaparova, D., Efremova, I.: Changing role of the university in innovation development. New challenges for Russian regions. Procedia - Soc. Behav. Sci. **214**, 359–367 (2015)
17. Schumpeter, J.A.: The theory of economic development. An inquiry into profits, capital, credit, interest and the business cycle. Translated from the German by Redvers Opie. Transaction Publishers, New Brunswick (U.S.A) and London (U.K.) (2008)
18. Severo, E., Guimarães, J., Dorion, E.: Cleaner production and environmental management as sustainable product innovation antecedents. A survey in Brazilian industries. J. Clean. Prod. **142**(1), 87–97 (2017)
19. Tomczak-Horyń, K., Knosala, R.: Evaluation of employees' creativity as a stimulator of company development. Procedia Eng. **182**, 709–716 (2017)
20. Uslu, T., Bülbül, I., Çubuk, D.: An investigation of the effects of open leadership to organizational innovativeness and corporate entrepreneurship. Procedia - Soc. Behav. Sci. **195**, 1166–1175 (2015)

Digital Economy as a Way to Ensure Economic Growth

E. B. Razuvaeva[1], N. V. Starun[2(✉)], and L. G. Elkina[3]

[1] Sterlitamak Branch of Bashkir State University, Sterlitamak, Russia
[2] Samara State University of Economics, Samara, Russia
starun@mail.ru
[3] Bashkir State University, Ufa, Bashkortostan Republic, Russia

Abstract. The study considers the essence of the digital economy; presents various approaches to the definition of the digital economy; highlights the problems of digitalization in Russia, the prospects of the digital coverage of all spheres in the Russian Federation, the digital economic development in Russia and its results. Russia with its scientific and intellectual capabilities, appropriate technologies has all the prerequisites for the further digital economic development. The study is devoted to widely discussed strategies of the digital economy and their influence on the increasing competitiveness and economic security of the country. The authors consider digital technologies in the economy, business, social sphere and public administration; present the results of the digital development strategy and the features of digitalization in Russia; carry out a comparative analysis of the contribution of the digital economy to Russia's economic development compared with the countries that are considered leaders in this field; note the state of the digital economy and the level of its digitalization in Russia; identify the priorities for the digital economic development and how it strengthens the position of our country. The authors present the costs for implementing the digital development strategy and the sources for its financing.

Keywords: Economic security · Information security · Information infrastructure · Digital economy · Digitalization · Digital state

1 Introduction

The concept of the digital economy appeared in 1995, when this concept was associated with the development of information technology. Over time, this term acquired a broader meaning and began to cover all spheres of life. In the Address to the Federal Assembly on December 1, 2016, Russian President V. V. Putin proposed to "launch a large-scale system program for the economic development of a new technology, the so-called digital economy" and at the same time "rely on Russian companies, scientific, research and engineering centers of the country". V.V. Putin pointed out that this is "a matter of national security and technological independence."

S. I. Ashmarina et al. (Eds.): ISCDTE 2019, LNNS 84, pp. 116–127, 2020.
https://doi.org/10.1007/978-3-030-27015-5_15

The digital economy involves the use of electronic technologies and services, processing and analysis of digital data in economic activity. In order to respect national interests and ensure national security, the Digital Economic Development Program in the Russian Federation involves the creation and implementation of priorities for technologies and services based on domestic developments. To digitalize the country, centers of the digital economy have been created, which will become the driving force of this Program. However, the implementation of such a large-scale project requires certain funds, which should have a positive effect. Like any transitional system, the introduction of the digital economy requires both expenditures and efforts to implement it, as well as necessary staff who has new thinking and who will contribute to the implementation of the outlined strategies.

The tool for implementing such a global transition to the digital economy will be information technologies that will require additional development. The problem is that we need the enormous coverage of the population and territories by this Program. This causes the need to bring information to the public and organizations, expand the base, increase the speed of information and many other problems.

To date, the digital economy was studied in many scientific works. The strategy for implementing the planned event is reflected in the Digital Economic Development Program in the Russian Federation until 2035. The collective monograph "The Digital Economic Development in Russia as a Key Factor in Economic Growth and Improving the Quality of Life of the Population" presents the results of research on digital technologies in social sphere and public administration.

The digital economic development was considered in articles, forums and conferences. For example, Kapranova in her article "Digital Economy in Russia: State and Development Prospects" discusses digital technologies, examines the state of the digital economy in Russia and the world and makes conclusions on the further realization of the digital potential of the economy [2].

Sagynbekova in the article "Digital Economy: Concepts, Perspectives, Development Trends in Russia" examines the history of the emergence and development of the digital economy, studies the concepts of the digital economy and analyzes the development of this area in Russia [4].

The collection of articles of Ivanov and Malinetsky "Digital Economy: Myths, Reality, Prospects" illustrates material on digital reality, risks and threats of the digital economy, digital needs of Russia [1].

The purpose of this study is to highlight the concept of the digital economy, to study the place of Russia in world digitalization, to analyze the development strategy of the digital economy of the Russian Federation until 2030.

The objectives of the study:

- To consider the concept of the digital economy;
- To justify the need for the digital economic development in Russia;
- To analyze the implementation of the digital strategy in the Russian Federation.

2 Methodology

The concept of the digital economy is revealed using the methods of the historical and logical development process, as well as the process of introducing the results of scientific and technological progress and information technology, computerization of all spheres of activity.

The study was conducted using the following methods:

- The method of scientific abstraction, which represents categories and concepts reflecting the most important aspects of the object under study;
- The method of analysis and synthesis, which allows a comprehensive approach associated with the introduction of the digital economy into our lives;
- Historical and logical methods, due to which economic processes associated with the emergence and further introduction of the digital economy into the national economic system are examined in unity and logically consistently;
- The method of induction and deduction, which allows describing the concept of the digital economy and its distribution in Russia.

3 Results

The main results of the Digital Economic Development Program of the Russian Federation should be:

1. Develop information infrastructure, i.e. development of communication networks and infrastructure for data storage and processing, development of a state digital platform for the provision of public services, the Internet of Things, provision of communications along federal road and railway lines, import substitution of communication links, access to digital services for citizens.

The Ministry of Communications and Mass Media of Russia will ensure the promotion of this direction, and the center of competence will be Rostelecom PJSC. The plan of activities in this area contains 350 proposals for the development of new-generation networks, infrastructure for storing and processing data, ensuring equal access of citizens to digital services in the field of healthcare and education. The total amount of financing is about 427 billion rubles. At the same time, it is planned to allocate 98.6 billion rubles from the federal budget, extra-budgetary funds - 328.5 billion rubles.

In 2024, 97% of households should have broadband Internet access at a speed of at least 100 Mbps. All public health facilities should be provided with broadband Internet access at a speed of at least 1 Mbit/s with satellite connection and at least 10 Mbit/s with fiber optic connection in 2018. In 2024, all educational institutions, as well as public authorities and local governments should be provided with broadband Internet access at a speed of at least 100 Mbit/s. By this time, all objects of the transport infrastructure should be provided with sustainable communication networks with the possibility of wireless data transmission. At the same time, in 2018 their share should be at the level of 96.5%. The plan also contains indicators for the development of fifth-generation networks (5G).

From 2018 to 2024, the number of reference data centers in the federal districts should be increased from 2 to 8. The capacity of Russian data centers should be 80 thousand seats in 2024. In the same year, Russia's share in the global volume of data storage and processing services should increase to 10%.

By the middle of 2019, the requirements for the state unified cloud platform by state and local governments should be normatively defined, and a plan to transfer their information systems and resources to the state unified cloud platform should be approved [6].

2. Provide information security that allows citizens to interact with the relevant authorities and to report illegal actions in the field of ICT. It ensures the control of processing and access to personal data, including in social networks. The Ministry of Communications and Mass Media of Russia will be responsible for this area, and Sberbank PJSC will be the center of competence.

A system of incentives should be created for the acquisition and use of computer, server and telecommunications equipment produced in Russia. Mechanisms encouraging the use of domestic software by all participants of information interaction have been created.

In addition, national standards for cyber-physical systems, including the Internet of Things, must be adopted. There is control over the processing and access to personal data, large user data, including in social networks and other means of social communication. National and regional computer incident response centers have been established. Following the results of the Program, a system of measures will be developed to support Russian manufacturers of ICT products and services who are patenting products abroad. Due to this digitalization, the share of the internal network traffic of the Russian Internet routed through foreign servers should decrease to 10% by 2024. The value share of the public sector and companies with the state participation of foreign software should fall from 50% in 2018 to 10% in 2024.

The share of subjects of information interaction (state authorities and local governments, companies with state participation) using security standards in cyber-physical systems and in the Internet of Things should increase from 10% in 2018 to 90% in 2024.

The share of citizens who have increased literacy in the field of information security, media consumption and Internet services should be 50% by 2024. The average idle time of state information systems as a result of computer attacks should be reduced from 65 h in 2018 to 1 h in 2024 [5].

3. Form research competencies, i.e. digital change of economic sectors and their individual subjects (blockchain, quantum computing and artificial intelligence). The Ministry of Communications and Mass Media of Russia will be responsible for this digitalization, and Rosatom and Rostec companies will be centers of competence.
4. Provide regulatory control and consolidation of the legal status of the Runet, i.e. creating a system of regulatory control of the digital economy, which will be based on digital technologies. This digitalization will achieve two goals. The first is to

eliminate key legal constraints and create separate legal institutions aimed at addressing the priorities of the digital economy. The second is to develop a permanent mechanism for managing change and competencies (knowledge) in the field of the digital economy regulation.

It is intended to create a "single digital environment of trust" and an environment of e-civil circulation, conditions for interaction between business and the state, conditions for the effective use of intellectual property, data storage and processing, and it is intended to introduce innovative technologies in the financial market. In this case, we need to digitalize such areas as legal proceedings, notoriety and collection of statements. It is supposed to establish basic concepts and institutions for legislation in the field of artificial intelligence, robotics. The Ministry of Economic Development will be responsible for this direction, and Skolkovo Foundation will be the center of competence.

As part of the unified digital environment of trust, it is proposed to unify the legal requirements for identification. In particular, it is necessary to define the conditions for identification and authentication using a "mobile" or "cloud" electronic signature, as well as using a mobile number and driver's license. There should not be legal restrictions on electronic signatures.

In addition, it is proposed to consolidate the general procedure for conducting remote identification using the Unified system of identification and authentication and confirmation of biometric personal data (face image, voice) in the biometric system. The unified digital environment of trust involves mass provision of telemedicine services, including in hard-to-reach areas.

The task of forming the sphere of electronic civil turnover, in addition to defining new and clarifying the existing requirements for transaction forms, offers and obligations, is to create a base of model contracts and sample terms of contracts used by counterparties in electronic form and provides for the possibility of self-fulfillment of the agreement.

The concepts of an electronic duplicate/image of a paper document, the conditions for recognizing the validity of an electronic document should be clarified, the structure of an electronic document should be defined. In addition, it is necessary to determine the requirements for the destruction of electronic documents. The scope of electronic civilian turnover also implies the right of insurance agents and insurance brokers to enter insurance contracts in electronic form.

To improve the conditions for collecting and processing data, we should eliminate the immediate legal restrictions that prevent the use of data, including geolocation data, and create legal conditions for using and processing data, the procedure for accessing data, and the responsibility for unauthorized access to it. In addition, it is planned to clarify the concepts of "personal data" and "biometric personal data", as well as to revise the rules for processing biometric data. It is necessary to eliminate the broad interpretation of concepts defining professional secrets, and to clarify the procedure for transferring banking secrets, communication secrets and medical confidentiality to third parties, with the consent of such subjects in order to process accumulated data and to achieve social, state and economic goals.

In addition, the procedure for disclosing information about the equipment used in the Internet of Things, timely connection to the Internet and safe operation of devices should be defined. In order to use the results of intellectual activity, we must eliminate legal uncertainty by including information as an object of civil rights. It is also planned to clarify the procedure for circulation of computer programs, to clarify the concepts of "program" and "software" and to ensure equal legal protection for various types of programs, regardless of the specifics of their creation, purpose and introduction into circulation.

It is planned to ensure the possibility of describing intellectual property in the form of digital models and enter the electronic form of issued security documents (patents and certificates). The list of measures to stimulate the digital economy includes projects to introduce lower taxation of income from intellectual property rights (Patent Box), as well as to clarify the parameters of companies applying the multiplying factor to R & D expenses and the acquisition of intellectual property rights when calculating the tax on profit.

Additionally, it is planned to optimize the use of reduced insurance premium rates for Internet companies and software development companies, to introduce special tax breaks on personal income tax for business angels, to simplify Russian tax residency for highly qualified individuals, and to expand the deduction of VAT from Russian companies with "export of electronic and IT services. It is also planned to work out the terms of taxation of Russian online retailers, stimulating the development of electronic commerce and to clarify the parameters of VAT taxation of services provided in electronic form by foreign sellers.

IT and telecom infrastructure objects are planned to be included in the list of possible objects of agreements stipulated by the law on PPP and the law on concession agreement. Barriers to digital models and virtual tests should be eliminated as part of standardization mechanisms including modeling the real parameters of products. It is planned to comprehensively reform the legislation in the field of certification in order to increase the competitiveness of Russian technologies and Russian business through the transition to electronic certification. At the same time, it is necessary to eliminate regulatory restrictions on the creation of electronic certification centers.

In the direction of legal proceedings, it is planned to eliminate legal restrictions on electronic document circulation, to eliminate the parallel functioning of federal and regional registries of enforcement proceedings, and to guarantee the participants of legal proceedings the right to remote participation using video conferencing systems.

Regarding the activities of notaries, it is planned to establish the possibility of performing a notarial act by producing a notarial document in electronic form and certifying the fact that the document was signed in the presence of a notary with a reinforced qualified electronic signature. It is also supposed to introduce a mechanism for remote notaries to perform certain notarial actions (certifying transactions, testifying the fidelity of copies of electronic documents and extracts from them, testifying the translator's signature on translated documents from one language to another, certifying the time of submission of electronic documents, and other activities). It is also planned to establish the legal status of electronic systems for fixing legal facts (including using distributed registry technologies, etc.) and to create a legal basis for an electronic system of alternative dispute resolution mechanisms (mediation and online dispute resolution) [7].

5. Prepare staff for the digital economy. As a result of this direction, a system for distributing personal digital certificates should be developed. There should be a mechanism for assessing the level of competencies, which will give advantages to people, who are entering higher education institutions (a digital equivalent of the TRP standards). Training and testing programs will be aimed at shaping the core competencies of the digital economy. We expect a system of standards for the competencies of the digital economy for all ages. In addition, the priority project "Digital School" should appear and the concept of a complex, alienable system of knowledge, software and hardware, which can be replicated in the regions of Russia, has been developed.

Educational organizations will have to start using personal profiles of students' competencies and provide personal development. The number of students enrolled in higher education programs in the field of IT should increase from 60 to 120 thousand students over several years.

Advanced training programs should be developed for teachers taking into account WorldSkills Russia standards for competencies that are a priority for the digital economy (at least 5,000 teachers must undergo advanced training programs each year). At least 15 cities of the country with high scientific potential should have accelerators on nurturing project teams in the interests of the digital economy. In addition, there should be 7 venture funds with the participation of universities, their graduates, partner companies to finance student startups in the digital economy at the first stage.

The country should have a digital bonus system, allowing students to receive rewards for various achievements (winning the Olympiad or sports competition, social activity, etc.) and to launch startups, to receive investments from venture funds, to go through acceleration programs and to carry out other types of business activities in the virtual environment.

A comprehensive educational and accelerated environment should emerge allowing a schoolchild, student, young scientist and specialist, regardless of their place of residence, to gain knowledge, skills and business experience, access to the advanced "end-to-end technologies" of the digital economy, access to the resources necessary for creating and developing a business, entry into communities and expert networks in technology and entrepreneurship. Associations and schools, universities and colleges, selected active educational institutions, business associations and individual companies, regions and selected active territorial entities are expected to be involved in the creation and development of this environment.

By the end of 2021, 20 advanced training programs should be developed in accordance with the competencies in demand in the digital economy, while 5 million people should be trained in programs [8].

4 Discussion

The term "Digital Economy" is a relatively new concept for our country. However, this term is tightly used in everyday life and extends to the spheres of economic activity, covers an increasing number of the population of Russia, which uses it. Many experts give the definition of this concept. Thus, in the Digital Economic Development Program of the Russian Federation, the digital economy is "the totality of social relations emerging from the use of electronic technologies, electronic infrastructure and services, technologies for analyzing large amounts of data and forecasting in order to optimize production, distribution, exchange, consumption and to increase social economic development of the state" [9].

The Information Society Development Strategy for the Russian Federation for 2017–2030 gives the following definition: "The digital economy is an economic activity in which the key factor in production is digital data, processing large volumes and using the results of the analysis, which can significantly improve the efficiency of various types of production, technology, equipment, storage, sale, delivery of goods and services compared with traditional forms of management" [2].

The digital economy, according to the World Bank, is "a system of economic, social and cultural relations based on the use of digital information and communication technologies" [11].

According to Ivanov, Doctor of Economics, a member of the Russian Academy of Sciences, "the digital economy is a virtual environment that complements our reality." That is, if you take, for example, Tinkoff Bank, you can agree that this is a kind of environment without a specific location, but with the help of which you can perform all those operations via the Internet that are usually performed in the offices of traditional banks [1].

Professor of RAS, Doctor of Technical Sciences, Malinetsky defines the concept of "the digital economy" from two sides - firstly, this is an economy based on digital technologies, which characterizes the field of electronic goods and services; secondly, it is economic production using digital technologies [1].

Thus, there is no consensus on the definition of the digital economy, however, it is clear that this is the economic development with the use of information technology, a kind of virtual environment where business activities are presented in electronic form.

Economy digitalization covers more and more areas of human activity. So, in the industry it is the introduction of robotic technologies, digital design and modeling of production processes, the implementation of industrial products via the Internet, recycling and processing. In agriculture, this is a transition to a new model of intelligent agriculture, which will be based on integrated automation and production robotization, as well as ecosystem modeling.

Electronic commerce, as an element of the new economic model, means an increasing number of participants in this process, in which sellers and consumers, manufacturers, business and the state are involved. It is growing as a world electronic trade turnover, which creates favorable conditions for the development of markets of individual countries and in particular, Russia [10].

In the field of communications, a promising solution is "a software-defined mobile network of a new generation with virtual implementation of network functions, unlimited scalable cloud resources and the possibility of operational analytics based on the concept of big data" [9]. This solution in the field of communications and telecommunications will allow you to avoid hacker attacks and analyze existing data.

In addition to the active development of electronic commerce, a significant increase is observed in the field of financial services. These are payments and transfers made via the Internet, asset management, blockchains and other types. The digitization of management should cover other areas of activity. The Digital Economic Development Program of the Russian Federation covers all areas of its application. Unfortunately, our country is lagging behind the world leaders in implementing the digital economy. Today, RAEC analysts estimate the contribution of the digital economy to the Russian economy at 2.1% of GDP, and the contribution of the mobile economy at 3.8% of GDP. The amount of the contribution is 4.35 trillion rubles or 5.06% of GDP [3]. Table 1 presents data on the contribution made by the digital economy to Russia's gross domestic product in comparison with other countries.

As can be seen from the presented data, almost all indicators characterizing the contribution of the digital economy to Russia's GDP are lower than other countries that are leaders in digitalization. It reduces the competitiveness of the country and may threaten its economic security. At the same time, it is natural to assume that such a state of affairs will allow foreign companies to ensure their competitiveness and strengthen their positions in relation to Russian enterprises.

Table 1. Contribution of the digital economy to Russia's GDP compared to other countries, 2016

Country	Share in GDP, %	Household spending on the digital sphere, %	Investment of companies in digitalization, %	Government spending on digitalization, %	Export of ICT, %	Import of ICT, %
USA	10,9	5,3	5,0	1,3	1,4	−2,1
China	10,0	4,8	1,8	0,4	5,8	−2,7
EU countries	8,2	3,7	3,9	1,0	2,5	−2,9
Brazil	6,2	2,7	3,6	0,8	0,1	−1,0
India	5,5	2,2	2,0	0,5	2,9	−2,1
Russia	3,9	2,6	2,2	0,5	0,5	−1,8

Source: compiled by the authors.

According to the estimates of McKinsey Global Institute, due to the digital economy there will be an increase in China to 22% of GDP, in the USA - 1.6–2.2 trillion USD, in Russia up to 8–10% by 2025. It will increase the country's GDP by 4.1–8.9 trillion rubles and the transition to the digital economy will become a major factor in economic growth.

The experience of the leading countries shows that the digital economic development increases the competitiveness of the country, provides lower prices, and provides a significant advantage for companies and consumers. In Russia, it is expected to introduce the digital economy in various sectors and areas in the period up to 2025. In medicine, the document flow should be completely in electronic form, in state structures - 95%, in the services sector of state bodies - 75% and so on. However, we need human capital and proper management to create an effective national economy, in addition to creating ICT.

The priority areas for the digital economic development are the construction of a smart home, the production of UAVs, the creation of a digital city, the creation of professional networks, and the significant use of 3D printers. The introduction of the digital economy in Russia is constrained by certain problems, among which are the lag in the development of information technologies, shortcomings in the training of qualified personnel and inadequate information security.

However, solving the problems associated with the introduction of the digital economy, you can achieve a certain economic growth and get the effect of the Digital Economic Development Program.

The government suggests in its Digital Development Strategy the opportunity to increase the effectiveness of this Program by following actions:

- Minimize corruption in the administrative system;
- Optimize taxes by calculating the individual tax burden;
- Provide public services through a single digital platform;
- Transit to the model of participatory budget (created with the participation of citizens).

The Economy Digitalization Program in Russia should be the basis for solving issues of public administration, the economy of the country, business, the social sphere and raising the standard of living of the entire population.

The country's government identifies several areas that need to be implemented as part of the Digital Economic Development Program of the Russian Federation.

It is supposed to carry out actions in the following directions:

- Information infrastructure;
- Information Security;
- Regulatory control;
- Personnel and education;
- Formation of research competencies and technological reserves.

All proposed activities require financial investments. Thus, 100 billion rubles will be allocated from the budget for Information Infrastructure, 22.3 billion rubles for Information Security, 0.9 billion rubles for regulatory control, and the formation of research competencies and technological reserves - 48 billion rubles [2].

Until 2020, it is planned to spend 521 billion rubles on the implementation of measures under the Digital Economic Development Program of the Russian Federation.

The main result of economy digitization should be a new economy, ensuring an adequate level of the country's competitiveness.

All the opportunities and directions of economy digitalization considered in the study will help to solve problems in the social sphere, overcoming the backlog of the leading countries, reaching a new level of development and economic growth.

5 Conclusion

Russia has all the prerequisites for realizing digital potential and accelerating the pace of digitalization. New technologies will affect the growth of the quality of life of the population, develop business and government and create new forms of socialization of people and their communications. Digitalization will create a synergistic effect and lead to a general growth of the Russian economy. One of the main obstacles to the digital economic development is the lack of coherent actions by the government, business and the scientific community.

The Digital Economic Development Program of the Russian Federation set not so many economic tasks that are associated with the development of the digital industry as a whole, as well as new ways of producing and bringing domestic technologies to global markets. The Program lacks some indicators: there is no share of e-commerce in turnover, no share of high-tech jobs in the overall structure of employment, no indicators of high-tech exports, or other indicators directly related to economy digitization. The Program did not include the issue of stimulating large companies, as well as medium and small businesses to actively implement digital innovations and increase investment in research and development. The Program is aimed at creating basic "services" for economic development - regulation, public services, IT infrastructure, which is also of great importance.

The development of the digital industry requires a developed domestic IT industry. In recent years, the government has provided incentives for the development of the IT industry. However, the digital economy needs the IT industry that will be able to produce and maintain IT products of the highest quality so that they can compete with foreign designs.

The Digital Economic Development Program of the Russian Federation has a lack of consistency; there is no close relationship with other government projects. As the economy digitalizes, problems in the protection of intellectual rights, as well as personal information, may arise related to the cross-border nature of information transfer and the operation of most services. Special attention should be paid to cyber security issues.

It is necessary to stimulate and support such areas as import substitution, export of information technologies, ensuring equal conditions for business activities with Internet companies in Russia, access and storage infrastructure, and promotion of all types of mass digital communications and services.

The technologies that can have the greatest impact on the economy are artificial intelligence technologies, big data analytics, cloud computing, Internet of Things, robotics, autonomous vehicles, production of customized products and 3D printers, social networks and other types of digital Internet platforms.

In the world, there is a process of strengthening the information component in ideology, politics and economy. That is why there is a need for a deep and systematic study of this process. The digital economy is not only the basis for creating new business models, qualitative changes in business models, the nature of business, its manageability and flexibility. It concerns the fundamental principles of civilization.

References

1. Ivanov, V.V., Malinetsky, G.G.: Digital Economy: Myths, Reality, Perspective. Moscow, IPM RAS (2017). (in Russia)
2. Kapranova, L.D.: Digital economy in Russia: state and development prospects. Econ. Taxes Law **2**, 58–69 (2018). (in Russia)
3. RAEC: The report "Runet Economy" (2018). http://raec.ru/live/raec-news/10192/. Accessed 27 Apr 2019. (in Russia)
4. Sagynbekova, A.S.: Digital economy: concepts, prospects, development trends in Russia. Theory Pract. Innov. **4**(28) (2018). http://www.tpinauka.ru/2018/04/Sagynbekova.pdf. Accessed 27 Apr 2019. (in Russia)
5. TDAdviser: Information security of the digital economy (2018). http://www.tadviser.ru/a/394110. Accessed 20 Apr 2019. (in Russia)
6. TDAdviser: Information infrastructure of the digital economy of Russia (2019). http://www.tadviser.ru/a/390240. Accessed 20 Apr 2019. (in Russia)
7. TDAdviser: Legal regulation of the digital economy (2019). http://www.tadviser.ru/a/383190. Accessed 20 Apr 2019. (in Russia)
8. TDAdviser: Staff and education in the digital economy (2019). http://www.tadviser.ru/a/399707. Accessed 20 Apr 2019. (in Russia)
9. The Digital Economic Development Program of Russia until 2035 from 28.07.2017. http://spkurdyumov.ru/uploads/2017/05/strategy.pdf. Accessed 14 Apr 2019. (in Russia)
10. Urmantseva, A.: Digital economy: how specialists understand this term. RIA Novosti, 16 June 2017. https://ria.ru/science/20170616/1496663946.html. Accessed 27 Apr 2019. (in Russia)
11. World Bank: World Development Report 2016 "Digital Dividends" (2016). https://openknowledge.worldbank.org/bitstream/handle/10986/23347/210671RuSum.pdf. Accessed 14 Apr 2019. (in Russia)

Strategies for Obtaining Added Value in Developing Technological Innovations

M. V. Simonova[(✉)] and N. V. Kozhuhova

Samara State University of Economics, Samara, Russia
est-samara@mail.ru

Abstract. A model for creating added value of a product is developed, which revealed the tendencies in changing the structure of creating added value of goods and services, caused by the widespread use of digital technologies, when the main share is increasingly accounted for at the stage of creating a product with new qualities. The distinction between the use of the concepts of labor productivity and productivity of an economic system is substantiated. Productivity of an economic system has a broader interpretation and includes the entire complex of economic processes, such as labor productivity, marketing, managerial productivity and the creation of new products with high consumer qualities. It is determined that the highest added value is provided by the products created with the use of digital technologies. It justifies the priority of high technology stage and the shift of the center of gravity in creating added value to the stage where the idea is formed, technologies are being developed and created. Decomposition of the goals and objectives of labor productivity at the enterprise level is carried out, and areas of responsibility are identified at various management levels. A model is developed that substantiates the aggregate growth in labor productivity and the level of investment in short, medium, and long terms with the use of traditional and digital technologies.

Keywords: Labor productivity · Added value · Innovations based on digital technologies · Productivity of the economic system · Strategies for obtaining added value

1 Introduction

The competitiveness of the Russian economy is determined to a great extent by the goods and services that are produced on the territory of the country and in those regions that have a certain specialization. One of the significant factors on which the achievement of specified indicators depends is labor productivity, the growth rates of which in recent years cannot provide the necessary growth. To make objective, comprehensive analysis of the current situation we need a deep scientific analysis of the subject of the study, that is, the economic nature of labor productivity, the place and the role of labor productivity in the production system, significant factors affecting productivity growth and the indicators necessary to analyze the dynamics of productivity, reasonable forecast and management.

S. I. Ashmarina et al. (Eds.): ISCDTE 2019, LNNS 84, pp. 128–136, 2020.
https://doi.org/10.1007/978-3-030-27015-5_16

Theoretical studies of labor productivity have rich historical roots since the beginning of the twentieth century, both in Russia [2] and in other countries [9], but so far the subject of research provides wide opportunities for discussion of different aspects characterizing labor productivity. For example, it is interesting to ask whether labor productivity can be considered a factor of production or its indicator [16], how to measure labor productivity [17], what determines the growth of labor productivity [10]. Most often we can note the confusion between the concepts of labor productivity and productivity of production system in general; the economic literature has an understanding of the distinction between these concepts [19] but at the mesoeconomic and macroeconomic management levels these concepts are often interchanged [4]. This leads to inadequate interpretation of reasons and consequences, makes the search of managerial solutions more complicated, reduces a serious economic problem to populist statements and waste of funds [12]. Separation of these concepts is necessary to highlight priority areas in state economic development programmes, both at the federal and regional levels; to focus efforts on priorities while maintaining strategic guidelines [3]. In the development of labor productivity, the selection of the most significant stages in the creation of the added value of a product or a service is becoming increasingly important. In the most general form, this process can be formalized as a model, which makes it possible to single out the stages of creating added value (Fig. 1).

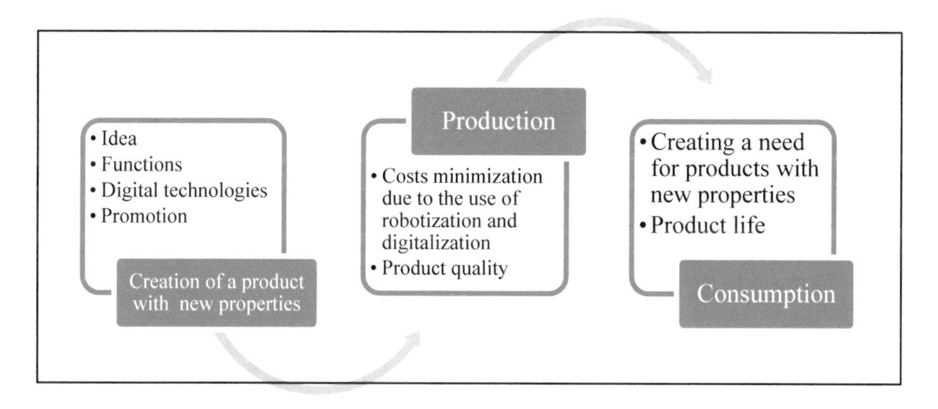

Fig. 1. The model of creating added value of a product (Source: compiled by the authors)

Modern Russian authors mostly have similar positions in the interpretation of the goals of increasing labor productivity: reducing the cost of products while improving their quality [8]. In general, we can agree with them, but we propose a broader interpretation – the highest labor productivity indicators can be obtained with a high added value of goods. This can be achieved on the account of new consumer qualities of the product, when its final value for the consumer is much higher than the total costs of the manufacturer, which in the current situation is reached by the use of digital technologies. Consequently, to increase labor productivity, it is necessary to create a qualitatively new consumer value of goods and services, shifting the center of gravity in creating added value from production to research and development centers, marketing, which, creating new products, simultaneously create the public demand for it.

2 Methodology

In the research such methods of economic research as modeling, the method of scientific abstractions, analysis and synthesis, systemic and functional approaches; special research methods, sociological and comparative methods were used.

3 Results

As the result of the research, it can be stated that in the strategic aspect most of added value creation with the development of digital technologies is shifted to the left block, shown in Fig. 1. In the short term, to provide the daily activities of enterprises, the main opportunities for creating added value fall on the average and right blocks, but its value will be significantly less than at the first stage. This understanding of the logic of obtaining added value allows making priorities in the achievement of short-term and long-term goals. The productivity of labor is mainly in the second block of creating added value directly at the enterprise. Increased labor productivity is possible in the short term if production costs are reduced. The most significant investments are needed to create products with high added value and with new consumer qualities (Table 1).

Table 1. Priorities when creating added value*

	Short-term goals	Medium-term goals	Long-term goals
Idea, design, marketing of a product			✓
Production	✓	✓	
Consumption		✓	✓

Source: compiled by the authors

Of course, at each of the stages of creating added value, it is necessary to analyze labor productivity, but these should be different indicators and parameters that require additional research, especially in the services sector and in the agricultural business sector [7]. For example, some studies point out the ambiguity of using labor productivity indicators in service companies. The increase of processes automation is also ambiguous here [15].

In general, the whole process of added value creation can be characterized from the point of view of productivity, but in this case one should speak about the productivity of the entire system and calculate it in total by macroeconomic indicators, that is, as the ratio of total income to total costs, as it is generally done in macroeconomic calculations and comparisons [13]. Scientific research and digitalization, on the basis of which new products and technologies are created, are becoming the main center of added value creation [1]. In the long term at this stage the main quantity and quality of jobs will be concentrated. At the same time, at the production stage, due to new automated and robotized technologies, the number of jobs will be reduced, the labor in this system changes its character, and the requirements to the quality of the workforce increase. However, in the short term, the stage of production where traditional technologies are used is decisive and it is necessary to look for ways to increase labor productivity here [6].

The growth rates of economic productivity in the world began to decrease at the turn of the 2000s [5]. The dynamics of economic productivity growth rates in G-7 countries in the period from 1973 to 2013 shows a significant slowdown in all countries [11]. This suggests that traditional technological solutions are exhausted and investments in these technologies can no longer provide the necessary economic growth. In Russia, productivity growth rates show a similar negative trend, they are significantly lower than that of the world leaders both by absolute indicators and by the indicators of decline [14]. According to the international electronic service "Trade Economy", the labor productivity in Russia increased by 1.5% in 2017 compared to the same period of the last year after decreasing by 0.3% in 2016. The productivity in Russia was on average 3.15% from 2003 to 2017, reaching a record level of 7.50% in 2006 and a record minimum of 4.10% in 2009.

It should be noted here that it is not the productivity of labor is being compared, but the productivity of the entire economic system, which we talked about earlier. This is a matter of principle, since the distinction between these concepts makes it possible to structure goals, to relate tasks, solution tools and achievable goals. If you distinguish productivity in industries using traditional technologies and those using advanced technologies, based on innovative solutions, you can see the difference in productivity growth rates when the increase of investments in fixed capital in traditional industries does not lead to the expected productivity increase.

The increase of the productivity of the entire economic system can be achieved, among other things, by the increase of labor productivity which needs to be distinguished. We should also understand a logical dependence of the economic system productivity and labor productivity, which are related but not identical concepts. Clarification of the application of these concepts is necessary for the logic of setting goals and objectives, reasons and consequences, for distinguishing related but different processes.

We understand labor productivity as the productivity of people's labor not taking into account the "past" labor, that is, the labor on the creation of technological systems. This separation allows us to consider the innovative process of creating new technologies a separate system, the management of which should be based on a different logic with different priorities, unlike that with traditional technologies. The defining of labor productivity in these parallel processes should be carried out according to different indicators and with different evaluation parameters because of significant differences in the nature of labor activity during the innovation and production processes.

During the production process exact observance of technological parameters is necessary, which directly affects the nature of the quality of the workforce.

If such a division is made, it becomes obvious that the goals and objectives of labor productivity must be decomposed (Fig. 2). The productivity of people's labor in the production system is determined by other factors - labor organization, production losses, capital productivity and motivation. So, it can be said that the productivity of the economic system depends on many factors, including the productivity of people's labor, but this is not the only factor. That is, labor productivity, as it is often formulated in various programs and projects, cannot be the goal of production, it is one of the indicators that determines the performance of the entire system, but not the only one. At the same time, almost everywhere, an increase in labor productivity is set as the ultimate goal of production, which replaces the real goal and does not allow, among other things, to achieve the necessary indicators of labor productivity.

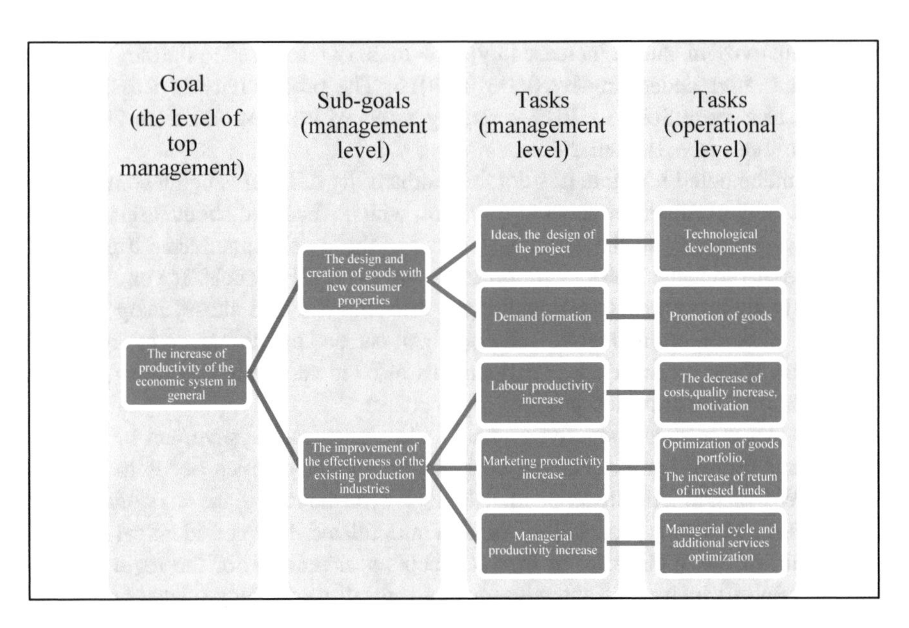

Fig. 2. Decomposition of goals to improve the productivity of the economic system (Source: compiled by the authors)

We proposed the system of decomposition of the strategic goal of improving the efficiency of the economic system, the task of increasing labor productivity is on the level of the managerial tasks, which is an integral part of the managerial goal of increasing production efficiency. Other equally important element in increasing production efficiency is the increasing of marketing and managerial productivity. The formation of an optimal portfolio of orders, the commercial return of invested funds, the optimal management and production structure play an equally important role in production efficiency as the increase in labor productivity plays directly in the production process.

With the increase of the efficiency of existing industries using traditional technologies, we can expect the increase of labor productivity of 5–7% during the first three years after the activities on increasing labor productivity at a particular enterprise are started (Fig. 3, Appendix A).

At the same time, it is necessary to accelerate the process of creating and developing an innovative technological basis for the production of goods with new consumer properties and the search for possible consumer groups, as well as the creation of motivation for such demand. The return from this process in the form of high labor productivity can be expected only after a sufficiently long period of producing ideas, technological developments, testing and transition to mass production. This period can take from 3 to 5 years, but then labor productivity and efficiency of the economic system can increase many times. It is necessary to take into account the uneven distribution of investments in these processes. At the initial stage, with the planning horizon of 3-5 years, the largest capital investments are required in the creation of new products with high added value and new consumer properties. During the same period, the increase in the efficiency of existing industries on the basis of the use of digital technologies can have a tangible effect and partially compensate these investments.

It is fundamentally important that these processes are parallel, since the center of gravity of added value, as far as capital intensity and duration are concerned, is transferred to the intellectual stage of high-tech production, and the income from creating such products is postponed to a later period, but the labor productivity is significantly higher. The first 3–7 years of high-tech products creation with high added value are accompanied by high costs, but the period of mass production is reduced due to robotization and automation. Thus, it is possible to follow the trend of changes in the structure of employment in the long-term period, which we talked about in previous works [18]. Our forecast for the mid-2000s showed an increase to the 20 s, the need for highly qualified and qualified specialists, mainly engaged in technologies, information and service, which practically corresponds to the current situation in the short term. The strategic forecast of the structure of employment in accordance with the dynamics of labor productivity, presented in Fig. 6, can be formulated as follows: reducing of the number of industrial personnel while increasing the number of people employed in research, engineering, information and service organizations. It seems important and necessary to develop the theoretical foundations of a new labor economy, the cost of robots production, perhaps wages and training of robots. Human participation in the production phase can be minimized; however, setting goals and objectives should in any case be made by the man.

4 Discussion

Determination of the real role and place of labor productivity in the system of factors of production, scientific substantiation of the concept contents allows identifying and applying specific tools to increase labor productivity in the form of proven and new methods, such as the use of scientific organization of labor based on the principles of careful production, reduction of unproductive costs, increasing of motivation and others. The dynamics of growth in labor productivity in two parallel processes of creating new products and increasing the efficiency of existing industries in Fig. 5 shows different growth rates in different time intervals.

To coordinate the parallel processes of creating new technologies and increasing labor productivity in production using traditional technologies, managerial and guiding actions are needed, which allow comparing and synchronizing the multidirectional actions of numerous participants and interested parts, both individuals and businesses.

5 Conclusions

Thus, as a result of the study:

- there was developed a model for creating the added value of a product, revealing the trends in changing the structure of creating added value of goods and services, when the main share of costs is more and more at the stage of creating a product with new qualities;
- the distinction between the use of the concepts "labor productivity" and "productivity of the economic system" is justified; it is determined that the concept "labor productivity" is not always validly applied; in most situations, it means "the economic system productivity" in accordance with the economic contents of the concept, which is broader and includes the entire set of economic processes, such as labor productivity, marketing, managerial productivity and the creation of new products with high consumer qualities;
- it is found out that the highest added value can be obtained in products made with the use of digital technologies;
- priorities for managing the process of added value creation is developed, which allows determining strategies for increasing labor productivity;
- to determine the contents of the concept "labor productivity", the decomposition of goals and objectives of labor productivity at the enterprise level was carried out, the areas of responsibility at different management levels were highlighted, which clearly shows the logic of performing tasks as the element of management;
- there was developed the model of the dynamics of growth rates of labor productivity and the level of investments in the short, medium and long terms with the use of traditional and digital technologies, justifying the use of different tools in different periods depending on the periods of achievement of the goals;
- there was formulated a long-term need for theoretical substantiation of a new labor economy in conditions of robotization, taking into account the production of robots.

Appendix A

See Fig. 3.

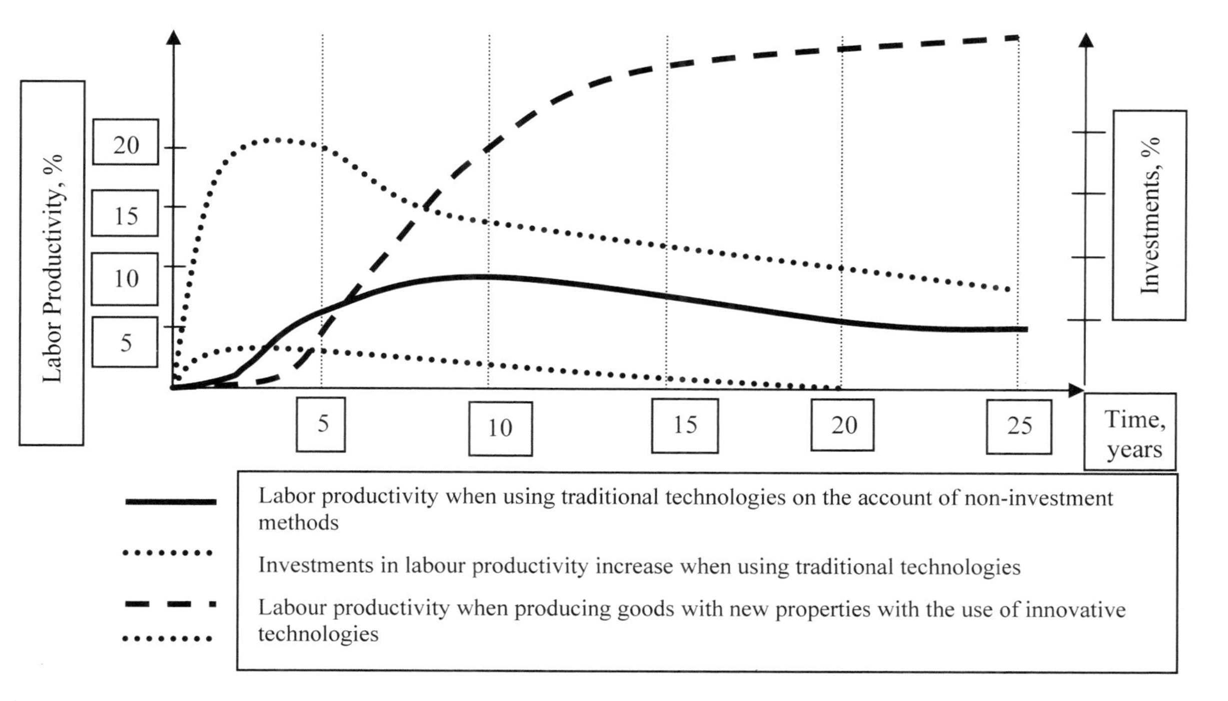

Fig. 3. Growth rates of labor productivity and investment costs when using traditional and advanced technologies (Source: compiled by the authors)

References

1. Acemoglu, D., Restrepo, P.: Artificial intelligence, automation and work. In: The Economics of Artificial Intelligence: An Agenda. NBER Working Papers 24196. National Bureau of Economic Research, Inc. (2018)
2. Barykin, E.: The ideas of economy, usefulness and productivity in the works of pre-revolutionary researchers of public finance and financial law. In: Troshkina, T.N. (ed.) Financial, Tax and Customs Law: Theory and Practice, pp. 11–46. HSE, Moscow (2014)
3. Ciccone, A., Hall, R.E.: Productivity and the density of economic activity. Am. Econ. Rev. **86**(1), 54–70 (1996)
4. Chistnikova, I.M., Antonova, M.V.: Theoretical aspects of research of factors of increase of labor productivity at the enterprise. Eur. Sci. **4**(5), 18–20 (2015)
5. Czumanski, T., Lšdding, H.: Integral analysis of labor productivity. Procedia CIRP **3**, 55–60 (2012)
6. Datta, D.K., Guthrie, J.P., Wright, P.M.: Human resources management and labor productivity: does industry matter? Acad. Manag. J. **48**(1), 135–145 (2005)
7. Duarte, M., Restuccia, D.: The role of the structural transformation in aggregate productivity. Quart. J. Econ. **125**(1), 129–173 (2010)
8. Fedchenko, A.A.: Methodical approaches to studying of labor productivity. Econ. Labor **3** (1), 41–62 (2016)
9. Grayson, C.J., O'Dell, C.: American management on the edge of the XXI century. Economics, Moscow (1991)
10. Jarkas, A.M., Bitar, C.G.: Factors affecting construction labor productivity in Kuwait. J. Constr. Eng. Manag. **138**(7), 811–820 (2014)
11. Knyaginin, V.N.: New technological revolution: Challenges and possibilities for Russia. Expert-analytical report. Moscow: The Centre of Strategic Development (2017). https://csr.ru/wp-content/uploads/2017/10/novaya-tehnologicheskaya-revolutsiya-2017-10-13.pdf. Accessed 23 Apr 2019
12. McGrattan, E.R., Prescott, E.C.: The labor productivity puzzle. Working Papers 694, Federal Reserve Bank of Minneapolis (2012)
13. Mikheeva, N.N.: The comparative analysis of labor productivity in Russian Regions. Reg.: Econ. Sociol. **2**(86), 86–112 (2015)
14. Russia Productivity. Trading Economics. https://tradingeconomics.com/russia/productivity. Accessed 23 Apr 2019
15. Rust, R.T., Huang, M.H.: Optimizing service productivity. J. Mark. **76**(2), 47–66 (2012)
16. Selezneva, T.O.: The essence of economy and the importance of labor productivity. Sci. Discuss.: Econ. Manag. Issues **5**, 91–101 (2015)
17. Shash, N.N., Borodin, A.I.: Indicators and methods of measuring labor productivity and the possibility of their use at enterprises. Sci. Notes Petrozavodsk State Univ. **3**(1), 96–101 (2015)
18. Simonova, M.S., Sankova, L.V., Mirzabalaeva, F.I.: Employment in innovation production networks: Regional sample. In: Mantulenko, V. (ed.) GCPMED 2018 - International Scientific Conference "Global Challenges and Prospects of the Modern Economic Development", The European Proceedings of Social & Behavioural Sciences EpSBS, vol. LVII – GCPMED 2018, pp. 1341–1348. Future Academy, London (2018). https://doi.org/10.15405/epsbs.2019.03.136
19. Syverson, C.: What determines productivity? J. Econ. Lit. **49**(2), 326–365 (2011)

Digitalization of Labor Regulation Management: New Forms and Content

V. A. Schekoldin, I. V. Bogatyreva(✉), and L. A. Ilyukhina

Samara State University of Economics, Samara, Russia
scorpiony70@mail.ru

Abstract. The article is devoted to the use of digital (information) technologies for the main types of work on labor regulation: the calculation of norms for production and new technological processes, the development of regulatory materials for labor regulation, the establishment of optimal standards of servicing and number with the use of economic and mathematical methods, processing of the results of working time costs study and the analysis of the state of labor regulation. The order of development of an automated labor regulation system is shown; the algorithm for calculating production norms and new technological processes is developed.

Keywords: Labor regulation · Labor norms · Regulatory materials · Digitalization · Information technologies · Norms calculation algorithm

1 Introduction

The development of information technology in conditions of digital economy leads to a gradual transformation of the content of labor process [9]. Consequently, the role of labor regulation, as an important element of production management, becomes more important. Enterprises are interested in reducing labor costs in production. In this regard, the requirements for the organization of work on labor regulation are becoming stricter. At present, enterprises are independent in the organization of labor regulation. The state guarantees assistance to enterprises in organizing labor regulation, creating legal and methodological support of work.

The Russian enterprises have gained some experience in using information technologies to solve problems in the field of labor regulation. The automated labor information processing systems existing at the Russian enterprises mainly concern the work on accounting of productivity and wages, while the work on automated rationing of technological processes, i.e. on setting standards for products and new technologies are underdeveloped and not methodically developed. This can be explained by the complexity and diversity of standardized works and the need to use a significant amount of regulatory information.

The relevance of the study of issues of digitalization of labor regulation is determined by the increasing role of labor regulation as one of the most important functions of production management in the digital economy; high complexity of the calculations caused by the complexity of production; the lack of use of automated systems in labor regulation at the enterprises.

S. I. Ashmarina et al. (Eds.): ISCDTE 2019, LNNS 84, pp. 137–143, 2020.
https://doi.org/10.1007/978-3-030-27015-5_17

The aim of the research was to study the content of labor regulation management in the conditions of digitalization and the use of digital technologies in labor regulation. In order to achieve this goal, we solved the following tasks: the system of managing labor regulation in the conditions of digital economy was presented, the main types of labor regulation work to be digitized were presented, the procedure for developing an automated labor regulation system was justified, the algorithm for calculating production norms and new technological processes was worked out.

2 Methodology

As the theoretical basis of the study there were taken the works of Russian and Western scientists on the problems of the development of digital technologies and their use in the regulation of labor. When solving the tasks, methods of quantitative and qualitative analysis as well as the methods of predictive analytics and work with Big Data were used. We studied the best practices of work automation in the field of labor regulation. The scientific validity of the results obtained and the formulated conclusions is determined by the use of monitoring of intelligent information systems, software and corporate information systems used by the enterprises in the regulation of labor.

3 Results

3.1 Forms and Content of the Labor Regulation Management in Conditions of the Digital Economy

Labor regulation management in the digital economy is aimed at ensuring the maximum coverage of labor processes by the labor norms, to provide high quality of the established labor costs standards, their progressiveness, soundness and equal intensity.

The objects of labor regulation management are: the actual process of developing and implementing norms for specific operations and work; providing norms for the labor regulation; updating (changing, revising) of labor costs standards; accounting and analysis of indicators of the state of labor regulation; systematic work on the improvement of the labor regulation.

The labor regulation at an enterprise in conditions of digitalization can be managed in the following forms: the establishment of planned indicators on the state of labor regulation; economic motivating based on established targets or on changes in the actual value of indicators; normative, information, methodological and documentary support of the labor regulation management (standards for labor regulation, methodical materials, instructions, regulatory documents).

Thus, it can be said that labor regulation management is a separate system and the part of the overall enterprise management system. The system of labor regulation at the enterprise should be regulated in the Provisions on the Labor Regulation System (or Standard) of an enterprise, which is approved by a local regulatory act of the enterprise which are approved by the local regulatory act of the enterprise.

The system of labor regulation in the digital economy is a set of interrelated elements, the content and effective functioning of which provides the effectiveness of the system as a whole (Fig. 1).

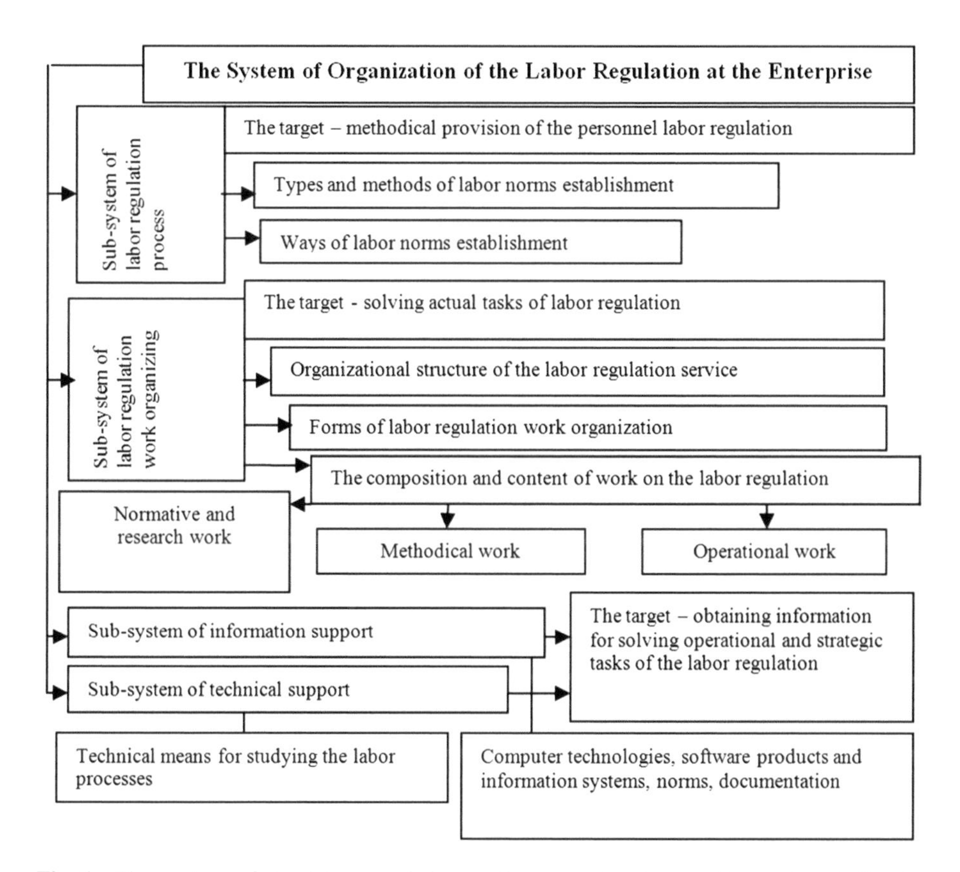

Fig. 1. The system of the labor regulation at the enterprise in conditions of digitalization (Source: compiled by the authors)

3.2 Automation of Work on the Labor Regulation

The object of automation in the field of the labor regulation at the enterprise is the work performed by the specialists in labor regulation. Depending on the level of management (enterprise, department) and the functional division of labor among specialists in the departments, tasks and functions in the field of labor regulation differ and are specified in the job instructions.

To establish the list of functions and tasks to be automated, it is advisable to conduct a survey of work performed by the specialists.

Practice and scientific research show that the main types of work on the regulation of labor, subjected to digitalization, are: the establishment of standards for the production of new products; new technological processes; the development of regulatory materials for the regulation of labor; the choice of the optimal standards of service and the norms of number with the use of economic and mathematical methods; processing of the results of the study of the working time costs and work processes and reporting on the state of the labor regulation and the quality of existing standards (Fig. 2).

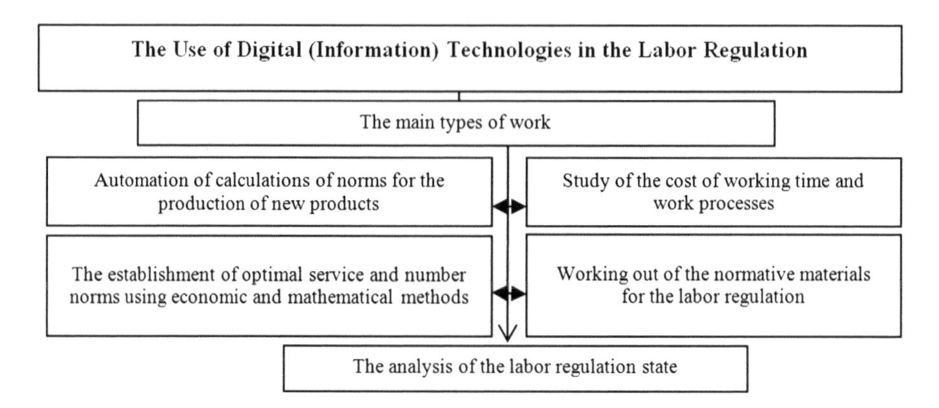

Fig. 2. Digital (information) technologies in the labor regulation (Source: compiled by the authors)

3.3 The Development of an Automated Works Regulation System

At the industrial enterprises, the work on the calculation of norms for technological processes of manufacturing products (automated rationing systems - ARS) has been automated to the least extent.

The experience of the enterprises shows that the following types of automated rationing systems can be distinguished: local (autonomous) systems, providing only the calculation of norms (technological processes are developed); the development of the technological process and the calculation of norms in one system (computer-aided design of technological processes - CAD-TP); traditional (one variant) system, i.e. the calculation of norms according to the established regimes and norms of time and multivariate system providing optimization of technological regimes and norms.

At a number of Russian enterprises, an autonomous automated system of rationing according to established technological regimes and time norms is the most prevalent (Fig. 3).

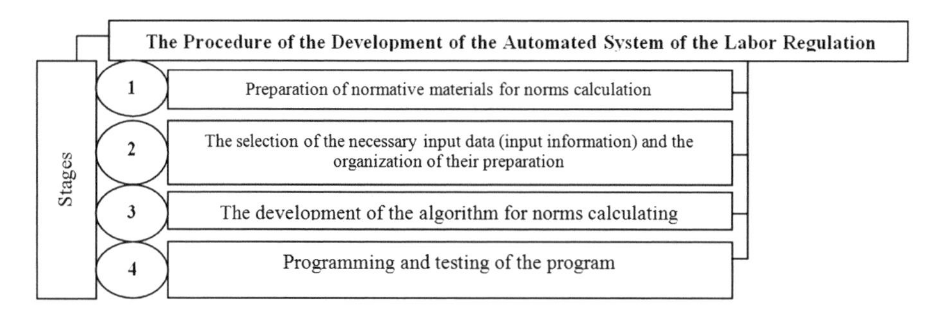

Fig. 3. The stages of the development of an automated regulation (Source: compiled by the authors)

The preparation of normative materials for the calculation of norms consists of the unification of normative tables (maps) from which time values are selected, if they are covered by a single algorithm for calculating and fitting tabular time values by normative dependencies (if they are absent in the normative materials).

The initial data for the calculation of norms generally include: information on manufactured products (parts); the composition of the technological operation (transitions) and the technological modes of processing; the description of the organizational and technical conditions of the operations; time standards; reference data for the calculation of labor intensity and tariffs.

From the point of view of using information in the calculation process, it is subdivided into a constant one used for rationing all the details of operations of a given type of work (time norms, tariff rates, etc.) and a variable one used each time when rationing a particular item of operation.

After clarifying the necessary initial data for the calculation of norms, forms are developed for presenting the initial information (input document). The content of this document is determined by the peculiarities of technological processes and equipment used.

The algorithm is based on a mathematical model for calculating norms.

$$tunit = (tb + ta) * \left(1 + \frac{am + ar}{100}\right), \quad (1)$$

where t_b – basic time for an operation;

t_a – additional time for an operation;

a_m, a_r – time for a working place maintenance and rest, % to the operation time.

It is rational to have a block structure of the algorithm: each block represents a separate stage in the calculation of norms. Figure 4 shows an example of an enlarged flowchart of an algorithm for calculating standards for turning work.

At this stage it is planned to develop a form for issuing calculation results. This form must comply with the standard regulatory documentation adopted at the enterprise (technical regulation map).

The last stage is the compilation and regulating of the program.

According to the developed algorithm, the program of calculation of norms is compiled. After testing the program is ready for use.

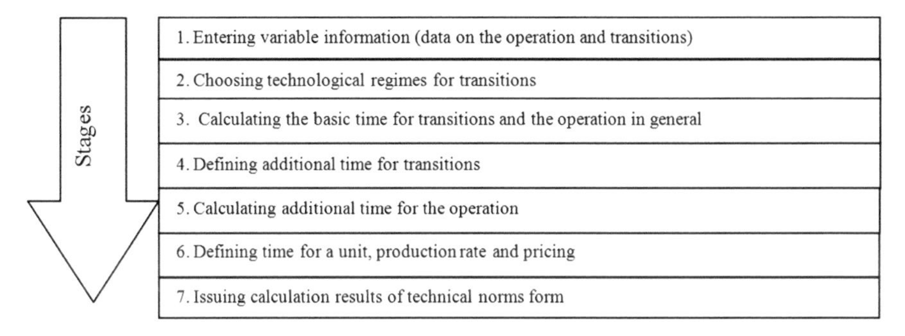

Fig. 4. Block diagram of the algorithm for calculating norms for turning work (Source: compiled by the authors)

In general, it can be said that the development and implementation of an automated labor regulation system leads to saving of working time both in production and in management spheres. If working time in production is not saved or it is difficult to determine it, then we can calculate only time savings in management.

4 Discussion

The scientific works of many Russian and foreign scientists are devoted to labor regulation at enterprises. The use of digital technologies in the microelement rationing of labor and their further development is viewed in the works of Sukhanova and Pikalin [10], Schekoldin [6], Miuskova and Kireeva [4], Maksimov [2], Malinin, Bakhtizina and Startsev [3]. The development of information technologies is a necessary element in improving the existing and developed software tools the working out labor norms [1].

In the 20th century, quite a lot of similar microelement systems were developed. Some of them are used today for modeling and forecasting individual elements of the production process. In foreign practice, the MOST method has become widespread. It is considered the most effective and inexpensive [5].

The development of IT-technologies in the regulation of labor reduces the complexity of the work of specialists engaged in labor regulation and increases the efficiency of using their working time [7]. Undoubtedly, modern computer technologies used to develop labor norms can improve the quality and reduce the complexity of their development. They speed up the process of calculating time norms and the development of new products as well as the reporting on the state of labor regulation and the quality of existing norms. Studies have shown that the automation of calculation processing leads to 5–6 times reduction of the time spent on performing calculations, and 2–3 times reduction of the time spent on paperwork. In this regard, it is necessary to consider the labor regulation at the enterprise as a process integrated into the activities of the company [8]. Thus, the labor regulation management in conditions of digitalization is particularly important.

5 Conclusion

The dynamism of modern production revealing itself in the frequent updating of product range, the improvement of equipment and the development of technology, increase the volume of work on labor regulation. There are tens of thousands of norms at medium-size enterprises. Studies show that specialists engaged in labor regulation in conditions of serial, small-scale production spend up to 70% of their working time to establish standards. Due to large volumes of information processed, labor regulation is a rather labor intensive process. Now, in the period of widespread use of computer information technologies, it has become possible to automate the main work on the regulation of labor, which will significantly reduce the time spent on manual processing of information. Working time is saved due to the transferring of a significant part of calculations in computer programs; reducing the time to refine the results of calculations due to a decrease in the number of errors; efficiency, i.e. quick access to the required information stored in databases; almost complete elimination of time spent on the preparation of working papers; streamlining and reducing workflow at the enterprise. Thus, the use of digital (information) technologies in the labor regulation makes it possible to reduce significantly the time and intensity of labor in the field of developing norms and standards and to solve the problems of improving labor regulation comprehensively.

References

1. Bychin, V.B., Novikova, E.V.: Labor rationing as an element of effective internal management in modern conditions. Russ. J. Labor Econ. **5**(1), 77–86 (2018). (in Russian)
2. Maksimov, D.G.: The emergence and development of microelement work measurement. Bull. Udmurt Univ. Ser. Econ. Law **1**, 68–71 (2014). (in Russian)
3. Malinin, S.V., Bakhtizina, A.R., Startsev, G.N.: Methods of work measurement in the coordinated system of modern production. Bulletin USPTU. Science, Education, Economy. Series Economy **3**(17), 90–101 (2016). (in Russian)
4. Miuskova, R.P., Kireeva, E.L.: Development and renewal of standard time by index method. Labor Rat. Setting Compens. Ind. **4**, 59–65 (2013). (in Russian)
5. Puvanasvaran, A.P., Mei, C.Z., Alagendran, V.A.: Overall equipment efficiency improvement using time study in aerospace industry. Procedia Eng. **68**, 271–277 (2013). https://doi.org/10.1016/j.proeng.2013.12.179
6. Schekoldin, V.A.: Automation of Work on Labor Standardization through Information Technology at Industrial Enterprises: Monograph. Publishing House of Samara State University of Economics, Samara (2013). (in Russian)
7. Schekoldin, V.A., Bogatyreva, I.V., Ilyukhina, L.A., Kornev, V.M.: The development of IT-technologies in labor standardization and quality assessment of standards: challenges and ways of their solving in Russia. Helix **8**(5), 3615–3628 (2018)
8. Shutina, O.V.: Business of work quota setting. Bull. Omsk. Univ. Ser. Econ. **2**, 19–25 (2009). (in Russian)
9. Simonova, M.V., Ilyukhina, L.A., Bogatyreva, I.V., Vagin, S.G., Nikolaeva, K.S.: Conceptual approaches to forecast recruitment needs at the regional level. Int. Rev. Manag. Mark. **6**(S5), 265–273 (2016)
10. Sukhanova, A.V., Pikalin, Y.: Microelement work quota setting. Sci. Bus.: Dev. Ways **12** (78), 50–52 (2017). (in Russian)

Digital Transformation of Tax Administration

M. A. Nazarov[(✉)], O. L. Mikhaleva, and K. S. Chernousova

Samara State University of Economics, Samara, Russia
nalogi_audit@mail.ru

Abstract. The relevance of the problem is due to the results of the analysis of the use of modern digital technologies in the practice of tax administration. Advanced technologies continue to develop and contribute to the modernization of taxation. The purpose of this study is to observe the possibilities of digital technologies for the organization of tax administration process and its transformation. The integration of advanced technologies allows to provide tax administration in real time regime. The assessment of the application of the latest technologies allowed to identify certain factors that affect the efficiency of tax administration processes. In the course of the study, the tasks of analyzing the effectiveness of the use of modern digital technologies in tax administration and identifying the key factors that influence its effectiveness were solved.

Keywords: Taxation · Tax administration · Digitalization · Digital technologies · Internet services

1 Introduction

Digital technologies, penetrating into all spheres of life, change the economy and management processes. Automation of processes changes the usual models of development. Taxation is no exception and is also subject to these modern trends. Nowadays, it is becoming increasingly difficult to track cash flows and this creates risks for tax revenues. In many countries, digitalization in the tax sphere is not developing as quickly as desired and does not allow the tax authorities to take advantage of modern technologies.

In Russia, developed and implemented digital technologies have allowed bringing tax administration to a high level and achieving a leading position in the world.

The purpose of this study is to observe the possibilities of digital technologies for the organization of tax administration processes. The assessment of the application of the latest technologies allowed to identify certain factors that affect the efficiency of tax administration.

2 Theoretical Background

There are different approaches to the definition of the term "tax administration". Bryzgalin understands tax administration primarily as procedural aspects of interaction between fiscal authorities and obligated persons [2]. From the point of view of

S. I. Ashmarina et al. (Eds.): ISCDTE 2019, LNNS 84, pp. 144–149, 2020.
https://doi.org/10.1007/978-3-030-27015-5_18

Maiburov tax administration is the process of management of tax production, implemented by tax authorities and other bodies (tax administrations) with certain powers in relation to taxpayers and payers of fees [3].

Tax administration broadly covers many areas of activity and regulation of the tax system of the state [4]. For normal functioning, the state needs funds that come from tax revenues. The importance of taxes is confirmed by their leading position in the budget revenues. During the period from 2013 to 2017, tax revenues to the Russian consolidated budget showed almost constant annual positive dynamics. In nominal terms, tax revenues increased by 58%. During the same period, in real terms, tax revenues increased by 19.9%. At the end of 2018, the consolidated budget received 21.3 trillion rubles, which is almost 4 trillion rubles (23%) more than in 2017. It is obvious that a significant contribution to the growth of revenues was provided by the improvement of tax administration.

The chief administrator of tax revenues is the Federal Tax Service of Russia. To achieve the goals and objectives set for the tax authorities, there is a whole range of activities related to:

- improvement of tax administration and tools to improve the quality of tax control;
- improvement of electronic interaction with taxpayers;
- implementation of measures to create a Federal government information system for the formation and maintenance of a Federal resource on the population;
- carrying out an experiment on the introduction of a special tax regime "Tax on professional income";
- improvement of tax authorities work on registration of control and cash equipment and control of its application;
- further work on improving information exchange between tax and customs authorities.

Modern development of tax administration is aimed at solving these and other problems.

The Fourth Industrial Revolution and the accelerated development of cyber-physical technologies lead to essential changes in national tax systems. The main areas in which taxation meets cyber-physical technologies are digitalization, robotization, M2 M and blockchain technologies [8].

A set of interrelated and shareable digital technologies include, in particular: modern robotics and artificial intelligence (robotics & AI), big data and analytics, cloud computing, industrial internet, the horizontal and vertical system integration, simulation, augmented reality, additive manufacturing and cybersecurity. European countries also widely implement digital technologies and risk-oriented approaches in the field of tax control [6, 7]. Digital innovations to improve tax administration have identified the need to analyze the impact of these changes on the efficiency of tax administration.

3 Results

Analysis of different approaches to the definition of tax administration allowed to consider it as the activity of the authorized bodies associated with the processes of management of taxes and fees.

The purpose of tax administration is to ensure budget tax revenues by combining the methods of tax regulation and control and using modern information technology. The Federal Tax Service uses a whole range of digital technologies in tax administration: automated control system (ACS) "VAT", a platform for the administration of online cash registers, AIS "Labeling", internet services, mobile applications, etc. (see Fig. 1).

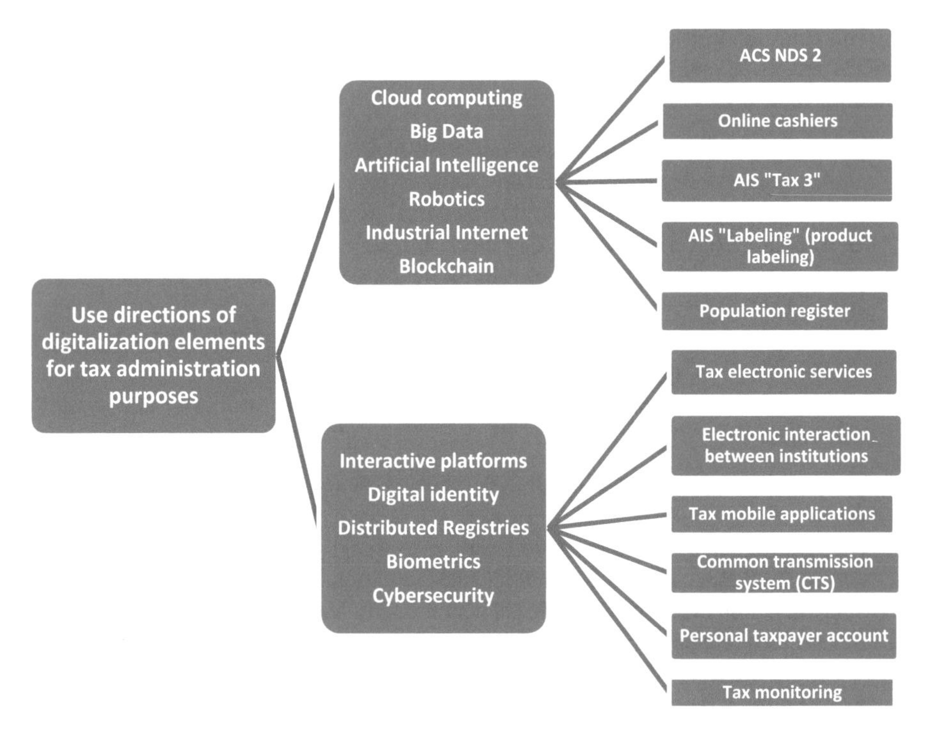

Fig. 1. Use directions of digitalization elements for tax administration purposes (Source: compiled by the authors)

The project "Big Data", the creation of AIS "Tax-3" involves full tax administration of legal entities and individuals (in the existing modules worked out the system of registration and accounting of taxpayers, administration of property taxes of individuals, departmental interaction, the subsystem of control of VAT chains), as well as a full cycle of inspections and debt collection.

Comprehensive analysis of financial and economic activities of taxpayers using publicly available criteria of tax risk allowed abandoning the total control. To improve the efficiency of tax administration, the tax authorities use a risk-based approach in their managerial control. Risk analysis tools became a key element to create the favorable tax environment [1]. For this purpose, various risk criteria are used, for example: tax burden, losses for several periods, high VAT deduction level, etc. The use of risk-based approach significantly reduced the number of on-site tax audits. Today, less than 1% of taxpayers is covered by such inspections. At the same time, in EU countries the tax authorities check in average 2–3%.

Automatic control over VAT system (Automatic Control System VAT) empowers Russian tax authorities through the use of tools of special control over the participants of the entire chain of product and money transfers [5].

Digitalization of tax administration contributes to the efficiency of tax administration. Since the beginning of the application of ACS "VAT-2" tax revenues to the budget system grew faster than the economy.

Thus, revenues to the Russian Federal budget for 2018 increased by more than 30,2% compared to 2017 and amounted to 11.9 trillion rubles according FTS. In 2017 the increase of tax revenues was 20% compared to 2016.

The introduction of ACS "VAT" provided a significant increase in the collection of VAT (VAT revenues to the Russian consolidated budget from 2015 to 2018 increased more than 1,5 times – from 2.45 to 3.6 trillion rubles). Since 2015 ACS "VAT" analyses claimed information in the tax returns about invoices taxpayers and identifies contradictions between the invoicing of taxpayers. The share of doubtful deductions for VAT due to ACS "VAT" decreased and amounted to less than 1%. This figure is one of the lowest in the world. For example, in the EU countries this figure varies on average from 3 to 7%.

The application of this control system forced the taxpayers to abandon the use of illegal schemes of VAT returns and operations with fictitious firms. In 2011, firms with signs of fictitious, were about 40% of the total number of legal entities, as of 2017, the figure fell to 7.3%. This information technology, on the one hand, provided transparency of economic operations, but on the other, only those taxpayers who, according to the results of the pre-test analysis, have increased tax risks of working with fictitious firms became the subjects of tax audits.

In order to control the turnover of goods from the manufacturer to the final user, a digital GIS technology "Labeling" was developed: a system for labeling goods with RFID tags (fur market) and QR codes (medicine market). It has been done to counter the illegal sales of these products. In 2016, the mandatory labeling of fur products was introduced. Only during the first year, more than 25% of the market participants of fur goods legalized their businesses. Since 2018, an experiment on the labeling of tobacco products and shoes labeling has begun, and from 2020, such a system will become mandatory for medicine.

At the same time, it should be noted that the introduction of RFID-marking of goods and further legalization of shadow turnover did not increase tax revenues. The use of a special tax regime (UTII) with the payment of a single special tax by taxpayers contributed to the withdrawal from taxation of legalized sales turnover of fur goods. As a result, there was a reduction in illegal trade, but fiscal interests were not achieved. In this regard, it is necessary to work out proposals to limit the use of UTII in the implementation of labeled products.

Another digital technology in Russian taxation is a new platform for the administration of cash registers, which allows you to see in real time regime how sales are going, what impact seasonal and time factors have on them, what commodity items are in demand and much more. Online cashiers can be registered on the website of the Federal tax service of Russia with the help of a special electronic service. The process is gaining momentum and at the end of 2017, 500 thousand taxpayers registered 1.6 million online cashiers (which exceeds the old cashiers by 40%), punched checks for 18.7 trillion rubles (including VAT—1.6 trillion rubles).

However, there are several problems with the implementation of the order of application of the new online cashiers. These include financial costs that significantly exceed the amount of tax deduction provided, as well as the presence of not full coverage of the Internet connection throughout the territory of the Russian Federation. This causes some difficulties in the implementation of legislation procedure for the use of online cashiers in doing business.

Another tool of tax administration was developed by the Federal tax service for reduction illegal business activities - "electronic" system of legalization of income of self-employed (individuals who do not have employees and independently sell goods, works, services without registration as an entrepreneur) through the mobile application "My tax" and automated payment of tax.

According to the Federal tax service in Russia about 25 million self-employed people, who are estimated by experts as the largest resources for the growth of tax revenues. The problem is also the access of such self-employed persons to social insurance (health and pension). A pilot project for testing this regime began in 2019 in several regions of the Russian Federation. It should be noted that only tax administration cannot solve this problem. It is necessary to make the mechanism of interaction between taxpayers and tax authorities simple and convenient, and new technologies will help to create a system of voluntary compliance with tax legislation.

Recently, the process of communication of tax authorities with taxpayers is carried out on the basis of new information technologies. Thus, the website of the Federal tax service presents interactive online services for different categories of users, personal taxpayers accounts of individuals and legal entities, with which it's possible to interact with taxpayers remotely. At the same time, even the use of interactive services of the Federal tax service when does not always make it possible to assess the tax risk with choosing counterparty.

In the sphere of tax administration of the largest companies, a special regime of interaction between taxpayers and tax authorities is provided - tax monitoring, which works on the basis of online access of FTS to their accounting systems. FTS can quickly track transactions and assess whether tax reporting is correct. Drill-down technology decodes the data of the declarations up to the primary documents. At the same time, the inspector's task is to assess the risks together with the taxpayer and eliminate such risks.

Digital tax administration technologies allow integration with other agencies, organizations and even countries to create a single information space for the use and exchange of information. Thus, the Russian Federation has joined the Common Transmission System (CTS) of the OECD, which technically will provide automatic exchange of information with foreign partners. The information obtained by CTS will allow to identify the presence of undeclared foreign financial accounts and assets, to compare the information provided by Russian citizens on controlled foreign companies. As a result, it is really possible to trace the scheme of financial flows and ownership of assets for tax control purposes. While using new digital technologies of tax administration, it is necessary to take into account their results.

4 Conclusion

Digital transformation should be seen not only as technology, but also as organizational changes, new competencies, and the introduction of improved business models and processes. The presence of certain factors that affect the efficiency of tax administration: improvement of tax legislation, legalization of income, identification of fictitious firms, etc., will contribute to the development of digital transformation.

The use of digital technologies in tax administration allows following an innovative way of its development, making it convenient for taxpayers and reducing the tax risks of the state.

References

1. Advokatova, A.S.: Development of tax control models as a factor reducing shadow economic processes under digitalization conditions. Econ. Taxes Law **11**(5), 136–145 (2018). https://doi.org/10.26794/1999-849X-2018-11-5-136-145
2. Bryzgalin, A.V.: Legal basis of tax administration. Econ. Law **3**, 3–17 (2007). (in Russian)
3. Mayburov, I.A.: Taxes and Taxation, 5th edn. Unity-dana, Moscow (2012). (in Russian)
4. Mishustin, M.V.: Information and Technical Foundations of State Tax Administration. Unity-dana, Moscow (2005). (in Russian)
5. Nazarov, M., Mikhaleva, O., Fomin, E.: Digital economy: Russian taxation issues. International Scientific Conference. Global Challenges and Prospects of the Modern Economic Development. In: Mantulenko, V. (ed.) The European Proceedings of Social and Behavioural Sciences, GCPMED 2018, vol. LVII, paper 129, pp. 1269–1276 (2019). https://doi.org/10.15405/epsbs.2019.03.129
6. Semerád, P., Bartůňkova, L.: VAT control statement as a solution to tax evasion in the Czech Republic. Procedia - Soc. Behav. Sci. **220**, 417–423 (2016). https://doi.org/10.1016/j.sbspro.2016.05.516
7. Tasalov, K.A.: Risk-oriented approach in the field of tax control in the United Kingdom. Nalog. – Taxon. **7**, 82–91 (2018). (in Russian)
8. Vishnevsky, V.P., Chekina, V.D.: Robot vs. tax inspector or how the fourth industrial revolution will change the tax system: a review of problems and solutions. J. Tax Reform **4** (1), 6–26 (2018). https://doi.org/10.15826/jtr.2018.4.1.042

Comparative Analysis of Automatized Systems for Management Processes Information Support

Yu. A. Pertulisov, E. S. Smolina(✉), and L. A. Vodopyanova

Samara State University of Economics, Samara, Russia
ekaterinsmolin@yandex.ru

Abstract. Changes in the economic conditions of the development of enterprises require the use of new methods in making management decisions. This leads to constant changes in the information support of each enterprise. Information support forms the basis of management activities at enterprises. Properly selected software will make it possible to automatize the implementation of information processes. Foreign automatized systems differ from domestic ones. The goal of the research is studying of automatized systems used in our country and abroad. Within the frame of the research it is necessary to solve the following tasks: to consider the theoretical foundations and the essence of information support at enterprises; to determine the peculiarities of automatized systems use in the management processes at the enterprises of the Russian Federation and the enterprises of foreign countries; to analyze automatized systems. The scientific novelty of the research is the comparative analysis of the automatized systems used in Russian and foreign companies and of the further development of these systems. The research methodology includes the following methods: theoretical analysis, used to summarize existing theoretical and methodological research in the field of automatized systems in management processes; synthesis, used to determine the characteristics of the automatized systems in management processes; comparison, used to identify differences and common features of the Russian and foreign software products. The main conclusions and results can be used for research in the field of information support of management processes.

Keywords: Information support · Automatized systems · Management process · Management decisions

1 Introduction

Nowadays information is one of the key resources of each organization. It helps to combine a lot of different kinds of knowledge into one. To combine all the information in one place, special software was created that allows you to combine a lot of information and makes it possible to make a right management decision in future. Information can be found in many sources and it is stored in various places. The main task of automatized systems is collection of information, processing of it, thematic unification and presentation in the form convenient for the user [7].

S. I. Ashmarina et al. (Eds.): ISCDTE 2019, LNNS 84, pp. 150–158, 2020.
https://doi.org/10.1007/978-3-030-27015-5_19

At present there are no restrictions concerning the type of information collected, as well as the types of media used in computer science. Software products which are used now make it possible to integrate various kinds of information and create a field of information resources. This process allows the user to get the necessary knowledge at a certain point in time.

2 Methodology

The theoretical foundations of information support of management processes are viewed in the scientific works of Russian and foreign scientists. The essence of the development of various forms and methods of management decision-making is based on the achievements of technical progress; it is associated with the development of computer science, the improving of legislation, receiving, processing and transmitting of information in various ways. The essence of project management research can be found in the works of Korneev [7], Trofimova [10], Kazakova [6], Webb, Gibson and Forkosh-Baruch [13], etc. These authors formulated certain theses, revealing the essence of information support of management processes.

3 Results

In modern conditions information plays an important role in formation of any enterprise. It is also an important tool for making management decisions. With the advent of a large amount of information, domestic and foreign developers every year try to create new means of storing information, they write new programs that help in organization's activities. Currently management information support works with the help of various automatized management systems. With the advent of a large amount of information, domestic and foreign developers try to create new means of storing information every year; they write new programs that help in the organization activities. Currently, management information support carries out work with the help of various automatized control systems [9].

An automatized control system is an information system designed for automatized control of various processes at an enterprise. The introduction of automatized control systems should be justified. The work of the system should lead to positive technical, economic and social results.

A certain number of general requirements are made to an automatized control system. To make the system work properly, the compatibility of all elements of this system should be provided. The system should also be adapted to changes in both the internal and external environment; it should be constantly developing.

In addition, it is necessary to take measures to protect the automatized system from illegal actions of the staff and from external influences that can lead to a partial or complete shutdown of the system. An unauthorized hacking of the system can also take place that will not lead to the system breakdown, but will cause information leak [10]. Automatized control systems are built on the basis of existing application programs, which make it possible to replace one program with another or to make changes in a program.

The software in the automatized control system makes a number of system requirements. The software is configured in such a way that when a single software product disables, it does not affect the operation of the entire system. It is also necessary to take measures to protect the system against errors when inputting and processing of information is made and to ensure the specified quality of the performance of the automatized system functions [4].

The software of the automatized system should allow the configuration of individual components of the system in such a way that does not affect the operation of the entire system.

At present enterprises are trying to deliver information to the consumer quickly. For this purpose distributed, multi-level automatized management systems are created. The examples of such systems are banking, tax, statistical and other services.

Currently financial management in any company cannot be performed without the introduction of special financial management software. Modern programs help in making management decisions, thereby eliminating various errors in financial accounting. The programs used in the organization help to provide new opportunities in the analytical work of the company. The software also provides the necessary information that contributes to the implementation of the task [5].

4 Discussion

Information support is a complex of various forms of documents, classifiers and regulatory basis. Information support also provides the keeping of a large amount of information that is used in the functioning of automatized systems. Information support implements the preparation of necessary information in individual management structures of the organization as well as in the enterprise system in general.

The main property associated with the adoption of management processes is directly contained in the information nature. The orientation towards the development of scientific and technical research in the organization is based on obtaining of the necessary information, performing the analysis of the information received. All necessary actions allow making right decisions on the modification of objects in the organization. If the information, which will be used for making management decisions, is objective and as accurate as possible, it will provide more accurate reflection of the real state of the managed object [3].

In any organization, regardless of the type of activity, information is the main element of the management process. It should provide the head of the organization with a good idea of the key tasks and the state of the system.

Before characterizing information in the management process as an object of labor, it is necessary to take into account a number of its features. First, information is used long. Although it is part of the finished product, it does not lose its consumer properties. This feature offers a certain individuality of its formation. Data bank is one of the first areas for combining a large amount of information in one convenient place. In future, the information in the databases will be periodically adjusted, updated and continues to be used at the enterprise [6].

Information is a special subject of labor, as it is capable of self-development. A sufficient amount of information gives a great opportunity to determine the direction of development of a managed object (an organization or an individual object) and to identify new links between individual classification groups of information.

Information, in dependence on the degree of preparation, can be divided into:

- Primary information. Such information includes a set of various data and indicators that describe specific aspects of the process and its elements.
- Secondary information [2].

Information can also be divided into models:

1. Processes or elements that describe the general state of the object.
2. Processes characterizing changes in various elements in the management process.
3. Common solutions with active focus [11].

To analyze information support, it is necessary to distinguish the following types of information:

1. Depending on the described processes. Such information includes production-economic, technic and technological, social and organizational information.
2. In relation to the managed object. External or internal production information.
3. According to the role in the management process. Such information includes directive, planned, regulatory and analytical information.
4. According to the degree of renewal and the order of receipt. Information can be constant or variable; it can be stored for a long time. It can also be cyclic, periodic and operative.
5. According to the degree of conversion. This type of information includes primary, derivative and generalized information [12].

Distribution of information plays an important role when organizing information support at the enterprise. It means that information can be divided into:

- command information;
- outgoing information;
- incoming information.

All this helps to take right decisions in the management process of the organization.

Economic information is the information that is created in the process of production and economic activities of the organization and management of these activities.

Economic information, in its turn, has several peculiarities:

1. Economic information has a discrete form of presentation.
2. It can be expressed both in alphabetic-digital and in digital form.
3. It is reproduced on different material carriers (hard drives, USB-drives).
4. Large volumes are processed within certain time limits which depend on specific functions.

5. Primary information which appears in one place is displayed in other control functions. So, this information undergoes various changes several times. This contributes to multiple rearrangements of the data.
6. With a small number of information processing operations, the volume is large.
7. Original data and calculation results are subjected to long storage on data bases.
8. One of the key requirements to economic information is its completeness, accuracy and timeliness [14].

The procedure of the formation of information support includes several steps:

(1) The description of the object state.
(2) Modelling classification links in databases and other software.
(3) The reflection of the dynamics of individual elements and processes in information models.
(4) The reflection of the dynamics, local processes links and the enterprise production in general.

To introduce software it is necessary to find out what company it is designed for: small, large or medium. In addition, it is necessary to clarify in which sphere the company works.

For a number of software products there is a basic set of requirements that should be used in financial activities:

- software should be appropriate for the system of accounting used in the organization;
- software should correspond to all modern requirements and provide all the information management processes need;
- software should have functions which completely correspond to the specifics of the organization work and profile;
- programs should keep records not only on the surface of the financial activities but also have an opportunity of more profound accounting in certain areas of production and the company in general.

There are many foreign companies (15–20) which supply Russia with the software for the sphere of financial activities. But only some firms can compete with them. This situation is stipulated by the fact that for solving complex tasks of management process enterprises are aimed at large firms that already have their own standards for financial accounting. Foreign software is designed for a quick introduction into the customer's company. Such systems require tuning to the following enterprise features:

- existing and modernized business processes;
- because of close ties with foreign partners there is a parallel reporting in accordance with Russian and international standards of financial accounting;
- specific complicated organizational structure;
- internal ideas of financial accounting, analysis, production and sales.

Foreign firms that supply software to the Russian market are built on the modern technological basis. Their software products make it possible to solve more complex problems at large enterprises.

But the main advantage of domestic software developers is that Russian software is completely oriented towards the domestic market. This is due to the fact that financial activities are significantly different from that of foreign partners, which in turn allows making right management decisions. At the same time, it is easier to adapt domestic programs to changes in the financial accounting legislation.

Also there is a tendency of the Russian financial system getting closer to the international one. Foreign accounting programs are used by those enterprises that keep records according to the international standards and regularly sign contracts with international companies. Such organizations use complex corporate systems in their work. One of such popular systems is the ERP-system.

Financial accounting in large organizations is conducted with the use of the latest ERP-systems which allow automatizing not only of financial accounting but of production management in general.

ERP system is a system for planning resources and reports of an enterprise that automatizes accounting and management. ERP-system is introduced into the organization in order to make the interconnection of all departments of the company in a computer system.

For the proper work of the ERP system a single database for all the departments of the organization is created.

Integration is one of the important stages in the automatizing of the work of any organization. As integration is an expensive process, it takes a lot of time [8].

ERP-system is multifunctional and it is rather complicated for automatizing of various processes. For better automatization, a large amount of data from various sources is required.

In order to make it possible for the organization to react to various situations quickly, all the necessary data is loaded into the system in time which allows making right management decisions.

Without the use of integrated ERP-system solutions making management decisions will be ineffective and the organization will not be able to reach a new level of the organization work automatization.

If we refuse to implement integration at the enterprise in various situations and rely only on the human factor, this can lead to a lot of mistakes, both in the financial activities of the enterprise and in making management decisions.

Each of companies depending on the type of activity uses different software. These can be domestic programs or foreign ones. Let us compare two programs of the ERP – system in order to understand which one is better: the domestic program 1C: Management of a manufacturing enterprise and the foreign program Oracle AIM [13]. Let us first view domestic software: 1C: Manufacturing Enterprise Management.

1C: Manufacturing Enterprise Management was developed by the company 1C which presents it as one of its main products. This program is completely created by the software developers of the company 1C and all the improvements, changes and implementation is made by the company.

The main task of this software product is to create the necessary configuration for a specific customer based on the basic configuration.

This is a complex system and after certain improvements the program will carry out the management functions of the organization of a certain type, for which the improvements were made. The advantage of this software product is that the program can be adapted to any type of organization activities.

One of the important disadvantages of the program is lack of modularity. That is to solve many problems you can create certain reports or perform processing. But if the documents or reference books are being changed, then absolutely all subsystems in the configuration should be changed.

Every year 1C changes the structure of its program. This is explained by the constant changes in the legislation and financial statements. The company is constantly improving the program to promote it on the market. This software product is one of the leaders of domestic software, which is later going to be used not only on the domestic, but also on the foreign market.

Let us view foreign software.

Oracle is one of the world largest business software developers. The main products of the company are:

- Database;
- Fusion Middleware;
- Applications

The main Russian customers of Oracle are: Ingosstrah, MTS, VTB-24 and Vympelcom.

The main program of Oracle, made for assistance in conducting of business is Oracle AIM (US) program. With the implementation of this system according to this methodology documentation is fixed according to special patterns on each stage. And each involves a set of documents prescribed in the methodology.

The main advantage of Oracle in comparison with 1C is that the set of documents can be used partially, for a certain stage. 1C methodology is worse than Oracle one in the regard that the documentation is not worked out in detail, but is constructed on state standards, which, in turn, have not been updated for many years.

The use of the V-shaped model in Oracle AIM involves a new version of the cascade model. For 1C: Manufacturing Enterprise Management the cascade model is also preferable.

The use of Oracle AIM program requires special knowledge and qualifications from the manager. Unlike Oracle AIM, 1C uses standard methodologies and life cycle models. On the one hand, special knowledge will allow the manager to act exactly as he needs. But on the other hand, ready-made patterns and recommendations can make the project more manageable [1].

General disadvantages and advantages of using Oracle AIM methodology are presented in Table 1.

Besides the assistance to the manager of the enterprise in making management decisions with the help of the software, Oracle is working on the creation of artificial intelligence. These will be Oracle Enterprise Resource Planning (ERP) Cloud and Enterprise Performance Management (EPM) Cloud programs. These programs should promptly provide the staff with necessary and timely information. Due to this, the managers and the financial director will be able to implement new business models at the enterprise.

Table 1. Disadvantages and advantages of Oracle AIM software

Advantages	Disadvantages
The rules for the documents are stipulated for each stage	The amount of the documentation is too large
The tasks for each item of processes and stages are worked out in detail	It is necessary to involve specialists in the implementation of this methodology
The manager can use all types of the project documentation as well as its parts	Because of the large amount of the documentation it is necessary to increase the project team. It can also increase the duration of the project
All changes and agreements are recorded in the project documentation	

Source: compiled by the authors based on [5].

The integration of "smart" processing, based on the rules within the framework of the Oracle Cloud Applications software, will make it possible to almost completely automate the execution of labor costly tasks which in future will greatly affect the financial activities of the enterprise. Artificial intelligence will allow verification and monitoring of compliance with the policies. Compliance with the policies will be achieved at the expense of faster reports, increased accuracy, as well as the simplification of the work of the end users.

New capabilities of artificial intelligence are going to contribute to reducing of the costs of the enterprise, to the opportunity to strengthen relations with key contractors due to the instant analysis of their profile and risks.

Also artificial intelligence helps to reveal hidden regularities in the data. This is necessary in order to provide the analyst with the necessary data that will help to deal with this or that situation in time; it will contribute to improving the quality of operating and financial decisions affecting business results.

Also, artificial intelligence is used to check users in the organization. It is created to make all employees fulfil security regulations and to protect confidential information from both internal and external threats.

Oracle ERP Cloud helps organizations to automatize the main business processes, to integrate scattered systems, align financial business processes with technological requirements and adapt to changing business requirements.

Artificial intelligence is being actively introduced by Russian companies, too. Many Russian business leaders believe that artificial intelligence has a positive effect on management activities.

5 Conclusion

In accordance with the analysis of software products of Russian and foreign companies, we can make a conclusion that Russian developers are mainly focused on the domestic market. This is a positive indicator, as all the necessary information and changes in the legislation affect the structure of the entire software product only to a small extent.

Foreign companies are more focused on the external market. The reason can be that many partners of the company represent another state. Thus, the software product allows you to solve the problems of two companies in different countries.

So, we can make a conclusion that developers of various companies have a goal to help companies in management processes. Software products are constantly being upgraded, they help to answer a particular question more accurately, draw a conclusion based on the data.

The development of software products will constantly develop. All this is connected with the development of techniques, as well as with the creation of artificial intelligence, which in future is going to be able to combine all software products into one. This will contribute to the interconnection of many companies in various countries in conducting of business.

References

1. Ashkanasy, N.M., Humphrey, R.H., Huy, Q.N.: Integrating emotions and affect in theories of management. Acad. Manag. Rev. **42**(2), 175–189 (2017). https://doi.org/10.5465/amr.2016.0474
2. Fetisova, O.V., Kurchenkov, V.V., Matina, E.S.: The Role of Network Companies in the Development of the Regional Consumer Market. VolGU, Volgograd (2013). (in Russian)
3. Gorbunova, M.L., Prikazchikova, Yu.V.: On the strategies of entering the foreign markets by Russian innovative non-resource companies. Russ. Manag. J. **14**(4), 49–80 (2016). (in Russian)
4. Greenberg, A.S., Korol, I.A. (eds.): Information Management. Uniti-Dana, Moscow (2016). (in Russian)
5. Gromov, G.R.: Information Technology Essays. Infoart, Moscow (2017). (in Russian)
6. Kazakova, N.A.: Modern Strategic Analysis. Urite, Moscow (2014). (in Russian)
7. Korneev, I.K.: Information Support of Management Activities. Masterstvo, Moscow (2017). (in Russian)
8. Mišút, M., Mišútová, M.: Evaluation of ICT implementation into engineering education. Paper presented at the International Conference on Advances in Information Technology (ICAIT 2013), Jeju Island, Korea (2013)
9. Smirnov, E.A.: Management Decisions. RIOR, Moscow (2016). (in Russian)
10. Trofimova, L.A.: Methods of Management Decisions Taking. Urite, Moscow (2016). (in Russian)
11. Rajnoha, R., Kadarova, J., Sujova, A., Kadar, G.: Business information systems: research study and methodological proposals for ERP implementation process improvement. Procedia - Soc. Behav. Sci. **109**, 165–170 (2014). https://doi.org/10.1016/j.sbspro.2013.12.438
12. Vuksic, V.B., Pejic Bach, M., Popovic, A.: Supporting performance management with business process management and business intelligence: a case analysis of integration and orchestration. Int. J. Inf. Manage. **33**(4), 613–619 (2013). https://doi.org/10.1016/j.ijinfomgt.2013.03.008
13. Webb, M., Gibson, D., Forkosh-Baruch, A.: Challenges for information technology supporting educational assessment. J. Comput. Assist. Learn. **29**(5), 451–462 (2013). https://doi.org/10.1111/jcal.12033
14. Zamecnik, R.: The measurement of employee motivation by using multi-factor statistical analysis. Procedia - Soc. Behav. Sci. **109**, 845–850 (2014). https://doi.org/10.1016/j.sbspro.2013.12.553

The Impact of Digitalization on the Economic Security Index of GDP

O. A. Naumova(✉), I. A. Svetkina, and T. A. Korneeva

Samara State University of Economics, Samara, Russia
naumovaoa@gmail.com

Abstract. To assess the state of economic security in Russia, both traditional and relatively new macroeconomic indices are used, which depend on changes in digital tools. With the help of macroeconomic indices, the level of economic security of any country is assessed. The study assesses the development of key economic security indices of the Russian Federation. Special attention is paid to the value of GDP and the influence of the digital economy on it. Different points of view on the volume of "digital" GDP and methods of its calculation are presented. The authors systematize views on the absolute and relative indices of the volume of the digital economy in Russia and estimate these values. The study is conducted to determine the share of the digital economy in the country's value of GDP, as well as the reasons for different values obtained, and GDP is used as a characteristic of economic security of the country.

Keywords: Digital economy · Economic security · Gross domestic product · Share of the digital economy in GDP

1 Introduction

The system of macroeconomic indices allows measuring economic security of the country. In addition, using macroeconomic indices, it is possible to measure macroeconomic results and analyze their prerequisites.

The state of economic security is measured by a list of criteria and indices determining the threshold values of the economic system. As the experience of many foreign countries testifies, the system loses the ability for dynamic self-development outside of these values, the competitiveness of the economy falls, corruption and criminality increase, the standard of living of the nation decreases.

Economic development, including complication and diversification of its processes, especially in the financial sector, and the new development of information systems and technological solutions based on them, form new challenges for the transformation of all relations in the economy and, as a result, legal and institutional changes. Speaking about the development of digitalization and the digital economy, it is necessary to consider the issue of economic security in two different aspects:

- In the aspect of digitalization and subsequent digitalization of the communication system between various actors;
- In the aspect of the digital economy itself.

S. I. Ashmarina et al. (Eds.): ISCDTE 2019, LNNS 84, pp. 159–164, 2020.
https://doi.org/10.1007/978-3-030-27015-5_20

One of the development features of the digital economy is its share in the country's GDP. This is the Economic Security Index*.

Gross domestic product (GDP) is the cost of all final services and goods that are produced during the year in the state, in all sectors of its economy for export, consumption and accumulation, using factors of production of the state, regardless of the nationality of economic agents.

The objective of this work is to study methods for determining the share of the digital economy in the country's value of GDP, as well as the reasons for differences in the values obtained, and GDP is used as an indicator of economic security of the country.

2 Methodology

The authors used such research methods as analysis, synthesis, description and comparison. Their application is determined by the theoretical nature of the study which included the following stages: formulation of the problem, analysis of information on this topic, comparison and description of different scientific views on the studied issues, synthesis of different approaches to the problem.

3 Results

The study collects research results of research and consulting companies in Russia and the world in the digital economy and presents quantitative indices of the digital economy. The authors assessed the reasons for significant differences in the values obtained using the methods of comparison, analysis, direct counting. As one of the reasons were methods of calculating the absolute volume of the digital economy, its cost measurement, as well as its share in GDP. We will measure the state of economic security of Russia using some macroeconomic indices (Table 1).

Based on the data in the table, we can talk about a slight increase in indices after the crisis in 2015. Both the Volume Index of GDP and investment in fixed assets are growing. Destructive indices, such as inflation rate and Labor Market Strains Index, on the contrary, are declining.

As a consequence of the fall in production and the investment recession, there should have been a further slowdown in economic growth. This was reflected in the Volume Index of GDP. After the crisis of 2015 and 2016, the decline in GDP was respectively −3.9% and −0.2%.

Russian scientists found out that all economic security indices, one way or another, are interconnected [12]. The deterioration of some indices leads to the deterioration of others, creating risks of economic security. In order to neutralize the risks of economic security, it is necessary to start with the economic restructuring, switch to the innovative economic model and modernize the system of economic management.

The innovative model of the economy directly depends on the growth of the cluster of digital goods and services and the digital economy as a whole. Thus, the growth of the digital economy will increase economic security of Russia, ensuring the growth of

Table 1. Some indices of economic security in 2013–2018 *

Indices	01.01.2015	01.01.2016	01.01.2017	01.01.2018	01.01.2019
Volume index of GDP, %	0,6	−3,9	−0,2	1,5	2,3
Industrial production index, %	101,7	96,6	101,1	101,0	102,1
Share of investment in fixed assets in GDP, %	19,7	17,8	20,4	25	27
Inflation rate, %	11,4	12,9	5,4	2,5	2,9
Labor market strains index, %	5,2	5,6	6	4,9	4,4
External debt of the Russian Federation, billion US dollars	599,9	518,5	511,6	518,8	453,7

Source: compiled by the authors according to Rosstat data [3].

key indices. For an adequate assessment of the growth rate of the digital economy, we need a unified approach to measure its volume. According to publications on the volume of the digital economy, we obtained the following results (Table 2).

Table 2. GDP volume of the "digital economy", billion rubles (2017)

Source of information	GDP volume of the "digital economy", billion rubles. (2017)
Information agency TASS	4300
The Boston consulting group – an international consulting company	1754
Russian association of electronic communications	4153
National Research University Higher School of Economics	2514
McKinsey & Company|Global management consulting	3229

Source: compiled by the authors.

It can be seen that the deviation of indices is very significant, in some values more than 2 times which indicates a radically different approach to determining the components of the digital economy. It is the main reason for the Volume Index of the "digital" economy. Some economists [8] include the turnover of the Internet and the automation of production (for example, technologies "smart home" and so on) in this concept. Another point of view [11] is that this is primarily the ICT field. In order to understand this question let us turn to the most frequently used definitions.

According to Gartner research company, "digital business is a new business model that covers people, business, things; scalable globally for the whole world through the use of information technologies, the Internet, and all their properties, suggesting effective personal service for everyone, everywhere, always [2]. There is also an opinion that the digital economy is a set of types of economic activity based on digital technologies and characterized by active introduction and use of digital technologies for storing, processing and transmitting information in all spheres of human activity [5]. The Boston Consulting Group is of the opinion that this is a sphere of economic activity that includes online consumption, the cost of building the infrastructure of this consumption. [4]. According to specialists of National Research University Higher School of Economics, the digital economy is the creation, distribution and use of digital technologies and related products and services [10].

Different approaches give different quantitative expressions. Table 3 presents the share of the digital economy in Russia's GDP and the global scale according to research data from various scientific schools.

We note a high share of the digital sector in the global economy and it is difficult to argue with the fact that it will grow from year to year. Russia is already living in the digital era: according to a number of Internet users, Russia ranks first in Europe and sixth in the world. For example, 60% of the population uses smartphones. This is more than in Brazil, India and Eastern European countries.

Table 3. The share of the digital economy in Russia's GDP

Source	Share in Russia's GDP
The World Bank	2,8%
Russian Association of Electronic Communications (RAEC)	5,1%
BCG	2%
OECD, National Research University Higher School of Economics, calculations by Stolypin Growth Economy Institute	3,6%
McKinsey	3,9%

Source: compiled by the authors on the basis of statistical data [1, 4, 6, 8, 9].

At the same time, Russia is not included in the group of leaders in the development of the digital economy in many indices - the level of digitalization, the share of the digital economy in GDP, the average delay in the development of technologies used in leading countries.

According to McKinsey, the potential economic effect from digitalization of the Russian economy will increase the country's GDP by 4.1–8.9 trillion rubles (in 2015 prices), which will be from 19 to 34% of the total expected GDP growth [6].

4 Discussion

The importance of macroeconomic indices is obvious, since they play a significant role in the country's economy. General macroeconomic indices include the most significant interdependent and specific parameters of the market economy, which give the most complete description of the state of the economy as a whole. The indicative system is aimed at assessing and predicting threats to national interests of Russia and the ability to repel these threats, by adapting government agencies to new conditions that arise in the modern world. To ensure economic security of a country, region, or industry, it is necessary to single out those security factors that have different outcomes and can act as potential threats that pose a danger to vital economic interests, both of the individual and of the state as a whole. The government of the Russian Federation has developed a strategy of economic security until 2030 [7], in which the prerequisites for the reliable prevention of its threats are formed.

5 Conclusion

The growth rate of GDP has a positive effect on the level of economic security of any country. The positive factors for the development of the digital economy include: strengthening control over various processes at different levels of objects and subjects; simplifying the interaction between the subjects of the economic space; reducing the cost of document management at different levels of management; saving time, etc.

We should also mention the additional threats to economic security associated with the development of the digital economy. There is a shadow digital economy that reduces GDP and increases digital fraud and penetration into the personal life of a person through digital tools. There is a digital divide between different levels of countries and the inaccessibility of digital benefits to developing countries. Jobs are reduced, some professions disappear. And most importantly, there is access to a huge amount of data and they are used by hackers for personal gain (each person with a gadget experienced an attack on social networks, by phone, on bank cards).

Due to differences in approaches to the definition of the digital economy in the structure of GDP and the impact on economic security, it is difficult to quantitatively and qualitatively assess the scale of its influence. There is always a risk, but it's not worth giving up one of the greatest benefits of the modern world.

References

1. All-Russian Academy of Foreign Trade: Monitoring of current events in the field International Trade No. 5 (2018). http://apec-center.ru/wp-content/uploads/2018/02/Monitoring_5_RFTA_APEC_OECD.pdf. Accessed 14 Feb 2019. (in Russian)
2. BitNovosti: Gartner: Programmable Economy will change everything (2015). https://bitnovosti.com/2015/12/11/gartner-says-programmable-economy-will-disrupt-global-economy. Accessed 25 Feb 2019. (in Russian)
3. Federal Service of State Statistics (GKS): Indicators for assessing the state of economic security of Russia (2018). http://www.gks.ru/wps/wcm/connect/rosstat_main/rosstat/ru/statistics/econSafety/. Accessed 12 Feb 2019. (in Russian)
4. Institute of Statistical Studies and Economics of Knowledge Research University HSE: The contribution of digitalization to the growth of the Russian economy (2018). https://issek.hse.ru/news/221125086.html. Accessed 14 Feb 2019. (in Russian)

5. Lastovich, B.: ICT infrastructure of the digital economy. Simple truths (2017). http://www.iksmedia.ru/articles/5434122-IKTinfrastruktura-cifrovoj-ekonomik.html. Accessed 14 Feb 2019. (in Russian)
6. McKinsey & Company: Digital Russia. New reality (2018). https://www.mckinsey.com/ru/our-work/mckinsey-digital. Accessed 14 Feb 2019. (in Russian)
7. Presidential Decree of 13.05.2017 N 208: On the Strategy of Economic Security of the Russian Federation for the period up to 2030. http://www.consultant.ru/document/cons_doc_LAW_216629/. Accessed 02 Feb 2019. (in Russian)
8. PWC: Perspectives development "Internet of Things" in Russia (2017). https://www.pwc.ru/ru/publications/the-internet-of-things.html. Accessed 14 Feb 2019. (in Russian)
9. RAEC: Runet summed up the year (2018). https://raec.ru/live/raec-news/10766/. Accessed 14 Feb 2019. (in Russian)
10. Suslov, A.B.: Digital economy: What and how to measure? (2018). http://www.gks.ru/publish/conf0918/suslov.pdf. Accessed 31 Jan 2019. (in Russian)
11. Vasilenko, N.V., Kudryavtseva, K.V.: The emergence of a new type of economy: the interdependence of its digitalization and servization. In: Babkina, A.V. (ed.) Digital Transformation of Economy and Industry: Problems and Prospects, pp. 67–91. Publishing House of Polytechnic University, St. Petersburg (2017). (in Russian)
12. Voronkov, A.N.: The impact of economic crises on the economic security of Russia. Econ. Res. Dev. **3**, 61–69 (2018). http://edrj.ru/article/11-03-2018. Accessed 31 Jan 2019. (in Russian)

Digital Transformation of Municipal Management Under Sustainable Development

A. A. Sidorov(✉), N. V. Lazareva, and N. V. Starun

Samara State University of Economics, Samara, Russia
sidorov120559@yandex.ru

Abstract. Digitization of environmental information and digitalization in municipal management can positively influence sustainable development of territories. The main documents of the rulemaking support of digital technologies and ICT in Russia are analyzed. The authors illustrate the effects of using digitization for the population of municipalities and consider the possibilities for digitizing environmental information in municipalities of the Samara region of the Russian Federation, a large and environmentally ambiguous region of Russia. The sources of the initial matrix for digitizing municipal environmental information are official public information, and the authors consider the appropriate categories and indicators, problems and ways of using them. It is proposed to transform the ideas of "smart sustainable city" into projects "smart sustainable municipality", to include digital technologies for sustainable development of municipalities in the strategies of socio-economic development of territories with specific proposals for solving local environmental goals, with indicators for their implementation.

Keywords: Digitalization · Environmental information · Municipal development

1 Introduction

Current changes in society are associated with the widespread use of digital technologies and information and communication technologies (ICT). According to the Global Connectivity Index (GCI), which measures the progress of 78 large countries on the path of digital transformation, it is clear that there is a clear correlation between digitalization and sustainable economic growth [4]. GCI shows that every additional $ 1 invested in ICT infrastructure can generate revenue of $ 3.70 in 2020, and in 2025 it determines a potential yield of $ 5. The increase in GCI only by 1 point corresponds to the growth in productivity by 2.3%, in innovation by 2.2% and in national competitiveness by 2.1%. The increase in investments in ICT by 20% leads to the increase in the country's GDP by 1%. In this ranking, Russia has improved its position: 2015 - 40th place, 2016 - 37th place, 2017 - 34th place, in 2018 - 36th place, it is assigned to the group of middle-countries (Adopter).

S. I. Ashmarina et al. (Eds.): ISCDTE 2019, LNNS 84, pp. 165–171, 2020.
https://doi.org/10.1007/978-3-030-27015-5_21

2 Methodology

Currently, in Russia, the majority of state and municipal services are provided in electronic form through state service portals. Digital technologies and ICT offer great opportunities to solve many municipal problems [10].

The purpose of the study was to assess the state of digitization and digitalization in the municipal government for sustainable development of sub-regions of the Samara region, a large region of Russia. The objectives of the study are to determine the effects of using digitization for the population of municipalities, to consider the possibilities, to find sources of forming the initial matrix for digitizing municipal environmental information and to identify digitalization problems in municipal sustainable development planning and ways to solve them.

The materials for the analysis are norm-setting documents, official statistical information and literary sources. The authors use methods of description, comparison, mathematical analysis, logical construction.

3 Results

From the point of view of the rule-making support of digital technologies and ICT, the following fundamental directives are in effect in Russia:

- "Strategy for the Development of Information Society in the Russian Federation for 2017–2030", approved by the Decree of the President of the Russian Federation № 203 of May 09, 2017 [13]. The document defines the goals, objectives and measures for the implementation of domestic and foreign policy of the Russian Federation in applying information and communication technologies aimed at the development of information society, forming a national digital economy and ensuring national interests and strategic national priorities;
- The national program "Digital Economy" for the period up to 2024, approved by the Decree of the Government of the Russian Federation No. 1632-p of June 28, 2017 [15]. It defines the goals, objectives, directions and deadlines for implementing the basic measures of state policy for the development of the digital economy in Russia, in which data in digital form is a key factor in production in all spheres of social and economic activity. The program has identified five basic areas, which include regulation, personnel and education, research competencies and technical reserves, information infrastructure and information security. A special subcommittee on the digital economy at the government commission on the use of information technologies has been created to operate the program.

The potential of digitization and digitalization is very high to reduce the load on the environment and increase the resilience of municipal ecosystems [8]. At the same time, they consider it is necessary to integrate environmental municipal goals built on EU directives with the technological capabilities of digitization in order to practically implement digital technologies in achieving the goals of environmental directives.

Digitization of environmental information at the municipal level in the Russian Federation can be justified by the action of Federal Law No. 131 "About General

Principles of Organizing Local Self-Government in the Russian Federation" of October 6, 2003, Article 15 "Issues of local importance of the municipal district" in paragraph 9 provides for inter-village events to protect environment [2].

Digitization of municipal environmental information and placing it on the website of the municipal administration are to:

- Simplify and clarify the results of environmental activities in the public understanding;
- Ensure the availability of environmental information;
- Influence the implementation of environmental policy targets;
- Transit to international environmental standards;
- Stimulate the modernization of production;
- Stimulate a constructive dialogue between the administration (government), civil society and business;
- Transit from exclusive use of norms of administrative law to economic and civil law;
- Provide reputational privileges for administration and business;
- Improve environmental education and upbringing of the population.

Let us consider the possibilities of digitizing municipal environmental information in municipalities of the Samara region of the Russian Federation. The region covers an area of 53.6 thousand km^2 in the middle reaches of the Volga River. The territory is characterized by a moderate continental climate, includes a large variety of conditions of the forest-steppe and steppe zones. The authors identified natural and anthropogenic environmental instability, an ambiguous situation in the sub-regions and unresolved diverse problems in land use, forest use, air pollution, water use, water supply and wastewater disposal, waste management [11].

The initial matrix for digitizing municipal environmental information can be compiled from publicly available information from the annually published Report of the Ministry of Forestry, Environmental Protection and Nature Management of the Samara Region "On the environmental situation in the Samara region" [7]. It contains gross and specific indicators of air and water pollution, waste management and other information in urban districts (10) and municipal districts (27). It is proposed to use the following categories and indicators (Table 1).

But at the same time, there is a need to solve many problems. So, if we refer to the data on the negative environmental impact in the municipal area Neftegorsk from the Report [7], we see two expectedly interrelated parameters. Article 11 mentions one licensed waste disposal facility, and the next Article 12 "Placed waste at licensed facilities" – has no information. One can only guess where the waste goes, why it does not go to the licensed facility and what the supervisor does? The answer is found in Article 13: the number of "Unauthorized waste disposal sites liquidated during the year" is 344 objects with a total area of 2.25 ha. Table 55 of the Report [7] "Specific environmental impact indicators for the territories of municipalities" has no complete information on the following parameters: "Number of vehicles (per 1000 population)" in 7 districts: Volga, Kinel, Krasnoyarsk, Neftegorsk, Pokhvistnevo, Privolzhsky, Syzran; "Discharge of polluted wastewater into surface water bodies (m^3/year/per 1 permanent resident)" in the Krasnoarmeysk district. At the same time, it is necessary to

Table 1. The original matrix of municipal environmental information

Categories	Indicators
Atmospheric air	The number of objects that have a negative impact on environmental protection systems is subject to regional environmental supervision (units), the number of objects having pollutant emissions (units), the number of stationary sources of pollutant emissions (units), the amount of emissions from stationary sources (thousand tons), emissions of pollutants (tons/year per inhabitant), emissions of pollutants (tons/km^2 per year), number of vehicles (pcs. per 1000 population)
Water basin	Discharge of wastewater into surface water bodies (million m^3/year), including not proper treated water (million m^3/year), number of water users having discharge of wastewater into surface water bodies (units), water withdrawals from natural water bodies (m^3/year per inhabitant), discharge of polluted wastewater into surface water bodies (m^3/year per permanent resident)
Territory	Forest cover (%), annual growth of tree plantations (%), share of protected areas in the total area of the territory (%)
Wastes	Unauthorized landfills and waste (units), licensed waste disposal facilities (units), the volume of waste disposal at licensed disposal facilities (thousand tons), the amount of waste generated (tons/year per inhabitant), the amount of waste generated tons/km^2 per year)

Source: compiled by the authors.

note a certain progress on the above index in comparison with the previous period. Thus, in the Report-2015, there was no data on the indicator "Discharge of polluted wastewater …" in 9 areas: Bogatovsk, Bolshaya-Glushitsa, Bolshaya-Chernigovska, Borsk, Elhovska, Isaklyi, Krasnoarmeysk, Pestravska, Shentalyi. Such restrictions make it difficult to plan the calculations and they reduce the objective implementation. On the basis of a set of indicators of environmental performance, it is possible to emphasize Borsk and Khvorostyansk (4 best parameters from 5), Kamyshlinsk (3 parameters from 5).

Along with this, we find significant discrepancies in the list of indicators given in the Report [7] and indicators provided for in the Strategy for Environmental Security of the Russian Federation for the period up to 2025, approved by the Decree of the President of the Russian Federation No. 176 of April 14, 2017 [12] in which much attention is paid to waste. So, out of a total of 18 indicators of environmental safety, 10 reflect the state of waste with differentiation into 4 hazard classes and the proportion of disturbed lands and others are taken into account. These indicators are not in the materials of the aforementioned State Report.

Russia also adopted GOST R ISO 37120-2015 "Sustainable Development of Communities. Indicators of Urban Services and Quality of Life" [4]. This standard is identical to the international standard (ISO 37120: 2014 "Sustainable Development of Communities - Indicators for Quality and Life, IDT)". The main indicators are applied in Section 8 "Environment": the concentration of fine suspended particles (PM2.5 and PM10), emissions of greenhouse gases in tons per capita. Section 16 "Solid Waste" represents the following main indicators: the share of the urban population provided with the service of regular removal of solid (household) waste, the share of municipal solid waste that is being processed. Section 20 "Wastewater" considers the proportion of the urban population provided with the sewage disposal service, the proportion of urban effluents that are not treated, the proportion of urban effluent undergoing primary, secondary and tertiary treatment. These indicators are not in the above regional and federal documents. Great difficulties arise in obtaining primary data due to the fact that in a large part of municipalities there is no environmental monitoring of water bodies, land resources, the qualitative component of emissions to the atmosphere, the biological diversity and others. It is obvious that it is necessary to unify the list of used indicators and to ensure the conditions for conducting environmental monitoring using unified methods.

Digital technologies and ICT can increase energy efficiency and reduce air emissions through the use of digital communications in transport, digital booking and placement of products, storage facilities, digital information in the choice of consumption, digital business planning and more [6]. The use of digital technologies, including those aimed at creating a favorable ecological environment, is mostly formulated in the framework of the projects "smart city", "smart sustainable city". This is such an agglomeration, which is able to meet the needs of modern residents, does not cause damage to other citizens and future generations [5]. Such projects are created on the basis of cooperation between private and state, industrial and municipal subjects, with the involvement of specialized universities or research institutes. Digitalization has great potential to support environmental benefits, but realizing potential requires thoughtful decisions from technological and political, regulatory, organizational and other measures [3]. At the same time, the budgets of many municipal districts in Russia are not able to finance infrastructure projects, for example, the construction and modernization of engineering networks, sewage treatment plants, water intakes, the depreciation of which reaches 90%. Cost projects are implemented exclusively with the involvement of regional budgets. Municipal-private partnership practically does not find distribution in this area.

Digital transformations of sustainable development of municipalities are summarized in the Strategies of Socio-economic Development of Territories. The assessment of the environmental factor in the Strategies of Socio-economic Development of the Regions of the Russian Federation showed its ambiguous state: no environmental factor (12% of documents), non-priority status (26% of strategies), lack of targets (73%) of environmental development [9].

The analysis of the Strategy for Socio-Economic Development of the Samara Region for the Period up to 2030 [14] shows that some environmental aspects are reflected in the SWOT analysis and in the strategic directions for the regional development. In the document, they are located after solving the problems of increasing the competitiveness of the economy through the development of its industries. The next section, improving the quality of life of the population is the improvement of the ecological situation, which has 12 pages of text. It is symbolic that this section is located between the sections of increasing the material well-being of inhabitants and the development of the education system and personnel support for economic growth. It is planned to have 10 directions of the regional environmental policy in order to improve all components of the environment. However, in reality, only 10 specific targets are expected to be implemented in 3 planned directions, the rest are declarative in nature.

The developed Strategy for Social and Economic Development of the Bezenchuk District of the Samara region for the Period up to 2030 implies the achievement of the most important goal of “ensuring environmental sustainability based on preserving and restoring the environment and introducing technologies for processing and recycling solid municipal waste”. But, at the same time, only declarative “expected outcomes” are outlined, although many specific indicators are given in the SWOT-analysis and PEST-analysis.

4 Discussion

“Digitization” and “digitalization” are two conceptual terms that are closely related and are often used interchangeably. In the literature, each term is considered in detail and it is argued that a clear distinction between them is of analytical importance [1, 16]. Digitization is considered as the material process of converting analog information streams into digital bits. Digitalization is defined as a way of restructuring many areas of social life around digital communication and media infrastructures.

5 Conclusion

Since there is the discrepancy between indicators, we need unify the methodology of GOST R ISO 37120-2015, the Strategy for Environmental Security of the Russian Federation, existing materials of the regional environmental ministry and digital transformation into a single list of indicators for assessing the environmental status at the municipal level. It is proposed to expand the capabilities of the municipal administration site so that residents can not only learn about the ecological condition of the area, but also report on the problems they face in everyday life, make suggestions, draw up queries. At the same time, the user must have the ability to trace the entire path of his appeal to the contractor, and the normative period for solving the question being asked must be determined. We consider it is reasonable to transform the ideas of a “smart sustainable city” into projects “smart sustainable municipality”. When developing plans for strategic development of municipalities, it is proposed to use a three-pronged socio-ecological-economic system approach in drafting policy documents. We should more fully take into account the potential of ecosystem services that have the prospect of their implementation in a specific territory. As part of digital transformation of sustainable development based on monitoring of a dynamic environmental situation, it is necessary to include specific proposals for the implementation of environmental goals that can be supported by appropriate technologies based on local conditions and regional and national environmental programs and projects.

References

1. Brennen, J.S., Kreiss, D.: Digitalization. In: The International Encyclopedia of Communication Theory and Philosophy, pp. 1–11 (2016)
2. Federal Law No. 131: About General Principles of Organizing Local Self-Government in the Russian Federation, 6 October 2003. http://www.consultant.ru/document/cons_doc_LAW_44571/. Accessed 12 Apr 2019. (in Russian)
3. Global Connectivity Index (GCI). https://www.huawei.com/minisite/gci/en/country-rankings.html. Accessed 12 Apr 2019
4. GOST R ISO 37120-2015: Sustainable Development of Communities. Indicators of Urban Services and Quality of Life. http://docs.cntd.ru/document/1200123370. Accessed 12 Apr 2019. (in Russian)
5. Höjer, M., Wangel, J.: Smart sustainable cities - definition and challenges. In: Hilty, L., Aebischer, B. (eds.) ICT Innovations for Sustainability. Springer Series Advances in Intelligent Systems and Computing, pp. 333–349 (2016). https://doi.org/10.1007/978-3-319-09228-7_20
6. Kramers, A., Wangel, J., Höjer, M.: Governing the smart sustainable city: the case of stockholm royal seaport. In: Proceedings of ICT4S 2016, Atlantis Press (2016). https://www.atlantis-press.com/proceedings/ict4s-16/25860372. Accessed 29 Apr 2019
7. Report of the Ministry of Forestry: Environmental Protection and Nature Management of the Samara Region On the environmental situation in the Samara region. http://www.priroda.samregion.ru/environmental_protection/state_report. Accessed 12 June 2019. (in Russian)
8. Ringenson, T., Höjer, M., Kramers, A., Viggedal, A.: Digitalization and environmental aims in municipalities. Sustainability **10**(4), 1278 (2018). https://doi.org/10.3390/su10041278
9. Shelomentsev, A.G., Belyaev, V.N., Ilinbaeva, E.A.: Environmental factor assessment in socio-strategies economic development regions. Bull. Corp. Law Res. Cent., Manage. Venture Capital Investment Syktyvkar State Univ. **1**, 169–179 (2014). (in Russian)
10. Shibaeva, N.A., Matveeva, A.P.: new format public sector services in digital economy. Innovative Develop. Econ.: Trends Prospects **1**, 263–270 (2018). (in Russian)
11. Sidorov, A.A., Lasareva, N.V., Firulina, I.I., Sapova, O.A.: Diagnostics of natural indicators of ecological safety of rural territories of the region. In: Mantulenko, V.V. (ed.) SHS Web of Conferences, Problems of Enterprise Development: Theory and Practice 2018, vol. 62, p. 15002 (2019). https://doi.org/10.1051/shsconf/20196215002
12. Strategy for Environmental Security of the Russian Federation for the period up to 2025, approved by the Decree of the President of the Russian Federation No. 176 of 14 April 2017. http://kremlin.ru/acts/bank/41879. Accessed 12 Apr 2019. (in Russian)
13. Strategy for the Development of Information Society in the Russian Federation for 2017–2030, approved by the Decree of the President of the Russian Federation № 203 of 09 May 2017. http://www.kremlin.ru/acts/bank/41919. Accessed 14 Apr 2019. (in Russian)
14. Strategy for Socio-Economic Development of the Samara Region for the Period up to 2030. http://economy.samregion.ru/programmy/strategy_programm/proekt_strateg/. Accessed 13 June 2019. (in Russian)
15. The national program "Digital Economy" for the period up to 2024, approved by the Decree of the Government of the Russian Federation No. 1632-p of 28 June 2017. http://government.ru/docs/28653/. Accessed 14 Apr 2019. (in Russian)
16. Tilson, D., Lyytinen, K., Sørensen, C.: Digital infrastructures: the missing IS research agenda. Inf. Syst. Res. **21**(4), 748–759 (2010)

Digital Technologies as a Tool for Solving Basic Industrial Problems in the Agro-Industrial Complex

E. P. Gusakova, A. V. Shchutskaya(✉), and E. P. Afanaseva

Samara State University of Economics, Samara, Russia
avs2020@yandex.ru

Abstract. Due to the widespread introduction of new digital technologies and platform solutions into production processes it is possible to solve the problems of food supply, to increase the production of competitive export-oriented agricultural products, and to increase the efficiency of agricultural and agro-industrial enterprises. The national project "Digital Agriculture" is to solve these problems. The authors hypothesize that digital technologies should cover all the main production processes in the agro-industrial complex and can become tools for solving basic production problems. The purpose of the study is to analyze the development of digitization of agriculture in Russia and the implementation of its elements in solving basic production problems, to give an overview of state programs for the development of digital agriculture, to identify prerequisites and obstacles in the development of digital agriculture, and to identify areas of digitalization of agriculture in Russia, taking into account advanced world and domestic practices. The authors use a variety of methods and techniques of economic research. The study shows that despite the lag of Russian agriculture in matters of digitalization from developed countries, the introduction of digitalization into the chain of agro-food products is actively gaining momentum. According to the results of the study, it is concluded that only with a systematic and integrated approach to digitalization one can achieve high results in the development of the industry and in achieving national development goals.

Keywords: Digital agriculture · Digital technologies · Mobile market · Precision farming · Robotics

1 Introduction

One of the most important tasks in the modern world is to ensure food security. According to UN forecasts, by 2050 the number of inhabitants on the planet will grow to 9.7 billion people. To feed such a number of people it is necessary to increase food production by 70%. To do this, agriculture must radically change - turn into a large-scale production capable of satisfying the growing food demands of the population. The transition to a new model of farming based on digital technologies is now being paid increased attention by leading international organizations and national governments.

S. I. Ashmarina et al. (Eds.): ISCDTE 2019, LNNS 84, pp. 172–179, 2020.
https://doi.org/10.1007/978-3-030-27015-5_22

The intensive introduction of digitalization and the Internet of Things into agriculture promises to turn the industry, less affected by IT, into a high-tech business due to the explosive growth of productivity and reduction of non-productive costs that are attributes of Agriculture 4.0.

In the message to the Federal Assembly on December 1, 2016, the President of the Russian Federation declared "to launch a large-scale system program for the development of the economy of the new technological generation, the so-called digital economy", which gave a start to digitalization in Russia. The program "Digital Economy of the Russian Federation" was adopted in 2017, the implementation of which is aimed at improving the competitiveness of the Russian economy, improving the quality of life of the population, ensuring economic growth and national sovereignty.

The departmental project "Digital Agriculture" was developed by the Ministry of Agriculture of the Russian Federation in 2018, incorporating it into the State Program for the Development of Agriculture and Regulation of Agricultural Products, Raw Materials and Food Markets. The goal of the project is the digital transformation of agriculture through digital technologies and platform solutions to ensure technological breakthroughs in the agro-industrial sector and to achieve a 2-fold increase in labor productivity at "digital" agricultural enterprises by 2024".

The project "Digital Agriculture" provides for a number of activities:

(1) Scale domestic integrated digital agro-solutions for agricultural enterprises: "Smart enterprise"; "Smart land use"; "Smart field"; "Smart greenhouse"; "Smart garden"; "Smart farm".
(2) Develop an electronic educational system "Knowledge Land" = "The 55th Digital Agrarian University", in which 55,000 specialists of domestic agricultural enterprises will train the competencies of the digital economy in 2019–2021;
(3) Develop an intellectual system of state support measures + a personal cabinet of subsidy recipients with the participation of the Agricultural Bank.

The relevance of the development of digital technologies in agriculture in Russia is not only in the urgent need to use the country's enormous resource potential in order to increase food exports, but also in striving to increase the efficiency of agricultural production through the use of digitalization to solve basic production problems in the agro-industrial complex.

2 Theoretical Background

To achieve the purposes of the study, general scientific methods of cognition and special methods were used: dialectical, abstract-logical, comparative analysis, research of economic and social processes based on statistical analysis, etc. The information base of the study is the data of the Federal State Statistics Service of Russia, the Ministry of Agriculture of the Russian Federation, information portal of infrastructure organizations dealing with the problem of the digital agriculture development, the Internet.

3 Results

The introduction of digital technology is a key factor in the growth of production volumes and business efficiency. The use of digital technologies in agriculture is actively developing throughout the world [3, 4, 6, 7, 10]. An analysis of scientific literature shows that the United States, Australia, Canada, and the EU countries occupy leading positions in the application of digital technologies in agriculture. Russia ranks 15th position in the world ranking of countries in terms of digitization of agriculture. Only 10% of arable land is processed using precision farming technologies, and elements of the Internet of Things are used by no more than 5% of Russian agricultural producers. About 70% of agricultural producers are working on extensive technologies [8]. According to some scientists, the transition of agricultural production in Russia to digital technologies will increase the potential output of agricultural products by about 1.8 times with the same expenditure of agricultural resources [5]. Digital technologies currently used in agribusiness can be grouped into four large groups, presented in Table 1.

Table 1. Digital technologies in the agro-industrial complex

Digital technologies	Applicable tools	Industry of application
Precise agriculture	Navigation and telemetry systems (systems of precise positioning of the unit in the field, parallel driving, yield mapping); remote sensing of the Earth (operative satellite images and aerial photographs), geo-information systems (GIS), differential fertilization, etc.	Crop production
Robots	Unmanned systems and aircraft (UAVs), drones for monitoring the field and harvesting, smart touch sensors, etc.	Crop production, animal production, food industry
IoT-projects (Internet of Things)	Peripheral equipment (sensors), communication channels (GPS/GLONASS satellite communication, etc.), IoT platforms/IoT applications allow you to automate the entire cycle of production and marketing of products	Crop production, animal production, logistics and product sales
Big Data	Approaches, tools and methods for processing huge amounts of data, allowing you to quickly make management decisions	Crop production, animal production, food industry, logistics and product sales

Source: compiled by the authors.

According to the study conducted by J'son & Partners Consulting, the most common digital technologies among American farmers are: a computer with high-speed Internet access, analysis of soil samples (used by 98% of respondents); yield maps, yield monitors, GPS navigation systems (about 80%); differential fertilization technologies and prescriptive cards (60%) [12].

In Russia, the largest agricultural enterprises show the greatest interest in innovative technologies in agriculture. For example, the agro-holding "Rusagro", processing almost 1% of all agricultural land of the country, actively implements forecasting models based on meteorological and vegetation data and optimizes the technologies used. The agro-holding aggregates large data on the development of all crops, detailed information on the operation of machinery, on the characteristics of the fields, on the state of the soil, on the applied technologies and weather conditions. The data comes from its own weather stations, as well as from weather services, satellite monitoring and GPS tracking, from monitoring and measuring sensors in the fields. Analyzing all the collected data, the agro-holding adjusts its production programs [14].

In the Samara region, Bio-Ton, the strategic enterprise in the field of agro-industrial production, has been engaged in digitization of land over the past few years and has been actively introducing digital farming methods in its fields. Tractors and self-propelled sprayers are equipped with AMS, Trimble CFX-750 DGPS precision farming systems and the Wialon transport monitoring system.

Due to the analysis of large data collected from sensors on agricultural equipment, agricultural producers are already able to double the output. Thus, according to the experience of using the Agrosignal system that controls the logistics of agricultural equipment with sensors in 150 farms with a total area of more than 2 million hectares, productivity can be increased by 100%, reducing losses by 50% and increasing the crop-producing by 10–15% [14].

According to the research of scientists of Kuban State Agrarian University, 28 regions (out of 40 surveyed) use elements of precise agriculture in Russia. The leaders in the number of farms using elements of precision farming are Lipetsk (812 farms), Oryol (108) and Samara (75) regions.

The scientists of this university, based on surveys of experts (representatives of the scientific and educational community, production, business and administrative bodies), identified the following most promising areas of digital technology: parallel driving systems, differential fertilization and spraying, creation of electronic field maps, information systems and monitoring.

The introduction of digital economy technologies, according to the Analytical Center of the Ministry of Agriculture of Russia, makes it possible to reduce the cost of agricultural production by at least 23%.

Today in Russia, the best performance in applying digital technologies has been achieved in the regions that are leaders in crop production (Krasnodar Territory, Lipetsk region, Rostov region, Republic of Tatarstan, Samara region). Their experience has to be used in other regions [9].

Animal production in Russia should gradually be transformed into "smart animal production" - an agricultural industry engaged in breeding farm animals, a feature of which is the introduction of new generation systems and technologies to automate animal care in order to increase production and reduce costs.

A "smart farm" is a fully autonomous, robotic, agricultural facility designed for the breeding of agricultural species/animal breeds (meat, dairy, etc.) in an automatic mode that does not require human participation (operator, breeder, veterinarian and etc.).

Such a farm independently analyzes the economic feasibility of production, consumer activity, the level of the region's general health (country, region, etc.) and other economic indicators using the necessary digital technologies (artificial intelligence, Internet of things, big data, neural networks and etc.). On the basis of such an analysis, the farm decides which species/breed of agricultural animal (with the specified qualitative and quantitative indicators) must be bred.

At present, the subjects of the agrarian sector are moving to digital, intellectual and robotic technologies, that is, to robotization of the industry. It should be noted that the average density of robotization in the world is 74 robots per 10 thousand workers in the economy as a whole. The level of robotization in different countries is as follows: in South Korea - 631 robots per 10 thousand workers, in Singapore - 488, in Germany - 309, in Japan - 303, in the USA - 189, in Russia – 3 [11]. In 2017 robotics were used in 28 regions and in 113 Russian agricultural organizations, mainly dairy products. The leaders of the robotic dairy farms in Russia are the following areas: Kaluga region - 37 units, Sverdlovsk region - 11 units, Republic of Tatarstan - 6 units, Kirov region - 6 units, Perm region - 4 units [2]. The largest in Russia is the robotized farm LLC Vakinskoe Agro of the Ryazan region, which gets up to 50 kg of milk from a cow per day. From 2006 to 2016, 393 units of robotics were introduced in the Russian Federation (the annual purchase was uneven and amounted to 12 units in 2006 and 59 units in 2016). While over the past 3 years the number of introduced units of robotics has decreased by 27%. The reason is that the farms often use imported robotics in the Russian Federation, but the change in the exchange rate and the rising cost of this equipment has made it difficult for many agricultural organizations to purchase it.

The number of manufacturers of robotic equipment for milking cows in the world is not too large. In Russia, DeLaval remains the leader in terms of installed milking robots in 2016 with a market share of more than 40%, LLC Leyli RUS with a share of Lely milking robots - 30%, GEA Farm Technologies - 20% and others. The launch of domestic robots - milkers by the company Promtekhnika-Privolzhie in 2016-2017 will reduce the starting barrier for dairy farms planning to robotize milk production.

4 Discussion

The development of intelligent farming and animal production is only gaining momentum. It is predicted that in the coming years the number of "smart" systems introduced into the process of breeding animals should increase significantly, as well as their use.

It is possible to increase the level of production and consumption of dairy products in Russia through the introduction of new technologies in animal production. In particular, it is necessary to develop farms with automated control systems, the parameters of which vary depending on the microclimate and the condition of animals on farms. Only such farms can improve the quality of milk to the extra class and ensure a stable growth of the milk productivity of animals.

The most important task is to create digital technologies that ensure the independence and competitiveness of the domestic livestock complex; attract investments; develop and introduce technologies that increase the milk productivity of animals up to

13,000 l/year; reduce the incidence with cow mastitis and therefore reduce the cost of antibiotics; develop and implement autonomous production technologies (without an operator), energy efficiency and energy mobility in the "Smart farm", and produce safe, high-quality, functional, food.

"Smart processing" is a new stage in the development of the food and processing industry, which relies on serious digitalization of agricultural production. Today, the main task is to transit the food industry to Industry 4.0 (introducing cyber-physical systems into food production) for enterprise's benefits, to implement BigData (technologies for collecting and processing large amounts of data) and cloud technologies for solving managerial decisions. This is the introduction of enterprise management systems (MES, EAM, LIMS, planning, logistics, etc.), new solutions in the process control system, the introduction of data centers in the food industry. The most robotic meat processing factory GC Cherkizovo in Kashira is an example of such systems. The production chain at the factory is supported by robots, not people. The industry uses philosophy 4.0: artificial intelligence and robots communicate with each other, thereby controlling the entire production chain. Engineers and IT specialists only maintain system performance [13].

In the current economic situation, digitalization can provide a reduction in cost and final food prices. At the same time, both the processes within the agricultural production cycle and the suppliers of raw materials, marketing, logistic and transport links should be "connected". At the same time, it is possible to restructure existing relationships and even exclude intermediate links located on the way to the consumer from the value chain.

A partial solution to this problem will be the creation of an alternative market for agricultural products, which will allow producers to reach customers directly, bypassing intermediaries, through electronic services and the INTERNET.

Foreign farmers have long been using digital technologies for operative communication with each other and with counterparties in order to conclude transactions, to obtain up-to-date information about the main price offers and sales markets, etc. The Esoko application in Africa, mobile farming in China appeared in the early 2000s. Today they provide tools for mass communication that are actively used by food market participants, non-profit organizations and the state.

Buryatia was a pioneer in introducing digital technologies in marketing of agricultural products and promoting local food to the domestic and foreign markets. The Mobile Market project was recognized as the best in the category "Development of Agriculture and the Creation of Comfortable Conditions for Business Development in Rural Areas" at the Russian Investment Forum in Sochi in February 2018. The goal of the project is to create a market for agricultural products, which allows producers to reach customers without intermediaries through cellular communications, and vice versa. The main stages of the project are to develop an automated information system, a mobile application, a database of sellers, a database of buyers, a call center and to organize product delivery. One of the effects of the project was to reduce the cost of realizable agricultural products by 20%.

The practice of "Creating a mobile market for agricultural products" was noted by the Agency for Strategic Initiatives as successful and replicable in other regions. Currently, the organization and promotion of the market for local agricultural products using mobile communications is being implemented in the Kaliningrad region, in the Stavropol Territory, the Republic of Dagestan and the Republic of Altai [1]. However, one should not idealize digitalization. The main problems hindering digital transformation are the lack of scientific and practical knowledge of modern agricultural technologies, the lack of financial opportunities for farmers to purchase new equipment and platforms, digital inequality, which is expressed in the absence of mobile communications in small localities and the Internet, the foreign origin of most of the resources used in Russia for communication and organization of digital services. The introduction of technology is only a tool, and the key factor in the development of the agro-industrial complex is quality management and human resources.

5 Conclusion

The use of digital technologies to solve basic industrial tasks in the agro-industrial complex will increase the efficiency and competitiveness of agricultural production, the use of the country's huge resource potential in order to increase food exports and develop non-oil food exports.

State programs for the development of digital agriculture stimulate the transition to digital intellectual and robotic systems and are an effective mechanism for the digital transformation of the agro-industrial complex, which has both economic and social prerequisites.

The prerequisites for digital transformation have already been formed due to the analytics of "big data", "cloud" technologies, cheap sensors, broadband mobile communications, artificial intelligence and the Internet of Things, service robotics.

Despite the lag of Russian agriculture in matters of digitalization from developed countries, the introduction of digitalization into the chain of agro-food products is actively gaining momentum. However, technologies are only a tool, and their implementation largely depends on a systematic and integrated approach to digitalization, which will allow achieving high results in the development of the agro-industrial complex and in the implementation of strategic national priorities.

References

1. Agency of strategic initiatives: The shop of correct solutions. https://asi.ru/store/. Accessed 15 Apr 2019. (in Russia)
2. Boiko, A.: Area for the introduction of milking robots in Russia. Robo Trends (2019). http://robotrends.ru/robopedia/geografiya-vnedreniy-i-planov-vnedreniya-doilnyh-robotov-v-rossii. Accessed 15 Apr 2019. (in Russia)
3. Fournel, S., Rousseau, A.N., Laberge, B.: Rethinking environment control strategy of confined animal housing systems through precision livestock farming. Biosyst. Eng. **155**, 96–123 (2017). https://doi.org/10.1016/j.biosystemseng.2016.12.005
4. Gan, H., Lee, W.S.: Development of a navigation system for a smart farm. IFAC-PapersOnLine **51**(17), 1–4 (2018). https://doi.org/10.1016/j.ifacol.2018.08.051
5. Korotchenya, V.M.: Russia and agriculture 4.0. Econ. Agric. Russia **6**, 98–103 (2018). (in Russia)

6. O'Grady, M.J., O'Hare, G.M.P.: Modelling the smart farm. Inf. Process. Agric. **4**(3), 179–187 (2017). https://doi.org/10.1016/j.inpa.2017.05.001
7. Otles, S., Sakalli, A.: 15 - Industry 4.0: the smart factory of the future in beverage industry. Prod. Manage. Beverages **1**, 439–469 (2019). https://doi.org/10.1016/B978-0-12-815260-7.00015-8
8. Semenov, S.A., Vasiliev, S.A., Maximov, I.I.: Features of implementation and application prospects of digital technology in agro-industrial complex. Bull. Chuvash SAA **1**(4), 69–76 (2018). (in Russia)
9. Shchutskaya, A.V., Afanaseva, E.P., Kapustina, L.V.: Digital farming development in Russia: regional aspect. In: Ashmarina, S., Mesquita, A., Vochozka, M. (eds.) Digital Transformation of the Economy: Challenges, Trends and New Opportunities. Advances in Intelligent Systems and Computing, vol. 908, pp. 269–279. Springer, Cham (2019). https://doi.org/10.1007/978-3-030-11367-4_26
10. Shimamoto, D., Yamada, H., Gummert, M.: Mobile phones and market information: evidence from rural Cambodia. Food Policy **57**, 135–141 (2015). https://doi.org/10.1016/j.foodpol.2015.10.005
11. Skvortsov, E.A., Skvortsova, E.G., Sandu, I.S., Iovlev, G.A.: Transition of agriculture to digital, intellectual and robotics technologies. Econ. Reg. **14**(3), 1014–1020 (2018). (in Russia)
12. The Internet of Things in Agriculture (Agriculture IoT/ AIoT): World experience, application cases and the economic effect of implementation in the Russian Federation. http://json.tv/ict_telecom_analytics_view/internet-veschey-v-selskom-hozyaystve-agriculture-iot-aiot-mirovoy-opyt-keysy-primeneniya-i-ekonomicheskiy-effekt-ot-vnedreniya-v-rf-20170621045316. Accessed 15 Apr 2019. (in Russia)
13. Tshemlyaev, A.: The largest sausage factory in Europe was opened in Kashira (2018). https://mosreg.ru/sobytiya/novosti/news-submoscow/krupneyshiy-v-evrope-zavod-po-proizvodstvu-kolbas-otkrylsya-v-kashire-2830. Accessed 15 Apr 2019. (in Russia)
14. Vartanova, M.L., Drobot, E.V.: Prospects for digitalization of agriculture as a priority direction of import substitution. Econ. Aff. **8**(1), 1–18 (2018). https://creativeconomy.ru/lib/38881. Accessed 10 June 2019. https://doi.org/10.18334/eo.8.1.38881. (in Russia)

Digital Technologies as a Factor of Expanding the Investment Opportunities of Business Entities

M. E. Konovalova[1(✉)], O. Y. Kuzmina[1], and S. A. Zhironkin[2]

[1] Samara State University of Economics, Samara, Russia
mkonoval@mail.ru
[2] Siberian Federal University, Krasnoyarsk, Russia

Abstract. The relevance of the study is due to the development of digitalization processes of economy, including its financial sector. The introduction of modern technologies contributes to the creation of new financial products, the emergence of fundamentally new mechanisms and ways implementing the investment process. In this regard, the article is aimed at studying the investment behavior of macroeconomic entities in the context of establishing a new economic model that creates conditions for the formation and operation of new tools and investment methods. The article is aimed at identifying and describing the features of the investment behavior of economic entities, the appearance of which is due to the transformation of motivational incentives to implement investment activities. The leading approach used in the research is systemic-dialectical, which involves a comprehensive consideration of the investment process as a systemic phenomenon. The article identifies the prerequisites for changing the investment behavior of macroeconomic agents, due to the emergence of new investment tools, such as crowd-funding, crowd-investing and crowd-lending. It is proved that the introduction of digital technologies expands the investment opportunities of the subjects, which contributes to more effective implementation of their economic and institutional interests. It is shown that the ongoing evolution of the social-and-economic system, caused by digitalization, changes the traditional understanding the types of economic activity of economic agents. The article materials are of practical value for assessing the level of investment potential of macroeconomic entities; for analyzing the process of introduction and development of new technological platforms.

Keywords: Investment process · Investment tools · Crowd-funding · Crowd-lending · Crowd-investing · Digitalization

1 Introduction

The formation and development of information economy creates the prerequisites for the transformation of traditional patterns of macroeconomic agents' behavior, their motivation and ways of implementing investment strategies are changing. New technologies allow business entities to use the digital system in the field of their activity [6]. This process can include data certification (implementation of large data storage

S. I. Ashmarina et al. (Eds.): ISCDTE 2019, LNNS 84, pp. 180–188, 2020.
https://doi.org/10.1007/978-3-030-27015-5_23

technologies), digitalization (conversion of information chains from analog to digital format), virtualization (splitting of reality into separate elements), and generation (re-programming and recombination). Not only consumption and communication models are transformed, but also investment models [5], new instruments for implementing investment strategies are emerging, new sectors of the economy are being created. The relevance of the study is due to the ongoing changes in the social-and-economic system, that are based on rapid development of information (digital) technologies, which changes the existing mechanism of the investment process. The problem highlighted in the article is related to the transformation of the investment behavior of economic entities, which currently have a greater choice of investment instruments and methods of their implementation. The purpose of the study is to identify the role of digital technologies as one of the factors expanding the investment opportunities of macroeconomic agents. The implementation of this goal involves the following tasks: studying the existing digital technologies that create new investment tools; identifying features of the investment behavior of business entities; analyzing the benefits of modern investment tools for lenders and borrowers.

2 Methodology

The main methodological approach used in the research is systemic-dialectical. The use of this approach makes it possible to dissect the investment process as an integrated system, the individual elements of which are in dialectical interaction. The relationship between them is deeply dialectical, since it reflects the contradictory unity of the individual components of the investment process as a system. The use of a systemic-dialectical approach allowed us to identify the features of the investment behavior of macroeconomic subjects in the conditions of economy digitalization.

3 Results

The most popular is the so-called crowd-funding, which is the interaction of individuals or legal entities, pooling their capital or other financial resources using the Internet, in order to implement a variety of projects. According to experts, the volume of operations that are carried out using alternative financing technologies (crowd-funding) will continue to grow.

Considering the process of expanding the range of investment tools through the use of new information technologies, one should pay the most attention to the mechanism of crowd-investing. Crowd-investing projects are implemented on P2P platforms. They are very diverse. These include, among other things, crowd-lending or public lending (P2P-lending), and a royalty model and joint-stock crowd-funding.

Like any other financial institution, the crowd-lending platform is a subject to risks, since even using the most advanced technology to collect data on borrowers, for example, Big Data technology, we cannot rule out the occurrence of any force majeure circumstances. Unsecured lending itself, while crowd-lending does not provide such a form of security as collateral, represents a highly risky type of financial activity.

Another risk is directly related to the activities of an administrator or manager of crowd-lending platforms. There are no regulatory standards regarding the timing of the funds raised in the administrator's accounts, the mechanism for paying debts if the platform operator is bankrupt. The procedure of protecting the crowd-lending clients from risks is not clear enough, if such exists at all. There is a threat of using the funds acquired by criminal means, as well as the risk of lacking the complete and reliable information about the financial solvency of the participants of crowd-lending model.

Platform risks can be partially compensated by non-rigid, but, nevertheless, substantial control by financial regulators. In the US and the UK the operation of P2P lending system is possible only if it is licensed. Platforms should report on the services they provide, their profitability, and access to this data should be open. Recently, they have intensively started talking about the need for introducing a risk management procedure, toughening capital requirements, which will make it possible to impart more regulated nature to crowd-investing activities.

The volume of loans provided by means of crowd-lending platforms is growing exponentially, but they are still not able to replace banking services completely. For example, the largest American Lending Club platform, assessing the potential of P2P services in the United States, concludes that at this stage, the market for crowd-lending services has a capacity of about 500 billion US dollars, while the volume of loans issued by retail units of American banks is 4 trillion dollars, which clearly demonstrates a small share of alternative financing. However, the growth rate of direct lending is amazing - 158% per year. A similar situation is observed in Europe: the total volume of the European market for alternative financing does not exceed 3 billion euros, of which about 1 billion euros accounted for P2P lending. The Russian market for alternative financing is currently in its establishment; according to the results of 2017, 30.3 billion rubles were issued online, the average loan amount is 13,800 rubles, which indicates a slight importance of new ways to attract investments. However, according to expert estimations made by Ernst & Young, by 2035 already 36.7% of all funds received in the form of loans will be accumulated through crowd-funding platforms. The volume of the market for alternative financing by the date indicated will amount to $ 178.6 billion, taking into account annual growth by an average of 51.2% [3].

Another form of crowd-investing is joint-stock crowd-funding. According to many researchers, this is the most popular form of public financing, providing for the investor's remuneration in the form of a share of property of an enterprise or a specific project. This type of investment attraction is based on a whole range of new tools that expand the investment opportunities of macroeconomic agents, including mining, securitization, and ICO (initial coin offering).

In the standard form, the ICO assumes the attraction of money from investors due to their involvement in the start-up, which is associated with the development and promotion of a service or platform, usually based on crypto-currency operations. At special information sites (crypto-currency forums), key financial, economic and technical information about the project (investment strategy, time of its implementation, description of the participants' team, specific activities related to the implementation of this project) appears. Using the block-chain technology, a certain number of unique ID - digital tokens is issued. Technically, tokens are issued by adding transactions to the block-chain with an indication of their number and assigning a unique identifier. Any token issued during ICO is contained simultaneously on many nodes, the record of it can neither be destroyed nor changed, which allows increasing its liquidity and ensure its safety.

Participation in ICO projects is quite risky. Thus, in the first half of 2018, the price of ICO tokens decreased by 66%, which indicates high volatility, and, consequently, the risk of investing in alternative sources of financing, holding back the development of this market. Over the same period, just $ 15 billion was raised using ICO tool. Only ten largest ICOs ensured the volume of growth in value, the rest were not profitable. The stability to this market is not added by the fact that so far no country has defined the legal ICO status, investors are not legally protected in case of a failure of the subject being financed. The anonymity of transactions and the possibility of using crypto-currency exchanges for money laundering and financing criminal projects also confuse. Nevertheless, we noted earlier that some attempts to regulate the procedure for conducting ICOs are being made. The Government of the Russian Federation has submitted for public discussion regulations that imply the legalization of the ICO sector in order to increase the investment activity of macroeconomic agents.

The block-chain technology, on which ICO is based, can be used by macroeconomic agents not only to obtain investment through a share in a start-up, but also when performing speculative operations in the crypto-currency market. As it is well known, to ensure safety and reliability in the block-chain, either PoW (work proof) or PoS (share proof) is used. When proving work, each of the nodes, on which the cryptographic record is fixed, is involved in solving a complex computational task, the result of which is the hash code of a new block recorded in the block-chain, which is accompanied by a fee in the form of receiving crypto-currency (mining). When using share proof, every block-chain network participant has the right to assure the newly received block with his electronic signature. This right is constantly transferred from one participant to another with a probability proportional to what share of a crypto-currency in the total volume the miner has. In this context, mining can be viewed as an activity aimed at either creating a crypto-currency or validating with the aim of obtaining remuneration in the form of crypto-currency. In any case, mining is an entrepreneurial activity, which implies the existence of not only income, but also costs. PoW miners have especially high costs.

Not only robot consultants have artificial intelligence, but also robots for high-frequency algorithmic trading, hybrid robots. Their development is associated with the growing need for customization of financial services, the need to reduce the role of intermediaries; however, at this stage their use is very limited. Robots can now be pointed to specific financial instruments; they do not perform well in volatile markets, in times of crisis, increasing the risks associated with cyber-security. The growth rate of using this kind of IT technologies in investment activity is very high, up to 70% per year. According to estimates of the National Research University "Higher School of Economics", by 2020, 5% of global investment, that is 2.2 trillion US dollars, will be made with the help of robot consultants [7].

The above analysis suggests that the introduction and application of information technology significantly expands the range of both opportunities and investment tools. It modifies the traditional mechanism of savings transformation into investments, which leads to the emergence of new models of investment behavior of macroeconomic entities.

4 Discussion

Any information technology is aimed at reducing costs in the process of making a decision. Collecting information, its processing, that is, sorting, ranking, eliminating zero information (information that repeats known characteristics or is useless to make a specific decision), structuring, and also, as a final step, its evaluation is not free for the individual, since it requires time and money. Information technologies, appearing in the form of certain tools, that is, in essence, acting as means of labor, facilitate the life of economic entities, accelerate economic processes, make them more structured and transparent. It is important to understand that in the conditions of transition from industrial to post-industrial society, information becomes, firstly, an independent factor of production, and secondly, an object of commodity relations, which allows considering information technologies not only as a means of labor, but also as a subject, that is, on the one hand, they act as an instrument of labor, while, on the other, a finished product. They not only help to make the most optimal decisions, but also become the goal of human activity.

The use of information technologies often has a negative impact on economic processes due to the lack of a full material base, institutional imperfections, and weak communication links of economic entities introducing technological innovations. Vulnerability is high due to the need to balance between information security and privacy. Nevertheless, despite the negative features of using the information technologies, they are becoming more common.

If the technological mainstream of the 90s was Internet technologies [9, 14], then in the zero years of the twentieth century these were mobile and wireless networks [1], cloud technologies, big data technologies [4] and other digital innovations.

The process of economy digitalization is interpreted differently by representatives of the scientific community. From the point of view of some scientists, the development of the digital economy is a separate stage in the development of society associated with the global introduction of digital technologies that change the existing paradigm of coordination of economic actors. Others, on the contrary, believe that digitalization is not a separate stage of development, but an integral feature of information economy itself, since digital technology is just a new way of presenting discrete transmission of information signals [12]. There are more sophisticated approaches that reveal the essence of digital economy. So, one should distinguish the levels of digitalization. The digital economy is based on the "digital sector": organizations that, using information technology, develop digital products and services. "Digital economy" itself is that part of economy represented by companies that have business models based on digital products and services. Under "digitalized economy" is understood the use of ICT in all sectors of economy.

As we see it, the current stage in the development of society is characterized by the transformation of the role of factors of production, as a result of which information (knowledge) becomes an independent and limited resource determining many economic phenomena. This process underlies the establishment of information economy as a separate stage in the development of society. Digitization, on the other hand, is a specific set of new tools that extend the functionality of macroeconomic agents.

Digital technologies in the financial sector demonstrate a special intensity of introduction, changing the traditional ideas of business entities about the ways of investing. Among the technologies that have this kind of impact, block-chain, Big Data (Big Data are structured and unstructured data of huge volumes and significant diversity), cloud technologies, quantum technologies, artificial intelligence and neural networks, robotization and cryptography should be distinguished. Let us consider each of them in detail.

Thus, users of the block-chain technology act as a "collective notary", which confirms the authenticity and legitimacy of the transactions made, and this means that it is almost impossible to change and falsify information.

The problem of reliability and security of data storage is becoming increasingly important in the context of exponential growth of mass data. It is a well-known fact that in the last ten years more information has been produced than in the entire history of the mankind. Permanent growth of information volumes caused the need to create Big Data technology, which allows storing, processing and analyzing huge amounts of data, which causes a change in the principle of their analysis. The human brain is not able to accumulate and process large amounts of information, more over to identify patterns and facts of future value for an economic entity, while Big Data technology is capable of processing huge amounts of data from completely different sources in completely different formats (structured, semi-structured and unstructured data). Like the block-chain technology, Big Data provides cost savings, although it's not so much about transaction costs as IT infrastructure and software costs [8, 10, 13]. Labor costs are reduced through the use of more modern methods of data accumulation, management, and the development of behavioral strategies.

Cloud technologies play the same role in saving costs. Being data processing technologies, they assume the provision of computer resources and facilities to the user as an Internet service. The consumer can easily use his own data without thinking about the infrastructure, operating system, software. The "cloud" is the whole set of hardware and software, which allows processing and executing client requests. The prototype of cloud technology is the Internet itself, which once again underlines the diversity of cloud computing used in the information environment.

Like all informational innovations, cloud technologies have some drawbacks, especially concerning the protection of information, its security and the availability of free access to the Internet. There are also institutional constraints associated with a poorly regulated system of interaction between company-user and cloud application companies. Nevertheless, despite certain constraints, the development of cloud technologies is an irreversible process.

Quantum technologies are also associated with computational issues. According to R.V. Dushkin, quantum technologies can be divided into the following areas: quantum information transfer, quantum sensing, quantum computer, and quantum calculation itself [2]. The most developed area today is quantum information transfer, since quantum communication channels have already been created, with the help of which quantum key distribution protocols are implemented, which implies a completely different degree of information protection.

The expansion of the scope of quantum information transfer is limited by the computing power of modern computers. It is impossible to implement the mathematical computational model of quantum physics, using usual architecture of modern machines, even in emulation mode, since the need for computing power increases exponentially with the number of qubits. The problem could be solved when creating a quantum computer. A universal quantum computer "in hardware" has not been developed yet; there are only its prototypes that work with a small number of qubits, which does not allow implementing complex quantum algorithms. If the humanity achieved that, the tasks of modeling complex systems and chaotic dynamic processes, ranging from weather modeling to exchange trading, would be effectively solved. But at the same time, it is important to understand that quantum technologies carry certain threats, in particular, breaking of existing information systems based on modern cryptography achievements. The consequence of such transformation will be almost complete destruction of available data secrecy. Under these conditions, the trajectory of further economic development is not so definitive.

The competition principle is the basis for modern economic system, i.e. the struggle between economic entities for the right to possess resources. A significant role in this process is played by the asymmetry of information, which from the point of view of institutional approach is a definite stimulus for economic development. The absolute transparency of economic sphere during the destruction of traditional cryptography technology will lead to the transformation of existing economic system. The system can be saved, but only under the condition of progress in the field of encryption technology, which will ensure the data secrecy even when using quantum computers. This scenario is considered to be the most likely by us.

The widespread use of quantum technologies and new crypto-operation tools can lead to the appearance of what is called artificial intelligence in the scientific literature, or rather, an artificial intellectual system with self-awareness [11]. Most often, artificial intelligence means the process of mapping a set of anthropomorphic tasks on a variety of analytical tools to simulate human behavior. This concerns the modeling of creative processes, the creation of an intelligent interface, new architectural products, within which the construction of effective intelligent systems, especially robotized ones, can be implemented.

The fifth and sixth technological modes led to the transformation of human needs by changing the market criteria for evaluating goods and services. Mass commercialization of technologies implies the existence of such criteria for robots as self-learning (the use of artificial intelligence) and autonomy (robots must surpass human abilities, evolve, and independently exist in a hybrid environment). Robots are very diverse mechanisms, ranging from automation and intelligent agents to trans-border essences and automated human versions. Such differentiation is explained by the variety of environments and methods of application, the degree of specialization, mobility, and the nature of technological operations.

Robotization, artificial intelligence, cloud technologies and neural networks, quantum technologies are widely used in the financial sphere. They changed the mechanisms of banking and insurance, exchange trade, investment activity beyond recognition, and that is why information technologies in economic literature are often called financial technologies. In our opinion, this is not entirely correct, since information technologies are used not only in the financial sphere, but also in other areas of life, for example, medicine, construction, management, and the sphere of intellectual law. The scope of information technology is not capable of replacing its informational nature, and, therefore, the use of the term "financial technology" seems to us inappropriate.

As it was mentioned above, information technologies lead to the transformation of not only objective reality, but also a subjective assessment of the processes taking place. Motivation, targets, behavior of business entities are being changed, especially when it's time to make investment decisions. Modern information technologies underlie the creation of fundamentally new financial instruments and mechanisms that allow investment agents to carry out economic activities. There is a change in traditional ideas about the methods and forms of investment, which expands the investment opportunities of subjects, transforms the ways of interaction between investors and borrowers, and expands the range of investment tools.

5 Conclusion

In the course of our study the following conclusions were made:

- in transition from industrial society to post-industrial one information becomes, firstly, an independent factor of production, and secondly, an object of commodity relations, which allows us to regard information technology not only as a means of labor, but also its object in the form of a finished product;
- it is worth noting the dual nature of IT technologies impact on investment processes. The leveling of information asymmetry, the transformation of economic sphere into an absolutely transparent one, repeatedly increases the possibilities for controlling and coordinating the activities of macroeconomic actors, but completely deprives them of their development incentives. Therefore, the introduction of IT technologies in economic sphere should not be accompanied by a loss of basic market economy principles, in particular, competition;
- considering the process of expanding the range of investment tools through the use of new information technologies, it should be noted that the introduction of alternative investment methods occurs against the background of transformation of investors' subjective assessment of processes; their motivation, targets, and, consequently, the behavior change. There is an increase in investment opportunities; there are new ways of capital accumulation, and new methods of making investment decisions.

References

1. Australia's digital economy: Future directions Canberra, Department of broadband, communications and the digital economy (2009). http://unpan1.un.org/intradoc/groups/public/documents/apcity/unpan039471.pdf. Accessed 6 May 2019
2. Dushkin, R.V.: Review of modern state of quantum technologies. Comput. Res. Model. **10** (2), 165–169 (2018). https://doi.org/10.20537/2076-7633-2018-10-2-165-179
3. Fintech course: Market development prospects in Russia (2018). https://www.ey.com/Publication/vwLUAssets/EY-focus-on-fintech-russian-market-growth-prospects-rus/$File/EY-focus-on-fintech-russian-market-growth-prospects-rus.pdf. Accessed 6 May 2019. (in Russian)
4. G20 Digital economy development and cooperation initiative (2016). http://www.g20.utoronto.ca/2016/g20-digital-economy-development-and-cooperation.pdf. Accessed 6 May 2019
5. Handel, M.: The effects of information and communication technology on employment, skills, and earnings in developing countries. Background paper for the World Development Report 2016, Washington, DC, World Bank (2015). https://blogs.worldbank.org/category/tags/wdr-2016?page=1. Accessed 6 May 2019
6. Heeks, P.: Information and Communication Technology for Development. Routledge, Abingdon (2017). https://doi.org/10.4324/9781315652603
7. ISSEK HSE: New financial technologies. Trendletter **11**, 1–4 (2016). https://issek.hse.ru/data/2016/12/16/1112515468/Layout.pdf
8. Khezrimotlagh, D., Zhu, J., Cook, W.D., Toloo, M.: Data envelopment analysis and big data. Eur. J. Oper. Res. **274**(3), 1047–1054 (2019). https://doi.org/10.1016/j.ejor.2018.10.044
9. Lane, N.: Advancing the digital economy into the 21st century. Inf. Syst. Front. **1**(3), 317–320 (1999). https://doi.org/10.1023/A:1010010630396
10. Mavragani, A., Tsagarakis, K.P.: Predicting referendum results in the Big Data Era. J. Big Data **6**(1), 3–10 (2019). https://doi.org/10.1186/s40537-018-0166-z
11. Penrose, R.: Shadows of the Mind. A Search for the Missing Science of Consciousness. Oxford University Press, Oxford (1994)
12. Popov, E.V., Sukharev, O.S.: Digital economy: "Irrational Optimism" of management and financing. Econ. Taxes Law **11**(2), 6–17 (2018). https://doi.org/10.26794/1999-849X-2018-11-2-6-17
13. Sohangir, S., Wang, D., Pomeranets, A., Khoshgoftaar, T.M.: Big data: deep learning for financial sentiment analysis. J. Big Data **5**(1), 3–22 (2018). https://doi.org/10.1186/s40537-017-0111-6
14. Tapskott, D.: The Digital Economy: Promise and Peril in the Age of Networked Intelligence. McGraw-Hill, New York (1994)

Participatory Budgeting in City of Prague: Boosting Citizens' Participation in Local Governance Through Digital Tools (Case Study)

E. Velinov[1,2(✉)], S. I. Ashmarina[3], and A. S. Zotova[3]

[1] Skoda Auto University, Mlada Boleslav, Czech Republic
emil.velinov@savs.cz
[2] RISEBA University of Applied Science, Riga, Latvia
[3] Samara State University of Economics, Samara, Russia

Abstract. The article analyzes the case study of realizing some local initiatives such as Participatory Budgeting (PB) for example and their promotion among the citizens with the help of various digital instruments. Participatory budgeting has another effect that the future appears as a strategic motto: growing political culture and educate citizens engaged in the resolution of their city. Thus, citizens are becoming important players in local politics and create a real civil society. PB is a democratic process that involves suddenly, that all members of the local community/municipalities are invited to be directly involved in decisions about the use of Public Finance (defined portion of the budget). The authors make the analysis of mechanisms, digital trends and other resources that help the citizens of Czech Republic use this local governance initiative and develop it significantly.

Keywords: Digitalization · Diffusion · Digital tools · Local governance · Participatory budgeting

1 Introduction

Digitalization nowadays is booming across cities worldwide as the era of Internet of Things, artificial intelligence, smart cities, connectivity and e-mobility is becoming contemporary phenomena in public administration [11]. Thus, the local governances are seeking to increase citizens participation in the daily life of the municipalities [12]. At the same time, the city councils are trying to promote smart cities concepts, which eventually reduce the pollution of the air in the cities, reduces the car traffic, reorganizes public transport routes, increase efficiency of commuting to and from work, etc. [1]. For example, Participatory Budgeting (PB) is a decision-making process in which citizens negotiate among themselves and with government officials in organized meetings over the allocation of new capital investment spending on projects, such as health care clinics, schools, and street paving [17]. It is an innovative budgeting practice as it promotes social justice through the allocation of more resources for poorer neighbourhoods, encourages citizens' direct participation in distributing resources

S. I. Ashmarina et al. (Eds.): ISCDTE 2019, LNNS 84, pp. 189–197, 2020.
https://doi.org/10.1007/978-3-030-27015-5_24

within each of the municipality's regions based on the mobilization of community members [5]. In addition, it can establish new accountability mechanisms to hold the elected politicians answerable to the inhabitants. In more successful cases, citizens gain authority to make important policy decisions, which have the potential to alter the basic decision-making process in politics. PB programs combine elements of direct democracy (i.e., direct mobilization of citizens in decision-making venues) and representative democracy (i.e., election of representatives).

The PB model has initially be used by the Brazilian local governments. Under the auspices of PB, citizens select a number of delegates who will be in their opinion best to promote their interests or the interests of the whole community [5]. These representatives then make decisions about matters relating to the entire local community, whilst their activities are subject to the citizens' scrutiny. This Brazilian model or similar model has been adopted by local authorities in both developed and developing countries. Scholars have used various frames of reference in exploring the adoption and practical use of PB. Amongst them, the theory of social capital [10] or theory of practice [8] are notable. Nevertheless, the introduction of new practice in another context is pinpointed as the diffusion of innovation. The theory of diffusion by Rogers (2003) envisages that a process of innovation passes through a series of phases before the reform becomes accepted practice [13]. Albeit PB has been propagated by various actors as a democratic innovative practice, its diffusion in various contexts is not explored through the work of Rogers (2003) [13]. Therefore, this study intends to explore endeavours of introducing PB in a Czech local government by relying upon the Rogers' theory of diffusion. In doing so, this study demonstrates how the discussion concerning the introduction of PB emerges and how the political authority has reacted to the endeavour of introducing PB [9].

2 Methods of Data Collection

This case study, based on document analysis, Prague Municipality website and FloweeCity.com website, explores how the discussion concerning the introduction of PB emerges and how the political authority has reacted to the endeavour of introducing PB in the Local Government of Prague, the capital city of the Czech Republic. One of the co-authors made several visits in August 2018 and February 2019 to Prague municipality with a view to gaining a preliminary insight into the PB and Digitalization processes there. The visit's findings were discussed amongst a group of two researchers. One of the researchers is living very close to the Prague town hall administrative area and he is aware of the accounting and budgeting practices prevailing in Prague local government.

The initial round's findings enabled us to comprehend an existing political struggle in the various stages of diffusing PB and Digitalization [15]. Having discussed the findings of our document investigation, we then sorted out the issues that should further be investigated and scheduled the second round of visits.

3 Empirical Findings

3.1 The Context of Czech Local Governments

The Czech Republic has a democratic system of government based on parliamentary democracy and free competition among political parties. The regions in Czech Republic are divided into seventy-six districts including three "statutory cities" (without Prague, which had special status). The districts lost most of their importance in 1999 in an administrative reform; they remain as territorial divisions and seats of various branches of state administration. A further reform in effect since January 2003 created 204 Municipalities with Extended Competence "little districts" which took over most of the administration of the former district authorities. Some of these are further divided between Municipalities with Commissioned Local Authority-"second-level municipalities". In 2007 the borders of the districts were slightly adjusted and 119 municipalities are now within different districts. The local government in the Czech Republic has two layers: 14 regions and 6 234 municipalities. These are self-administrated units; people elect their representatives for regional and municipal government. Until December 31, 2002 there were also 76 district offices serving as local branches of the government. In the Czech Republic, local autonomy means that certain matters are administered by an authority other than the state government (public corporations – municipalities or regions, as well as other authorities). These authorities are relatively independent in this activity. They are not subordinated to state administration bodies (e.g. ministries), which supervise only the legal aspects of public corporations' work. Citizens come into contact most often with local autonomy that includes municipalities, regions and the Capital City of Prague. Operation of local government authorities is regulated by laws on local autonomy, such as Act No. 128/2000 Coll., on Municipalities, Act No. 129/2000 Coll., on Regions and Act No. 131/2000 Coll., on the Capital City of Prague, which primarily govern these authorities' separate powers. Currently the capital city of Prague is divided into 22 districts, which are led by mayor and district councils. In each Prague district the council members are elected by the residents and the council members along with the mayor have four years mandate. The local government of Prague has special status among the twenty-two districts. The local government of Prague is governed by council assembly of eleven members, who are representatives from different political parties. The mayor of Prague is only the representative of Prague municipality and she has limited decision making rights. The council assembly members are elected by the citizens every four years and the council assembly is the highest governing body of Prague local government. Under the council assembly is situated the city council consisting of sixty-seven council members from the above-mentioned political parties and independent members. The city council of Prague is held once per month and it is public, which means that each resident of Prague can attend the city council meetings every month. All members of council assembly and city council of Prague are listed on the official website of Prague municipality www.praha.eu and all the agendas from the council assembly and city council meetings are available online as well. Local government authorities' separate powers include matters which are in the interest of the municipality (region) and citizens of the municipality (region), unless stated otherwise by applicable laws in regard to specific situations. Typically, this involves management of municipal property. A municipality can serve as an example to show how control of local government authorities works. Other local government authorities work likewise. Running of the municipality is mainly affected by the members of its council, as well as of course by its citizens. In practice citizens have the right to express their opinion on drafts of the budget and final balance sheet (and of course to examine them). In addition, citizens may request that the community board or local council discuss a specific issue. When the required number of signatures is achieved, the community board or local council must discuss the relevant issue within 60 or 90 days, respectively. The same rights belong to owners of real state property located within the municipality. Similarly, each member of a municipality council has the right **to submit proposals** to be discussed at meetings of the council, community board, committees and commissions. In addition, each council member can **make enquiries, comments and suggestions and request information** from almost anyone involved with the municipality (e.g. heads of budgetary institutions, such as museums, theatres, etc.). A crucial point is that s/he must receive a response within 30 days.

3.2 The Stages of Introducing Participatory Budgeting in Prague Prior Conditions for the Participatory Budgeting and Digitalization

Participatory Budget is an essential step on the road to participatory democracy. Thus, such a governing body, where certain process belongs to the people and not to the political establishment. Participatory Budget is one of the effective tools of self-government, which is closer to decision-making within the meaning of subsidiarity as close to the people affected by such decisions most affect. Participatory Budgeting in the Czech Republic in most of its forms gives citizens the opportunity to influence crucial funding streams in municipalities, regions and sometimes even within individual countries. Digitalization has been promoted by Prague City Council by launching special website on open data news on Prague called FloweeCity.cz. This initiative of Prague municipality comes with the fact that Prague is trying to become more digitalized and more and more residents are using mobile apps to bring up ideas on how to improve their place for living [16]. It allows the residents of Prague to get informed with the entire daily life of Prague. It makes Prague's data more open and the residents digitalized [2].

Participatory budgeting has another effect that the future appears as a strategic motto: growing political culture and educate citizens engaged in the resolution of their city through using social networks and smart cities services. Thus, citizens are becoming an important player in local politics and create a real civil society [3].

PB is a democratic process that involves suddenly, that all members of the local community/municipalities are invited to be directly involved in decisions about the use of Public Finance (defined portion of the budget). According to the Czech Act on Local Governments, the decision must be always approved/confirmed by the council. Ordinary citizens, in this case are not only familiarized with the budget as imposed by law on budgetary rules (218/2000 Coll.) but even they are informed through the form of official boards and at the city council meetings. According to the Local Government Act citizens not only have an advisory role, but they are also entrusted with executive power based on the condition that the public budget is in the area/district, where they live. Furthermore, they are competent in attending meetings and the citizens possess "autonomy" in terms of decision making which has beneficial effects for the society.

Knowledge of Participatory Budgeting
It is envisaged that the knowledge concerned with the respective innovative practice plays a vital role in the diffusion of innovation [6, 13] or at least to raise a debate concerning such innovative practice [7]. This is no exception in the context of Czech Republic. The existence of PB as a practice of strengthening democratic governance and its potential advantages have become the main trajectory of promoting the PB discussion in the Czech cities is the awareness of its existence and its advantages.

In the Czech Republic the non-profit organizations which are focused on propagating PB among the citizens are developing different materials on PB and are trying to disseminate it among the citizens in order to increase the awareness and knowledge on PB among all the stakeholders. Several years ago Agora organization created the website www.participativnirozpocet.cz, which is a source where the citizens can find all basic necessary information on PB and they can get familiar with the fundamentals of PB how it works, where it works and what benefits could bring to them. Moreover, citizens and local governments get more acquainted with how PB works and where PB has been implemented around the world. This increase the level of awareness among all stakeholders and in the recent years it triggered the necessity of talking and discussing on PB best practices from abroad and their eventual implications in the Czech Republic as well. In Prague city council from year 2012 until now several times citizens have stood up in front of the city council members and the mayor of Prague where they openly have informed and talked about possible implementation of PB in Prague municipalities.

Persuasion for the Adoption of Participatory Budgeting
Various knowledgeable actors would strive to persuade the decision-making authority in the adoption of innovative practices that they deem important for the survival of a specific group or institution [13]. Particularly, those who are aware of global trend are likely to be in the forefront of making convincing speeches concerning innovative accounting and budgeting practices [6]. In the Czech Republic, the citizens several times have been endeavouring to convince the politicians in the local government of Prague to introduce PB and to demonstrate that PB could bring positive results for the residents and the municipality. For instance, in 2012 the citizens and the non-governmental organizations Agora and Alternativa Zdola from the Czech Republic have been working on documents and case studies showing that PB adds value for all stakeholders in the local level of public governance. They have been referring to best practices from the US, Canada, Spain and especially Poland in order to prove that PB is a trajectory of positive changes for the future of the place where we live.

Also, the non-profit organizations Agora and Utopia have been successfully supported the local government of Bratislava in Slovakia, where in 2013 the PB has been introduced successfully and it has been working for several years now. Furthermore, 'Agora' and 'Alternativa Zdola' have implemented successfully in several towns in the Czech Republic. Agora Central Europe organizes this year in March a big conference named 'Good Municipality of 2016-participation, communication and transparency' in the residence of the current mayor of Prague in order to popularize PB and to demonstrate an innovative method of citizen participation. The aim of this conference is not only popularizing the PB but also that non-profit organizations named Agora, CPKP, Partnership and Green Line would like to address as many as possible representatives from the public sector on all level of policy making- district, region and republic level. Thus, these non-profit organizations would like to sign a pact with the stipulated representatives from the public administration and they would like to make sure that the citizens will be involved and engaged in the decision making process in the public administration. This pact is supported with another two official documents called methodology and standards. This way of persuasion of the local governments to include citizens in the decision making process has been inspired by similar activities in the UK (organization Compactem) and Norway (IdeBanken). For the moment in the Czech Republic this pact is perceived only as a tool but Agora and their partners would like to convert it into valid official document for citizens' participation in the public administration. In the last two years, the political party 'Pirates' has allocated 0,4 million Czech crowns in year 2014 and 1 million Czech crowns in year 2015 from its budget on PB which explains to certain degree the emerging importance of PB for finding future dialogue between the local government of Prague and its citizens.

Decision, Implementation and Confirmation of the Participatory Budgeting
The stage of decision is the stage in which all arguments, i.e. supportive or opposing are taken into consideration and the powerful decision making authority discloses its opinion [4, 6, 13]. The decision could be yes for the proposed innovative practice or no for it. In some contexts, the top leaders may consider to postpone the introduction of innovation. As evident by the conducted interviews with the council members and administrators, the ruling party's politicians are not willing to implement the whole PB idea. The elected representatives can see PB as a threat to their power and domination or merely an idealistic model in reality that cannot function successfully [8]. Indifference or opposition to the introduction of PB is also a clear standpoint of the answers to the question of whether PB should be introduced or not. In April 2014 officially in Prague District 7 has been introduced PB by the majority of the district council who were at that time socialists (KSCM socialist political party). The district council of Prague 7 has decided to allocated 1 million Czech crowns to be allocated by the citizens. One of the district council members of Prague 7 that time was from the socialist and she decided to take a risk and give opportunity to citizens to make decision on the budget of the district because she considers the residents as well informed and agents of knowledge for the best of the place where they live. Her project involved citizens and it was for building play fields for the kids in the area. This attempt became successful not because of her political party belonging but because the project was transparent, well-structured and different stakeholders were participating in it. Since then in Prague district 7 has been working PB successfully.

In 2013 in Prague District 10, the citizens have been participating for first time in the history of the district in allocating 21 million Czech crowns for refurbishing of public building in their district. Since then, PB has been successfully introduced in Prague 10 district. Besides that, the citizens have increased the scope of the project and they have extended the project not only to the building's refurbishment, but also for reconstruction of the sidewalks and reducing the noise in their streets. So PB showed in this district that the proactive citizens think globally and broader about the place where they live and that PB is only for cities/towns where the citizens think broader and are extremely proactive to propagate different projects and where the district council members are risk takers. Since 2013 Prague municipality has introduced different model of PB by allowing a citizen from the society to become members of grant and selection committee in Prague municipality which is a positive gesture from the municipality towards its residents. However, the willing citizens should his or her application online and the Prague local government decides which citizen will occupy a spot in the respective committee.

In Sept 2015 in Prague District 22 for first time in the history 600 000 Czech crowns have been allocated as a part of PB and the citizens have been voting for the budget of 2016. The implementation stage of PB is often claimed to be the most difficult stage of the reform process. At this stage, the actors, who are against the reforms, become more active and strive to alter, modify, and reinvent the very essence of the proposed reforms [6]. In year 2015 the Prague Institute of Planning and Development, led by Petr Hlavacek (member of the democratic party ODS), has started to develop official manual in Czech language on PB with the ultimate goal to propagate, support and show to the citizens what is PB, its structure, functioning and implementation in the practice. The Prague Institute of Planning and Development (IPR Prague) is the body in charge of developing the concept behind the city's architecture, urbanism, digitalization, development and formation [14]. It is an organisation funded by Prague and represents the city in spatial planning matters. IPR Prague is also in charge of the important task of processing geographical data and information, both for applied research and for the creation of supporting documentation that is important for the development of the city (particularly the Prague Analytical Land Use Documentation). Pehe (2016) explains the passivity of citizens on public matters as follows: "Probably most important reason is the fact that the Czech Republic is related to the political sphere of citizens society, which is relatively inactive and weak. It is still a new experience for the most Czech. In the former state of Czechoslovakia, there were any signs of independence because these democratic principles were put down. The result was that the Czechoslovaks and free civil society began to build again until after the fall of communism mode“ [10]. Many habits normalization period, however, understandably survive even a decade after the fall Communist regime. The gap between "policy" and the citizen was so deep that it still has not been fully overcome. The author [10], however, sees hope in young people who are stigmatized standardization traumas, and which is already evident that they are much more active in the expression of their civic stance emphasizes that almost all large public events in recent were work just by young people. It is therefore possible that young people in the Czech Republic can also hope for the introduction of PB in future years.

4 Discussion and Conclusions

Scholarly works have increasingly focused their attention on the public sector services innovation under the rubric of new public management reforms. Albeit it is vital to explore such reform endeavours in the local governments of Czech Republic because of its socialist legacy as a member country of former Soviet Union, scholars have largely neglected reforms of the country's local governments. As such, this study makes a particularly interesting contribution by pinpointing how the political leadership of Prague City Council endeavours to impede and delay the demand of the community to introduce PB and Digitalization that enables them to publically raise their voice concerning the allocation of public funds. In addition, this study underscores the usefulness and strength of Rogers' diffusion theory to explore and pinpoint the existing struggle in the process of adopting innovative accounting and budgeting practices.

Prior conditions such as weaknesses of existing practices particularly instigate to import new practices that would contributes to superseding at least most of them [6, 13]. Our study demonstrates that context specific issues such as the lack of co-operation between the citizens and corruption related accusation concerned with the elected politicians have contributed to emerging the debate dealing with the necessity of introducing PB. In this respect, the two non-government organizations and opposition political parties have shown the keen interest of introducing PB. They have raised their voice in the council meeting and have explained the importance of adopting PB. In addition, a petition consisting of citizens' signature has also been submitted to the mayor as a strategy to persuade the political leadership in favour of adopting the innovative budgeting model. Nevertheless, the ruling political party and its leadership in the council is hesitant to rely upon PB as they perceive it to be based on the socialist ideology.

The municipality's political leadership has implicitly indicated that they are not willing to adopt PB in this near future but the council would strive to provide more and more opportunities for citizens to raise their voice with respect to particularly strategically important issues such as construction projects. Similarly, the council is leaned to co-operate with the non-government organizations with a view to developing a report, which would underscore advantages and disadvantages of this budgeting practice. This will enable the political leadership to delay the introduction of PB and to show that they will consider adopting PB sometime later. Nevertheless, just focusing attention on advantages and disadvantages or challenges in in the implementation of PB would envisage that they are likely to be interested of comprehending disadvantages or challenges as such comprehension can be deployed in further postponing or making counter arguments against the emerging request for the adopting of PB. Therefore, we argue that the diffusion of PB in the Prague Municipality has not moved beyond the decision stage.

References

1. Barreto, L., Amaral, A., Baltazar, S.: Urban mobility digitalization: towards mobility as a service (MaaS). In: 2018 International Conference on Intelligent Systems (IS), Funchal - Madeira, Portugal, pp. 850–855. IEEE (2018). https://doi.org/10.1109/is.2018.8710457
2. Beutelspacher, L., Mainka, A., Siebenlist, T.: Citizen participation via mobile applications: a case study on apps in Germany. Int. J. Electron. Gov. Res. (IJEGR) **14**(4), 18–26 (2018). https://doi.org/10.4018/IJEGR.2018100102

3. Bernhard, I., Norström, L., Snis, U.L., Gråsjö, U., Gellerstedt, M.: Degree of digitalization and citizen satisfaction: a study of the role of local e-government in Sweden. Electron. J. e-Gov. **16**(1), 59–71 (2018)
4. Broadbent, J., Guthrie, J.: Public sector to public services: 20 years of "contextual" accounting research. Account. Audit. Account. J. **21**(2), 129–169 (2008). https://doi.org/10.1108/09513570810854383
5. Célérier, L., Cuenca Botey, L.E.: Participatory budgeting at a community level in Porto Alegre: a bourdieusian interpretation. Account. Audit. Account. J. **28**(5), 739–772 (2015). https://doi.org/10.1108/AAAJ-03-2013-1245
6. Ezzamel, M., Hyndman, N., Johnsen, A., Lapsley, I.: Reforming central government: an evaluation of an accounting innovation. Crit. Perspect. Acc. **25**(4–5), 409–422 (2014). https://doi.org/10.1016/j.cpa.2013.05.006
7. Vinod Kumar, T.M.: Smart environment for smart cities. In: Vinod Kumar, T. (eds.) Smart Environment for Smart Cities. Advances in 21st Century Human Settlements. Springer, Singapore (2020). https://doi.org/10.1007/978-981-13-6822-6_1
8. Kuruppu, C., Adhikari, P., Gunarathna, V., Ambalangodage, D., Perera, P., Karunarathna, C.: Participatory budgeting in a Sri Lankan urban council: a practice of power and domination. Crit. Perspect. Account. **41**, 1–17 (2016)
9. Musso, J., Weare, C., Bryer, T., Cooper, T.L.: Toward strong democracy in global cities? social capital building, theory-driven reform, and the Los Angeles neighbourhood council experience. Public Adm. Rev. **71**(1), 102–111 (2011). https://doi.org/10.1111/j.1540-6210.2010.02311.x
10. Pehe, V.: Socialism remembered: cultural nostalgia, retro, and the politics of the past in the Czech Republic, 1989–2014, Doctoral dissertation, UCL University College London (2016). https://core.ac.uk/download/pdf/79523595.pdf. Accessed 5 June 2019
11. Prins, C., Adams, M.: Digitalization through the lens of law and democracy. In: Prins, C., Cuijpers, C., Lindseth, P.L., Rosina, M. (eds.) Digital Democracy in a Globalized World, pp. 3–23, Elgar Law, Technology and Society Edward Elgar, Cheltenham (2017)
12. Richter, C., Taylor, L., Jameson, S., Pulgar, C.P.: Who are the end-user(s) of smart cities? a synthesis of conversations in Amsterdam. In: Coletta, C., Evans, L., Heaphy, L., Kitchin, R. (eds.) Creating Smart Cities, pp. 121–130. Routledge, Abingdon (2018)
13. Rogers, E.M.: Diffusion of Innovations, 5th edn. The Free Press, New York (2003)
14. Schnoll, H.J. (ed.): E-Government: Information, Technology, and Transformation: Information, Technology, and Transformation. Routledge, New York (2015)
15. Srivastava, L.: Digitalization: structure and challenges. J. Account. Financ. Mark. Technol. **2**(3), 22–26 (2019). http://management.nrjp.co.in/index.php/JAFMT/article/view/334. Accessed 5 June 2019
16. Tian, J., Li, H., Chen, R.: The emerging of smart citizen concept under smart city environment. Aper presented at the Proceedings of the International Conference on Electronic Business (ICEB), pp. 739–742, December 2018
17. Velinov, E., Maly, M.: Participatory budgeting in prague magistrate: empty hopes or realistic future? Publication name, pp. 230–235 (2016)

The Possibilities of a Paperless Company Concept

P. Šuleř and V. Machová(✉)

Institute of Technology and Business, School of Expertness and Valuation, Okružní 517/10, 37001 České Budějovice, Czech Republic
machova@mail.vstecb.cz

Abstract. The review examines the possibilities of a paperless company concept, and explored mainly resources available on Scopus and Web of Science databases. The researcher found out that companies are still using traditional ways of doing things despite innovation of new technologies like the 3D printing, and AI. A few percentage of companies like health insurance firms and technology outfits are moving towards becoming completely paperless, whereas a majority of establishments combine the two concepts (paper and paperless) for record keeping and other operations. It is crystal clear that the paperless concept has not been fully realized. However, most companies, especially in the developed world, have a great potential in fully becoming paperless in the near future. Relatively, in order to go paperless, companies should organize all files, convert, and scan new documents, and most importantly should develop robust documentation management software. There is thus, no doubt for a future research in the subject matter using neural networks to accurately predict the future.

Keywords: Paper · Paperless · Company concept

1 Introduction

The idea of paperless concept is not new, but still in transition. In fact, paperless company concept is referred to as a myth. This concept was talked about by Pake [5] in his article. He envisioned offices in the future, especially desk job workers having computers that keep documents in files saved on the desktop. In his paper, he quoted Giuliano [1] who foretold that the utilization of paper for documentation will diminish over time. Yet, some school of thought alleged that businesses have increased their use of papers over the years since then. Nowadays, those predictions can be seen and experienced in our work places. Though not completely. Paperless company concept application depends upon the sector of a business and department of operation. For instance, heads of companies and CFO's can be more paperless than administrators down the management hierarchy.

A paperless company concept is a concept if adopted in companies will totally eradicate or significantly reduced the use of papers in a company setting. The concept is achieved mainly by changing document into digital form which is environmentally friend, efficient, and can help considerably reduce operation cost.

S. I. Ashmarina et al. (Eds.): ISCDTE 2019, LNNS 84, pp. 198–202, 2020.
https://doi.org/10.1007/978-3-030-27015-5_25

Traditionally, papers are used on a daily basis by variety of companies around the world to circulate information at work place. Regrettably, paper documents are sometimes not accessible to some workers, not user-friendly, negative impact on the environment, and unreadable owed to variety in font sizes. Decades ago, some organizations have anticipated principles that would revise the use of papers to determine the possibilities of the paperless concept. Paperless documentation is also an unconventional means of information dissemination. The possibilities of paperless company concept are expected to truly change the way information is being managed today.

In February, 1999 Mitsubishi Automotive Engineering Company in Japan received the ISO9001 quality certification and the environmental management certification (ISO14001) in February, 2000 for using a paperless company concept. During this period the paperless concept was very unpopular for engineering company in Japan. Traditionally, the application of a quality system thoughtfully requires an increase in company paperwork. Fortunately, Mitsubishi surprised all by reducing large amount of paperwork in adopting a paperless system. Implementation of the paperless concept enable the company to save one million six hundred thousand pages within two years of application of the concept, which resulted to saving thirty million yen each year, comprising labor cost, sending and filing of papers as in the case of the paper concept [4].

2 Results

A decade ago, a Swiss company presented an update on its defense programs. The organization reduced their country's defense posture as well as the national defense budget by adopting paperless company concept. Paperless operations, especially in logistics support increasingly outsourced their activities to industries. This move lead to an increasing presence of the organization in logistic domain by developing an internet portal that relates to paperless concept to enhance cost verification, evaluation, and maintenance valuation, thereby eliminating the use of paper.

According to Haggith [2], Joshua Martin, executive of the Environmental Paper Network in the United States concurs that there is a far reaching recognition that most paper items are made from dead trees, however he is confident that the idea of the 'Don't print' message is confusing. Be that as it may, some companies are tapping into the desire for digitalization by linking the move from paper to web based documentation, thereby saving trees. Recently, there is a campaign to go paperless, focusing on financial organization clients and by utilizing a research by Javelin Strategy and Research to advance natural advantages coming about because of a move from paper to online transactions.

To experiment the possibility of paperless company concept, Ruchalski [7] organized a 12-weeks research to proof the concept of paperless documentation. Evaluated the Health Information Designs/Thomson Healthcare and Etreby Computer company for electronic delivery of PPIs in 6 chain and 4 independent community pharmacies, with 5 pharmacies evaluating each system. He found out that Pharmacists esteemed the two frameworks to be open and easy to understand. Drug specialists from the sample size felt that paperless PPIs were challenging to peruse and that printing takes excessively long time to complete. In that same year Slaney [8] had a different standpoint. He carried out a paperless company concept internal project by Adobe which lead to the creation of Portable Document Format (PDF) with the primary intent that file format of documents could be easily sent within and among companies regardless of their operating systems.

Digitalization has become a key strategy for the sustainability of many companies around the world. In recent years, organizations and companies are thriving to attaining Industry 5.0 where jobs positions are available only for specific period of time (GIG economy), and traditional full time workers are diminishing leading to a change in company operations. From company view point, overtaking another company and maintaining a dominant position is important to getting benefit which enables them to continue in business. On the other hand, from the workers stand point, digitalization of operations is indispensable to face new business GIG economy, and workers who would like to save their jobs in companies should be compelled to innovate or be able to quickly adapt to innovative ways of doing their work to stay workable.

The application of industry 5.0 such as digitalization and mobile marketing should be transitioned to a business model that provide revenue logistics digital business so that such businesses can be more evident [6]. There is a possibility that company agreement with customers can be paperless within operating systems. The concept of paperless companies through digitalization is an unquestionable requirement that will help them to cope in this technology era. Corporations may move to end up becoming smart players by taking no advantage, but simply focus around arranging paperless contracts with clients. On a fundamental level, firms should not claim resources, rather should be ready to provide services.

2.1 Evaluation of Paper and Paperless Company Concept

On one hand, it is believed that long paper document is easier to read and preferable by some people. However, productivity and profit maximization is diminished in companies that completely use papers as the only means of documentation. Also, paper documents are difficult to find on shelf and can easily be misplaced, time consuming to manually look for documents, and can occupy space in an office. On the other hand, paperless concept increase mobility, reduce response time of companies to customers, and are more secure. Nonetheless, the risk of hacking, toughness to filter through documents, expensiveness and time consuming to convert existing paper document into digital, frequently updating of software/hardware, and malware attacks and network crashing are critical challenges associated with the paperless company concept. The above mentioned challenges if not properly taken care of, may short down completely paperless companies. It can therefore be inferred that all companies now or in the near future should ensure they have technology in their portfolio to remain competitive in the world of business.

2.2 Opportunities Associated with Paperless Concept

Major changes in the business community has taken place in recent technology era. Conventional management concept has been proven very difficult to survive in innovative companies. The increasing interdependence of world economies, growing scale of international trade and flow of international capital with the rapid spread of technology has left companies with no option, rather, to think about innovative ways of management to match up the global competiveness; hence the need for adoption of paperless concept cannot be over-emphasized. In such situation the use of cutting edge management ideas, innovative technology management style has turned into an unavoidable pattern of plan of action, while the relating company information re-engineering is essential for the data management transformation. The paperless company concept is believed to enhance efficient resources integration, reduce cost of operations, and enable faster communication in industries.

Strategic and rational thinking maintain and keep organizations in a superior position. Some people believe the notion that going paperless is hard, and others think that the application of the concept is possible, but should not affect company operations and the transition should also not happen overnight.

Research conducted at the government level comparing Nigeria with the UK in terms of adopting paperless concept reveals that the concept is vital in implementing e-governance in government agencies. The findings revealed that, the concept is beneficial in so many ways; save paper, boost productivity, and make information sharing easier. In the developing world, an increase in the use of internet among citizens will enhance usage of e-governance. E-governance on one hand leads to improvement in revenue collection and at the same time contribute to reducing deforestation [3].

So, the innovation of new technologies and with numerous benefits associated with the concept, it is definite that there is a greater possibility that the concept will be universally accepted and utilized. In addition, the 100s of billions of e-mail circulating each day around the world, which was not possible 50+ years ago, and the rapid technological development shows a clear path towards the future of paperless company concept.

3 Conclusion and Future Work

The concept is not completely in practice. However, the appearance of cloud computing has made paperless concept more visible nowadays. Consequently, at present, there is no exact figure showing the percentage of companies that are completely paperless. With the increasing advocacy to save our environment from global warming due to deforestation, amassed urge of companies for large profit maximization, and the competiveness in innovation, it very well may be affirmed that the paperless concept will overtake the current trend. So, limited literature in this topic suggests the need for future research on the possibilities of the paperless company concept incorporating neural networks to accurately predict the future.

References

1. Giuliano, V.E.: The office of the future. Boomberg Business Week (1975). https://www.businessweek.com/stories/1975-06-30/the-office-of-the-futurebusinessweek-business-news-stock-market-and-financial-advice. Accessed 10 June 2019
2. Haggith, M.: Is paper free really green? PPI Pulp Paper Int. **52**(11), 13–16 (2010)
3. Iyoha, F., Jinadu, O., Ayo, C., Gberevbie, D., Ojeka, S.: E-government adoption and environmental bonuses: a study of Nigeria and United Kingdom. In: Decman, M., Jukuc, T. (eds.) Proceedings of the 16th European Conference on e-Government, pp. 91–97. Academy Conferences Ltd., Reading (2016)
4. Ohishi, H.: Complete paperless intranet process management system for ISO9001 quality and 14001 environmental management systems. SAE Technical Papers, pp. 1–10 (2002)
5. Pake, G.E.: The office of the future: an in-depth analysis of how word processing will reshape the corporate office. Boomberg Business Week (1975). https://www.bloomberg.com/news/terminal/O1OYOI6S972A. Accessed 20 May 2019
6. Rahman, N.A.A., Muda, J., Mohammad, M.F., Ahmad, M.F., Rahim, S., Fernando, M.-V.: Digitalization and leap frogging strategy among the supply chain member: facing GIG economy and why should logistics players care? Int. J. Supply Chain Manage. **8**(2), 1042–1048 (2019)
7. Ruchalski, C.: The paperless labeling initiative: a proof-of-concept study. Ann. Pharmacother. **38**(7–8), 1178–1182 (2014)
8. Slaney, S.: A brief history of adobe's portable document format. New England Printer Publ. **67**(3), 24 (2004)

Digitalization as a Driver of the New Economic Order and Regional Development

Regional Digital Maturity: Design and Strategies

E. K. Chirkunova, G. A. Khmeleva(✉), E. N. Koroleva, and M. V. Kurnikova

Samara State University of Economics, Samara, Russia
galina.a.khmeleva@yandex.ru

Abstract. The paper is aimed at substantiating regional digital maturity, its understanding, structure and measurements. The study reviews the terminology genesis of digitalization, reveals the necessity to transfer from digital economy to the understanding of digital maturity, disclose the structure of regional digital maturity including the rate, depth and coverage of the digital transformation of a local economy and community. The authors have analyzed the digital maturity data of 85 Russian regions. The results show that digitally matured regions are among the most successful Russian subjects. The study contributes to the body of knowledge of regional competitiveness. The authors have developed the strategies of regional digitalization and the mechanism for their integration in the world economic system.

Keywords: Region · Digital transformation · Digitalization · Innovation · Maturity · Integration

1 Introduction

Today's world is in the midst of the greatest informational and communicational revolution in the history of mankind. The distribution of digital economy expands the horizons and opens up new possibilities for territorial development especially for ambitious young people [4]. The development of the digital economy concept has resulted in the current understanding of ICT-based goods and services as components of digital economy. Governments and businesses use ICT services to facilitate information processing and communication [9]. The Russian Association of Electronic Communication (RAEC) views digital economy as markets where value added is created by digital technologies [10].

In his message to the Federal Assembly in 2016, the President of the Russian Federation proposed to 'launch a large-scale systemic programme for economic development' as a response to global challenges thus determining the strategic path for national development.

Recently, the phenomenon of digital transformation of territories has been investigated in detail. Digital economy has been identified as an inevitable stage of regional development, with traditional productive sectors and service industries being transformed substantially by the penetration of informatization and digitalization of economic processes [3], the existing markets being changed and the new ones developed [1], new approaches to management decisions being worked out. The changes in the

S. I. Ashmarina et al. (Eds.): ISCDTE 2019, LNNS 84, pp. 205–213, 2020.
https://doi.org/10.1007/978-3-030-27015-5_26

need for IT by businesses and community could not but have impact on the market of Internet use by population, organizations encompassing fixed broadband and mobile access, the speed of the Internet, the use of social media and technologies for e-business [2]. In order to measure such changes, the authors introduce the term of regional digital maturity, estimate the digital maturity based on Russian regions, reveal the connection with competitiveness. The paper forms the basis for choosing the strategy of regional digitalization based on regional digital maturity.

2 Theoretical Background

2.1 From Digital Economy to Digital Maturity

The theoretical background for regional digital maturity can be found in many scolars' works on a wide range of issues. The papers by Bell [6] and Galbraith [13] have formulated qualitative parameters of post-industrial community development, among which are science- and technology-intensive production techniques based on informatization and the concentration of the most employed in the services sector. M. Porter suggested that scientific and technological progress particularly in the IT-sphere is among the most important developments being able to change the laws of competition [21].

The ideas of Toffler [26] are connected with the notion of an information society that was further named as a post-industrial community, with information as its critical resource as Castells [7] stated in his research.

The term 'digital economy' was first suggested by Tapscott in 1995, when he showed the Internet being able to alter radically the relations in society and business environment with rising new generation using digital technologies [25].

We can observe the fundamental change in the world and the creation of new business models, i.e. digital transformation in response to the penetration of digital technologies such as cloud computing, mobile Internet, social media and big data [12]. As a result, the companies from various industries try to assess their capacity in a new era [14]. Not long ago, this phenomenon was named digital maturity in scientific publications [5, 24]. Digital maturity is widely viewed in case of enterprises [27] both of high-tech sector and traditional industries [23].

However, the rates of mastering digital goods and services, new technologies by enterprises, the levels of using digital services result in territorial digital maturity – the one of cities and regions. The authors think that it may be justifiable to refer digital maturity to regions while assessing the level of digital technologies spread, the use of digital services by people and businesses, and consequently the level of digital economy development.

2.2 Approaches to Measuring Digital Economy and Digital Community Development

Currently, there are not any common approaches to measuring digital economy. However, in general, one can distinguish the ones used by national statistical offices and professional associations (organisations) in the ICT sphere.

As for our research, we have taken into consideration the European and Russian experience, and the main sources of information were the Federal Statistics Service (Rosstat) data, the database of the official governmental statistics – the Unified Interdepartmental Statistical Information System, as well as the results of the information society survey (2018) by the Ministry for Digital Development, Communications and the Media of the Russian Federation, the Federal Statistics Service, Higher School of Economics [15].

As a result, the estimation of digital economy through regions was conducted in two ways: the level of the Internet penetration and the level of digital economy development (see Fig. 1).

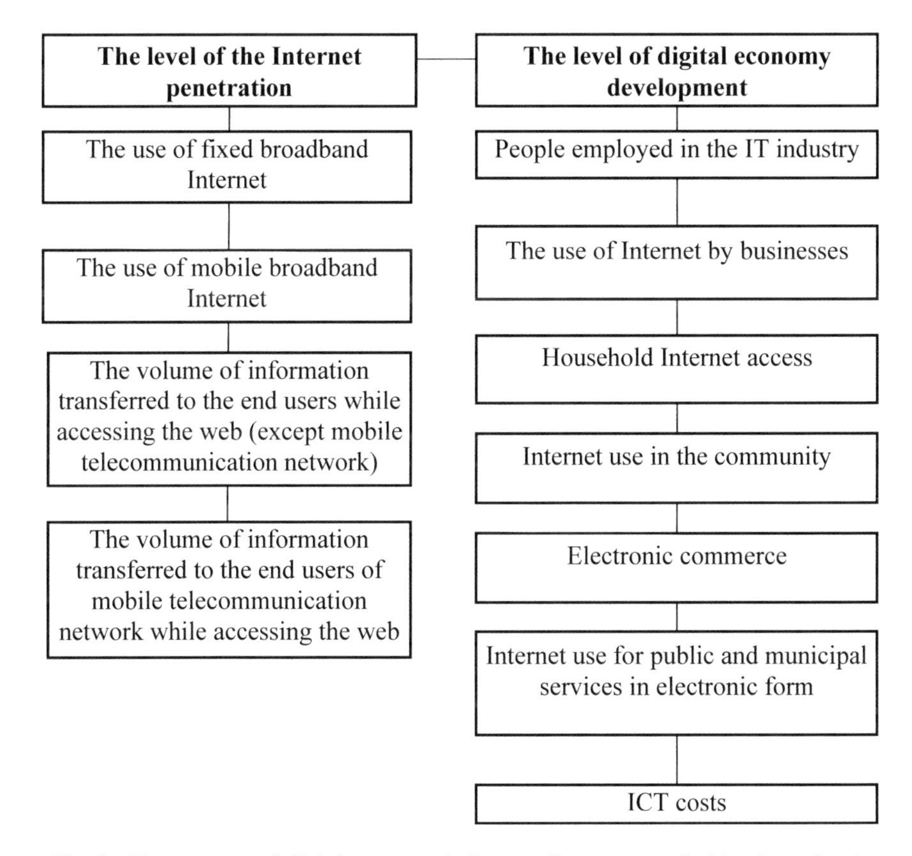

Fig. 1. The structure of digital economy indicators (Source: compiled by the authors).

The level of the Internet penetration establishes a framework technical condition while opening opportunities for the information exchange between as many people as possible that is why the data on the number of Internet users and traffic volume were analyzed.

Digital economy development is estimated using the indicators of the Internet penetration into day-to-day life and the business of companies. Apart from the employment in the IT industry and ICT costs, the data on the Internet use in the community and by businesses, electronic commerce, public services characterize technological, organizational and management changes taking place within economy and society closely associated with rapid ICT development and spread.

Considering increasing ICT employment, the number of economic activities newly accessible due to digital technologies and therefore their growing economic significance, digital economy measurement is a primary objective to make the right political choices, develop a tax policy and distribute resources in Russian regions.

2.3 Research Questions and Research Design

The introductory review of the literature has the fundamentals of regional digital maturity. In order to develop the types of regional digitalization strategies, we have considered the terms of digital economy and digital maturity. Despite the significant contribution made by these scolars, they do not define 'regional digital maturity'. However, it is important to keep in mind that different types of regional digitalization strategies should reflect different levels of regional digital maturity. Furthermore, we summarized the approaches to measuring digital economy and digital society development used to estimate regional digital maturity in order to verify our theoretical findings. We used the integral indicator method based on the indicators of Fig. 1 in order to calculate the level of regional digital maturity. The indicators were normalized to convert them into a comparable form. The proposed set of indicators was tested for 85 Russian regions according to the data from 2017. The comparison of regional digital maturity levels enabled to develop the types of regional digitalization strategies.

Our contribution initiates a new understanding of digital maturity with regard to regions and creates types of innovative digitalization strategies.

Based on the problem described and its current understanding, this research would answer the following questions:

What is regional digital maturity?

What benefits could regions derive from digital maturity?

What is the methodology of achieving digital maturity by a region?

We attempted to use a theoretical and an empirical strategy in our research. The theoretical study collected data on substantial characteristics of digital maturity phenomenon. The empirical examination was used to prove practical application of theoretical knowledge and develop the types of regional digitalization strategies.

3 Discussion

Digital transformation in regions occurs in all spheres through the large-scale use of digital and information and communication technologies aiming at increasing efficiency and competitiveness. In Russia, 83.7% of the population and 66% of organizations of the companies used the Internet in 2017. It is widely assumed that regional economic digitalization is characterized by the level of the Internet penetration and its impact on the developments in various sectors of the economy (the structure of personnel demand and employment, electronic commerce, public and municipal services) [8]. However, the main features of digital transformation are not only the Internet but also the full spectrum of Industry 4.0 technologies (big data, Internet of Things, virtual and augmented reality, 3D printing, blockchain, autonomous robots). The Internet is the way to 'revive' new tehnologies, and in that sense its penetration and use by households and businesses may characterize the level of regional digital maturity.

The calculations have shown that Russian regions differ significantly in terms of the Internet penetration level, digital economy development level and as a result digital maturity level (see Table 1).

Table 1. The differences in digital maturity levels between Russian regions (2017)

Regions	Internet penetration (digital divide)	Digital economy development	Digital maturity
The leading regions	**3,13** (the Pskov Region)	**4,27** (the Moscow Region)	**6,77** (the Moscow Region)
The outsiding regions	**0,093** (the Republic of Ingushetia)	**0,99** (the Republic of Daghestan)	**1,59** (the Republic of Daghestan)
Average	**1,51** (~the Tyumen Region)	**2,52** (~the Astrakhan Region)	**4,03** (~the Republic of Mordovia)
Median	**1,44**	**2,55**	**3,92**
Standard deviation	**0,55**	**0,66**	**1,01**

Source: calculated and worked out by the authors.

Though ICT have emerged as a critical development instrument affecting many aspects of the work, regional digital divide is reflected in disparities in access to ICT and their use [19, 20].

There remains digital divide in Russian regions. The authors believe that its depth depends on spatially differentiated factors and is particularly apparent in geographic ranges of rural areas. In Russia, digital divide throughout subregions is not taken into sufficient consideration and is still subject to research. The main development vector of digital economy in Russia is expected to have been provided by the largest mega cities and populous regions of the European part of the country. Regional digital maturity development levels are sufficiently different and indirectly dependent on the levels of economic development among regions, their average per capita income, consumer prices and other 'offline' components [11]. Moreover, behavioural patterns that are more dynamic in Moscow and the Moscow Region may also affect digital maturity.

The industrialized regions such as the Tyumen and the Moscow Regions have more active advantage of digital economy (Fig. 2).

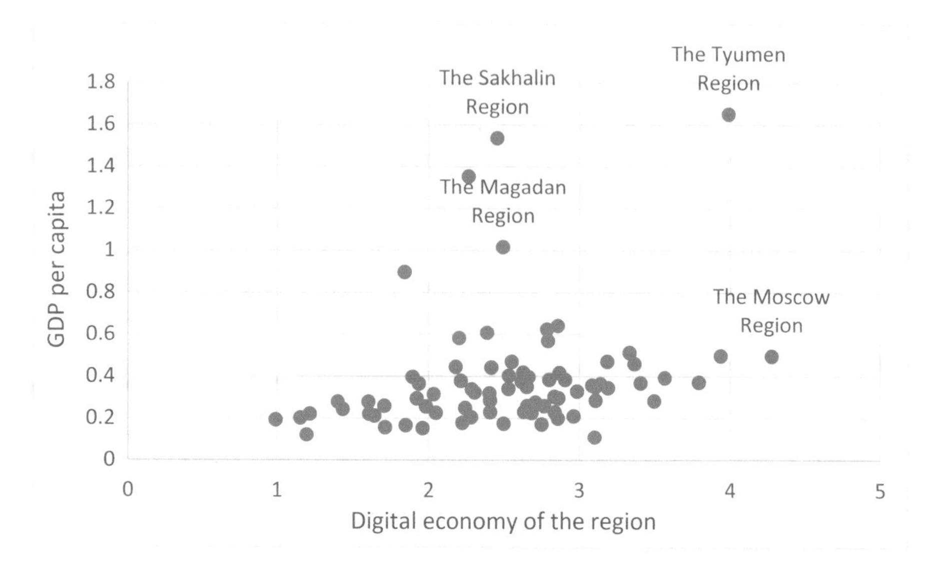

Fig. 2. The generalized model of GDP and digital economy. Note: the Moscow Region includes the data on the city of Moscow. (Source: calculated and worked out by the authors)

It is not yet necessary to count on digital economy for the regions with the extractive industries (the Sakhalin Region, the Magadan Region), as they are satisfied with the welfare of mining and marketing of natural-resource commodities.

Thus, digital maturity largely depends on the level of digital economy development. The Internet penetration only creates opportunities for regions which can then make full use of them for prosperity of economy.

There was an uneven development of electronic commerce throughout regions in 2017. The variability in the proportion of organizations placing orders for goods (work, services) over the Internet ranged from 18.1% in the Republic of Dagestan to 50.7% in the city of St Petersburg. The organizations in regions geographically remote from the country centre are more active in electronic commerce. Thus, digital maturity may perhaps contribute to reducing geography interference [18].

There has been no data in Russian regions yet that digital transformation would contribute to productivity growth impacting employment in various sectors [16].

Smart cities, Housing and Public Utilities and transport digitalization are among promising areas in Russian regions.

The acceleration of digital maturity rates would promote the welfare of a region. This is especially relevant to resource-poor regions. Often, such regions have more difficulty in developing digital economy because of the general situation of long-term underdevelopment. Regions may be suggested various strategies aiming at digital maturity according to their differences (primarily scientific and technological capacities and a productive base):

digital technological accelerator – the strategy of creating their own digital technologies (fostering growth of investments in IT-sphere, governmental support of

both startups and scaleups – the companies successfully growing for some years [19];
digital cooperation – the strategy aimed at technological activity of creating their technologies in a region including the digital ones, active cooperation in creating such technologies and attracting technologies from other regions and countries;
propulsion technologies – the strategy of attracting digital technologies from abroad and other regions.

Regional authorities should adhere to the following guidelines for developing digital maturity:

1. To increase investment in digitalization and strongly encourage the use of opportunities offered by public-private partnerships in terms of investment development plans and investment funds [8].
2. To ensure the reliance on scientific advances for expanding competitive knowledge-based industries through digitalization of industrial enterprises and creating innovative centres (centres of excellence in technologies) at universities [17].
3. To provide support to industry, new business development and smart specialisation which will serve local interests in the context of digitalization [22].
4. To develop digital skills of regional citizens and contribute to expanding broadband Internet access [2], especially in rural areas.

4 Conclusion

The paper is aimed at revealing the genesis of digital transformation phenomenon and provide a better understanding of digital maturity. We have showed that though the spread of the Internet is the engine of digital economy it does not automatically involves creating all advantages of digital economy for regions.

Today, the regions of high digital maturity benefit from digital economy despite digital divide. We have proposed three strategies of developing regional digital maturity that found the basis for further discussions on regional digital transformation and integration in the world economic system.

Acknowledgments. This paper is an output of the science project No. 19-510-23001 (Application 2019) “Modeling of processes of individual regions integration in the world economic system” of the Russian Foundation for Basic Research and the Foundation “For the Russian Language and Culture” in Hungary.

References

1. Adapting industrial policies to a digital world for economic diversification and structural transformation. UNCTAD. https://unctad.org/meetings/en/SessionalDocuments/cimem8d5_en.pdf. Accessed 23 Apr 2019
2. Akerman, A., Gaarder, I., Mogstad, M.: The skill complementarity of broadband Internet. Q. J. Econ. **130**(4), 1781–1824 (2015). https://doi.org/10.1093/qje/qjv028

3. Ashmarina, S.I., Kandrashina, E.A., Dorozhko, J.A.: Digitalization as a source of transformation of value chains of telecommunication companies using the example of PAO megaphone. In: Ashmarina, S., Mesquita, A., Vochozka, M. (eds.) Digital Transformation of the Economy: Challenges, Trends and New Opportunities. Advances in Intelligent Systems and Computing, vol. 908. Springer, Cham (2020). https://doi.org/10.1007/978-3-030-11367-4_57
4. Atkinson, R.D., Castro, D., Ezell, S.J.: The digital road to recovery: a stimulus plan to create jobs, boost productivity and revitalize America. In: The Information Technology and Innovation Foundation, Washington, DC (2009). https://www.itif.org/files/roadtorecovery.pdf. Accessed 23 Apr 2019
5. Becker, J., Knackstedt, R., Pöppelbuß, J.: Developing maturity models for IT management – a procedure model and its application. Bus. Inf. Syst. Eng. **1**(3), 213–222 (2009). https://doi.org/10.1007/s12599-009-0044-5
6. Bell, D.: The Coming of Post-industrial Society: A Venture in Social Forecasting. Basic Books, New York (1973)
7. Castells, M.: The Internet Galaxy: Reflections on the Internet, Business, and Society. Oxford University Press, Inc., New York (2001)
8. China's digital transformation: The Internet's impact on productivity and growth. McKinsey Global Institute (2014). https://www.mckinsey.com/~/media/McKinsey/Industries/High%20Tech/Our%20Insights/Chinas%20digital%20transformation/MGI%20China%20digital%20Executive%20summary.ashx. Accessed 23 Apr 2019
9. Barefoot, K., Curtis, D., Jolliff, W.A., Nicholson, J.R., Omohundro, R.: Defining and measuring the digital economy. Working Paper Bureau of Economic Analysis, USA (2018). https://www.bea.gov/system/files/papers/WP2018-4.pdf. Accessed 23 Apr 2019
10. Digital Economy of Russia: Association of Electronic Communications (RAEC) (2018). https://raec.ru/activity/analytics/9884/. Accessed 23 Apr 2019
11. Digital Russia: A new reality (2017). http://www.tadviser.ru/images/c/c2/Digital-Russia-report.pdf. Accessed 23 Apr 2019
12. Fitzgerald, M., Kruschwitz, N., Bonnet, D., Welch, M.: Embracing digital technology. MIT Sloan Management Review (2013). https://sloanreview.mit.edu/projects/embracing-digital-technology/. Accessed 23 Apr 2019
13. Galbraith, J.K.: The New Industrial State. Houghton Mifflin, Boston (1967)
14. Gannon, B.: Outsiders: an exploratory history of IS in corporations. J. Inf. Technol. **28**(1), 50–62 (2013). https://doi.org/10.1057/jit.2013.2
15. Information Society in the Russian Federation: Statistical collection. In: Sabelnikova, M.A., Abdrakhmanova, G.I., Gokhberg, L.M., Dudorova, O.Yu., et al. (eds.) HSE, Moscow (2018). https://issek.hse.ru/news/234186392.html. Accessed 23 Apr 2019
16. Khmeleva, G., Tyukavkin, N., Bulavko, O., Prosvetova, A., Egorova, K.: The regional allocation of labor resources in the context of economic restructuring and renewal: using a case from Russia. Probl. Perspect. Manag. **15**(3), 377–393 (2017). https://doi.org/10.21511/ppm.15(3-2).2017.07
17. Khmeleva, G.A., Agaeva, L.K., Chirkunova, E.K., Shikhatova, E.E.: Russsian Universities: the innovation centres of digitalization in the region. In: Mantulenko, V. (ed.) GCPMED 2018 - International Scientific Conference Global Challenges and Prospects of the Modern Economic Development, The European Proceedings of Social & Behavioural Sciences EpSBS, GCPMED 2018, vol. LVII, pp. 1728–1740. Future Academy, London (2019). https://doi.org/10.15405/epsbs.2019.03.175
18. Krugman, P.: Increasing returns and economic geography. J. Polit. Econ. **99**(3), 483–499 (1991). https://pr.princeton.edu/pictures/g-k/krugman/krugman-increasing_returns_1991.pdf. Accessed 23 Apr 2019

19. Lee, C.-Y.: Geographical clustering and firm growth: differential growth performance among clustered firms. Res. Policy **47**(6), 1173–1184 (2018). https://doi.org/10.1016/j.respol.2018.04.002
20. Liu, H., Fang, C., Sun, S.: Digital inequality in provincial China. Environ. Plann. A: Econ. Space **49**(10), 2179–2182 (2017). https://doi.org/10.1177/0308518X17711946
21. Porter, M.E.: Competitive Strategy: Techniques for Analyzing Industries and Competitors. Free Press, New York (1980)
22. Randall, L., Berlina, A.: Governing the digital transition in Nordic Regions: the human element (2019). http://www.nordregio.org/publications/governing-the-digital-transition-in-nordic-regions-the-human-element/. Accessed 23 Apr 2019. https://doi.org/10.30689/r2019:4.1403-2503
23. Remane, G., Hanelt, A., Wiesböck, F., Kolbe, L.M.: Digital maturity in traditional industries – an exploratory analysis. In: Twenty-Fifth European Conference on Information Systems (ECIS), Guimarães, Portugal. Research paper (2017). https://www.researchgate.net/publication/316687803_DIGITAL_MATURITY_IN_TRADITIONAL_INDUSTRIES_-_AN_EXPLORATORY_ANALYSIS. Accessed 23 Apr 2019
24. Schwer, K., Hitz, Ch., Wyss, R., Wirz, D., Minonne, C.: Digital maturity variables and their impact on the enterprise architecture layers. Probl. Perspect. Manag. **16**(4), 141–152 (2018). https://doi.org/10.21511/ppm.16(4).2018.13
25. Tapscott, D.: The Digital Economy Anniversary Edition: Rethinking Promise and Peril in the Age of Networked Intelligence. McGraw-Hill, New York (2014)
26. Toffler, E.: Shock of the Future. AST, Moscow (2002)
27. Valdez-de-Leon, O.: A digital maturity model for telecommunications service providers. Technol. Innov. Manag. Rev. **6**(8), 19–32 (2016). https://doi.org/10.22215/timreview/1008

Information Society Development in Regions of the Russian Federation

N. V. Kulikova, N. P. Persteneva(✉), and T. V. Ruslanova

Samara State University of Economics, Samara, Russia
persteneva_np@mail.ru

Abstract. One of the top priority tasks for most states nowadays is the transition to digital economy. To assess the level of development in the sphere of information and communication technology (ICT), the indexes published in the annual reports of International Telecommunications Union, are used. In 2017 Russia occupied 45 position in the International Rating of ICT development, which is positions lower than in the previous year. The position of the Russian Federation (RF) in this rating is largely influenced by the level of digitalization of its regions. In general, information society development in the RF is a herculean task due to significant variations in the levels of ICT use in various regions of the country. The authors of the paper carried out analysis of information society development in the regions of the RF in the period between 2015 and 2017. The results of the analysis can be effectively used in complex estimations of information society development levels and be a useful tool in administrative decision-making in the sphere of digitalization of regional economy and economy of the Russian Federation on the whole. As a result of this policy, Russia's international information society development ranking is supposed to improve.

Keywords: Digital economy · Knowledge-based society · Information society development index · Regions

1 Introduction

On the global level, modern society is characterized by the ever-increasing demand of the stakeholders for accurate and complete information. This brings forth the changes in all spheres of life, from social and cultural to economic and industrial ones. Thus emerges the new kind of society, currently known as "information society" or "knowledge-based society".

The Russian Federation is no exception from the global trend. The issue is widely discussed in contemporary works of the leading Russian economists. However, the present state of research in the field is characterized by the distinct lack of a unified, consistent approach to the content and methodology of measuring the level of development of information society. This explains extensive number of issues concerning both devising a comprehensive definition of "information society" and the development of evaluation methodology and approaches. This situation requires further research.

S. I. Ashmarina et al. (Eds.): ISCDTE 2019, LNNS 84, pp. 214–224, 2020.
https://doi.org/10.1007/978-3-030-27015-5_27

The Ministry of Digital Development, Communications and Mass Media of the RF has produced a version of methodology for evaluating the level of information society development, available from open source. We believe that, despite being consistent and well-rounded, the approach has some significant drawbacks. One of them is disregard for such important factor as the level of ICT development in various spheres of public life, especially in education. Education has always served as the ground for personality development in accordance with the most up-to-date social requirements. Excluding such important factor as education from the methodology for estimating information society development may lead to inaccurate estimation on both regional and federal level. This explains the necessity of further research in the field.

2 Methodology

The authors applied both general scientific and purely statistical methods and approaches. Institutional approach was used to analyze organizations, specializing in estimating the level of digitalization in a society.

General methods include:

- historical method, that proved the inevitable nature of ongoing digitalization as the basis of post-industrial society;
- comparative method, that allowed to discover current lack of reliable methodology for estimating the level of ICT development in the regions of the RF.

Statistical methods include:

- summary and classification method that allowed to process initial data;
- graphical and table methods used for displaying the resulting data;
- absolute, relative and median values that allowed to define the quantitative characteristic of an economic phenomenon;
- time series analysis and forecasting used for studying the data dynamics;
- index method used to summarize the results of data analysis.

The authors used «Gretl» and «QuickMap» for data processing.

3 Results

At present, Rosstat does not allocate any specific markers of information society. Thus, the authors created a system of such markers that allows to calculate information society development index (ISDI) more accurately on the basis of other markers, available from governmental open source.

To create this updated index, the authors selected 50 markers that reflect the readiness of a region for the transition to information society. The system of these 50 markers was then used to create the information society development index (ISDI) that reflects the actual use of ICT in a particular region during a particular period. The index can be used to monitor and compare the level of information society development in various regions of a country.

As for the ISD estimation methodology, the Ministry of Digital Development, Communications and Mass Media of the Russian Federation has proposed a methodology for calculating ISD Index for the regions of the RF. It applies indexes that can be subdivided into the following sub-indexes:

- ICT Use in areas of priority;
- Factors of information society development.
- To calculate the level of information society development, the following formula is used:
- ISDIRRF = 1/3*IF + 2/3*IU(1),

where:

ISDIRRF – Information Society Development Index in Regions of Russian Federation;
IF - Factors of Information Society Development Sub-index;
IU - ICT Use in Areas of Priority Sub-index.

Authors of this paper included one more sub-index into the core of ICT Use Sub-index – ICT Use in Education (ICTUE). It is, by our firm belief, crucial for more accurate data analysis.

ICTUE includes such markers as:

- number of PCs used for educational purposes per 100 students of state and municipal educational institutions;
- percentage of educational institutions with broadband Internet connection (out of the total 100 percent of educational institutions in the particular region);
- number of PCs used for educational purposes that are integrated into local area networks (LANs) per 100 students of state and municipal educational institutions of higher professional education.

The authors applied the methodology suggested by the Ministry of Digital Development, Communications and Mass Media of the Russian Federation to calculate the ISD Index for all the regions of the RF and create a comprehensive rating of the regions, ranked in accordance with the level of information society development. Due to the fact that the authors of this paper included the ICT Use in Education Sub-index into the core of the ICT Use Sub-Index, data differs from that calculated according to the officially suggested methodology. Figure 1 illustrates the discrepancies in ISD Index, calculated according to both methodologies, and applied to the regions of the RF in 2016.

The graph clearly demonstrates that the resulting values of ISD Index obtained by applying the two methodologies differ significantly for every region of the RF. On average, the discrepancy between the results is 27,2%. This can be attributed to the fact, that the authors updated the officially recommended methodology by including some important markers, previously unaccounted for. The authors believe, that this comparison proves that the methodology of the Ministry of Digital Development, Communications and Mass Media of the Russian Federation will benefit from the update, suggested and empirically tested by the authors. Besides, the Ministry-recommended methodology has not yet been applied to two regions of the RF, namely, the Republic

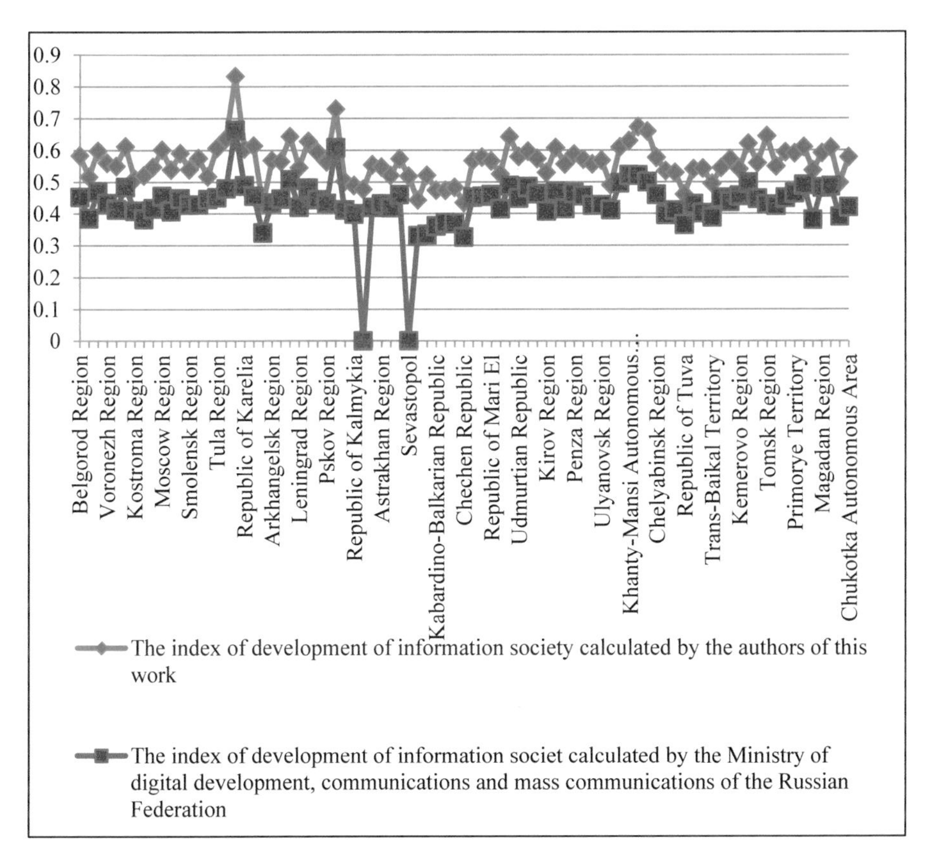

Fig. 1. Comparison of ISD Index in the regions of the RF in 2016 (Source: compiled by the authors)

of Crimea and the city of Sevastopol. The authors of this paper have also taken the liberty to calculate the ISD Index for those regions, alongside with the other regions of the RF.

The results of ISD Index calculation are later used to create a comprehensive rating of the regions of the RF, ranked in accordance with the level of information society development. The rating can be a crucially important tool for making administrative decisions on the regional and federal levels, concerning encouraging the use of ICT in a particular region (e.g. by increasing local ICT development and deployment budget) and the overall goal of expediting the transition of the RF to knowledge society.

The results calculated according to the authors' version of methodology, were also used to map the regions of the RF by the value of ISD Index in 2016 (Fig. 2) and 2017 (Fig. 3). Comparative analysis of the map model showed that in 2017 13 regions moved from the category of outsiders (=lowest ISD Index values) to the category of regions with average values of ISD Index.

The map model also shows that in 2017 there were several leading regions, namely: the City of Moscow, the City of Saint Petersburgh, Yaroslavl region, Tomsk region and

Fig. 2. Regions of the RF ranked by ISD Index in 2016 (Source: compiled by the authors using the program QuickMap)

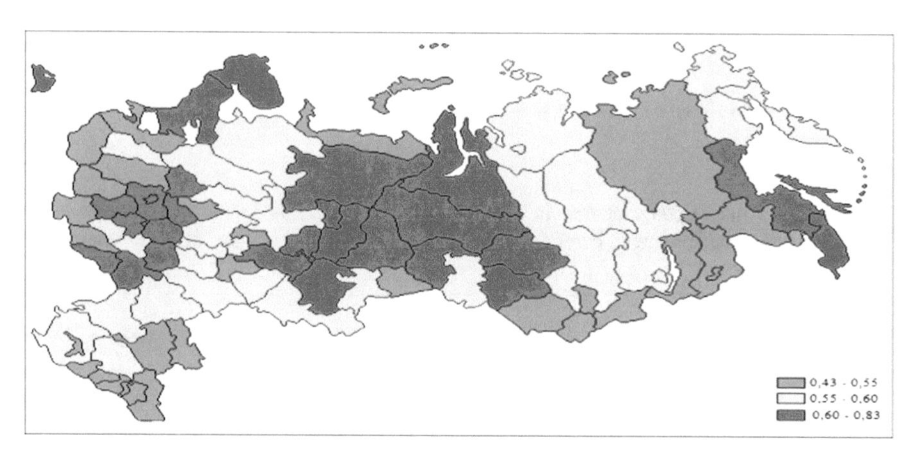

Fig. 3. Regions of the RF ranked by ISD Index in 2017 (Source: compiled by the authors using the program QuickMap)

Khanty-Mansi Autonomous Okrug. These regions have highest ISD Index values both in 2017 and in 2016.

According to information on Rosstat's official website, another important factor that affects the readiness of a region for information society developments the number of researchers based in this region. This marker also has its leaders: the City of Moscow (137,8 researchers per 10000 persons) and the City of Saint Petersburgh, (126,7 researchers per 10000 persons) due to the fact that the main scientific centers are based in those two locations. The lowest positions in the number of researchers per 10000 persons are occupied by Kostroma region (1,9 researchers per 10000 persons) and Yamal-Nenets Autonomous Okrug (1,5 researchers per 10000 persons). Mari El Republic and Zabaikalskiy Krai are also among the regions with lowest number of researchers: 3,5 and 5,2 researchers per 10000 persons, respectively.

To evaluate the dynamics of information society development in the regions of the RF, we compared the results of ISD Index calculations for 2016 and 2017. Figure 4 gives the results of comparative analysis of ISD Index values for the regions of the RF with maximum positive and maximum negative dynamics during the selected period (2016–2017). Let us note that maximum value increase in 2017 was 0,13 points and maximum value decrease was 0,07 points.

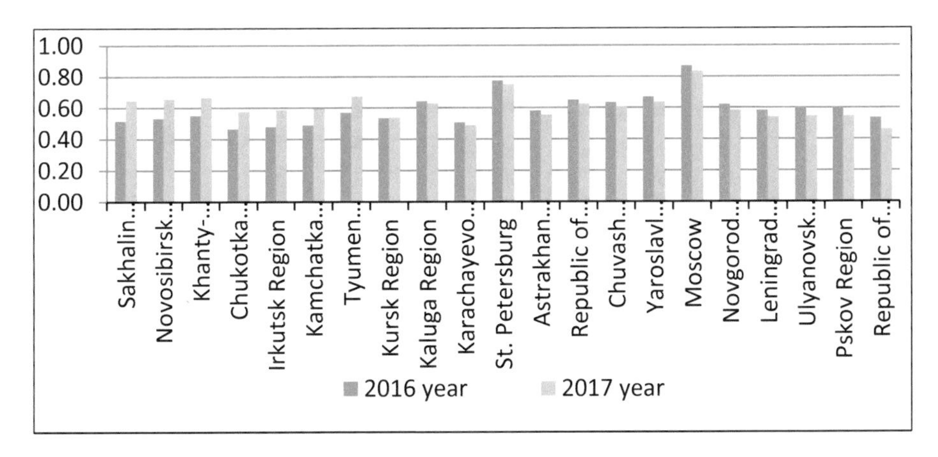

Fig. 4. Comparison of ISD Index values for regions with maximum positive and maximum negative dynamics in 2016–2017 (Source: compiled by the authors)

Regions with maximum value rise include: Sakhalin region, Khanty-Mansi Autonomous Okrug, Chukotka Autonomous Okrug and Irkutsk region. Increase in ISD Index values in those regions was mainly due to the increase in the number of PCs per 100 employees in state healthcare organizations; increase in the number of companies that actively use the Internet; increase in the number of cultural establishments that have their own website. Such rise in particular markers of ISD Index in these regions can be attributed to the respective increase in local ICT budget in 2017, compared to 2016.

Regions with slight decrease in ISD Index values include: Kursk region, Kaluga region, Republic of Karachaevo-Cherkessiya, City of Saint Petersburgh, Astrakhan region, Republic of Tatarstan, Chuvashia Republic, Yaroslavl region, City of Moscow, Smolensk region, Murmansk region and Mari El Republic. However, these regions in 2016 already had much higher ISD Index values than the regions which demonstrated significant rise in the index value in 2017. Thus we can conclude that the insignificant drop in ISD Index value in 2017 did not imply any detrimental effect to the actual level of information society development in those regions.

Unfortunately, ISD Index value is calculated on the basis of 50 markers and sub-indexes whose statistical data is only available in the period from 2015, and statistical data for 2018 has not yet been published for all the markers and sub-indexes. That is why it is currently possible to estimate the dynamics of ISD Index values in the regions of the RF only in the period from 2015 to 2017. The results of estimate are shown in Fig. 5.

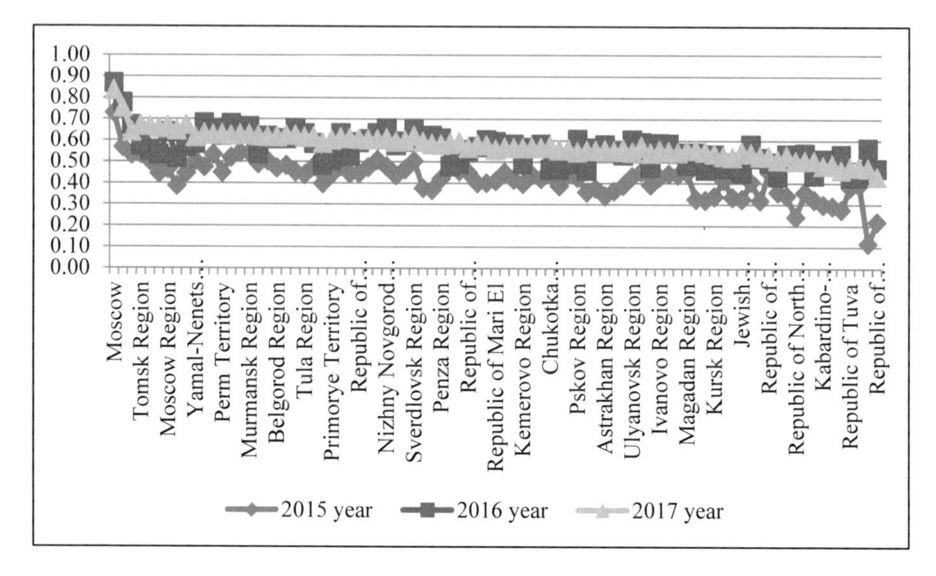

Fig. 5. ISD Index in RF regions in 2015–2017 (Source compiled by the authors)

Bar graph analysis reveals yearly increase in ISD Index values. The rise in values in 2017 compared to 2016 was 2,7% on average, and compared to 2015 was a staggering 37,4% on average. Besides, in 2015 there was a significant difference in ISD Index values among the regions of the RF, whereas in 2017 the value is virtually equal for all the regions in our research. This undoubtedly proves that there is positive dynamics in the information society development throughout the RF.

Short time series of collected ISD Index values (2015–2017) deems forecasting information society development unnecessary, because the forecast is bound to be unreliable. The authors therefore forecasted the dynamics of the values of markers and sub-indexes with maximum share in the calculation of ISD Index value.

One of those markers that affect significantly the current state and the rate of information society development in the RF is the share of companies that use ICT in their operations. The marker includes all the companies operating in all spheres of economy: primary and secondary sector companies, healthcare organizations, financial institutions, retail and wholesale enterprises, construction companies, companies working in tourism sector, cultural establishments, etc.

The marker in question is based on the following factors: number of PCs, servers and LANs, use of email and the Internet, and the share of companies that have their own websites. Table 1 contains the forecast values for the share of companies that use email in their operations:

In our view, the best-case scenario is more likely one because the yearly increase of ISD Index has been as high as 4.4% in the period from 2015 to 2017, which represents linear trend.

Table 1. Forecast values for share of companies that use email for 2018 and 2019, %.

Year	Forecast type	Forecasted value	
		Point Forecast	Interval Forecast (95%)
2018	Worst-case scenario based on quadratic trend	83,11	80,89–85,31
	Best-case scenario based on linear trend	91,77	89,20–94,35
2019	Worst-case scenario based on quadratic trend	79,73	77,53–81,94
	Best-case scenario based on linear trend	96,26	93,00–99,53

Source: compiled by the authors.

4 Discussion

There is a number of theories concerning information society. The term itself was first mentioned in the works of F. Machlup in 1962. The concept of information society stems from that of post-industrial society – the more universal one that its predecessors, because it describes a wide range of social and cultural changes and shows the vector of social development from pre-industrial through industrial to post-industrial society. The concept of information society, in its turn, focuses on the degree of information and communication technology (ICT) development and regards industry and use of scientific, technical and other kinds of information as the main driver of social development.

Often the terms "information society" and "post-industrial society" are used interchangeably, however this is not always correct: some post-industrial countries, like Canada, cannot claim to have a well-developed information society. Whereas in Japan and the USA information society is a fact of life.

Virtually every country that enters the information stage of development, has a national program of transition, like Finland, Russia, China and India, for example [6].

In Russia the concept of information society development is described in the following documents on the federal level: "Strategy of development of information society in the Russian Federation in 2017–2030" [12] and State Federal Programs "Information Society" [11].

We believe that information society as a statistical unit implies a mass phenomenon that characterizes the stage of societal development, where such factors as availability of reliable and complete information and use of state-of-the-art information and communication technologies (ICT) are crucial for further development of human beings.

Thus, there is a distinct global trend in modern statistics: issues, concerning estimations of digital economy and the degree of digitalization of society are becoming the focal point of research and Russia is no exception. However, at present there is lack of empirical research in this area in RF. Let us name a few of the most prominent and relevant ones. The works of Değerli, Aytekin contain statistical estimate of information technology development in the context of diffusion of innovations theory [3]; Vershinskaya, Alekseyeva compiled a rating of countries based on the readiness for information society [13]; Chinayeva reviews the experience of estimating advanced digitalization in a range of countries [2]. Digital economy development worldwide is the focus of research in the works of Kirshin, Mironova., Pachkova (2015) focus on BRICS countries [4]; Schlichter and Danylchenko focus on Romania, Cyprus and Estonia [9]; Njoh focuses on countries in Africa [8].

There is a range of works with comparative approach to the issue where the situation with information society in Russia is compared to that in other countries. Klochkova calculated the global rating of the RF based on the ICT Use Sub-index [5]; Smirnov and Mulendeeva apply the ICT Use Sub-index to evaluate the prospects of strategic alliances between the RF and the leading countries of the world [10].

On the regional level, Lapina carries out factor analysis of the level of regional readiness for transition to information society (excluding Moscow and Saint Petersburgh) [7]; Volkova focuses on geographical diversity of ICT Development Index (IDI) in the regions of the RF [14]; Adamadziev and Rabadanova research the dynamics and variations of the Information Society Indices (ISI) throughout the regions of the RF [1].

We still firmly believe that the number of research on complex statistical description of the current state of digital economy in the regions of the RF is insufficient.

5 Conclusion

Upon completion of the research, the autors came tot he following conclusions:

(1) The authors proposed an updated version of methodology for calculating ISD Index in the regions of the RF. Taking into consideration crucial importance of education for digital economy development, we included a new Sub-index into ISD Index formula, namely – the ICTU in Education Sub-index. It consists of a number of markers. The discrepancies in ISD Index values, calculated according to the methodology proposed by the Ministry of Digital Development, Communications and Mass Media of the Russian Federation and the values obtained by applying the revised methodology, suggested by the authors, reached 27,2%. This points out the fact that the markers included in ICTU in Education Sub-index have significant effect on ICT development.

(2) Dynamic analysis of the differences in regional ISD Index values demonstrates that where as in 2016 the leading regions were mostly those in the European part of Russia and the Ural, in 2017 appeared four more areas with high level of ICT Use: North-Western regions, South-Eastern regions around Moscow, the Ural/West Siberia and the Far East. Regions of East Siberia and North Caucasus show lowest ISD Index values with no positive dynamics.

(3) Key factors for increased ISD Index value in the RF are: increase in the number of PCs in healthcare organizations and increase in the number of cultural establishments that have their websites. Thus, it is the digitalization of social sphere that becomes the driver for ICT development in Russia.

(4) To understand current processes in this area, it is necessary to consider not only the differences between regional ISDI values, but also the speed of ICT development. We compared data for 2016 and 2017: the highest growth of ISD Index value (approximately 0.13 points) was recorded in the Northern and Eastern parts of the country – Novosibirsk region, Khanty-Mansi Autonomous Okrug, etc. The fastest fall, in general quite insignificant (approximately 0.07 points) was seen in the regions of European part of Russia (Mari El Republic, Kursk region, etc.) However, the original value in the regions that showed negative dynamics was very high, so it is incorrect to classify them as ICT development outsiders.

(5) ISD Index value in the RF was first calculated as recently as 2015, thus the time series is too short to make any reliable forecasts based on the collected data. The only marker of ICT development with sufficient time series for forecasting is the number of companies that use email, so the authors used it to predict the best-case scenario based on linear trend and the worst-case scenario based on quadratic trend for 2018-2019. The calculations show that the best-case scenario forecast with average annual increase of 4.4% is more likely. Forecast value for 2018 is 91.77%, and for 2019 it is 96.26%.

One of the goals of "Strategy of development of information society in the Russian Federation in 2017–2030" is the development of statistical tools for accurate measurements of its outcomes and for monitoring the markers of its achievement level. We believe that the updated ISD Index we proposed can be of use in achieving this goal and can become one of the key indicators for strategic documentation on the regional and federal levels.

References

1. Adamadziev, K.R., Rabadanova, R.M.: Evaluation of level of informatization of regions of Russia: dynamics, regional differentiation. Fundam. Res. **4**(2), 462–466 (2013). (in Russia)
2. Chinayeva, T.I.: Information and communication technologies and development of digital economy. Economist **6**, 61–67 (2018). [in Rus.]
3. Değerli, A., Aytekin, Ç., Değerli, B.: Analyzing information technology status and networked readiness index in context of diffusion of innovation theory. Proc. – Soc. Behav. Sci. **195**, 1553–1562 (2015)
4. Kirshin, I.A., Mironova, M.D., Pachkova, O.V.: Index assessment of readiness of the countries of BRICS group for information society. Proc. Econ. Finan. **24**, 318–321 (2015)
5. Klochkova, E.N.: Russia in the world information community. Questions Stat. **8**, 66–75 (2016)
6. Kostina, A.V., Khorina, G.P.: Information culture in the concepts of information society. Philos. Cult. **4**(52), 14–20 (2012)
7. Lapina, L.E.: Factor analysis of ratings of subjects of the Russian Federation on the index of readiness for information society. Anal. Model. Econ. Soc. Process.: Math. Comput. Educ. **20**(1–2), 96–104 (2013)
8. Njoh, A.J.: The relationship between modern Information and Communications Technologies (ICTs) and development in Africa. Util. Policy Energy Proc. **50**(C), 83–90 (2018)
9. Schlichter, B.R., Danylchenko, L.: Measuring ICT usage quality for information society building. Gov. Inf. Q. **31**(1), 170–184 (2014)

10. Smirnov, V.V., Mulendeeva, A.V.: Use of information and communication technologies: Russia in comparison with India. China US Econ. Anal.: Theory Pract. **18**(2), 308–326 (2019)
11. State program: Information society (2011–2020), approved by order No. 2161-p of 2 December 2011. https://digital.gov.ru/ru/activity/programs/1/. Accessed 04 Mar 2019. (in Russia)
12. Strategy of development of information society in the Russian Federation in 2017–2030, approved by the decree of the President of the Russian Federation from 09 May 2017 № 203. http://www.consultant.ru/document/cons_doc_LAW_216363/. Accessed 04 Mar 2019. (in Russia)
13. Vershinskaya, O.N., Alekseeva, O.A.: International indices of countries readiness for the information society. Proc. Inst. Syst. Anal. Russ. Acad. Sci. **61**(2), 19–25 (2011). (in Russia)
14. Volkova, N.N.: Index of ICT development of Russian regions. Econ. Entrepreneurship **4** (93), 1305–1309 (2018). (in Russia)

The Impact of the Digital Economy on the Development of the Stock Market in Russia

A. V. Vaulin and E. V. Pogorelova(✉)

Samara State University of Economics, Samara, Russia
jour.ru@gmail.com

Abstract. In the article, the author assesses the degree of influence of the development of the digital economy on the stock market in Russia. The author carries out a comparative analysis of the stock market indicators for 10 years and describes new mechanisms for working with securities that appear in the digital economy. In addition, the article describes methods and methods that have emerged due to the digital economy, which can increase the involvement of individuals in the economic development of the Russian stock market.

Keywords: Currency · Digital economy · Electronic digital signature · Exchange · Non-cash transactions · Private investment · Stock market · Securities

1 Introduction

In a market economy, the stock market is an important mechanism for attracting monetary resources for investment, modernization of the economy, stimulation of production growth. The Russian stock market belongs to emerging markets. Since its formation - since the early 1990s - quite a long time has passed, the market saw both crises and a period of stable growth. But in comparison, for example, with the market of the United States, the Russian stock market just started and is still very young. The Russian securities market and the associated financial and foreign exchange markets are gradually acquiring features typical of developed countries with a market economy, pursue the same goals, perform the same functions and use similar mechanisms. Statistical analysis of the main indicators of the dynamics and structure of the Russian stock market allows us to assess its general state, identify the main trends of further development, and compare the domestic market with the markets of other countries.

Modernization of business processes on the domestic stock market is an integral part of the reform of the Russian economy. This is due to the deep connection between the phenomena affecting the main sectors of the economy and their reflection in the financial sphere. The Russian securities market today is one of the fastest growing sectors of the domestic economy.

The key task to be solved by the securities market in Russia, from the point of view of the author, is to ensure a flexible intersectoral redistribution of investment resources, the inflow of national and foreign investments into Russian enterprises, the formation

S. I. Ashmarina et al. (Eds.): ISCDTE 2019, LNNS 84, pp. 225–232, 2020.
https://doi.org/10.1007/978-3-030-27015-5_28

of the necessary conditions to stimulate the accumulation and transformation of savings into investments.

Here there is a need to consider business processes in the stock market, under which the author proposes to understand the totality of consecutive and technologically related operations for the provision of financial products and (or) the implementation of a specific type of professional activity in the securities market [4].

Today, the problem of digital solutions in the stock market is one of their actual, as, in the opinion of the author; this part of the economy will become the strongest driver of the development of the entire economy of the country.

At the same time, digital transformation has already begun. This cannot be overlooked, since it affects all sectors of the economy and social activities, production, health, education, finance, transport, etc. [7]. The formation of the digital economy in Russia actually began with a message from President Vladimir V. Putin, in which he stressed the need to form a new web economy to improve the efficiency of industries through information technology [9]. Every day the spread of IT-solutions aimed at performing simple tasks, becomes larger and captures new, larger territories and markets. Information technology already densely entered all spheres of human activity and human knowledge. In particular, the use of information technologies in the business environment significantly expands the range of financial instruments and gives new opportunities to participants in the stock market.

The electronic information revolution and its product - electronic and digital economies also change the form of the organization of economic relations, institutions and organizations in the global space of a market (essentially capitalist) economy.

Information as the main factor of production in the form of modern technologies of the VI technological order (in this case ICT) has opened great opportunities for qualitative economic growth through the following tools and factors:

- Firstly, unlimited commercial sites on the Internet, the development of Internet trading, financial (stock and currency) exchanges;
- Secondly, reducing the size of companies for successful competition in markets, the development of horizontal management systems and the emergence of virtual enterprises (companies) and organizations, also called "cybercorporation";
- Third, reusing the same physical, labor and other resources to provide various services within the cloud infrastructure of the enterprise, specialized regional clusters of the digital economy and the digital ecosystem;
- Fourthly, the limited scope of operational activities is limited only by the size of the Internet [5].

It is also necessary to take into account the emerging trends in the world towards convergence of various types of financial systems (primarily the role of banking and non-banking financial institutions). In any case, it is already obvious today that Russia is in the process of intensively forming its own model of the market, which must absorb all the best that has already been created in the world.

This model can include any trading systems used on various exchanges and in various market segments, but it must be built on the basis of a single information space and market transparency. It does not exclude the use of various models for corporate, state and municipal securities.

Another problem that, in the opinion of the author, will help solve new digital technologies is an increase in the involvement of the population in the activity of the stock market and the formation of its own investment capital.

For example, in the US, according to a Gallop poll - 52% had assets in the form of securities. In Japan, the same figure is 39% of the total population of Japan. In Russia, according to the Moscow Stock Exchange, only 0.77% of the country's population participate in trading on the Moscow Stock Exchange as individuals. This indicator in Russia is 2 times less than in India. For a country that claims to be a world financial center, these figures are very low, so the government of the Russian Federation has developed a number of measures to attract the population to the domestic stock market [10].

Further, in the article the author considers how the digital economy influences the development of the stock market in Russia and the level of involvement of the country's population.

2 Methodology

To assess the impact of digital technology, data for several years were analyzed.

Based on the analysis, the author was convinced that in the period from 2007 to 2017 the number of clients of the Moscow Stock Exchange group increased significantly.

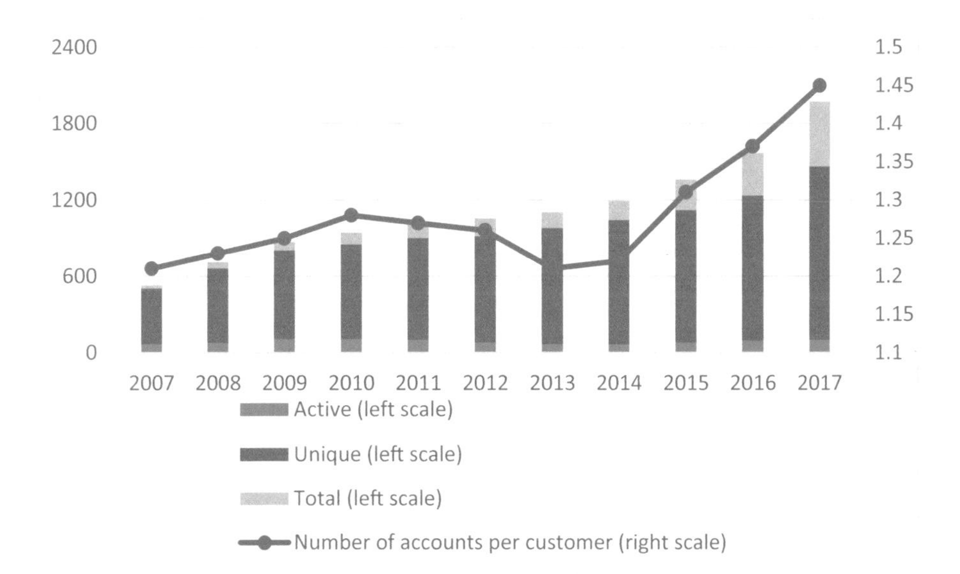

Fig. 1. Clients of the Moscow Exchange group, thousand people. Active (left scale) Unique (left scale) Total (left scale) Number of accounts per customer (right scale). (Source: authors, compiled on the base of NAUFOR calculations [8])

Figure 1 shows the number of customers registered in the Moscow Exchange group. The total number of clients of the Moscow Exchange group shows a long-term trend towards growth. In 2017, the growth in the number of unique customers accelerated to 19.2% (9.8% in 2016). At the end of the year, their number was 1360.8 thousand - 219.3 thousand more than in 2016.

For the period 2007–2017, the number of unique customers grew by 10.9% (CAGR). An important characteristic of the investor base is the number of so-called active investors, that is, those of unique customers who carry out at least one transaction per month. It should be noted that the niche of active clients in the general client mass is traditionally small in both relative and absolute terms. Thus, in the period 2007-2011 active customers averaged 15% of the number of unique customers. In 2012–2017 years this ratio has decreased to 7.7%. In the interval of 2007–2017, the number of active customers increased by 3.5% (CAGR). In 2017 there is an increased activity of investors: the average annual number of active customers was 101.8 thousand - an increase of 8.7% compared to the year before [8].

Such an increase in activity can be directly related to the development of digital technologies, namely the emergence of remote account opening, the execution of non-cash transactions through online systems, etc. At what increases not only the number of institutional investors.

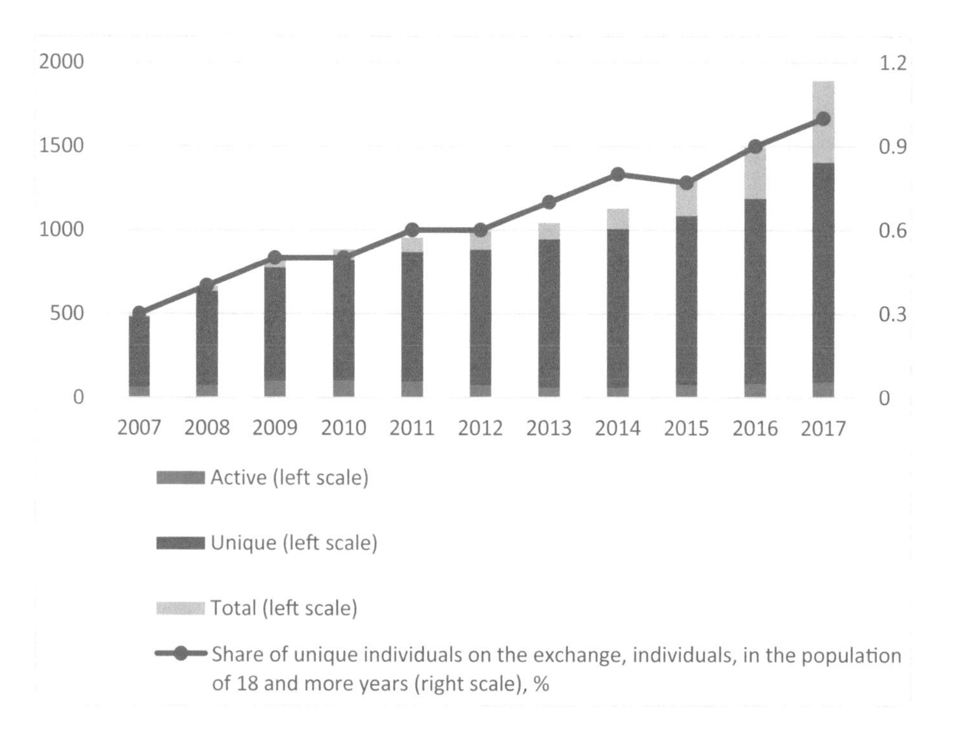

Fig. 2. Individuals residents - clients of the Moscow Exchange group, thousand people. Active (left scale) Unique (left scale) Total (left scale) Share of unique individuals on the exchange, individuals, in the population of 18 and more years (right scale), % (Source: authors, compiled on the base of NAUFOR calculations [8])

Figure 2 shows the number of clients of the Moscow Exchange group - individuals - residents. In the total number of unique clients, individuals - residents make up 96.4%. The number of unique clients of the Moscow Exchange group of resident individuals is constantly growing. In 2017, the number of such customers increased by 18.8% and amounted to 1310.3 thousand people. On a national scale, this is a very small amount - about 1.1% of the able-bodied population [8].

Author shows that with the advent of technology to work in the stock market interest in this area has increased significantly. At the same time, the share of participants remains minimal, which means that the industry's potential is enormous.

Confirmation that the development of digital technologies leads to an increase in the involvement of stock market participants illustrated in Fig. 3:

Figure 3 is a graph of the change in the number of financial transactions performed using remote technologies. Therefore, in 2007, remote channels accounted for only 8% of all banking and brokerage operations. In 2017, this figure is close to 60%.

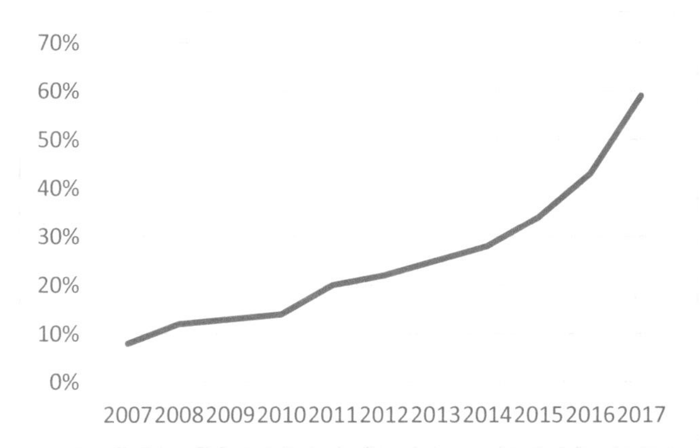

Fig. 3. Number of financial transactions performed through digital technologies, in % (Source: authors)

3 Results

As a result of the research, the author found that there is a direct correlation between the development of digital technologies and the involvement of the population in the stock market. In addition, not only the convenience of new applications attracts new users.

In the article "Calendar anomalies in the Russian stock", the author studies abnormal results on the Russian stock market and concludes that transaction costs significantly influence the calculation of yield on the stock market [2]. It is through lower costs that brokerage firms and banks can give lower rates for working in the stock market. It turns out that the reduction of tariffs along with the improvement of the distance channels of work substantially increases the interest in the stock market. In addition, this leads to an increase in the number of participants and turnover on the Russian stock market.

The results of the research also formed the author's opinion about the transparency of the financial system. In the article "State-owned enterprises in the Russian market: Ownership structure and their role in the economy", an in-depth analysis is made of the strong state participation in the activities of many public companies. This, in turn, has a negative effect on the development of a market economy [1]. Based on the data from this article, the author concludes that the development of digital technologies will result in the transparency of all business processes. Therefore, the state's share can be reduced because of the need for internal control.

4 Discussion

The theme of digitalization of financial services becomes one of the main themes between participants of the stock market: "Digitalization is narrowing the gap in financial knowledge between residents of cities and remote corners of Russia, the Director of business development "BCS Premier" BCS financial group Anton Grabarov: "Today, almost any citizen can, without leaving home, to buy and sell securities, to exchange currency, to open deposits, to manage your personal Finance, etc. recently, this seemed fantastic," he says.

Banks began to actively launch mobile applications for the purchase of shares with a simple and intuitive interface, convenient functionality and affordable tariffs for the purchase of securities.

Users of investment services got rid of the need to use special terms ("candles", "glass", options, futures, derivatives, bonds, funds, etc.), as well as to install terminals for buying/selling securities for trading on the Internet."

In the investment market, there is a need for a broker with a minimum "entry threshold", says Emeshev: "even the cheapest online intermediaries have a high trade cost, which makes purchases for small amounts too expensive. Banks have overcome all these difficulties and showed customers that it is easy to earn".

"The Digital channel has a huge potential. From there, we get a significant portion of new customers. In June of this year, we launched the mobile application "VTB My Investments" for individuals, combining products of brokerage services and services of the management company", - said the representative of the press service of VTB. The application has installed more than 24 thousand customers, of which about half are active customers with a total daily turnover of more than 3 billion rubles.

"To date, more than 30% of all applications for transactions with shares are processed through digital channels. The volume of attraction to mutual Funds through the online channel for the first half of 2018 exceeded the volume of attraction for the whole 2017 by 72%. The number of registered users since the beginning of the year is twice as much as at the end of last year" [6].

5 Conclusion

In one of the articles on the topic of neoindustrialization of the economy, the author writes: "… the problem of social prosperity of Russian people has increased especially because of the deindustrialization of the national economy as a negative consequence of unregulated market reforms. In turn, the prospects of achieving the European level of social prosperity in Russia can appear only under the conditions of neo-industrialization of the economy. In General, government efforts should focus on the development of a special form of social prosperity - the prosperity of the network based on recent progress [3]." Only the development of the digital economy will help Russia to reach the same level as the developed countries and achieve social and economic development.

In his study, the author showed that with the advent and development of the digital economy, the growth of the Russian stock market has accelerated significantly. This is reflected not only in an increase in turnover on the stock exchange, but also in an increase in the involvement of the population. Thanks to new technologies, such as online banking, wire transfers, digital signature, etc., the stock market opportunities become available even to remote regions. The author sees the following steps to further increase the involvement and financial literacy of the population:

- Improvement of the legal framework to improve financial transparency
- Implementation of broadband Internet access in all regions of Russia;
- Creation of the possibility of remote opening of brokerage and Bank accounts without the need to visit the office
- Use of digital technologies to train the population to work with finances

The author believes that first it is necessary to work out the legal framework, which will be able to fully is regulate all technological processes associated with the opening, maintenance and closure of brokerage and Bank accounts. The author is also convinced that the successful implementation of digital technologies can significantly increase the involvement of people in the stock market to at least 10% of the total population.

References

1. Abramov, A., Radygin, A., Chernova, M.: State-owned enterprises in the Russian market: ownership structure and their role in the economy. Russ. J. Econ. **3**, 1–23 (2017)
2. Caporale, G.M., Zakirova, V.: Calendar anomalies in the Russian stock market. Russ. J. Econ. **3**, 101–108 (2017)
3. Gasanov, M., Zhironkin, S.: Neo-industrial base of network social prosperity in the Russian economy. Proc. Soc. Behav. Sci. **166**, 97–102 (2015). https://doi.org/10.1016/j.sbspro.2014.12.490
4. Ivanov, M.E.: Increase of efficiency and modernization of business processes on the Russian securities market. Finan. Anal.: Probl. Solutions **24**(66), 30–33 (2011)
5. Ivanov-Shits, A.K., Aityan, SKh: Integration of Russia into the world economy and the globalization of stock markets. Bull. MGIMO-Univ. **6**, 1–9 (2009)

6. Information Agency of Russia TASS: Private investors increase investments in the stock market of the Russian Federation, saving themselves from low rates (2018). https://www.finanz.ru/novosti/aktsii/chastnye-investory-uvelichivayut-vlozheniya-v-fondovy-rynok-rf-spasayas-ot-nizkikh-stavok-1027460550. Accessed: 3 Mar 2019
7. Makarov, V.L., Bakhtizin, A.R., Burilina, M.A.: Prospects of the digitalization of modern society. Econ. Manag. **11**, 4–7 (2017)
8. National Association of Stock Market Participants (NAUFOR) (2017) Russian stock market, Events and Facts (2017). http://www.naufor.ru/download/pdf/factbook/ru/RFR2017.pdf. Accessed 3 Mar 2019
9. Negroponte, N.: Being Digital. Alfred A. Knopf, New York (1995)
10. Semenova, O.: Involvement of the population in the investment process (2017). https://bcs-express.ru/novosti-i-analitika/vovlechennost-naseleniia-mira-v-investitsionnyi-protsess. Accessed 3 Apr 2019

Trends in Optimizing the Formation of Consolidated Reporting in Holding Companies in the Context of Global Digitization

V. P. Fomin and E. S. Potokina(✉)

Samara State University of Economics, Samara, Russia
Potokinal@mail.ru

Abstract. In the environment of a rather abrupt change in Russia's economy development stages the issue of the influence of the form of functioning of a business on the financial indicators of its sustainability and development is attracting more and more attention. The study of the problems of holding companies functioning in various areas of the economy is of great research interest from the point of view of both theory and practice. The article examines the ways to optimize the electronic acquisition, processing, storage and application of financial, managerial and tax accounting data of holding companies. The tendency to further integration of financial and management accounting data has been revealed. A research of the current state of accounting software systems in holding companies has been carried out, discrepancies have been found out between the existing techniques of accounting data formation and the requirements of the digital economy. The analysis of the structure of the in-house electronic document interchange has revealed the need for introducing "smart digitization". For the purposes of this article publications of corporate reports of the largest holding companies were studied and the influence of non-financial information on the disclosure of consolidated statements was established. The requirements to reduce the consolidated financial statements publication time, provide their depth and fullness run into collision with the fragmented use of applied solutions. As a result of the study, a trend has been identified for the integration of accounting systems and consolidation into a single information platform or service center.

Keywords: Accounting · Consolidated reporting · Digitization · Electronic document management · Holding companies

1 Introduction

Holding companies are widely developed in European countries and in the West in almost every sector: production, industry, construction, trade, banking, insurance, etc. In order to enhance survival, sustainability and further development many modern organizations on their own combine their resources and transform into holdings. Being an important factor for operational and strategic decisions making by investors and shareholders the elimination of information asymmetries in the disclosed statements

S. I. Ashmarina et al. (Eds.): ISCDTE 2019, LNNS 84, pp. 233–242, 2020.
https://doi.org/10.1007/978-3-030-27015-5_29

affects the position of a holding on the stock market. Holding companies comply with high standards of reporting information disclosure and strive for maximum information transparency [1].

While examining the topic of trends in optimizing the formation of financial reporting indicators in holding companies in the environment of digitization of economy, we should note that this area of research is dynamically developing, new tools, technologies and resources are developed and need to be researched and tested in practice.

In the book "Accounting Theory" Hendriksen and Van Breda (1992) concluded that the accounting development was stimulated by technical progress, describing the future of accounting as follows: "The accounting registers will be replaced by databases, some of those will be financial information… users have acquired instant access to account information of any level of generalization, as well as the ability to define and modify data presentation formats" [3].

The important role of the functioning of a business or a business conglomerate structure has been pointed out by a lot of researchers: Shitkina [12], Voltchik, Bereznoy [13] and Savranskaya [11] who formulated the main advantages of holding structures in the form of a synergistic effect from the pooling of resources and the ability to plan and redistribute the tax burden within enterprises, thereby optimizing the consolidated financial statements.

The relevance of this topic is confirmed by the annual scientific and practical conferences and the validation by the Government of the Russian Federation in 2017 of the program "Digital economy of the Russian Federation". There is an increasing amount of studies by Russian researchers devoted to the problem of preparing and transferring of accounting disciplines to the knowledge economy and introducing new digital platforms. However, not many works are devoted to integrated research and overcoming new challenges on the way to modernization and adaptation of corporate financial accounting in the global digitization period.

2 Methodology

The study employs the analytical method, the method of interdisciplinary synthesis and the logical method. The main research methods:

First of all, we applied the method of analysis of the main factors influencing the formation of a corporate accounting system in holding companies. At the next stage we monitored the adaptability of information and technology systems for the formation of consolidated reporting to the expectations of investors and disclosed statement end users by means of interdisciplinary synthesis from the point of view of accounting, management and information technology.

At the final stage basing on the observation method (i.e. the perception of economic processes in their real form) and the logical method we predicted trends in the optimization of accounting systems used to consolidate financial statements based on integration, the best fitness to business needs and changes in the regulatory structure.

3 Results

3.1 Resources for Enhancing the Performance of the Process of Acquisition, Processing, Transmitting and Controlling Accounting Data in Holding Companies

The information process in the system of corporate financial and tax accounting is the process of acquisition, processing, accumulating, storing, searching and distributing accounting information and reported data.

In an automated information system, the process of managing accounting data is carried out by means of technical applied solutions, special data processing methods in order to generate information for a specialist-user. The tax code obliges taxpayers to run tax registers, as well as to submit electronically a declaration on certain types of taxes.

In Russia the market of software for automated accounting offers solutions of several key software-companies. As a rule, various types of records are kept on the same information field, the data recorded in accounting programs are processed in parallel in accounting, tax and management accounting.

Professor Fomin [2] states in his research that “to solve modern managerial tasks much more information is required than the existing reporting is able to provide. The absence of required data and misrepresentations in information sources lead to the necessity to adjust the information technologies mechanism, as well as to appropriately adapt the scientific apparatus of economic analysis, which implies formation of new demanded in practice qualitative characteristics of economic entities and their quantitative expression in figures”

In 2010–2017 there was a rapid development of information technology in the field of accounting, meanwhile the development was aimed at the creation of sector solutions – production, construction, retail, vehicles. At the same time, the market of services for enterprises is maintained at the proper level and includes setting up the program, updating the program, servicing, consulting, and training sessions on keeping records in this program. This work is aimed at the formation of special competences of accountants. Special competences are connected with a certain professional activity area [9].

In accordance with the approved program of the economy digitization, further development of the information space requires developed technologies, infrastructural environments, regulations, and qualified personnel.

In practice, holding companies are established in order to consolidate the results of the economic activities performed by a group of entities for increasing the competitiveness of their products, goods, services and maximizing profits by dividing functions, which ultimately leads to their total tax payments decrease. That is a typical strive for obtaining greater profitability at a lower cost. This goal is achieved due to transfer (internal) prices, which, unlike market ones, allow to compensate for the losses of some divisions of the holding company with the profits of other subsidiaries. As one of the features of transnational holding companies functioning we consider the sublimation of various management cultures and mentalities in general, mutual penetration and the imposition of different approaches in accounting through the concentration of capital, the desire to adapt the entity to changes in the external environment and their internal controllability [6].

The process of reporting data generation is preceded by the stages of collecting, transmitting and controlling accounting information. Below we will highlight some features of these processes in holding companies:

- horizontal exchange of a large number of internal reporting between the enterprises that are part of the holding company;
- vertical exchange of information and reporting data which goes from subsidiaries to the parent organization;
- "short" time of closing the period before issuing the monthly management reports, which accelerates the final indicators control process and may cause the reporting data quality decrease;
- advanced modelling of the monthly, quarterly, yearly results for determining the financial deliverables and tax burden of each intercompany enterprise and the parent organization itself. Detection of the plan/fact analysis deviations from the predicted indicators.
- because of insufficient standardization of management reporting forms, most of this type of reporting is not generated in applied accounting programs having built-in blocks of control ratios and reconciliation of indicators between financial and tax accounting, but in Excel programs. This means that accountants have to maintain another database or manually transfer accounting and analytical data from one program to the another;
- adaptation to the requirements of international accounting standards of the parent company determines the need for subsidiaries to maintain a large number of additional management and financial reporting registers, taking into account the cost and quantitative deviations associated with differences in accounting standards, which increases labor costs;

A way to improve technological effectiveness of electronic processing of accounting data can be the elimination of digital inequality between enterprises of the holding, as well as improving digital literacy. For many holding structures geographical branching is typical.

Head organizations tend to be situated in large cities and federal centers, manufacturing and service facilities may be located in rural areas. This causes differences in infrastructure and the level of qualification of personnel responsible for the formation of accounting and analytical data, which slows down the process of document circulation, information exchange and making decisions ultimately affecting the reporting data formation.

The evidence from practice shows that at the stage of active business development or business processes diversification the problem of building effective document circulation between the companies of the group is becoming more acute. Together with standardized incoming and outgoing documents, corporate ones are also generated. The introduction of an internal electronic document management system in holding companies will reduce bureaucracy, establish direct contact between the sender and the end user of data, and increase the speed of information exchange between legal entities integrated into a holding company.

3.2 The Adaptability of Accounting Programs to the Expectations of Investors and Disclosed Statement End Users

Today's financial environment could not be better characterized than by the saying "time is money." Accountants, financiers, analysts face the task of reducing the time of issuing analytical information while the quality of the data provided is still important. Managers and officials in charge want to be able to respond quickly and make decisions based on reporting data. The speed of response to informational changes is the key priority.

The increasing volume of the accounting data must be checked, classified and input into the system. For transnational holdings or holding companies, the correct classification of business operation items is of paramount importance. Errors in the primary classification lead to the reporting data misinterpretations [8]. One should not ignore the requirements of owners and investors for the corporate accounting - increasing its analyticity, visualizing reports in the form of graphs, diagrams, availability of reports at any time from any location of the investor.

Despite the availability of more enhanced technical and financial resources, the process of providing accounting programs for holding companies is considered more complicated. One way or another, tax and management accounting should be implemented. Accounting of the parent company is generally maintained according to the standards of RAS and IFRS, which necessitates the integration of the subsidiaries' accounting programs into the corporate space with the option of transforming the reporting date into IFRS.

One of the ways to transform the RAS data into IFRS statements is the manual adaptation of the reported data using MS Excel spreadsheets, which is performed in several stages. Each intragroup enterprise forms a turnover balance sheet in the standard accounting program, analysis of subcontume accounts separately in turnovers with group companies and third parties, translates this data into an individual report in MS Excel, sends it to the parent company, where the reports are checked, combined and consolidated. The consolidated data is re-classified with the help of mapping by IFRS criteria. This kind of transformation has significant drawbacks - reports in MS Excel spreadsheets are checked using control ratios, introducing any changes in previous reporting periods may lead to a distortion of the initial balances. There is no opportunity to see the specification of the reporting data in MS Excel, since they are collected from scattered files. This method of forming consolidated statements is suitable for enterprises that use standard accounting programs 1C: Enterprise and 1C: Accounting. The analysis allows us to assume that the presence of custom blocks to reflect business operations and the formation of accounting and analytical information is insufficient, which makes the process of forming analytical data for managing corporate taxation inflexible. The resulting reporting information does not take into account the requirements for visibility and visualization mentioned above.

A more reliable way is to conduct parallel accounting in the information system, in programs that support accounting under RAS and separately under IFRS. Relatively new applications for accounting automation in holding companies (for example, 1C Accounting Corp.) are characterized by IFRS accounting support, automation of parallel accounting objects the accounting policies for which differ from those of RAS. For large holdings, along with accounting, a treasury, payment management, finance, budget management, reporting under IFRS, management accounting, production planning, production scheduling, inventory management under production orders, etc. have been introduced. Increased time and labor costs caused by the necessity to perform the same operations in different programs at once are a natural disadvantage of this automation.

The functionality of "1C: ERP" applications developed for holding companies allows to automate several types of accounting (accounting RAS, tax and management accounting, IFRS) using the transformation method. The transformation method involves preparation of statements under IFRS through by introducing adjustments into the turnover balance sheet statements under RAS (reclassification of reporting items, estimated adjustments). The disadvantage of this method is that during transformation it is necessary to create a number of tables reflecting correction data from Russian articles to the corresponding IFRS indicators.

From the point of view of the information usefulness for the investor, the consolidated statements have the following main advantages over the individual reporting in a group company [5]:

- the notes to the consolidated financial statements show the management structure/ownership of the group;
- the consolidated statements allow to estimate the amount of «overpayment» for the acquisition of subsidiaries (goodwill reporting item);
- the capital of the consolidated company reflects the share of non-controlling shareholders which is the part of the retained earnings and reserves that is not owned by the shareholders of the parent company;

The intragroup transactions between the group companies are eliminated, as are intragroup remaining balances. Consolidated statements solely reflect the results of operations with third parties, therefore we can exclude the possibility of a «paper» increase in financial results (for example, due to the sale of assets at inflated prices between companies of the group) and balance currencies (accounts receivable and payable between companies of the group for the purchase and sale of assets overpriced).

In the digital information space holding groups actively interact with the external environment in the face of regulatory bodies. They submit reports to the Inspectorate of Tax Inspection, Social Insurance Fund, Pension Fund of Russia, Federal State Statistics Service, Federal Service for Alcohol Market Regulation; besides, they exchange primary documents, participate in tenders, confirm the 0% VAT rate in electronic form.

Declarations are submitted to tax inspections in electronic form through a special electronic document management operator. The platform for the exchange of electronic documents was launched in 2011. Also, in 2011 the Ministry of Finance adopted the Decree №.50n «On approving the procedure for issuing and receiving invoices in electronic form via telecommunication channels using electronic digital signature» .

For holding companies, submission of tax and financial statements in electronic form allows: to save time, to reduce technical errors; increase the speed of updating reporting formats. Besides, it guarantees the delivery of documents, prevents unauthorized reviewing and adjusting the documents, allows to issue in electronic form the certificate of budget settlements status and other documents.

The software systems «SBIS» (by «Tensor»), «Kontur-Extern», «Astral Otchet», «Sprinter» (by «Taxikom») and others – are designed to interact with accounting programs, which allows to submit reports prepared in electronic format to any tax inspection of the country. It is convenient for the companies inside the group because it allows carrying out preliminary control of declarations and making the necessary adjustments before finalizing the final management reporting and data consolidation for the holding company.

The complex follows up control ratios not only within a single declaration; it also checks the relationship between the declarations and statements. For example, «SBIS» system allows to check the correspondence between the indicators of revenue in the VAT declarations and incomes in the income tax declaration; in case there are discrepancies in the data, the system will issue a warning.

Amendments to Law № 54-FZ «On the application of the cash register equipment» [10] set the task of online transfer of fiscal cash register data to tax organs. Electronic document operators play the role of fiscal data operators. They provide services for connecting cash registers in the tax service, connecting cash register to the Internet, changing the pattern of work of holding companies with technical support centers.

Most holding companies, especially those placing their shares on the stock exchange or planning to go IPO, undergo a mandatory annual audit. The settlements with the budget are also subject to auditing. The intragroup companies accordingly request from the tax authorities information on the status of budget settlements. In case of discrepancies identified they make adjustments to their statements or send out notifications confirmed by the necessary documents (declarations, payment orders) for introducing corrections in the budget settlement card.

The taxation transfer into the digital environment poses technological challenges for their processing. The new «Big data» [7] system follows up all the Value Added Tax transactions. This caused the need to upgrade the existing IT infrastructure of regulatory bodies. However, holding structures also have to conduct the audit of the state of information and digital facilities and infrastructure.

4 Discussion

Analysis of these factors showed that in the current operation conditions holding companies need information technologies with fundamentally enhanced orientations and aimed at the integration processes of accounting systems. To solve this problem one can introduce a single information platform or information processing center for the enterprises of the holding group.

We can mark the methods of optimization the formation of consolidated reporting:

- recommend for group enterprises the usage and accounting into a single, unified program or the unified information platform
- development of accounting policy and accounting subconto that maximally approximates the requirements of national standards of the group's enterprises;

- development of accounting policy and chart of accounts according to IFRS;
- development and approval of the methodological basis for individual reporting of the group's enterprises under IFRS, considering industry specifics;
- the unification of the individual reporting forms;
- obtaining full information about the current and future activities of the group's companies;
- corporate seminars and round tables.

Analyzing the processes of implementing accounting information technologies in large holding companies, one should highlight a number of advantages which include a more flexible system for entering accounting data, customizable analytics, the option of maintaining a multi-level system of revenues and expenses, meeting the need for automation of cash flow management and control over budget execution and investment.

The features of the issuing of accounting and analytical information in holding companies affecting the trends of accounting are determined. A greater volume of information transmitted and the accelerated pace of business operations processing in holding companies in comparison with other enterprises cause a tendency to justifiably invest in digital technologies to increase accounting transparency. The immediate task is to convert more internal business processes into electronic form, to switch to the non-documentary collection of primary documents, which is very important for holdings with their intercompany transactions.

It has been established that adaptation to the requirements of international accounting systems, the formation of requests from foreign audit organizations require additional forms of financial and management reporting. In this regard, we can consider the development of the branch in management accounting as "intermediate" between RAS and IFRS and, therefore, the standardization of electronic registers for the formation of analytical and reporting data. Strengthening the impact of non-financial information on financial accounting in the future will lead to the formation of new forms of reporting data with a deeper level of disclosure [4].

It has been established that development takes place both in taxation itself and in applied technological processes, the effectiveness of tax administration largely depends on the quality of the acquisition, processing and efficient use of global information arrays. It has been determined that the tax authorities, in carrying out the functions of tax control entrusted to them, are oriented towards identifying tax dodgers and discard the policy of total control. It was determined that the state, in the face of regulatory bodies, is shaping a tendency to further reject paper documents, develop "cloud technologies", digital infrastructure environments, increase digital literacy and enhance information security.

5 Conclusion

As a result of the study, prospects for the development of electronic processing of financial and tax reporting indicators in holding companies in the environment of digital space were identified. The authors of the study come to the conclusion that investments should be made in data processing centers, systems for acquiring and storing accounting and analytical information.

In the future holding companies will be able to overcome digital inequality and increase digital literacy within subsidiaries. The necessity has been determined to develop horizontal and vertical communications in holding companies, with the classifying accounting programs into the accounting systems and accounting management systems and the transfer of basic management functions to the parent organization.

The conclusion is formulated about the necessity to increase the efficiency of accounting and financial teams of holding companies by means of launching "smart" digitization of basic business processes, where one intellectual program could model, forecast and recommend the most profitable actions of another. The need for further scientific research of management accounting and management reporting forms standardization has been identified.

The forecast for further digitization of tax accounting of corporate business will make the remote electronic interaction of corporate business with tax authorities more transparent.

The increase in the openness of tax authorities using an electronic data processing system with a powerful potential for automating tax administration has been marked, the information base and analytical capabilities of inspections have been increased; new automated methods of work with taxpayers were introduced which minimize direct communication of the tax inspector with the accounting department of the legal entity, meanwhile the monetary component of the effectiveness of tax audits has been increased.

These recommendations will reduce labor costs in the process of acquiring and processing accounting and analytical data, improve the quality of automatic generation of reporting data, reduce the time to find the necessary information, improve the quality of management decisions based on the generated analytics, facilitate the formation of registers on differences between RAS and IFRS in the process of financial audits.

In the future, new approaches to the creation of a modern digital space will contribute to the dynamic and innovative development of the economy, improving work efficiency and increasing the competitiveness of companies.

References

1. Amiram, D., Owens, E., Rozenbaum, O.: Do information releases increase or decrease information asymmetry? New evidence from analyst forecast announcements. J. Acc. Econ. **62**, 121–138 (2016). https://doi.org/10.1016/j.jacceco.2016.06.001
2. Fomin, V.P.: Methodology of formation and analysis of balanced indicators of economic entity development. Dissertation, Samara State University of Economics (2008). (in Russia)
3. Hendriksen, E.S., Van Breda, M.F.: Accounting Theory. Irwin (1992)
4. Kim, O.: The IFRS adoption reform through the lens of neoinstitutionalism: the case of the Russian Federation. Int. J. Acc. **3**(15), 345–362 (2016). https://doi.org/10.1016/j.intacc.2016.07.001
5. Krimpmann, A.: Principles of Group Accounting under IFRS. Wiley, West Sussex (2015). https://doi.org/10.1002/9781119044826

6. Lin, K.Z., Cheng, S., Zhang, F.: Corporate social responsibility, institutional environments, and tax avoidance: evidence from a subnational comparison in China. Int. J. Acc. **52**(4), 303–318 (2017). https://doi.org/10.1016/j.intacc.2017.11.002
7. Matveeva, T.V.: Big data technologies in tax administration (2016). https://www.nalog.ru/rn77/news/activities_fts/5979856/. Accessed 3 Mar 2019
8. Nobes, C.: Lessons from misclassification in international accounting. Br. Acc. Rev. **50**(3), 239–254 (2018). https://doi.org/10.1016/j.bar.2017.08.002
9. Obuschenko, T.N.: Define the basic, special and the key professional competence. In Osipova, L.Ya., Orlova, L.V., Belyaeva, L.N., Sabirova, G.T. (eds.) Actual Problems of Economics, Management, Marketing, Collection of Scientific Materials of Conference in Samara University of Management, pp. 184–188. SIU, Samara (2017)
10. Order of the State Duma of 22.05.2003 N 50-FZ: About the application of the cash register equipment. http://www.consultant.ru/document/cons_doc_LAW_42359/. Accessed 3 Mar 2019
11. Savranskaya, M.V.: Management of industrial holdings based on reproduction approach. Dissertation, Khyban State University (2008). (in Russia)
12. Shitkina, I.S.: Holdings: Legal regulation of economic dependence. Management in groups of companies. Wolters Kluwer, Moscow (2008). (in Russia)
13. Voltchik, V.V., Bereznoy, I.V.: Interests of groups and quality of economic institutions. Econ. Vestn. Rostov State Univ. **5**(2), 62–66 (2007). (in Russia)

Special Economic Zones as Instrument of Industry and Entrepreneurship Development

A. V. Streltsov, G. I. Yakovlev, and N. V. Nikitina(✉)

Samara State University of Economics, Samara, Russia
nikitina_nv@mail.ru

Abstract. The features of the organization and advantages of special economic zones necessary to attract mobile industrial capital of entrepreneurs from different countries in certain areas are considered. The problems hindering the wide development of special economic zones (SEZ) for the Russian economy, with the allocation of economic, administrative, regulatory component are revealed. The expediency of development of the system of division of risks of the current activity between residents of SEZ, their managing companies, regional and Federal authorities is shown. It is necessary to develop both incentive measures for resident enterprises and a mechanism for monitoring their functioning in the SEZ, which provides for the regulation of the profile of their residents ' activities, assessment of their impact on the level of competition in the market, elimination of cases of oppression and closure of competing domestic enterprises, dependence of the Russian industry on transnational companies.

Keywords: Integration · Risk · Resident · Special economic zones

1 Introduction

The most important potential tool for the integration of domestic industrial enterprises into global reproduction chains are special economic zones (SEZ), which have long been used in many countries, the most successful in China, Taiwan, India, Brazil, Mexico, the United States, Singapore. Many experts note a fairly rapid increase in the number of these zones. Volkova notes that in the early 1990s. such zones around the world have already been created more than 1000 [17], representing more than 25 varieties that differ depending on the characteristics of the home country, where up to 30% of the world trade turnover of products and services is carried out. In Russia, the pace of development of these zones is far behind that of the leading countries. Against the background of "normal" conditions of activity in the national economy, the SEZ format is distinguished by an increase in innovation and entrepreneurial activity of resident enterprises. The emergence and development of these zones is associated with the strengthening of the processes of integration and internationalization in the world economy. The economy of any country can function effectively only when a combination of a wide variety of types of economic organization, a multi-layered structure and freedom of business. It is impossible to find one universal form, equally effectively

S. I. Ashmarina et al. (Eds.): ISCDTE 2019, LNNS 84, pp. 243–251, 2020.
https://doi.org/10.1007/978-3-030-27015-5_30

implemented in different types and models of entrepreneurship inherent in different countries.

2 Methodology

Foreign researchers have been studying the nature of the activities of special economic zones for a long time and they put forward several concepts that underlie the existing approaches and methods of organizing production in local territorial entities. The importance of research in this area is confirmed by the scientific provisions set forth in the works of such scientists as Varnavsky [16], Sapir [9], Reznikov [8], Kashbraziev [3], Hernández and Pedersen [2], Strange and Magnani [11], Laplume, Petersen and Pirs [5], Starostina [12]. The attention of modern researchers focuses on such problems as structural shifts in cross-border production chains, value creation and multinational production, which required appropriate clarification of the methodological and categorical apparatus.

In domestic literature this topic is still widely unexplored. It is no coincidence that there is currently no single classification and typology of special economic zones. In the works of Smorodinskaya and Kapustina [10], Streltsov and Yakovlev [13], signs of classification of FEZs are by four main criteria: by nature of activity; by degree of integration into the world and national economy; by type of activity; by nature of ownership.

Describing the current state of world economic relations, Kondratyev speaks about a new model of international integration and the change of the main players, the disintegration of the material component of the value chains, stimulated by the technologies of industry 4.0, the differentiation of ways of growth of developed economies and increasing the share of services in international trade [4]. At the same time, Streltsov points out that SEZs are able to ensure the implementation of various goals for the development of the national economy, to improve the efficiency of a particular zone, it is necessary to clearly define its goals. They should form requirements to investors and a set of benefits for the constituent entities [14].

When organizing the SEZ, it is especially important to take into account three organizational and technical components:

(1) location of the territory;
(2) availability and condition of production, social and transport infrastructure;
(3) the set and timing of benefits.

The last factor is particularly important for SEZ residents. As a rule, the benefits are differentiated depending on the implementation of a number of requirements, primarily on the volume of investment of firms-applicants in the zone. A significant characteristic in favor of SEZs is also the establishment of a special customs regime in them.

3 Results

Developing countries often see their integration into global value chains as an opportunity to increase the value added of products and enhance competitive positions by improving the environment for international business and attracting foreign investment. Over the period 1995-2009, the level of countries ‘ involvement in global value chains increased by an average of 5-10%, and today this trend continues [6]. About 40% of the OECD countries ‘ exports are value added generated abroad. Since 1995. South Korea, India and China, whose participation index (GVC Participation Index1) ranged from 10 to 20%, improved their positions in the global value chain most notably. The average share of value added of services in the gross exports of OECD countries and their partners has also increased during the 15 years under review [7]. The largest economies of the EU (Germany, Great Britain and Italy), as well as India and the USA showed the greatest growth of this indicator, in gross exports of which the added value of services occupies an average of 40–50%. For Russia, this figure remained almost unchanged – at the level of 30%.

With the aim of embedding the global in the reproduction chain in our country SEZs were established in the early 1990-ies. In the initial period, this process was more image-political than economic in nature. In the context of the break-up of economic relations of enterprises after the collapse of the USSR, the disorganization of national economies, the emergence of problems of confidence in the power of the SEZ did not fulfill the functions assigned to them for the technical development of industry and the production of competitive products.

The second stage of development of the SEZ began with the 2nd half of the 2000s. OS-new their formation was the Federal law of the Russian Federation of 22.07.2005 № 116 - FZ “On special economic laws in the Russian Federation” [1]. The SEZ management Agency was established as the governing body. However, the insufficient elaboration of this concept of the SEZ development strategy also did not allow them to meet their expectations. Therefore, it is reasonable to 2016 was discontinued on the establishment of SEZ and territories of priority development, and management functions of the SEZ was re-given to regional authorities. In 2016. the decree of the Government of the Russian Federation terminated the activities of 8 SEZs recognized as ineffective. According to the control Department of the President of the Russian Federation, from 2006 to 2016 186 billion rubles were spent on SEZs, of which 24 billion were not mastered, and the amount of tax and customs revenues amounted to only 40 billion rubles [11]. At the same time, 525 residents (104% of planned installations) have already been registered in the SEZ, 95 companies with foreign capital, with a total investment of more than 221 billion rubles [15].

Due to the specifics of the Russian economy SEZ is divided into:

- industrial and production, located in the territories of large industrial regions, the main competitive advantage of which are price factors due to the optimization of production costs;
- technical and implementation, the main purpose of which is the generation, development and commercialization of innovative developments, digital technologies;

- tourism and recreation, involving extensive development of entrepreneurship in all areas of tourism;
- port and logistics, allowing to increase the volume of foreign trade activities.

The analysis of the peculiarities of functioning and development of Russian SEZs allowed to systematize the identified problems and offer directions of their solution (Table 1).

Table 1. Problems in the operation of domestic SEZs and ways to solve them

№	Problems in SEZ activities	Problem description	The way to resolve the problem
1	Provision of tax benefits and formation without prejudice to the revenues of regional budgets	Benefits are the main driving element in attracting investors to SEZs - the greater their value, the more attractive this SEZ becomes for the investor. In modern conditions of severe financial restrictions, the need to fulfill social obligations against the background of macroeconomic instability, both regional and Federal authorities do not go to increase benefits. And without this, domestic SEZs lose in investment attractiveness to many SEZs in the world, especially in comparison with the countries of South-East Asia	A balanced approach to the nomenclature and procedure for granting benefits, the use of public-private partnership mechanisms. If these benefits are formed at the expense of the Federal budget, then at the highest levels of the hierarchy of economic management it will be easier to compensate for the shortfall in income due to the high diversification of their sources of income. The lack of benefits can be compensated by a large organizational support from local/regional authorities, improvement of business infrastructure and business culture
2	Insufficient amount of investment attraction	Investments, especially foreign ones, go where there is low risk, stability and sustained growth. Unstable rates of economic growth in the Russian Federation, the economy's dependence on the global hydrocarbon market, the conditions of sanctions restrictions do	Creation of favorable investment and business climate, personal work with each investor, assistance in localization of production, development of stable social and economic situation, increase of incomes of the population, provision of guarantees,

(*continued*)

Table 1. (*continued*)

№	Problems in SEZ activities	Problem description	The way to resolve the problem
		not provide an acceptable low level of risk. There is no possible compensation in the form of increased return on invested capital	insurance create attractive conditions for investments
3	Lack of a capacious sales market and availability of resources	There are low and unstable growth rates of the Russian economy. Foreign investors when choosing SEZs to place their investments, in addition to benefits, are guided mainly either by a large sales market or by the availability of cheap resources. Currently, in the global practice there is a sufficient number of SEZs with cheaper resources, especially labor, than in the Russian Federation	A huge reserve for accelerated economic growth lies in the Russian society itself, its current unclaimed intellectual and business potential, reserves for increasing labor productivity, which requires a radical improvement in the business climate, the accelerated development of small and medium-sized businesses, the turnover of which grows to 10% annually
4	Insufficiently developed SEZ management system both at the Federal and regional levels	Russian SEZs are insufficiently specialized, compete with each other, which in General, reduces their overall competitiveness in comparison with the SEZs of other countries. The practice of successful domestic SEZs shows that the effectiveness of their activities directly depends on the level of management and "supervision" by the authorities	It is important that the regulation of their activities is carried out in addition to the Federal functional ministries (departments), at the level not lower than the Governor of the region or even at a higher level, and their activities should be accompanied by authorized agencies
5	Increased competition and displacement of domestic producers from the market	Very often, foreign investors do not try to create new innovative enterprises within the SEZ, but seek to duplicate already existing	Need a Mature and balanced industrial policy to foreign investors rather tried to create an SEZ is a new innovative enterprises, without

(*continued*)

Table 1. *(continued)*

№	Problems in SEZ activities	Problem description	The way to resolve the problem
		production in the domestic industry. They create enterprises that are more often focused on the intermediate consumer, in priority with foreign participation in the capital, which has either a large market share or high profitability. There is a dependence of the enterprise – intermediate consumer from transnational corporations. The increase in the scientific and technological level of production of the host country is also very controversial	prejudice to existing domestic industry production and not imported into Russia outdated equipment and technology, to crowding out of domestic producers. In addition, foreigners often implement an effective strategy of penetration into the target market – "on the shoulders of a competitor", acquiring its production assets, and promoting their successful brands to the detriment of scientific schools, innovative developments of Russian scientists and designers

Source: compiled by the authors.

The identified problems have the consequence that the already implemented fairly large budget funds invested in the creation and development of the SEZ have led only to separate, point results, and in the regions and so are industrially developed. For the depressive territories in which it was supposed to use the SEZ as "points of growth", they almost did not justify themselves. In this regard, the proposed directions include tax incentives, improving macroeconomic conditions, creating the necessary infrastructure, further harmonization of technical and management standards with the world, ensuring the stability of the regulatory sphere, creating attractive conditions for potential investors [18]. Their implementation will make it possible to use the SEZ as an effective tool for the development of industry and entrepreneurship.

4 Discussion

The analysis of the activity of the only SEZ in the Samara region – the Tolyatti SEZ established on August 12, 2010 is indicative from the point of view of revealing the features of successful SEZ practice. As of 2016 in the SEZ "Tolyatti" functioned 6 production LLC "Nobel automotive RUSIA", LLC "This automotive Rus", LLC "AtsumeteToyota tsuse Rus", LLC "hi-Lex Rus", LLC "of edsha Tolyatti", "praksayr Samara". Of the 6 named 5 industries produce parts and components for the automotive industry, for the pipeline of JSC "AVTOVAZ". During their work, all companies have jointly created 574 jobs.

Currently, it is planned to start work in the SEZ a few more residents who are already working on the creation of enterprises on the territory of the SEZ with different degrees of readiness: LLC "Togliatti paper factory" (planned 150 jobs), LLC "Solo-filmz" (planned 50 jobs), LLC "Porcelain" (planned 575 jobs), LLC "Samara plant of medical products (planned 85 jobs), LLC "Prodmash-Composite" (planned 150 jobs). In the future, the arrival of other investors is planned.

The results of the 6-year operation of the SEZ show that it was not possible to diversify the economy of the city on the basis of the creation of the SEZ (5 out of 6 resident enterprises belong to the automotive industry, as well as VAZ), the number of jobs created is extremely small compared to the planned. Output is also significantly lower than expected. From the point of view of the goals set for the SEZ as a whole, they have been unattainable for 6 years so far. Regarding the increase in the innovative level of production, it can be noted that, for example, the equipment of LLC "Edsha Togliatti" is not new, as it was transported to the SEZ "Togliatti" from the Czech Republic, from the Czech division of the German company "Edsha holding GmbH". The size of the exports of SEZ residents is negligible. According to the assurances of the management of LLC "Edsha Togliatti" the company is only studying the possibility of export, but even if it is decided on its purpose, it is required that the products of the enterprise should not create competition to other units of "Edsha holding GmbH".

If to analyze the composition of the planned location of the new re-sidenav area, all of them are domestic enterprises. Their products are focused on the domestic market, and they can not have any influence on the position of Russian industry in the world markets. In addition, a more detailed motivational analysis is needed: these enterprises would plan projects for the creation of production capacities in the absence of the SEZ format. If so, their investment in SEZs is a motive for probable tax evasion and, accordingly, the loss of the regional budget.

5 Conclusion

Taking into account the problems of the effective functioning of SEZs, it is necessary to begin again to develop the concept of their creation to ensure the activities, which should take into account both the competitive advantages and disadvantages of individual operating SEZs, and more clearly identify their specialization.

- A special system should be developed to attract investors to the SEZ, which should operate at the Federal level, with a mechanism to exclude competition of individual SEZs for investors.
- To increase the investment attractiveness of the SEZ, it is necessary to form benefits at the expense of both local and Federal taxes, and also use other, non-tax factors.
- The objectives of the SEZs should be clearly defined. If the goal is to increase the participation of domestic enterprises in the global reproduction chains, then the main condition for residents should be put forward - an increase in exports.

- Residents' compliance with the agreed requirements should be more closely monitored, and the profile of their activities should be developed in advance. The arrival of a foreign investor should lead to an increase in the efficiency of the domestic economy, and not to the bankruptcy of some enterprises and an increase in dependence on transnational companies – other Russian enterprises.

References

1. Federal law of the Russian Federation: On special economic laws in the Russian Federation ot 22 July 2005, №116 - FZ. http://www.consultant.ru/document/cons_doc_LAW_54599/. Accessed 02 Apr 2019). (in Russia)
2. Hernândez, V., Pedersen, T.: Global value chain configuration: a review and research agenda. BRQ Bus. Res. Q. **20**(2), 137–150 (2017). https://doi.org/10.1016/j.brq.2016.11.001
3. Kashbraziev, R.V.: Aims of international industrial cooperation. J. Econ. Law Sociol. **4**, 39–42 (2015). (in Russia)
4. Kondratyev, V.B.: New stage of globalization: features and prospects. World Econ. Int. Relat. **62**(6), 5–17 (2018). https://doi.org/10.20542/0131-2227-2018-62-6-5-17. (in Russia)
5. Laplume, A.O., Petersen, B., Pirs, J.M.: Global value chains from a 3D printing perspective. J. Int. Bus. Stud. **47**(5), 595–609 (2016). https://doi.org/10.1057/jibs.2015.47
6. OECD, WTO: OECD-WTO: Statistics on Trade in Value Added. OECD, Paris (2013)
7. OECD, WTO, UNCTAD: Implications of Global Value Chains for Trade, Investment, Development and Jobs. OECD, WTO, UNCTAD, Paris (2013)
8. Reznikov, S.N.: Imperatives and determinants of current and future restructuring of global supply chains: conceptual aspect. Bull. VEGU **5**(73), 38–48 (2014). (in Russia)
9. Sapir, E.V.: An integration model of the Russian region: methodological basis and stages of formation. Bull. Udmurt Univ. **26**(6), 65–74 (2016). (in Russia)
10. Smorodinskay, N., Kapustina, A.: Free economic zones: world experience and Russian prospects. Econ. Issue **12**, 126–140 (1994). (in Russia)
11. Strange, R., Magnani, G.: Outsourcing, offshoring and the global factory. In: Cook, G., McDonald, F. (eds.) The Routledge Companion on International Business and Economic Geography, pp. 1–26, Routledge (2017)
12. Starostina, M.I.: Performance issues of special economic zones in Russia. Russ. Foreign Econ. Bull. **8**, 13–30 (2016). (in Russia)
13. Streltsov, A.V., Yakovlev, G.I.: Peculiarities of conducting business activity in special economic zones of Russia. Econ. Entrepreneurship, **19**(4), 895–906 (2018). https://doi.org/10.18334/rp.19.4.38973. (in Russia)
14. Streltsov, A.V., Yakovlev, G.I.: The formation of special economic zones for innovative development of national economy. Econ. Entrepreneurship **2**(91), 968–972 (2018). (in Russia)
15. The Department of regional development: Ministry of economic development publishes a report on the results of the functioning of special economic zones. Official website of the Ministry of economic development of the Russian Federation (2016). http://economy.gov.ru/minec/about/structure/depOsobEcZone/2017030705. Accessed 03 Apr 2019. (in Russia)
16. Varnavskii, V.G.: International trade in value added terms: methodological issues. World Econ. Int. Relat. **62**(1), 5–15 (2018). https://doi.org/10.20542/0131-2227-2018-62-1-5-15. (in Russia)

17. Volkova, E.S.: Investments in special economic zones as an instrument of social and economic development of Russia. In: Dezhkin, V.N., Trunin, V.I. (eds.) Problems of socio-economic development of Russia, pp. 122–126. Publishing House St. Petersburg, State University of Economics and Finance, St. Petersburg (2008). (in Russia)
18. Yakovlev, G.I.: Peculiarities of import substitution strategy implementation in Russian industry. Bull. Samara State Univ. Econ. **5**(127), 59–64 (2015). (in Russia)

The Transformation of the Customer Value of Retail Network Services Under Digitalization

D. V. Chernova[1](✉), N. S. Sharafutdinova[2], I. I. Nurtdinov[3], Y. S. Valeeva[4], and L. I. Kuzmina[5]

[1] Samara State University of Economics, Samara, Russia
kafedra-ks@yandex.ru
[2] Institute of Management, Economics and Finance, Kazan Federal University, Kazan, Russia
[3] Kazan State Medical University, Kazan, Russia
[4] Kazan Energy University, Kazan, Russia
[5] Kazan Cooperative Institute, Russian University of Cooperation, Kazan, Russia

Abstract. The study presents theoretical and methodological studies of structural changes of retail network services under digitalization. Based on the analysis of literary sources and activities of Russian trade networks, the structure of the customer value of the service is proposed, which is supplemented by a new element for Russia. It is implemented only by large federal trade networks and it is an ecosystem or lifestyle, carrying out customer feedback through a mobile application, sites. In the American and European customer markets, we can notice new participants in trade services as a result of digital transformation. This is due to the fact that modern manufacturers and other IT trading platforms offer a wide range of services and involve customers. All this made it possible to propose a forecasting flowchart for a customer's way of receiving trade services under digitalization. An important direction of digitalization is the processing of Big Data analytics. In Russia, only 30% of networks use Big Data at all stages of retail. A classification of digitalization trends in Russian retail is presented as intellectual marketing, digitalization of operations, a must win strategy. Innovation activity allows the customer to be involved in the joint creation of the customer value to satisfy his needs, to provide and implement the customer value quickly and geographically available.

Keywords: Big Data · Customer value · Digitization · Innovation · Retail network services · Transformation

1 Introduction

In the sphere of traditional services, which the retail trade belongs to, the service and technological revolution marked the beginning of a set of large-scale structural reforms that predetermined the formation of network forms for the commercial space and the increase in its level of innovation, informatization and knowledge-intensiveness.

S. I. Ashmarina et al. (Eds.): ISCDTE 2019, LNNS 84, pp. 252–260, 2020.
https://doi.org/10.1007/978-3-030-27015-5_31

According to a study conducted by the Redis Business Class, PwC, Microsoft, 41% of retailers surveyed [5] are already in the process of digital transformation, and another 16% plan to begin its implementation in the coming year.

The trends in the development of retail services in the USA and Europe structurally and gradually affect the customer value of services of Russian companies with the emergence of new trade formats - Internet commerce, including online commerce, online transactions, mobile offer trading and retail platforms. Multi-channel strategies emerge and new participants emerge creating a trade service, for example, Amazon [5], whose share of online trading in 2017 was 4% of the total retail trade in the United States [13].

Determining trends in customer behavior in the retail industry are the reduction in the number of purchases and the purchase of less expensive products. Thus, over the past six months, the average check has decreased by 4.6%. This study combines the research of foreign scientists on the analysis of digitalization of American and European retail networks and the implementation of innovations in the activities of Russian network organizations (Porter [10], Kozlova et al. [6], Shangina [11], Dolinina et al. [2], Sharafutdinova and Valeeva [12], Nurtdinov [8]).

New vectors for the development of trade services as a result of digitization and a detailed study of work on this topic have predetermined the need to study and summarize the methodological aspects of the structure, the composition of participants in the customer value of the trade service, the practical aspects of innovation, including Big Data.

The objects of the study are retail networks. The subject of the study is economic relations arising while transforming the customer value of the retail network service. The main purpose of the study is a theoretical and methodological study of structural consequences of digital transformations of the customer value of retail network services.

2 Methodology

The research method includes the following steps:

1. Summarize and present the value-oriented development of retail network services in modern conditions through the structural model of the trading service and its components.
2. Propose the concept of the customer value of the trade service under digitalization using the example of foreign retail networks.
3. Identify and summarize the applied and implemented digital innovations, the application of Big Data in the activities of retail networks and how they affect the customer value of the service.

3 Results

In studies of the next technological revolution, which is usually called digital, much attention is paid to the relationship between digital and organizational transformation of economic actors. Implementing the ability to analyze Big Data, the Internet of Things and innovative technologies will require a renewal of economic institutions, leading to fundamental innovations. The analysis of this area at the corporate, industrial and transnational levels often relies on the concept of value networks proposed by Porter [10]. The important position of structuring the value of trading services, based on the theory of creating the customer value by Levitt [7], remains in the understanding of the trading service as mandatory and additional that affects the customer value (Fig. 1). Based on the generalization of approaches to the customer value levels and taking into account the development of retail, the authors proposed the following description.

It is important to note that the customer value of retail network services in Russia is updated by such an element as a potential customer-oriented level. This level of value can only be provided by large federal networks that finance large projects of the IT-industry aimed at communicating with customers and processing Big Data in order to expand and improve the customer value of the service. Thus, the Alphabet of Taste supermarket was the first among Russian food retailers to launch a Chinese service to pay for Alipay purchases.

The Chinese application WeChat is an example of a digital ecosystem that includes trade and is used by more than 1 billion people. It includes the functions of messenger, social network, ordering city services, a multimedia and news hub, as well as an electronic wallet. Currently, X5 Retail Group is developing a retail ecosystem development strategy for 15 years. A unified marketing strategy, a unified organizational structure and a technological landscape for the service ecosystem are assumed. Therefore, when e-commerce and online sales confirm their right to exist and develop in Russia, the same merger procedure will be launched at the level of business processes. Direct communication with the customer at his points of contact with the brand

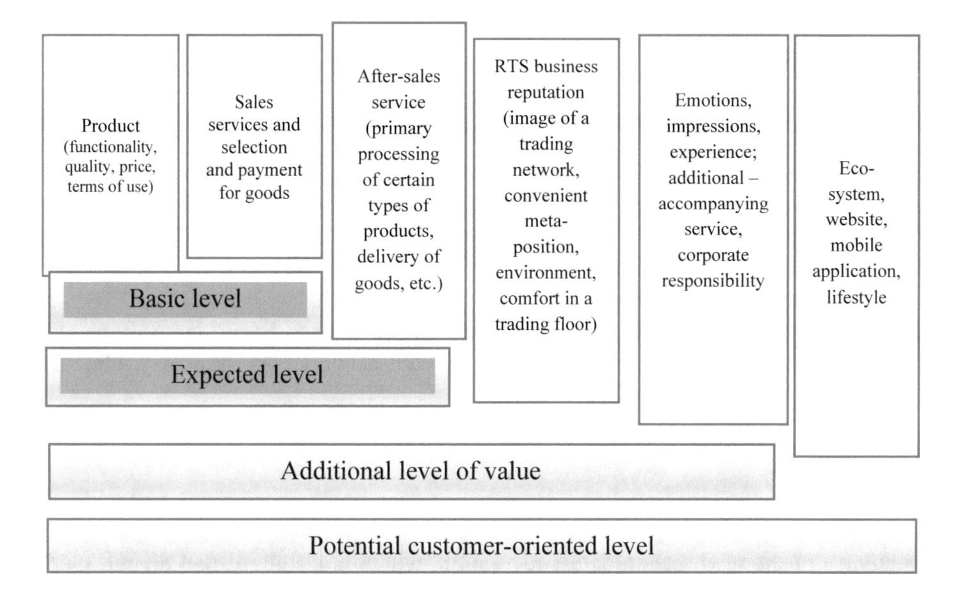

Fig. 1. Customer value of the RTS service for customers at the level of the retail network (Source: compiled by the authors)

and working with a large array of data gives a huge advantage in terms of improving the service. Price still remains the main factor in the decision to purchase. But according to a recent study by analysts from Kibo (2018 Customer Trends Report - Attracting an Informed Customer), buyers increasingly prefer buying experience. In this case, priority is given to a combination of quality goods and services.

Digital transformation as a source of the additional value, automation, personalization of services, changes in the environment or eco-system, interaction, transparency and control combine many activities and processes. The main prerequisite for their changes is that customers are more likely to interact with those network participants who best create the value for their requests. It is important to consider the ability to create the value beyond the actual purchase (for example, the impact on product use, the exchange of values or user experience).

In the retail sector there are 4 fundamental areas of digitalization:

- Work with customers, including all stages from the interest in the brand to the time of purchase;
- Ensuring operational efficiency, including staff performance, work with products, price tags, display of goods, etc.;
- Logistics and control of product deliveries;
- Control of IT infrastructure and security system.
- New entrants or online platforms are a definite challenge for retailers, reducing interest in the traditional interface.

The interface is the source of information for the customer, the emerging concept of retail trade, information flows about goods, consulting on sales.

Currently, manufacturers (brands) are trying to directly interact with the customer. They are able to create powerful branded ecosystems that interact with customers through Internet applications or platforms, direct sales, interaction programs and loyalty programs, as well as offer personal communications that create completely new values and make brands empirical. For example, sportswear manufacturer Adidas plans to control 60% of the brand's global retail space by 2020 due to a significant increase in the number of its own stores, the concept of "store in store" [4]. Digital platforms are digital intermediaries that effectively connect external producers/sellers with customers, thereby ensuring value-added interaction. Their goal is to facilitate the exchange of goods, services or social currency [9]. Examples of platforms are Alibaba, Wish, Ebay and Amazon Marketplace, which are currently among the most expensive in the world [3]. As a result of these combined changes, the competition for the interface with respect to the customer increases and affects the change in the value: producer → seller → customer.

The traditional network, as a rule, does not allow each group of components to directly interact with the end user. This trend currently affects, first of all, the sales transaction and the immediate process that leads to it, but ultimately will affect the constant interaction with the customer during the time of using the whole product or using the service.

Our approach is to describe the system and the whole network with new sources of the customer value, which arise through digitization and affect customer preferences and satisfaction. Digitalization affects the different stages of the decision-making process by customers (Fig. 2).

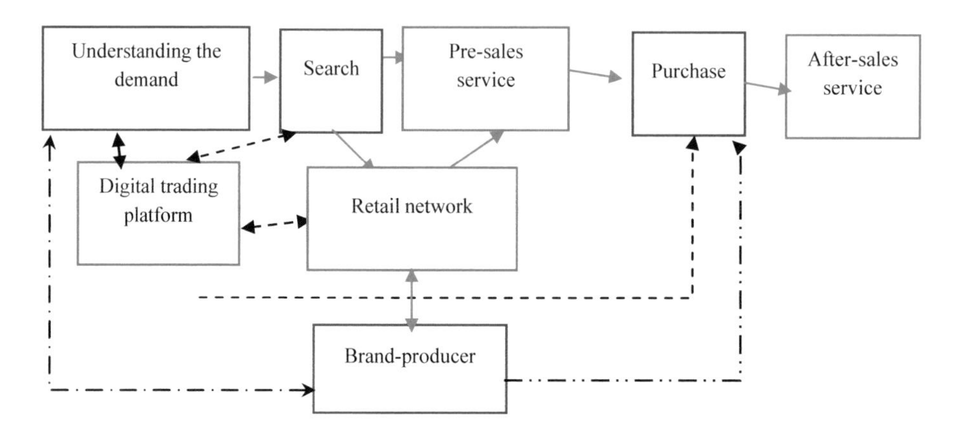

Fig. 2. A predictive block diagram of the customer receiving a trading service under digitalization (Source: compiled by the authors)

The proposed predictive model is based on studying the features of digitalization in international retail networks in the US and Europe. It is found out that interaction with customers within the framework of a trade service can be changed due to a new network member who will provide the greatest customer values. Digital trading platform can be the main source of information and organization of trade, therefore, its interaction with the retail network will be observed.

An online store, an application for a smartphone, IT device can serve as a digital platform, which will be the basis for informing and advising customers and participating in interaction.

Thus, digital transformation has a systemic structural effect on the value of a trade service, while managing interactions with the end customer through the brand-producer, platform, or retailer.

Despite the fact that the domestic retail sector is striving to become more technological, enterprises are developing front-office technologies (CRM systems for creating and optimizing loyalty programs) and back-office solutions (ERP, geo-information systems, BI tools, for example, to collect information about visitors over Wi-Fi), these enterprises do not have enough opportunities for large-scale implementation and restructuring of the principles of work.

In the Russian market, not all retail networks have the opportunity to receive and process Big Data in full due to limited technical resources and a small percentage of the population who use online services. Table 1 presents trade networks which applied Big Data technologies.

Table 1. Bid-data tools used by retail networks

Applied	Are to be applied	Are not applied and not to be applied
Lenta	X5 Retail Group	Auchan Groupe
City supermarket (ABC of taste)	Metro Group	O'Key
Gloria Jeans & Gee Jay	M.Video	Euroset
Megaphone	El Dorado	Computer center DNS
Yulmart	Svyaznoy	Seventh Continent

Source: compiled by the authors based on [1]

The analysis of Big Data is directed not only at new visitors to the retail network - the behavior of customers who are already retailer's customers is also evaluated. So, it is recommended to regularly examine data on the purchase history, attendance, paying particular attention to users who have not visited it for a long time and have not bought anything. These actions will help prevent customer churn. For example, to motivate them to buy by offering a personal discount or bonus points for the purchase of certain goods.

At each stage, you can take advantages of modern technological innovations. There are new ways to interact with the buyer in all three areas. So it is important to focus on those methods that provide the most effective return.

Big Data analytics is now applied at all stages of the retail process:

- Determining the location for the opening of a next store;
- Developing popular products;
- Forecasting the demand for these products;
- Price revision;
- Defining the target audience which might be interested in these products;
- Defining the best ways of customer orientation.

Personal income, customer loyalty, interest rates - this is only part of the factors affecting demand. Some of the most significant factors in modern retail trade are related to fluctuations in customer preferences and the level of demand, an increase in online operations and an increase in customer mobility. This has led to a decrease in brand loyalty.

Improvement of existing or creation of new values for customers may arise at different stages of the retail service. For example, retail networks can use face recognition systems to identify customers and adapt interactions with them, to form a proposal according to their mood (correlation with the data of purchase history, their preferences). In addition to the sensory value, which can be created using the aesthetic atmosphere of the sales area and retail network infrastructure, a personalized approach can provide emotional values (for example, shopping pleasure), cognitive values (for example, stimulating inspiration and motivation to learn new recommendations) or relative value (for example, a sense of personal appreciation) that goes beyond cost. Among the key digital trends in Russian retail, let us highlight intellectual marketing, digitalization of operations and a must-win-strategy (Fig. 3).

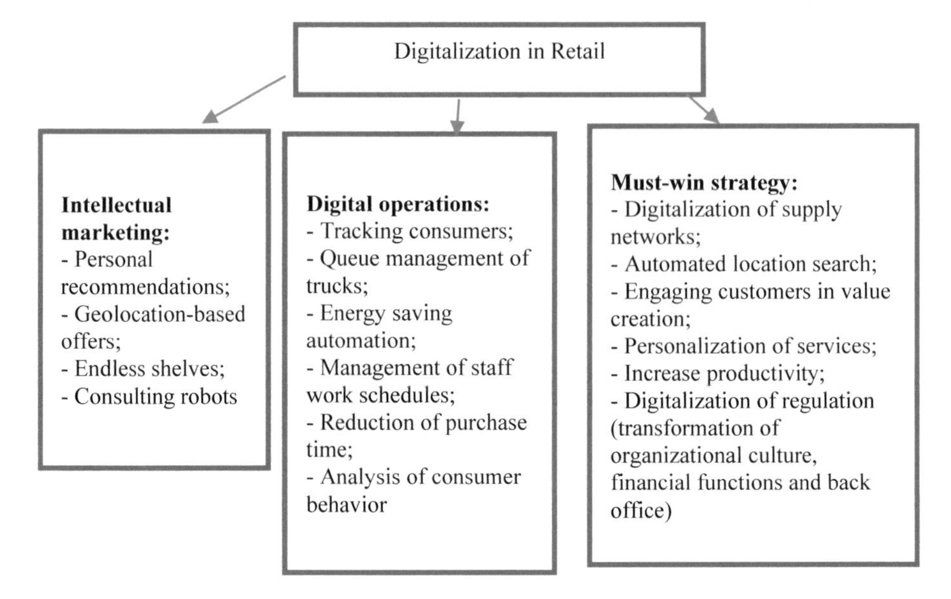

Fig. 3. Digitalization in retail (Source: compiled by the authors)

Intellectual marketing involves changing the traditional approach to standard operations and processes – for example, introducing electronic price tags or geolocation services for personalized offers. Digitizing operations – for example, automating logistics processes, tracking customer movement around the site – is designed to reduce labor, electricity, temporary losses, and so on. Implementation of the must-win-strategy - which includes digitalization of supply networks, automated search for locations and the development of omnichannel - will reduce logistics costs, increase revenue growth and choose the best geolocation for business needs. Retail networks are constantly finding new, innovative ways to extract data from an ever-growing amount of structured and unstructured information about the behavior of their customers. Customer focus allows for revenue and profitability of sales through the use of personalized approaches, online services.

4 Discussion

At each stage of the development of retail services, you can take advantage of modern technological innovations. The emergence of new ways to interact with the customer allows you to focus on those methods that provide the most effective returns. Each organization has set goals in the area of profit and loss, and retail trade is one of the most complex industries from this point of view. There is always the potential to optimize operations, change or increase market share. Taking advantage of current opportunities to streamline business processes, retailers will be able to significantly affect the financial result. Strategic directions include increasing information security (risk reduction) and changing processes to quickly respond to new market challenges.

The retailer's current key goals include increasing demand, attracting customers, and analyzing their behavior and preferences. It is necessary to adhere to three fundamental strategies to achieve these goals: customer focus, optimal infrastructure and effective use of relationships with business partners and suppliers. When customers explore the range and make purchases, retailers can use various opportunities to interact with them and get important information. One of the factors contributing to the success of sales is the ability to timely provide the necessary information to the customer, on the basis of which he can make a purchase decision. Decisions in this area are guided by three main tasks: formation of demand, stimulation and analysis.

5 Conclusion

In the future, retail networks will mainly sell products characterized by a large search for information. This transition will lead to smaller impulse purchases in the store and fewer opportunities for cross-selling - activities that are crucial for the profitability of the trading network. Therefore the need to transform network services is predetermined, since in the future customers will be able to buy anywhere and be satisfied. It will be important to introduce new digital POS solutions (for example, by installing tablets, digital price tags, in-store navigation applications, etc.), effectively integrating POS to organize customer travel for shopping.

New technologies continuously modify the way customers interact with retailers. Current customers pay special attention to values of products and services offered, therefore retailers should form personalized, interactive interaction, encouraging interest and emphasizing respect for preferences of each customer, whatever channels and forms of interaction he chooses.

Innovations in the field of mobile application platforms simplify customer interaction. Retailers can provide smart-phone apps to shoppers to create and use virtual shopping lists; search for information about the services of the store; use geographic information and navigation services, etc. Using Big Data in retail, retail networks are becoming more customer-oriented and can provide the necessary quality services through personalization of offers and improved marketing strategies.

References

1. Analytical statement. Retail-barometer (2019). https://www.pwc.ru/en/industries/retail-consumer/retail-barometr.html. Accessed 30 May 2019
2. Dolinina, O.N., Kushnikov, V.A., Pechenkin, V.V., Rezchikov, A.F.: The way of quality management of the decision making software systems development. In: Silhavy, R. (ed.) Software Engineering and Algorithms in Intelligent Systems. CSOC2018 2018. Advances in Intelligent Systems and Computing, vol. 763, pp. 90–98). Springer, Cham (2019). https://doi.org/10.1007/978-3-319-91186-1_11
3. Ernst & Young GmbH: Digitalriesen überholen Industrie – US-Internetkonzerne sind wertvollste Unternehmen der Welt (2018). https://www.ey.com/de/de/newsroom/news-releases/ey-20180629-us-internet-konzerne-sind-wertvollste-unternehmen-der-welt. Accessed 30 May 2019

4. Kell, J.: Now sporting giants Nike and Adidas are pushing the future of retail. Fortune (2016). http://fortune.com/2016/12/14/nike-adidas-retail-future. Accessed 11 Feb 2019
5. Keyes, D.: Amazon Captured 4% of US Retail Sales in 2017. Business Insider (2018). https://www.businessinsider.com/amazon-captured-4-of-us-retail-sales-in-2017-2018-1. Accessed 30 May 2019
6. Kozlova, O.A., Sukhostav, E.V., Anashkina, N.A., Tkachenko, O.N., Shatskaya, E.: Consumer model transformation in the digital economy era. In: Popkova, E., Ostrovskaya, V. (eds.) Perspectives on the Use of New Information and Communication Technology (ICT) in the Modern Economy. ISC 2017. Advances in Intelligent Systems and Computing, vol. 726, pp. 279–287 (2019). Springer, Cham. https://doi.org/10.1007/978-3-319-90835-9_33
7. Levitt, T.: Marketing success through differentiation – of anything. Harvard Bus. Rev. (1980). https://hbr.org/1980/01/marketing-success-through-differentiation-of-anything. Accessed 30 May 2019
8. Nurtdinov, I.I.: Leasing as a tool for business development in modern conditions. Financ. Anal. J.: Probl. Solut. **11**, 41–46 (2012). (in Russian)
9. Parker, G.G., Van Alstyne, M.W., Choudary, S.P.: Platform Revolution – How Network Markets are Transforming the Economy and how to Make them Work for You. W. W. Norton & Company, New York (2016)
10. Porter, M.E.: Competitive Advantage: Creating and Sustaining Superior Performance. Free Press, New York (1985)
11. Shangina, E.I.: The triad of vitruvius in the modern world. In: Cocchiarella, L. (ed.) ICGG 2018 - Proceedings of the 18th International Conference on Geometry and Graphics. ICGG 2018. Advances in Intelligent Systems and Computing, vol. 809, pp. 2095–2107 (2019). Springer, Cham. https://doi.org/10.1007/978-3-319-95588-9_187
12. Sharafutdinova, N., Valeeva, J.: Quality management system as a tool for intensive development of trade organizations. Mediterr. J. Soc. Sci. **6**(N1S3), 498–502 (2015). https://doi.org/10.5901/mjss.2015.v6n1s3p498
13. Thomas, L.: This chart shows how quickly Amazon is ‘eating the retail world. CNBC (2017). https://www.cnbc.com/2017/07/07/amazon-is-eating-the-retail-world.html. Accessed 2 Feb 2019

Digitalization of the Public Sector of the Regional Economy

T. M. Tarasova, L. V. Averina(✉), and E. P. Pecherskaya

Samara State University of Economics, Samara, Russia
alv94@ya.ru

Abstract. The article deals with the effectiveness of staffing and promotion of staff in public sector and digital enhancement of the above-mentioned issues based on qualitative and quantitative analysis of Samara region's public sector. Research covers the period from 2015 until and including 2017. Authors characterize social environment in public sector regional economy, determine challenges in staffing and training and the ways of their solution. The structure and strength of a personnel reserve should reflect the current and prospective differences of the HR management needs depending on: leave vacancies; staffing; local degree of the turnover rate; perspective and current plans development; the needs of the quantitative structural work of the subdivisions of the governing body in HR; the reserve ratio, in accordance with which the number of trained candidates from the candidates pool in the appropriate position is guaranteed; importance of digital technologies, including electronic access of staff and candidates to their portfolio. By results of the research, the authors offered a way of such digitalization, that can be the basis for purposeful process of staffing, training and retraining of public sector's staff aiming at development of regional economy.

Keywords: Government entity · Public officer · Staff · Digital technologies · Document database

1 Introduction

Staffing in public administration is an activity aimed at recruiting professionally trained employees of all the authorities who, according to the modern requirements, are able to effectively carry out the tasks and functions of the state and municipal bodies within the framework of the law and official powers, involving the use of various legal acts concerning mechanisms and technologies of staffing.

In assessing the main personnel of the state administration, the analysis of the quantitative and qualitative capabilities of municipal indicators is applied [1].

Quantitative indicators:

- number and proportion of employees, who have one or another type (level) of education;

The proportion of senior officials is considered according to age groups, law requirements, number of municipal employees, etc.

S. I. Ashmarina et al. (Eds.): ISCDTE 2019, LNNS 84, pp. 261–268, 2020.
https://doi.org/10.1007/978-3-030-27015-5_32

Qualitative methodological indicators:

- qualification,
- education level,
- business skills of municipal employees, etc.

The structure and strength of a personnel reserve should reflect the current and prospective differences of the current management body's personnel needs [5] and should be determined as mandatory depending on: leave vacancies in the development of the RF; staffing; local degree of this turnover rate taken into account; development of perspective and current plans only for the development of the results of the management body; the needs of the quantitative structural work of the subdivisions of the governing body in the personnel both for the closest period to support and for the distant stand out the prospect of norms; the reserve ratio, the positions representing a quantitative testing indicator to maintain, in accordance with which the number of candidates trained by the municipal reserve in the appropriate position is guaranteed [13].

2 Methodology

Employee security of Samara region administration had been considered. Research covers the period from 2015 until and including 2017. The authors analyzed gender composition and age, seniority (working experience) and educational levels of staff (public officers). The research methods are: system analysis, polling and interviewing.

3 Results

The security of the public sector staff of the Samara region is considered in Table 1 below (Table 1).

Table 1. The security of the public sector staff of the Samara region.

Indicators (employee)	2015	2016	2017
Amount of staff	4893	4752	4292
Replaced by employees	4212	4115	4178
Percentage of regular staff, %	83,2	84,6	97,4
The number of positions of staff	4214	4121	4056
Replaced by employees	4103	4096	3941
Percentage of regular staff, %	87,9%	89,1%	91,8%

Source: compiled by the authors.

Thus, having reviewed the data for three years, it can be noted that the staffing level of employees of Samara Region Government has decreased throughout the entire presented period. Compared to 2015, the number of employees in 2016 decreased by 3% (which amounted to 141 people). The change in staff number in 2017 compared to

2016 was 10% (that is, 460 people less). The ratio of 2015 to 2017 was 12%, namely, 601 people.

Also the downward trend has the number of public officers (civil service) posts. Analyzing the data of the presented table, we can conclude that the number of civil service posts in the Government of Samara Region in 2016, as compared to 2015, decreased by 2% (93 people). Compared with 2016, in 2017, the number of posts decreased by 1.5% (which amounted to 65 people). Thus, comparing the two extreme years, we note that the number of civil service posts in the Government of Samara Region in 2017 decreased, compared to 2015, by 3% (158 people). Gender-efficient composition of public officers (public sector employees) is shown in Fig. 1 below.

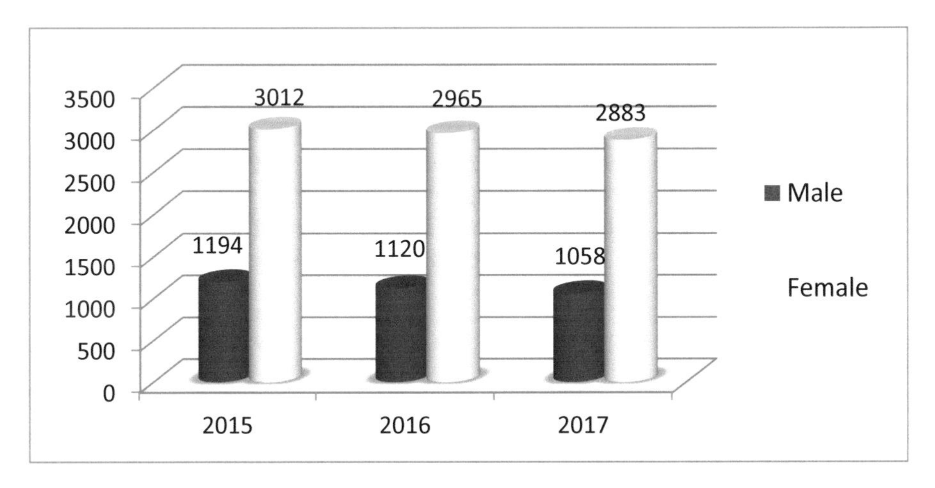

Fig. 1. Gender-efficient composition of public officers (public sector employees) (Source: compiled by the authors)

An analysis of the labor structure if public sector employees are quantitative in the Samara region by the highest representation of male and female employees indicates a gender imbalance. The analysis shows domination in amount of female staff. Such tendency has been observed for the last 3 years. Thus, in 2015 the number of full-time women working in the Samara region Government was 3012 people, and in 2017 the number of jobs decreased to 2,883 people. The number of men who work in public service is as follows: in 2015 amounted to 1,194, the amount of men slightly decreased up to 1,058 in 2017. Next, we study the age structure of public sector employees of the Samara region (Table 2).

Table 2. The age structure of public sector employees of the Samara region, people

Years of study	Age up to 30	Age from 31 to 40	Age from 41 to 50	Age from 51 to 60	Older than 61
2015	812	1745	1432	456	36
2016	750	1682	1254	586	64
2017	632	1477	1123	654	55

Source: compiled by the authors.

The age structure at stake confirms that public sector employees have a predominance of high level in the age group of 31 to 40, fewer public sector employees established in the age groups between 41 and 60.

At the same time, a significant reduction in the number of public sector employees is observed among the employees up to 30 years old and from 51 to 60 years old, which for the target number is related to the retirement of public sector employees and a decrease in the number of employees' motivation.

It is also necessary to pay much attention to such an aspect as a professional education of employees in public sector [7]. It is imperative that a large-scale successful development of the public sector employees and it's reforming according to the management tasks of the region are necessary and, consequently, employees at stake should have skills of professionally trained staff. Authors suggest, that employees with the necessary knowledge and skills are persons capable to performing their professional duties, solve social and economic problems of the local community at the proper level.

Also authors pay special attention in our study to the period of service, which is related to the quality indicators of public sector employees. Period of service - the length of service in years, duration of the work (service) of different categories of executives and specialists taken into account by the government in accordance with federal legislation and legislation of the subject of the Russian Federation.

The period of service is the basis for:

- filling the corresponding position;
- assignment of qualification level.

In addition, the experience gives the right to:

(1) obtaining necessarily a monthly allowance to the official salary for long service;
(2) the provision of annual additional leave;
(3) the payment of a lump-sum monetary reward upon retirement to the state pension;
(4) monthly supplement to the state pension;
(5) other social benefits and guarantees provided for employees by the legislation of the Russian Federation.

Thus, the public sector employees composition of period of service (seniority) in the Samara region in the period from 2015 till 2017 is shown in Table 3.

Table 3. The seniority of the public sector staff of the Samara region

Seniority	2015	2016	2017
Up to 1 year	279	215	160
From 1 to 5 years	650	612	544
From 5 to 10 years	1002	910	846
From 10 to 15 years	1086	1021	918
Over 15 years	1487	1502	1473

Source: compiled by the authors.

As we can see in a Table 3, a greater number of positions are held by employees with seniority of over 10 years. The number is made up of new employees who work up to 1 year and from 1 year to 5 years, their total amount had been 929 in 2015, 827 means a person equal in 2016 and 704 people by 2017. It can be noted that, even summing up groups of people up to 1 year and from 1 year to 5 years, the number does not exceed other groups. Perhaps this is due to the unpopularity of the desire to work in state bodies among the young generation.

The negative side of staffing level of the governing bodies worsens the conceptual management of social and economic development of the region. To improve the efficiency of personnel administration, it is necessary to further increase the attractiveness of the public service for young employees who have education in public administration sphere [11].

Below is a comparison of the specializations in education as a percentage of the educational spheres of public sector staff in the Samara Region (Fig. 2).

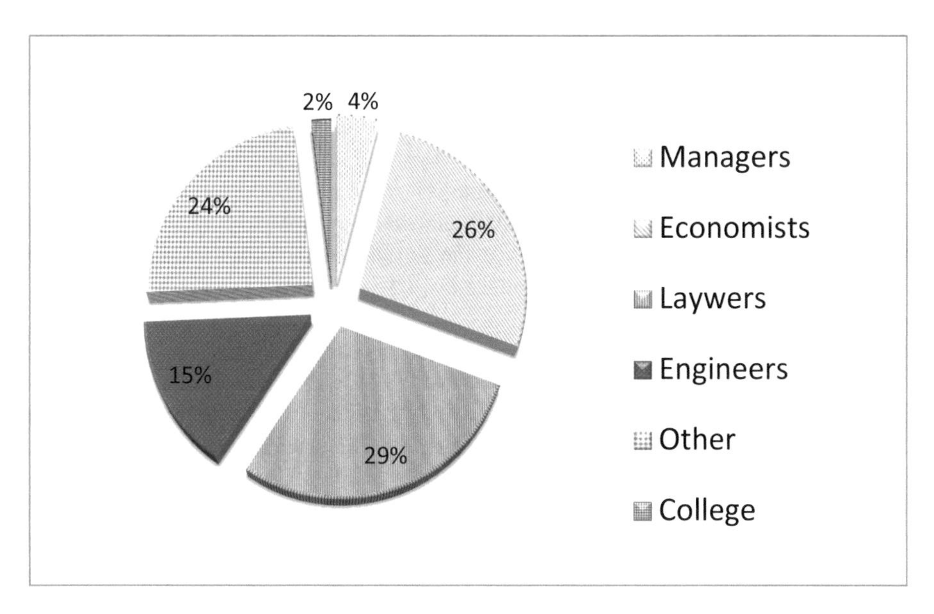

Fig. 2. Percentage of the educational spheres of public sector staff in the Samara Region (Source: authors)

It should be noted that most of the employees have a higher education level, which is a prerequisite for admission to the state service efficiency [14]. At the same time, only an insignificant part of employees has specialized education in the field of management, only 4% of the total number, which amounts to 159 people. The educational structure of employees is dominated by economists (26% of the total number, which was 1048) and lawyers (29% of the total number, which amounted to 1132 people). Also, employees with a different specialty prevail.

4 Discussion

Due to the study there were eliminated some challenges that regulation in public sector administration should resolve. Thus, there are unresolved issues of state and municipal service, the legislation referred to the competence of the Federation. There are problems affecting the passage of civil and municipal services and the solution of which is necessary in order to improve the government bodies activity.

The first problem is the absence on the territory of the Samara region of a regulatory legal act concerning the relationship between the class officials of the civil service of the Samara region, the federal state civil service, the classes of officials of the state civil service of another subject of the Russian Federation, the classes of officials of the municipal service, military and special ranks, the classes of officials of justice, the rank of prosecutors. This problem manifests itself in the transition to the civil service in the Samara region from other subjects of the Russian Federation or in the transition from the federal civil service and may affect the right of a civil servant to bonus and retirement payments, which is the main means of his material support and stimulates professional performance in a civilian position [4]. The solution to this problem will be the creation and consolidation in the territory of the Samara region of a regulatory legal act regulating the relationship of class officials of the state civil service of the Samara region, the federal state civil service, class officials of the state civil service of another subject of the Russian Federation, class officials of the municipal service, military and special ranks, class officials of justice, class officials of prosecutors. Also, it is possible to issue a draft decree of the President of the Russian Federation, which will indicate the criteria for the ratio of ranks and the standard version of the ratio of different ranks to solve this problem everywhere.

The second problem is related to the implementation of the legal provision stated in clause 11 of Article 14 of the Federal Law No. 79 "On the State Civil Service in the Russian Federation", which implies that a public officer has the right to additional professional education in the manner established by this Federal Law and other federal laws, as well as Clauses 3 and 5 of Article 62 of the Federal Law No. 79 "On the State Civil Service in the Russian Federation", which imply that the additional professional education of a public officer lasts throughout the period of the civil service, and such training is carried out as necessary but not less frequently than once in every three years [6]. In fact, these standards are not fully implemented. It is not always possible to realize the rights of each employee to improve their qualifications, since the amount of funds provided by the regional budget for educational purposes is limited. To solve the second problem, there are several options. One of them is to increase funding for public service agencies, additional cash flows will go to the conclusion of a larger number of contracts for the provision of services to obtain additional professional education.

The third problem is related to labor optimization. Currently, there is a need in the Administration of the Governor of the Samara Region to develop and implement software that would automate the recording of information about their income, property and property obligations, as well as information about income, property and property obligations of their spouse and minor children [2, 5, 8–10].

The software could reduce the time required to fill out the form, keeping the last year's version, for example, if the employee has not changed the real estate during the year, it does not need to be re-entered and described. Also, the program could automatically check the basic errors when filling [3, 12, 15]. On the basis of this software, it would also be possible to create electronic access of employees to their personal affairs in order to simplify the exercise of the right established by paragraph 8 of Article 14 of the Federal Law No. 79 "On the State Civil Service in the Russian Federation".

5 Conclusion

The retraining and improvement of public sector employees, taking into account the implementation of the principle of lifelong learning, maximally ensures that the future level of training corresponds to the challenges of the modern society and is relevant in the context of the socio-economic realities of present day. The solution to this problem is of paramount importance for the success of the reforms being carried out at the present time. A high level of professional training of employees is a necessary element of the system of effective public administration.

References

1. Axelrad, H., Luski, I., Malul, M.: Behavioral biases in the labor market, differences between older and younger individuals. J. Behav. Exp. Econ. **60**, 23–28 (2016). https://doi.org/10.1016/j.socec.2015.11.003
2. Batley, R., Mcloughlin, C.: The politics of public services: a service characteristics approach. World Dev. **74**, 275–285 (2015). https://doi.org/10.1016/j.worlddev.2015.05.018
3. Blank, J.L.T.: Measuring the performance of local administrative public services. BRQ Bus. Res. Q. **21**(4), 251–261 (2018). https://doi.org/10.1016/j.brq.2018.09.001
4. Dobre, E., Petrascu, D., Botina, M.: Risks management profile in local public services administration. Procedia Econ. Financ. **27**, 204–208 (2015)
5. Enke, J., Kraft, K., Metternich, J.: Competency-oriented design of learning modules. Procedia CIRP **32**, 7–12 (2015). https://doi.org/10.1016/j.procir.2015.02.211. Paper Presented
6. Federal Law No. 79: On the State Civil Service in the Russian Federation. http://www.consultant.ru/document/cons_doc_LAW_48601/. Accessed 27 July 2004
7. Garbett, D., Ovens, A.: Being self-study researchers in a digital world: an introduction. In: Garbett, D., Ovens, A. (eds.) Being Self-Study Researchers in a Digital World. Self-Study of Teaching and Teacher Education Practices, vol. 16. Springer, Cham (2017)
8. Juliani, F., Oliveira, O.J.: State of research on public service management: identifying scientific gaps from a bibliometric study. Int. J. Inf. Manag. **36**(6), 1033–1041 (2016). https://doi.org/10.1016/j.ijinfomgt.2016.07.003
9. Klimuk, V.V., Piskunov, V.A., Pecherskaya, E.P., Tarasova, T.M.: Methodical approaches to assessing the economic security. In: Mantulenko, V. (ed.) GCPMED 2018 - International Scientific Conference "Global Challenges and Prospects of the Modern Economic Development", The European Proceedings of Social & Behavioural Sciences EpSBS. GCPMED 2018, vol. LVII, pp. 283–290. Future Academy, London (2018). https://doi.org/10.15405/epsbs.2019.03.29

10. Lindgren, I., Madsen, C.Ø., Hofmann, S., Melin, U.: Close encounters of the digital kind: a research agenda for the digitalization of public services. Gov. Inf. Q. (2019 in press). Corrected Proof, 12 March 2019. https://doi.org/10.1016/j.giq.2019.03.002
11. Molotkova, N.V., Kulikov, N.I., Kudryavtseva, Yu.V., Pecherskaya, E.P.: Innovative technologies as a new social challenge in the labor market. In: Mantulenko, V. (ed.) GCPMED 2018 - International Scientific Conference "Global Challenges and Prospects of the Modern Economic Development", The European Proceedings of Social & Behavioural Sciences EpSBS. GCPMED 2018, vol. LVII, pp. 1259–1268. Future Academy, London (2018). https://doi.org/10.15405/epsbs.2019.03.128
12. Paskaleva, K., Cooper, I.: Open innovation and the evaluation of internet-enabled public services in smart cities. Technovation **78**, 4–14 (2018). https://doi.org/10.1016/j.technovation.2018.07.003
13. Pecherskaya, E.P., Klimuk, V.V., Tarasova, T.M.: Analysis of European approaches to improving the life of the population through the implementation of a mechanism of independent assessment of qualifications. In: Ashmarina, S., Vochozka, M. (eds.) Sustainable Growth and Development of Economic Systems. Contributions to Economics, pp. 249–257. Springer, Cham (2019). https://doi.org/10.1007/978-3-030-11754-2_19
14. Pecherskaya, E.P., Averina, L.V., Kochetckova, N.V., Chupina, V.A., Akimova, O.B.: Methodology of project managers' competency formation in CPE. IJME-Math. Educ. **11**(8), 3066–3075 (2016)
15. Shirahada, K., Ho, B.Q., Wilson, A.: Online public services usage and the elderly: assessing determinants of technology readiness in Japan and the UK. Technol. Soc. (2019). https://doi.org/10.1016/j.techsoc.2019.02.001

Assessment of Quality of Services of Public Transport in Digital Economy

A. R. Rakhmatullina(✉), A. N. Sivaks, and E. P. Pecherskaya

Samara State University of Economics, Samara, Russia
sseu_ar@mail.ru

Abstract. A research purpose - development of conceptual provisions and development of practical recommendations about assessment and improvement of quality of services of city public transport in digital economies. Implementation of an effective objective of a research demanded the solution of the following tasks: determinations of tendencies of forming of urban transport environment; identifications of problems and perspectives of development of services of city public transport; analysis of a state and development of system of public transport; formations of system of indicators of sustainable development of public transport; developments of methodical approaches to determination of service quality of city public transport; quality evaluations of services of transportation and service on city bus transport; developments of actions of the strategic plan of improvement of quality of services of public transport. Object of a research are such subjects of system of public transport as transport operators, consumers of services (passengers), bodies of the public and municipal administration. Research questions are organizational and economic relations developing in the course of rendering services of city public transport.

Keywords: Quality assessment · Transport services · Transport process · Quality standard · Public transport · Quality of services

1 Introduction

City public transport plays an important role in development of regional and municipal economy. All types of economic activity in city agglomerations feel the need for timely, reliable and ecologically safe transportations of passengers as manpower. The social importance of city passenger transport consists in providing to inhabitants broad access to the organizations and organizations of social security, health care and education, and also in ensuring higher mobility for elderly citizens, physically disabled people and children.

From the point of view of city mobility, public transport is more effective, than the privately owned vehicle, on use of road space and the consumed energy. For example, the bus transporting 40 passengers occupies only 2,5 times more spaces, than the privately owned vehicle transporting, as a rule, no more than 4 people, and the same bus consumes only 3 times more fuel, than the car.

Rationally organized system of public transport is easy and convenient in use, fast, safe and available. The fact that it can combine in itself (himself) several modes of

S. I. Ashmarina et al. (Eds.): ISCDTE 2019, LNNS 84, pp. 269–280, 2020.
https://doi.org/10.1007/978-3-030-27015-5_33

movement of passengers by trams, buses, trolleybuses and the subway acts as key feature of public transport. Modern information and communication systems allow users to have timely and available information on schedules of public transport, rules of transportation, forms of its payment and the transport operator.

In our days small transport operators perform passenger traffic with low quality of services, dangerous practice of driving and fatigue of vehicles while public transport offers higher level of service. Availability of the uniform municipal organization which plans network routes and defines quality of service, allows to achieve correspondence of need for services of public transport to the offer.

Development of city public transport in the Russian Federation is characterized by availability of the numerous problems connected with insufficient level of passenger service, their mobility with availability of transport services. These problems in many ways are defined by a state of disrepair of road and transport infrastructure, high fatigue of vehicles, lack of evidence-based transport planning in city conditions, unprofitability of municipal transportations of the population. In this regard development of conceptual provisions on quality evaluation of services of city public transport and practical recommendations about increase in the service level of passengers is urgent.

Theoretical and methodological bases of a research of the service industry to which activity of city public transport belongs, are stated in works of such foreign and domestic scientists as Adams et al. [1], Albalate and Bel [2], Anbarci, Escaleras and Register [3], Axhausen [4], Bartholomew and Reid [5], Battellino [6], Bayley, Curtis, Lupton and Wright [7], Bristow, Enoch, Zhang, Greensmith, James and Potter [8], Bureau [9], Granroos [10]. However in works of domestic and foreign scientists such important issues of quality evaluation of services of city public transport as forming of urban transport environment, establishment of system of indicators of service quality and development of actions of the strategic plan of its increase were insufficiently fully investigated. All this defined relevance of a subject of dissertation work, the purpose and research problems, its theoretical and methodological bases.

Scientific novelty of the research consists in the following:

- the economic, environmental, infrastructure and managerial problems interfering development of services of city public transport are defined;
- methodological approach to forming of system of objective and subjective indicators of service quality of transportation and service is offered;
- functions are proved and the process model of strategic and operational management of services of city public transport is developed;
- the system of target indicators of sustainable development of public transport including indicators of improvement of quality of transportations and service is developed;
- the method of determination of dynamics of sustainable development of city transport on the basis of calculation of the generalizing index is offered;
- the mathematico-statistical interrelation of satisfaction of passengers and service quality of city public transport is for the first time established;
- the technique of quality evaluation of services of public transport with use of economic-mathematical modeling is developed.

2 Methods

The research of service quality of transportation of passengers (objective parameters) and service quality (subjective parameters) on city bus transport was conducted in the order recommended by methodical provisions on carrying out inspection of use of public transport.

1. Preparation of carrying out statistical inspection of service quality of transportation and service on bus routes. At this stage the method of carrying out a research and forming of a statistical array based on polls and questioning of passengers of city bus routes No. 45, 56 and 70 was defined.
2. Preparation of questions of questionnaires of inspection of service quality of transportation and service, holding polls of passengers (respondents).
3. Processing of results of questioning of passengers. Processing of mark estimates of respondents of separate components of service quality of passengers on city bus routes was performed by the technique developed by the author according to which mark estimates of passengers were normalized for convenience of calculations from 0,1 to 1,0 or final mark assessment was normalized.
4. Mathematico-statistical modeling of estimates of passengers of quality of service on city bus routes.
5. Transfer of coefficients of a variation (Vi) in coefficients of ponderability of separate indicators of service quality of passengers on city bus routes in integral quality (wi) was performed on the basis of the fact that the some of variation of a qualitative character, the higher it the importance and, therefore, ponderability.

This circumstance can be expressed as the following mathematical dependence which allows to calculate ponderability coefficients:

$$w_i = \frac{V_i^{-1}}{\sum_{n=1}^{8} V_i^{-1}}, \quad \sum_{n=1}^{8} w_i = 1,000. \tag{1}$$

Calculation of the general service quality of passengers on city bus routes is carried out in Table 1.

Table 1. Service quality evaluation on city bus transport in Samara in 2018

Index	GPA (x)	Variability index (V)	Weight coefficient (w)	Multiplication (w · x)
1. Convenience of stopping points	8,0	0,248	0,115	0,92
2. Degree of comfort and design of salon	7,5	0,210	0,136	1,02
3. Politeness and courtesy of personnel	7,9	0,235	0,122	0,96
4. Quality of control of bus	7,0	0,196	0,146	1,02
5. Sense of security and reliability of the trip	6,7	0,179	0,160	1,07

(continued)

Table 1. (*continued*)

Index	GPA (x)	Variability index (V)	Weight coefficient (w)	Multiplication (w · x)
6. Convenience of trip payment	8,1	0,254	0,114	0,92
7. Level of information and communication support of the route	8,4	0,280	0,102	0,86
8. Convenience of feedback with carrier	6,2	0,272	0,105	0,65
Total ($\sum$)	-	-	1,000	7,42

Source: compiled by the authors.

According to data of Table 1 the general indicator of the expected and apprehended quality of service on city bus transport which is value judgment of his passengers made 7,42 points, in a rated type - 0,742.

3 Results

In a management system services of city public transport the important place is taken by strategic management, in particular, the strategic planning including goal setting and development of strategy in improvement of quality of services and the service level of passengers. Increase in level of their service on city public transport has significant effect on the social and economic purposes of municipal economy and transport operators.

There are political, legal, economic and social sales problems of a positive impact of development of services of city public transport on achievement of goals of social and economic policy of bodies of municipal management.

There is the following classification of the purposes in area of service quality of public transport:

- strategic objectives which are applied to all organizations of public transport. They are, as a rule, included in policy of the organizations and declare reference points in the field of service quality of passengers;
- the purposes characterizing workmanship of specific objectives and degree of satisfaction of passengers. They are applicable to all functional types of activity of the organizations of public transport and are responsible for service quality;
- the purposes in area of quality of the rendered services of city public transport which are connected with improvement of separate parameters of consumer properties of services;
- purposes in area of quality of processes and transactions of services of transportation of passengers, and also additional service.

It is possible to mark out the following features of a space arrangement of Samara:

- the urban space area is created as the circular sector with dominance of the radial and linear directions of road and transport infrastructure and transport system along the Volga River;

- that does not correspond to perspective requirements of development of public transport, road and transport infrastructure and leads the low level of complexity of building of the urban area and availability of its private (individual) building about the Volga River to jams (traffic jams) on roads;
- remoteness of certain sites of the urban area from each other, orientation of road and transport infrastructure along the Volga River and backwardness of cross highways is created by problems of further development of system of city public transport;
- many sites of the urban area (individual building, production and warehouse sites, waste grounds and green zones) have a sign of space and composition heterogeneity though certain sites have the signs of integrity and uniformity forming policies;
- on the urban area there were several pronounced policies, such as historical center of the city, Bezymyanka area and MEGA shopping Center, radio center territory - the venue of the FIFA World Cup.

4 Discussion

Objective indicators of service quality of city public transport of Samara for 2012–2019 are provided in Table 2.

Table 2. Objective indicators of service quality of city public transport of Samara

Index	Year							
	2012	2013	2014	2015	2016	2017	2018 (estim.)	2019 (forecast)
Average filling of salon in rush hours on means of transport, passenger/m^2 of the free floor space								
Bus	4	4	4	3	3	3	3	3
Trolleybus	4	4	4	3	3	3	3	3
Tram	5	5	5	4	4	4	4	4
Average interval in rush hours, min.								
Bus	6–7	6–7	6–7	6–7	6–7	6–7	6–7	6–7
Trolleybus	3–4	3–4	3–4	3–4	3–4	3–4	3–4	3–4
Tram	5–6	5–6	5–6	5–6	5–6	5–6	5–6	5–6
Average interval in the interpeak period, min								
Bus	8–10	8–10	8–10	8–10	8–10	8–10	8–10	8–10
Trolleybus	6–8	6–8	6–8	6–8	6–8	6–8	6–8	6–8
Tram	7–9	7–9	7–9	7–9	7–9	7–9	7–9	7-9
Average trip time	25,5	25,5	25,5	25,5	25,5	25,5	25,5	25,5
Rolling stock with the low and lowered floor level, %								
High and extra high-capacity buses	22,8	32,2	41,5	50,8	57,6	64,5	76,3	88,1

(continued)

Table 2. (*continued*)

Index	Year							
	2012	2013	2014	2015	2016	2017	2018 (estim.)	2019 (forecast)
Coefficient of release of the rolling stock of the municipal enterprises for the working days								
Bus	0,630	0,660	0,700	0,730	0,750	0,750	0,750	0,750
Trolleybus	0,780	0,780	0,780	0,780	0,780	0,780	0,780	0,780
Tram	0,828	0,828	0,828	0,828	0,828	0,828	0,828	0,828
Samara subway	0,760	0,790	0,800	0,830	0,820	0,820	0,820	0,820
Coefficient of release of the rolling stock of the private operators for the working days	0,620	0,670	0,70	0,730	0,750	0,750	0,750	0,750

Source: compiled by the authors.

As appears from data of Table 2, for the considered period there was a reduction of average filling of insides of buses and trolleybuses from 4 to 3 people/m^2 of the free area of a floor, insides of trams - from 5 to 4 people/m^2. The average interval of the movement of all types of public transport made on buses of 6–7 min., trolleybuses - 3–4 min., to trams - 5–6 min. The average interval of the movement of all types of city public transport did not change, as well as the average time of the passenger in way to the destination.

Information sources for objective assessment of services of transportation of passengers on city bus routes are data of federal and regional statistics, the Ministry of Transport and Roads of the Samara region, Department of transport of the lake Samara, transport operators, in particular MT Passazhiravtotrans. Basic data about indicators of service quality of transportation of passengers on city bus transport in 2017 and 2019, and also their rated values are provided in Table 3.

Table 3. Indicators of service quality of transportation of passengers by buses on city routes of Samara

Index	Unit	2017	2019	Standard
1. Fullness of insides of buses	m^2/passenger	0,15	0,17	0,20
2. Coefficient of filling of buses: - Regular daytime - On rush hours	Relation	0,30 0,85	0,29 0,75	0,28 0,73
3. Regularity of the movement on a route	%	93,1	96,0	98,0
4. Density of route network	km/km^2	3,8	2,6	2,5
5. Average interval of the movement on a route	min	11	8	6,5
6. Coefficient of route changing	Relation	1,9	1,6	1,4
7. Total time for a trip	min	55	48	40

(*continued*)

Table 3. (*continued*)

Index	Unit	2017	2019	Standard
8. Accomplishment of the schedule of the movement	%	93,1	96,0	98,0
9. Availability of buses with the low and lowered floor level	%	64,5	88,1	-
10. Share of buses with a capacity more than 18 passengers	%	67,2	70,1	-
11. Bus fatigue	%	70,0	60,0	-
12. Emission of harmful substances	Tons	4,0	2,9	2,5
13. Reduction of number of the road accidents to previous year	Percent points	5,0	5,0	-

Source: compiled by the authors.

Methodological approach to assessment of objective service quality of transportation in 2017 consists in comparison of each private indicator of quality of this year with the standard or, in case of its absence, with an indicator of planned 2019. It means that rated value of a private indicator of service quality of transportation is defined by the relation of its size in 2017 to the size of 2019 or to the standard if higher numerical value of the standard corresponds to higher quality of service. On the contrary, if higher numerical value of the standard corresponds to lower quality of service, then value of the standard is divided into value of this indicator of quality.

Designing of the general objective indicator of service quality of transportation of passengers by city bus transport is connected with determination of ponderability of each private indicator in the general indicator of quality. In work the method of determination of coefficients of ponderability, available to practical calculations, based on rank assessment of importance and importance of each of indicators of service quality of transportation of passengers in system of indicators is used.

The general objective service quality of transportation of passengers on city bus routes can be calculated as the amount of works of rated values of private indicators of quality on their coefficients of ponderability. Calculations of rated values of private indicators of quality, their coefficients of ponderability and general objective service quality of transportation of passengers by public buses of the city district Samara, except for private operators, are provided in Table 4. There is a problem integral (subjective and objective) quality evaluations of services of city public transport which is that many of objective parameters of services of transportation cause some subjective parameters of service quality.

Table 4. Quality evaluation of services of transportation on city bus transport in Samara

Index	Standard	Rank	Weight	Multiplication (col. 2 col. 4)
1. Fullness of insides of buses	0,75	8	0,066	0,05
2. Coefficient of filling of buses on rush hours	0,86	9	0,055	0,05
3. Regularity of the movement on a route	0,95	4	0,110	0,10
4. Density of route network	0,66	11	0,033	0,02
5. Average interval of the movement on a route	0,59	10	0,044	0,03
6. Coefficient of route changing	0,74	12	0,022	0,02
7. Total time for a trip	0,73	6	0,088	0,06
8. Accomplishment of the schedule of the movement	0,95	5	0,099	0,09
9. Availability of buses with the low and lowered floor level	0,73	7	0,077	0,06
10. Share of buses with a capacity more than 18 passengers	0,96	13	0,010	0,01
11. Bus fatigue	0,86	3	0,121	0,10
12. Emission of harmful substances	0,63	2	0,132	0,08
13. Reduction of number of the road accidents to previous year	0,95	1	0,143	0,14
Total	-	-	1,000	0,81

Source: compiled by the authors.

So, for example, total costs of time for trips and fullness of insides of buses in a chain of causes and effect relationships cause degree of comfort of passengers, and reduction of number of the road accidents creates a sense of security and reliability of journey of passengers. The integral quality evaluation of services of transportation and service on city bus transport combining its subjective and objective parameters can be given on the basis of expert evaluation of their contribution to integral assessment. Then, integral service quality of transportation and service on city bus transport will make 0,78, and calculation looks as follows: $0,6 \cdot 0,81 + 0,4 \cdot 0,742$.

The received assessment of objective service quality of transportation of passengers and subjective quality of service on city bus routes should be considered average as there are considerable reserves of improvement of quality - 0,22 (22%).

Insufficiently high degree of satisfaction of passengers of city public transport (bus routes) demands carrying out strategic planning of its development. Strategic planning and development of the program of development of services of city public transport on the basis of improvement of quality of service of passengers includes several stages of consecutive transformation (reforming) of the existing transport system of the municipality.

The following is among the main directions of development of services of city public transport.

1. Optimization and integration of transportations by different types of city public transport.
2. Optimization of use of road space for the purpose of providing a priority of public transport.
3. Choice of parameters of increase in power of a bus fleet, improvement of a design of buses, rationalization of a payment system of journey.
4. Increase in availability of city buses to socially unprotected citizens and physically disabled people.
5. Ensuring proper accomplishment of rules of journey of passengers and increase in internal efficiency of modal system.
6. Use of information and communication technologies, including means of communication with the driver and feedback with the transport operator.
7. Application of modern economic-mathematical methods and models of adoption of management decisions by transport operators and municipal authorities.

The important management decision in activity of municipal transport operators of bus transportations is optimum distribution of limited financial resources of the organization and subsidies of the budget for the separate directions of improvement of quality of service of passengers and to the actions implementing them. The theory of usefulness considered at the consumer choice forms a methodological basis of decision making on distribution of financial resources. According to this methodological approach, each of the directions of improvement of quality of service of passengers is characterized by a certain indicator of the usefulness from the point of view of the passenger or importance (importance). The coefficient of importance (importance) of this direction of improvement of quality of service (a) is offered to be determined as the attitude of its coefficient of ponderability (w) towards rated value of average mark assessment by passengers of the expected and apprehended quality (x):

$$a = \frac{w}{x}. \tag{2}$$

Then usefulness or importance of improvement of quality of services of city bus transport in its main directions will make:

- arrangement of stopping points - 0,144;
- increase in comfort and design of salon - 0,181;
- motivation of politeness and courtesy of personnel - 0,154;
- improvement of quality of control of bus - 0,209;
- growth of safety and reliability of journey - 0,239;
- journey payment reforming - 0,141;
- increase in level of information and communication providing a route - 0,121;
- improvement of feedback with carrier (personnel) - 0,162.

Problem definition of optimum distribution of financial resources in the directions of improvement of quality of service on city bus transport includes:

- target function of f(x) which defines the decreasing usefulness of an investment of each additional ruble of financial resources in service improvement of quality;

- matrix of usefulness of each direction of improvement of quality of service (a);
- a matrix of initial and equal distribution of means in the separate directions (x);
- the optimality criterion of target function (Maximize) maximizing the general usefulness (importance) of all directions of improvement of quality of service;
- restriction on total amount of distributed financial resources (g), equal 50 million rubles;
- matrix of optimum distribution of financial resources (h0);
- maximum cumulative useful effect of f(x0).

The computer solution of a problem of optimum distribution of financial resources in the directions of improvement of quality of service on city bus transport has the following appearance:

$$a = \begin{pmatrix} 0.144 \\ 0.181 \\ 0.154 \\ 0.209 \\ 0.239 \\ 0.141 \\ 0.121 \\ 0.162 \end{pmatrix} \quad x = \begin{pmatrix} 6.25 \\ 6.25 \\ 6.25 \\ 6.25 \\ 6.25 \\ 6.25 \\ 6.25 \\ 6.25 \end{pmatrix} \quad g = 50 \quad f(x) = a \cdot x^{0.5} \tag{3}$$

Given

$$\sum x \leq g \quad x0 = Maximize(f, x)$$

$$x0 = \begin{pmatrix} 4.346 \\ 6.854 \\ 4.968 \\ 9.153 \\ 11.957 \\ 4.162 \\ 3.065 \\ 5.495 \end{pmatrix} \quad f(x0) = 3.455 \quad a \cdot x0^{0.5} = 3.455$$

Thus, the largest size of financial resources (11,957 million rubles) it has to be allocated for actions for such direction of improvement of quality of service on city bus transport as safety and reliability of expectation of vehicles on stopping points, landings of passengers, their transportations and disembarkation.

5 Conclusion

The expected social and economic effect of implementation of the directions of improvement of quality of service of passengers of city bus transport consists in decrease in number of the road accidents and accidents by 6%, in increase in turnover of a bus fleet by 15%, in reduction of the actual run for 20% (more than 3 thousand rubles of economy on one vehicle a month), in decrease in fuel consumption by 7% (more than 4 thousand rubles of economy on one vehicle a month).

Theoretical value of work consists: in establishment of tendencies and factors of forming of urban transport environment; in identification of the economic, environmental, infrastructure and managerial problems interfering development of services of city public transport; in a formulation of author's determinations of concepts "services of public transport", "satisfaction of passengers of public transport"; in justification of methodological approach to forming of system of objective and subjective indicators of service quality of transportation and service; in the proof of mathematico-statistical interrelation of satisfaction of passengers and service qualities of city public transport.

Practical value consists in justification of functions and development of process model of strategic and operational management of services of city public transport, development of the system of target indicators and a method of determination of sustainable development of public transport, development of a technique of quality evaluation of services of public transport with use of economic-mathematical modeling.

The basic theoretical and methodical provisions of the thesis, the received results of a research were reported and discussed at university, interuniversity, All-Russian and international scientific and practical conferences. Practical developments of the author are used in activity of transport operators and bodies of municipal management of the city district Samara.

References

1. Adams, M., Cox, T., Moore, G., Croxford, B., Refaee, M., Sharples, S.: Sustainable soundscapes: noise policy and the urban experience. Urban Stud. **43**(13), 2385–2398 (2006)
2. Albalate, D., Bel, G.: Tourism and urban public transport: holding demand pressure under supply constraints. Tour. Manag. **31**(3), 425–433 (2009)
3. Anbarci, N., Escaleras, M., Register, C.A.: Traffic fatalities: does income inequality create an externality? Can. J. Econ. **42**(1), 244–266 (2009)
4. Axhausen, K.W.: Social networks, mobility biographies, and travel: survey challenges. Environ. Plan. **35**(6), 981–996 (2008)
5. Bartholomew, K., Reid, E.: Hedonic price effects of pedestrian- and transit-oriented development. J. Plan. Lit. **26**(1), 18–34 (2011)
6. Battellino, H.: Transport for the transport disadvantaged: a review of service delivery models in New South Wales. Transp. Policy Spec. Issue Int. Perspect. Transp. Soc. Exclus. **16**(3), 123–129 (2009)
7. Bayley, M., Curtis, B., Lupton, K., Wright, C.: Vehicle aesthetics and their impact on the pedestrian environment. Transp. Res. Part D: Transp. Environ. **9**(6), 437–450 (2004)

8. Bristow, A., Enoch, M.P., Zhang, L., Greensmith, C., James, N., Potter, S.: Kickstarting growth in bus patronage: targeting support at the margins. J. Transp. Geogr. **16**, 408–418 (2008)
9. Bureau, B.: Distributional effects of a carbon tax on car fuels in France. Energy Econ. **33**(1), 121–130 (2011)
10. Granroos, C.: Service Management and Marketing: Customer Management in Service Competition, 3rd edn. Wiley, Hoboken (2009)

Analysis of Innovative Activity of Russian Oil and Gas Companies in the Context of World Experience

E. S. Smolina[1(✉)], E. Yu. Kuzaeva[1], O. B. Kazakova[2], and N. A. Kuzminykh[2]

[1] Samara State University of Economics, Samara, Russia
ekaterinsmolin@yandex.ru
[2] Bashkir State University, Ufa, Bashkortostan Republic, Russia

Abstract. Despite the important role of the oil and gas industry in providing the state with financial resources, there is currently insufficient attention to innovations in this branch that leads to technological backwardness of Russian companies in the world economy. Therefore, there is a need to form an effective system of innovation management. In this regard, the consideration and comparison of the experience of innovative activities of foreign and domestic oil and gas companies is relevant. In this article, a sample of the largest oil and gas companies in Russia was made based on the results of the analysis of revenue and its profitability in 2017. The review of significant innovations and management systems of innovative activity is carried out on the chosen enterprises. Then, the authors assessed intangible assets and R&D results on the basis of annual financial reports for 2015–2017 and gave the comparative characteristic of innovative activity of domestic and foreign oil and gas companies on the basis of unit costs of R&D in relation to the volume of sold products and on the basis of the share of R&D costs in the revenue of companies. On the basis of this analysis, the authors formed conclusions about the innovative activity of the largest oil and gas companies in Russia and recommendations for its improvement.

Keywords: Innovations · Innovative activity · Intangible assets · Oil and gas companies · R&D results

1 Introduction

In modern conditions of globalization, to ensure high rates and quality of economic growth in Russia, improve its competitiveness and efficient use of resources, the main task is to develop innovative activities of enterprises in all sectors of the national economy, including the oil and gas complex, which occupies a special place in the Russian economy.

The oil and gas industry of Russia is an important element of the world energy market and part of the global system of all energy supply, it also forms a competitive production complex, which currently not only fully meets the needs of the country in

S. I. Ashmarina et al. (Eds.): ISCDTE 2019, LNNS 84, pp. 281–289, 2020.
https://doi.org/10.1007/978-3-030-27015-5_34

oil, gas and petroleum products, but also, thanks to its active export activities, creates a significant share of foreign exchange revenues to the federal budget.

Without innovation technologies it is impossible to develop new promising oil resources – the Arctic shelf, Eastern Siberia, the deep horizons of Western Siberia. Therefore, it's important to form an effective system of innovation management, to analyze and compare the experience of innovative activity of foreign and domestic oil and gas companies. The subject of this study is the innovative activity of Russian and foreign companies. The research is aimed at analyzing the existing innovative activity of oil and gas companies in Russia, as well as offering recommendations to increase their innovative activity based on the analysis of domestic and foreign experience.

Objectives of the study are:

- to carry out a sample of the largest oil and gas companies in Russia for 2017 based on the results of the analysis of revenue and its profitability;
- to review significant innovations, innovation management systems of these companies;
- to evaluate intangible assets and R&D results on the basis of annual financial reports for 2015–2017;
- to carry out a comparative description of the innovative activity of domestic and foreign oil and gas companies on the basis of the unit cost of R&D in relation to the volume of sales and on the basis of the share of R&D costs in the revenue of companies.

2 Methodology

In this study, the following methods of statistical analysis are used: absolute and relative statistical values, time series, sample. The topic of innovative activity of companies is considered in the works of Belozertseva [1], Ratner and Mikhailov [7], Nikulina and Miroshnichenko [5], Brem, Nylund and Schuster [2], Leten, Belderbos and Van Looy [4], Shams, Alpert and Brown [9], etc. The authors also analyzed the rating of RBC (Russian Business Consulting), the rating of the largest companies in Russia and annual reports of oil and gas companies.

3 Results

According to RBC survey of the 500 largest companies by revenue, 55 enterprises are representatives of the oil and gas industry, which is 11% [8]. The largest of them, with revenues in 2017 of more than 208 billion rubles, are 11 enterprises (Table 1).

The largest and most important in the industry are 7 oil and gas companies that are in the top 20 of largest enterprises and have revenues of more than 500 billion rubles. Data analysis the financial statements of companies for the year 2017 shows that the main holders of intangible assets (IA) are Gazprom, Rosneft and Transneft, and Gazprom has a 2.4 and 6.4 times larger amount of intangible assets in value than Transneft and Rosneft respectively, but only in 2017 the IA share in the assets exceed 0.1% (Table 2).

Table 1. Ranking of the largest oil and gas companies by revenue in 2017

№	Place in the rating	Company	Revenue (billion rubles)	Profit (billion rubles)	Profitability of revenue (%)	Share of sales among the 500 largest companies (%)
1	1	Gazprom	5966	997	16,71	9,53
2	2	Lukoil	4744	208	4,38	8,43
3	3	Rosneft	4134	201	4,86	6,71
4	9	Surgutneftegaz	1006	−62	−6,16	1,63
5	13	Transneft	818	233	28,48	1,33
6	16	Tatneft	580	106	18,28	0,90
7	18	Novatek	537	265	49,35	0,77
8	23	Joint stock oil company «Bashneft»	494	52	10,53	0,83
9	39	Sakhalin Energy	305	58	19,02	0,61
10	53	Slavneft	215	29	13,49	0,37
11	55	Novy Potok	208	4	2,09	-
In the sum:			19007	2091	11,00	31,11

Source: compiled by the authors based on [6, 8].

Table 2. Analysis of the dynamics of the value of intangible assets of the largest oil and gas companies in 2015–2017 (%)

Company	Share of intangible assets in the Assets			Growth rate of intangible assets		Share of intangible assets among the 7 companies		
	2015	2016	2017	2016	2017	2015	2016	2017
Gazprom	0,063	0,092	0,108	155,65	125,27	24,4	31	60,3
Lukoil	0,020	0,017	0,019	98,08	105,84	1,1	0,9	1,5
Rosneft	0,260	0,222	0,023	103,24	11,09	64,5	54,5	9,4
Surgutneftegaz	0,017	0,015	0,016	106,46	103,48	1,6	1,4	2,3
Transneft	0,205	0,343	0,549	186,34	142,14	7,3	11,2	24,7
Novatek	0,001	0,001	0,001	78,24	101,03	0	0	0
Tatneft	0,053	0,057	0,064	118,08	128,11	1	0,9	1,9
For 7 on average	0,116	0,126	0,078	-	-	-	-	-
Assets	-	-	-	113,09	103,84	-	-	-
Intangible assets	-	-	-	122,26	64,5	100	100	100

Source: compiled by the authors based on data from financial statements of companies.

The analysis of the costs on research and development shows that in 2017 Rosneft and Gazprom have 91.3% of the results of scientific research among the largest oil and gas enterprises, and in 2017, Rosneft's "scientific asset" exceeded Gazprom's one, and the science intensity of fixed assets by Rosneft is 4.9 times higher (Table 3).

Table 3. Analysis of the dynamics of R&D results of the largest oil and gas companies in 2015–2017 (%)

Company	R&D Results/fixed assets			Growth rate of R&D		R&D share among 7 companies		
	2015	2016	2017	2016	2017	2015	2016	2017
Gazprom	0,055	0,055	0,037	100,5	74,4	52,2	51	37,4
Lukoil	0,297	0,228	0,207	84,9	94,5	0,6	0,5	0,5
Rosneft	0,271	0,251	0,316	96,9	133	39,1	36,8	48,3
Surgutneftegaz	0,023	0,028	0,018	133,2	67,1	2,4	3,1	2,1
Transneft	0,216	0,288	0,217	264,4	120,5	0,9	2,3	2,7
Novatek	0	0	0	-	-	-	-	-
Tatneft	0,271	0,232	0,305	130,9	148,6	4,9	6,2	9,1
For 7 on average	0,161	0,155	0,157	-	-	-	-	-
R&D	-	-	-	102,7	101,5	100	100	100

Source: compiled by the authors based on data from financial statements of companies.

Currently, the main R&D directions of Russian oil companies are associated with the development of new fields, a decrease in unit production costs. The largest foreign energy companies today have a slightly different emphasis in R&D [20]. Research and development are often associated with new energy sources, with the creation of prerequisites for an adequate response to challenges and risks of the transition from a hydrocarbon-based economy to alternative renewable energy sources [10].

4 Discussion

Table 1 shows that the importance of the oil and gas industry in the country's economy is quite high: the 500 largest enterprises in the country have 31.11% of sales. The largest are Gazprom, Lukoil, Rosneft, Surgutneftegas, Transneft and Tatneft. In addition, three leading oil and gas companies are separated by a large difference in revenue of 3000 billion rubles.

There is a tendency that the oil companies with the amount of revenue up to 1000 billion have a great variation of profitability – 0–20% (with the possibility of increasing profits, as it is by Novatek and Transneft), while the largest companies (Lukoil and Rosneft) stabilize by around 4–5% (with the exception of the Gazprom profitability with revenue 16,71%).

Of the seven oil and gas companies studied in the annual report for 2017, only six enterprises explicitly stated about the ongoing innovations. According to the annual report of PJSC Gazprom, the company is trying to achieve ambitious production goals due to high technological level and system development of its own technical solutions. The main factor that makes the company the main generator and consumer of innovations is the implementation of significant oil extraction projects in the Arctic, the shelf of the Okhotsk Sea, Eastern Siberia and the Far East [13].

As part of the approved Program of Innovative Development until 2025, technological development is expected in the areas of search and exploration of hydrocarbon deposits, increase in the development of existing fields, the development of new fields and deposits on the continental shelf, improvement of the transportation efficiency, storage and processing of gas, as well as the sale and use of gas [14]. So, the company assumes technological development in all parts of the technological chain operating in the oil and gas industry with an emphasis on the gas dominant in its activity.

As part of its innovative activities, PJSC Gazprom cooperates with both European companies and entities of the Asia-Pacific region, as well as with an extensive network of supporting universities in Russia [3].

Control over the effectiveness of a large volume of R&D work with a volume of 6.3 billion rubles in 2017 (which is 1.57 times less than in 2016) is carried out by PJSC Gazprom through the R&D Commission engaged in the development of agreed proposals for their organization and implementation. Accordingly, the company's intellectual property management system is constantly being improved on the basis of the best international and domestic practices. The result of this system operating and innovative activity was the possession of 2269 subsidiaries rights and 991 patents by December 31, 2017. In 2017, 9% and 14% of them respectively were protected, and 227 patent applications were submitted, which is an additional increase of 10% [13]. This ratio means a significant breakthrough in the performance of innovative activity.

Only 405 objects (17.8%) were used in the production, but the economic effect is more than 7 billion rubles, which is 0.7% of the profit specified in Table 1. That means a slight effect from the commercialization of intellectual property. In addition, by 2017, PJSC Gazprom had implemented a number of measures for import substitution and localization of production, the total economic effect of which is estimated at 10.9 billion rubles [13], which is 1.5 times more than the effect of using the results of intellectual activity mentioned above, and it is 1.09% of profits according to Table 1.

Thus, PJSC Gazprom, as the largest oil and gas company, carries out innovative activity in all key areas of production, transportation and processing of gas and oil, however, the most important task for the company in 2017 was the import substitution of foreign technologies.

In the annual report of Lukoil in 2017, the priority development direction was the improvement of the oil extraction ratio and operational efficiency, its results were the increase in the share of horizontal wells up to 28.6% and the extraction of 27% of oil with the use of technologies increasing oil recovery of a layer [14].

Lukoil also carries out scientific and technical work on the development of Bazhenov Fractured Reservoirs and cooperates with universities. The innovative activity of the company is more focused on the improvement of oil extraction technologies, as well as the production of high-viscosity oil.

Innovations of Rosneft are implemented within the framework of the investment program and are aimed at the drilling development in the existing fields and the introduction of new ones. As part of the shelf development, the company is working on the creation of marine equipment and realizing of the import substitution program [15].

The innovative activity of PJSC "Surgutneftegas" also refers to all the main stages of the enterprise activity, but the most important area is to increase the efficiency of hydrocarbon extraction and commissioning of oil deposits with hard-to-recover reserves, which gave an economic effect of 10.2 billion rubles [11]. According to Table 1, it is 16.45% of the registered loss and 93% of the economic effect of Gazprom's innovations.

In the field of the intellectual property protection in 2017, OJSC "Surgutneftegas" protected 9 objects: 3 utility models, 3 databases, 1 trademark, 1 invention, 1 software [17]. PJSC Transneft carries out innovative activities within the framework of the adopted Program of Innovative Development focused on the development of pipeline transport of oil and oil products with the addition of measures for the period 2017–2021 for the development of the intellectual property management system, as well as cooperation with universities and scientific organizations.

PJSC "Transneft" is carried out the reorganization of the management system in the area of intellectual property and intangible assets, where an automated control system "Edison+" is introduced, 233 documents on the result of the intellectual activity were received in 2017, but among implementations for 13250 rationalization proposals and 17720 "Kaizen proposal" there are only 51 inventions, 25 utility models and 15 computer programs. The economic effect of using the intellectual activity results was 18,998 billion rubles, 93% of which is accounted for inventions and utility models, it is 17.9% of revenue and 2.7 times more than the similar effect estimated by PJSC Gazprom.

As a result of this review, we conclude that the economic effect of the intellectual property use is at the level of 7–19 billion rubles, that represents 10–20% of the profits of leading enterprises, but for oil and gas companies with revenue less than 500 billion rubles it can become a significant driver of development.

On average, among the 7th of the studied companies the share of intangible assets in assets has decreased from 0,13% in 2016 to 0.07% in 2017, however, it is important to note that Transneft's share of intangible assets in the assets increased up to 0.55% and has exceeded the value of Gazprom (Table 2). In contrast to these companies, Novatek and Surgutneftegas have a low value of IA, corresponding to 0.001% and 0.02% of the asset value, respectively.

If in 2016 the growth rate of the intangible assets value of 122.2% has exceeded the assets value growth of the studied companies by 9.17 percentage points, in 2017, the growth of assets was only 3.84% and intangible assets decreased by 35.5%. On the background of increasing of IA by Tatneft, Transneft and Gazprom, Rosneft reduced the intangible assets value in 2017 by 89% (Table 2).

It can be concluded that the largest oil and gas companies are characterized by a small share of intangible assets, but the most innovative and active of them are increasing it, which should form a long-term source of profit for them [12].

According to Table 3, it can be noted that «the knowledge intensity of fixed assets» (the ratio of the cost of R&D results to fixed assets) is gradually decreasing –

from 0.161% in 2015 to 0.157% in 2017 – on the background of Rosneft increasing the volume of the knowledge intensity, the share of Novatek and Surgutneftegas is negligible. At the same time, Tatneft reached the value of 0.3 by the indicator of science intensity, Transneft and Lukoil – the value of 0.21, but among the leading oil and gas enterprises the cost of R&D results of Lukoil is negligible – 0.5%.

Thus, Transneft and in particular Tatneft (occupying 4.9% of R&D among seven companies in 2017) are intensively increasing scientific assets, which can form for them intellectual property objects in the future (mainly patents for utility models and inventions), while the leading enterprises Gazprom and Lukoil are "more mature in technological terms" and have embodied most of the "scientific assets" in specific results of the intellectual activity. The only exception is Rosneft which started increasing the "science intensity" in 2017.

In the case of oil and gas companies, the innovative activity can also be estimated based on unit costs for R&D in relation to the volume of sales or in relation to the volume of hydrocarbon extraction [18]. For comparison, the unit R&D costs of the oil and gas company Exxon Mobil in 2015–2017 amounted to 5.3–6.3 USD/tonne of oil equivalent, by Shell these costs are 7.1–8.5 USD/tonne and by Rosneft – 0.7–1.2 USD/tonne.

Currently, the leading Russian oil and gas companies try to finance R&D at the expense of own means formed on the basis of net profit and depreciation [19]. Among Russian oil companies, PJSC Rosneft significantly increased its R&D expenditure. By the end of 2017, the company spent 0.8% of revenue on R&D, which corresponds to the level of many foreign oil and gas companies. At the same time, PJSC Gazprom and PJSC Lukoil continue to show limited demand for R&D: expenditures on the innovative activity were about 0.3% of revenue in 2017. This is 2–3 times lower than the corresponding indicators of the world's leading manufacturers in this branch (Statoil – 0.7%, Chevron – 0.6%, ExxonMobil – 0.5%).

5 Conclusion

Based on our research, we formulate the following conclusions.

From the 11 largest oil and gas companies in terms of revenue, two echelons can be distinguished: the "big three" – Gazprom, Lukoil and Rosneft, with revenues of more than 4,000 billion rubles per year, and the "second tier" – Surgutneftegas, Transneft, Tatneft and Novatek with revenues of 500–1000 billion rubles. The profitability of revenue of the "big three" with the exception of Gazprom is around 4–5%, while the "second tier" has 0–20% with the possibility of losses.

The analysis of the financial reports of enterprises for 2017 shows that the main holders of intangible assets are Gazprom, Rosneft and Transneft. The economic effect from the intellectual activity results usage is about 7–19 billion rubles, which is less in comparison with the increase in energy efficiency and import substitution.

The share of the IA value in assets and R&D results to fixed assets is relatively low (at the level of 0–1%), which is explained by the high fund intensity of the industry and the predominance of work on the search for new deposits over technological development.

In terms of R&D costs, Russian companies today lag behind foreign organizations in relation to the volume of hydrocarbon extraction [16]. Among the Russian oil companies, Rosneft significantly increased its R&D expenditures.

On the basis of these conclusions, we can give some recommendations on increasing the innovative activity of oil and gas companies.

(1) oil and gas enterprises of the "big three" need to coordinate their efforts in the field of development of intelligent wells technologies, carry out work on the development and updating of databases and domestic computer programs for the implementation of oil and gas activities;
(2) companies of the "second tier" need to form a joint network of universities and research centers, designed to accelerate research work in the field of improving the efficiency of individual processes and distribute risks of innovations;
(3) in addition to the energy efficiency programs, oil and gas companies need to adopt a program for increasing the knowledge intensity of their activities, containing specific goals to increase the share of the intellectual property to 0.5–1% of assets, and the results of R&D – to 0.3–0.4%;
(4) Russian companies should consider an opportunity for implementing joint projects involving foreign companies. Effective use of new technologies created on the Russian scientific and industrial base, formation of joint ventures with foreign partners for the implementation of oil and gas projects, involving the creation and implementation of innovative technologies, are important and productive development directions. Joint ventures with foreign capital create opportunities for access to new technologies tested in other countries.

In our opinion, the implementation of these measures will allow the largest oil and gas companies in Russia to significantly improve efficiency and follow the trends of the technological development formed in the international market.

References

1. Belozertseva, O.V.: Prospective use of innovative technologies in petroleum industry in Russia. Int. J. Appl. Fundam. Res. **8–3**, 502–505 (2015). (in Russian)
2. Brem, A., Nylund, P.A., Schuster, G.: Innovation and de facto standardization: the influence of dominant design on innovative performance, radical innovation, and process innovation. Technovation **50–51**, 79–88 (2016). https://doi.org/10.1016/j.technovation.2015.11.002
3. Lendel, V., Moravčíková, D., Latka, M.: Organizing innovation activities in company. Procedia Eng. **192**, 615–620 (2017). https://doi.org/10.1016/j.proeng.2017.06.106
4. Leten, B., Belderbos, R., Van Looy, B.: Entry and technological performance in new technology domains: technological opportunities, technology competition and technological relatedness. J. Manag. Stud. **53**(8), 1257–1291 (2016). https://doi.org/10.1111/joms.12215
5. Nikulina, O.V., Miroshnichenko, O.V.: Comparative analysis of features of financing of innovative activity of oil and gas companies in the world economy. Financ. Anal.: Probl. Solut. **32**(314), 23–39 (2016)
6. Rating of the largest companies in Russia-2017 in terms of sales. https://expert.ru/ratings/rejting-krupnejshih-kompanij-rossii-2016-po-ob_emu-realizatsii-produktsii/. Accessed 27 May 2019. (in Russian)

7. Ratner, S.V., Mikhailov, V.O.: Diversification of the project portfolio of oil and gas corporations as a way to maintain strategic competitiveness. Econ. Anal. Theory Pract. **13**, 11–20 (2017)
8. RBC 500 Rating: All business in Russia. https://www.rbc.ru/rbc500/. Accessed 01 Apr 2019. (in Russian)
9. Shams, R., Alpert, F., Brown, M.: Consumer perceived brand innovativeness: conceptualization and operationalization. Eur. J. Mark. **49**(9–10), 1589–1615 (2015). https://doi.org/10.1108/EJM-05-2013-0240
10. Silkin, V.Yu.: Russian oil and gas industry innovation policy: problems of overtaking development. Energy Policy **6**, 46–54 (2014). (in Russian)
11. Surgutneftegas Annual Report. https://www.surgutneftegas.ru/investors/reporting/godovye-otchety/. Accessed 01 Apr 2019. (in Russian)
12. Taherparvar, N., Esmaeilpour, R., Dostar, M.: Customer knowledge management, innovation capability and business performance: a case study of the banking industry. J. Knowl. Manag. **18**(3), 591–610 (2014). https://doi.org/10.1108/JKM-11-2013-0446
13. The Annual Report of Gazprom. http://www.gazprom.ru/investors/disclosure/reports/2017/. Accessed 27 May 2019. (in Russian)
14. The Annual Report of LUKOIL. http://www.lukoil.ru/FileSystem/PressCenter/121348.pdf. Accessed 01 Apr 2019. (in Russian)
15. The Annual Report of Rosneft. https://www.rosneft.ru/Investors/statements_and_presentations/annual_reports/. Accessed 01 Apr 2019. (in Russian)
16. Tomczak, T., Vogt, D., Frischeisen, J.: Wie Konsumenten Innovationen wahrnehmen Neuartigkeit und Sinnhaftigkeit als zentrale Determinanten. In: Hoffmann, C.P., et al. (eds.) Business Innovation: Das St. Galler Modell. Springer (2016). https://doi.org/10.1007/978-3-658-07167-7_12
17. Transneft Annual Report. https://www.transneft.ru/u/section_file/35361/otchet_interaktivnaya_razvorot.pdf. Accessed 01 Apr 2019. (in Russian)
18. Wang, Q., Cui, X., Huang, L., Dai, Y.: Seller reputation or product presentation? An empirical investigation from cue utilization perspective. Int. J. Inf. Manag. **36**(3), 271–283 (2016). https://doi.org/10.1016/j.ijinfomgt.2015.12.006
19. Yavarzadeh, M.R., Salamzadeh, Y., Dashtbozorg, M.: Measurement of organizational maturity in knowledge management implementation. Int. J. Econ. Commer. Manag. **III**(10), 318–344 (2015)
20. Zwick, T., Frosch, K., Hoisl, K., Harhoff, D.: The power of individual level drivers of inventive performance. Res. Policy **46**(2017), 121–137 (2017). https://doi.org/10.1016/j.respol.2016.10.007

Internet-Marketing in the Sphere of Higher Education

S. Ziyadin(✉) and A. Serikbek

Al-Farabi Kazakh National University, Almaty, Kazakhstan
sayabekz@gmail.com

Abstract. Long before this time, universities adhered to a certain strategy, which implied special criteria for admission, but at the same time limited marketing approaches and communications, which significantly influenced the methodology itself and had a unidirectional nature of development. With the gradual introduction of various changes in the technological sphere, higher educational institutions strengthened their positions, rethinking their strategies, began to adapt more to changing conditions and began to apply marketing approaches to meet the needs of clients in information and communication. The changed approach was obligatory for survival of the universities in the demand-side market. The article discusses the essence of the use of marketing in the field of educational services of higher education institutions, also analyzes the means and tools of marketing, allowing the university to consolidate its position in the market and increase competitiveness. The features of the promotion of educational services and the potential of marketing in the field of higher professional education are being advanced.

Keywords: Marketing · Digitalization · Communication · Online marketing · Social media

1 Introduction

At the present stage of development, the system of higher professional education in the Republic of Kazakhstan is increasingly undergoing changes, primarily related to the transition of educational institutions to a commercial basis. If in the early 2000s education and business were viewed as completely disparate spheres, now these are two interrelated processes. Every year it is more and more possible to observe the formation of market relations in the provision of this type of services.

The active number of growing non-state educational institutions creates a situation where universities are forced to "fight" for each applicant. In each constituent entity of Kazakhstan, a multitude of educational organizations are being created both on a state and a commercial basis. The presence of the same areas of training and educational programs significantly reduces the perception and memorization of the educational institution of a potential entrant.

Particularly relevant to the issue under consideration in modern times, when to a large extent there is a reduction in government funding of many educational institutions of higher vocational education [4]. To attract the so-called customers, universities have

S. I. Ashmarina et al. (Eds.): ISCDTE 2019, LNNS 84, pp. 290–297, 2020.
https://doi.org/10.1007/978-3-030-27015-5_35

to conduct a special policy. Strengthening the position in the market of educational services and improving competitiveness requires the use of marketing tools and tools by the university.

The marketing of educational services is understood as an effective tool for the market positioning of an educational organization, combining components of psychology, sociology, conflict management and other disciplines and promoting the development of partnership and social communications. One of the main elements of the marketing of educational services is the analysis of the needs of potential applicants and the technology of its satisfaction by promoting this type of service on the market [17].

Today, the demand for educational services has great prospects for development. According to the studied data, it was revealed that the country requires at least one million small and medium-sized businesses for the normal functioning of a market economy, which means that the need for specialists is increasing [1]. Consequently, it is necessary to introduce marketing in the field of education to form the educational services market.

The main types of marketing activities of the university include:

- research of demand for products and services of higher educational institutions;
- determination of the price of educational services;
- development of measures to stimulate sales of products and services of the university;
- market segmentation, identification of consumer needs in the market of educational services.

A feature of marketing in education is that services are less tangible. The main task of promoting the services and products of the university is the improvement of quality, as well as the creation of new areas of training, programs on an ongoing basis and the maintenance of their life cycle [15].

To date, the possibility of promoting educational services is widely represented. Marketing tools are able to provide awareness to the entire audience, which is oriented organization offering higher professional education. Various communication technologies are used, a number of actions are thought out. The simultaneous use of all advertising tools allows you to increase the effectiveness of marketing educational activities.

Of the many available means of marketing communication, we identified the most suitable for the promotion of educational services. These included: advertising and public relations, direct marketing, exhibition activities of a higher education institution and the organization's presentation on the global Internet [5]. But do not forget that marketing activity is conditional and involves a set of tools that may include elements of each separately.

The most famous educational marketing tool for a university is advertising. By advertising an educational institution, we understand the system of distribution in any form of educational services and products, as well as a system that is designed to shape an educational organization's image and interest in a certain circle of people.

Advertising is divided into text, online advertising, television, outdoor, souvenir and other types. However, advertising of educational institutions, as a rule, is standard.

The content is information about the availability of certain areas of training, the number of budget places, the duration of training and, of course, a high level of teaching.

Despite the fact that advertising is considered the most popular way to promote goods, in the field of education it loses its value and is inferior to other communication means of marketing in education.

Public relations is one of the key marketing tools in the framework of the activities of a higher education institution, since the meaning of this tool is in two-way communication with both consumers (applicants) and the society as a whole [11].

The goal of public relations is to form public opinion about the university in the minds of potential customers. Using this tool, an educational organization should instill the essence of the idea to the entrant, do everything possible to make the idea universal, so that the existence of the institution becomes interesting for each client [14].

Each event held within the framework of this promotion tool is aimed at creating and enhancing the prestige of the university. These include the open day, the anniversary of the university, press conferences, publication, problem discussions and others. Communication with the public allows reaching mutual understanding with consumers, forming the ideas and values of an educational institution on the basis of reliable information.

Not less popular element of marketing in the field of education is direct marketing, that is, direct interaction with the consumer (applicant) in order to build relationships [18]. If there is a positive impression about the university, then there is trust and the likelihood that they will choose this, and not another educational institution.

A distinctive feature of direct marketing is a wide promotion method, which includes both personal sales and other means of communication. Direct marketing in the activities of the university began to be applied not so long ago. First of all, it is connected with the establishment of direct contacts with applicants.

The areas of direct marketing in the field of education include direct sales marketing, that is, mailing advertising leaflets, organizing personal meetings of the university representative office and applicants, a database - and relationship marketing aimed at establishing long-term relationships with consumers of educational services [2].

The next key element of marketing in education is the organization of the exhibition activities of the university, which has recently been actively used as a promotion of educational services. The advantage of this marketing tool is the opportunity to work with interested parties directly through the organization of fair-trade exhibition activities, combining all types of educational services offered, personal contacts and the opportunity to learn more about the educational institution [3]. And, of course, the most popular way to promote products and services in any area is marketing in the global Internet. Electronic means of communication of higher education institutions include at least four areas of marketing activities:

(1) the creation and maintenance of the university website;
(2) advertising on the Internet;
(3) posting information about the services of an educational organization on industry portals;
(4) University blogging.

The corporate site is the "face" of the university, which should be of interest to all visitors who have fallen on it. With the help of the site, the client should receive all the necessary information about the university: the student - the schedule and news, the applicant - a description and availability of all educational programs, deadlines and the like.

A modern educational website is a communication structure that unites all levels of external and internal interactions, presenting them on the Internet. But despite the breadth of the presented potential of marketing funds, the possibility of promoting the educational services of a higher educational institution by most institutions is not fully utilized or is completely excluded due to funding restrictions. Reluctantly investing in marketing, the representation of the university shows its skepticism about the promotion of services [6].

It must be recognized that marketing in education is still used limitedly. Conducted PR-events are unsystematic in nature and most often boil down to advertising appeals. The reason for this comes from the definition of services and features of their promotion.

A service is an event offered by one of the parties, intangible and not leading to the acquisition of something [12]. Educational services are understood as an aggregate of knowledge and skills, aimed at meeting the various needs of the individual, society and the state, as a whole. Also, educational services are long-term in nature and have a high degree of responsibility. By purchasing them, the consumer does not know what he will receive at the end, what will be the result of his training [13]. As a result, the university in order to attract applicants and convince the feasibility of acquiring educational services presents the parameters of the services offered very clearly, using diplomas, licenses, certificates, and so on. The consumer presents the offered services as not conserved. It is common for a person to forget the information received, and the knowledge gained in the course of scientific and technological progress becomes obsolete. That is why the marketing of educational services should combine both external and internal PR factors.

Thus, to understand the essence of marketing in the field of education, we analyzed its various elements. Namely, the subjects of marketing relations, the problems of marketing of educational services and its content.

At present, marketing is increasingly being introduced into the system of higher education, which, in the first place, satisfies social needs, turns from a costly sphere into a significant factor in the development of the economy [7]. Today, to establish a clear position in the educational services market, the university is creating certain technologies, methods and ways to maintain competitiveness, so the use of marketing tools in the field of education is not in doubt.

The need for analysis, management, planning and control of demand, quality of educational services and products stimulates the representation of the university to use PR-methods in strategic development.

Proper use of marketing, taking into account the interests of the market and society as a whole will ensure the long-term well-being of the university.

2 Results

Social networks are channels of information transfer, the way we share our lives through pictures, words, videos. This is a kind of transmission of our attitude towards other people, objects, programs, events, etc. [9]. This is what unites us and what shows how we share information with each other.

Social network transmission forms are divided into 10 categories: publishing tools, sharing tools, discussion tools, social networks, micro-publishing tools, social aggregation tools, live broadcast, virtual worlds, social games, and multiplayer online games (MMO) [8].

Higher education tools that are in demand: publishing tools such as blogs and wikis, video and slide show sharing tools, discussion tools such as forums, social networks, and micro-publishing tools like Twitter (Fig. 1).

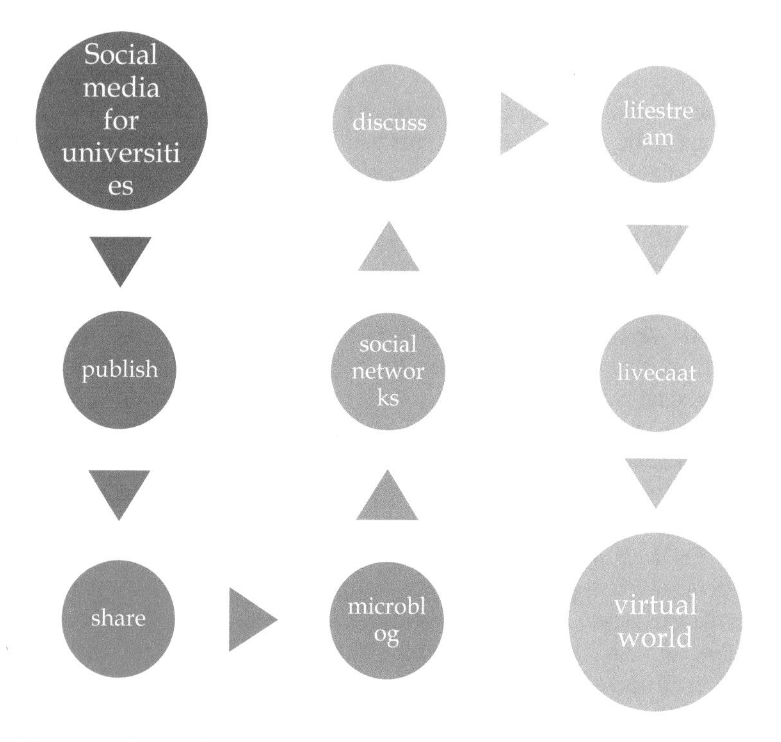

Fig. 1. Social media usage for universities (Source: compiled by the authors)

According to official demographic data, which show that 33% of Facebook users are between the ages of 18 and 24, 30.8% are between the ages of 25 and 34, and the third age group, representing 16.6%, are people of the age from 13 to 17 years. All of this suggests that a significant part of the university community is dependent on social networks, which makes Facebook one of the popular means of communication (Fig. 2).

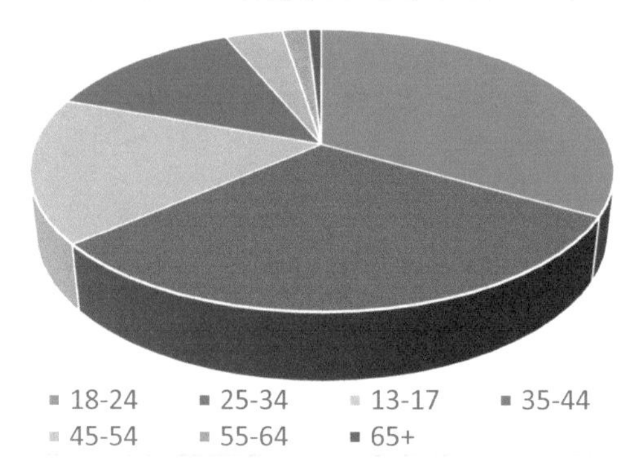

Fig. 2. Age groups using Facebook (Source: compiled by the authors)

Social networks offer their content to provide information for universities that gather around themselves a potential audience in the form of students and teachers, also allow teachers to communicate with already enrolled students and thereby increasing the image and reputation of the university [10]. The profile of a university or faculty in such a network could benefit from an informal language and support an image of an institution that is close to students, who understands their needs and expectations and is interested in knowing their opinions. Intercommunication in this way in a social network is useful for maintaining relationships with former students and for creating a graduate community, which is also an important promotion tool, since potential students often search for information in graduate communities.

As for private universities, they are less interested in this: 16% have a Facebook account, 7% have a Youtube account and only 2% have a Twitter account. None of the private universities use LinkedIn. This trend is the opposite of the situation for universities and colleges in the United States, where they implement communication on social networks faster than state institutions (Fig. 3).

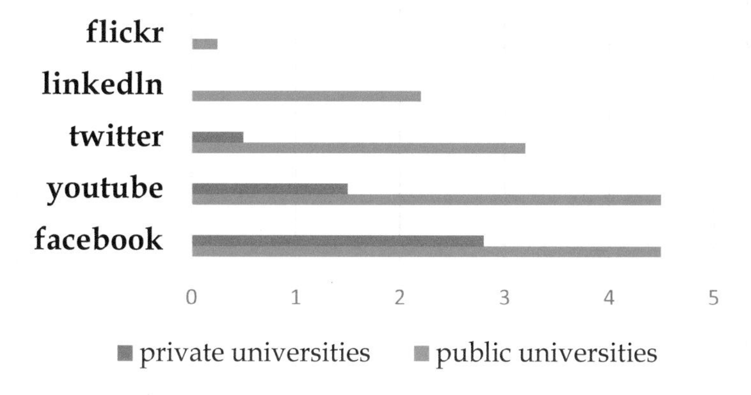

Fig. 3. Comparative use of social media instruments between public and private universities (Source: compiled by the authors)

The social networks accounts are more popular among students in admission periods because of getting the new information regarding the admission process or something else, the results and accommodation options.

3 Conclusion

Internet marketing is available to any educational institution. The space of the Internet, unlike the print or broadcast, is practically unlimited, and the search and access to information is carried out almost instantly.

Applying the method of promoting the university in social networks, information about it acquires a confidential nature and is distributed on the principle of "word of mouth". Deep targeting when using social networks and E-mail marketing, makes it possible to focus the advertising campaign on a specific segment of consumers. Interactive communication allows you to quickly respond to questions and feedback from consumers, maintain dialogue. The relatively low cost of Internet marketing is an indisputable advantage. Information on the website of the educational organization is constantly updated and has the priority of instant response. A professionally and personally sound teacher has a number of effective ways to communicate with the co-workers and even the authorities and efficiently convey them the actual working conditions and future opportunities for the school [16].

In the presence of a significant variety of simple ways to evaluate marketing actions using surveys, questionnaires, analysis of data counters. Traditional methods of collecting information most often show the effectiveness of site traffic, its communicative properties, but at the same time do not evaluate the quality of the presented Internet resources.

Using the Internet technology to explore preferences in issues such as the quality of teaching or administration, the university will face deliberately false statements. Such a reason as an "unfaithful" student "thrown in" may aim at informational deception. This forces marketing to be deliberate, to clearly plan the results of the university's marketing service.

Thus, it can be said that one of the main tasks of Internet marketing of educational services is the creation, development and maintenance of a relevant, constantly filled website, a kind of "calling card" of a university in the spirit of corporate traditions, as well as its positioning in the external environment.

Acknowledgements. The authors thank all anonymous reviewers for their helpful remarks. This research was supported by the BR05236689 "Diversification of tourism in the Republic of Kazakhstan in the transition to a digital economy: strategies and implementation mechanisms" scientific and technical program.

References

1. Al-Rahmi, W.M., Zeki, A.M.: A model of using social media for collaborative learning to enhance learners' performance on learning. J. King Saud Univ. Comput. Inf. Sci. **29**(4), 526–535 (2017). https://doi.org/10.1016/j.jksuci.2016.09.002
2. Arnett, D.B., German, S.D., Hunt, S.D.: The identify salience model of relationship marketing success: the case for non-profit marketing. J. Mark. **67**(2), 89–105 (2003)
3. Baldwin, G., James, R.: The market in Australian higher education and the concept of student as informed consumer. J. High. Educ. Policy Manag. **22**(2), 139–148 (2010). https://doi.org/10.1080/713678146
4. Cavazza, F.: Social media landscape (2008). https://fredcavazza.net/2008/06/09/social-media-landscape/. Accessed 1 Apr 2019
5. Cetin, R.: Planning and implementing institutional image and promoting academic programs in higher education. J. Mark. High. Educ. **13**(1–2), 57–75 (2004). https://doi.org/10.1300/J050v13n01_04
6. Chapleo, C.: Interpretation and implementation of reputation/brand management by UK university leaders. Int. J. Educ. Adv. **5**(1), 7–23 (2004). https://doi.org/10.1057/palgrave.ijea.2140201
7. Constantin Bratianu Ramona Diana Leon: Strategies to enhance intergenerational learning and reducing knowledge loss. VINE, **45**(4), 551–567 (2015)
8. Conway, T., Mackay, S., Yorke, D.: Strategic planning in higher education: who are the customers? Int. J. Educ. Manag. **8**(6), 29–36 (1994). https://doi.org/10.1108/09513549410069202
9. D'Andrea, V., Stensaker, B., Allison, J.: Images and identity in the branding of the university – exploring the symbolic and cultural implications. In: Stensaker, B., D'Andrea, V. (eds.) Branding in Higher Education. Exploring an Emerging Phenomenon, pp. 36–55. EAIR, Amsterdam (2007)
10. Doyle, P.: Value-based marketing. J. Strat. Mark. **8**(4), 299–311 (2011). https://doi.org/10.1080/096525400446203
11. Ivy, J.: A new higher education marketing mix: the 7Ps for MBA marketing. Int. J. Educ. Manag. **22**(4), 288–299 (2008). https://doi.org/10.1108/09513540810875635
12. Kotler, P., Fox, K.: Strategic Marketing for Educational Institutions. Prentice Hall, Upper Saddle River (1995)
13. Lauer, L.D.: Advancing Higher Education in Uncertain Times. Council for Advancement and Support of Education, New York (2006)
14. Noel-Levitz et al.: Focusing Your E-recruitment Efforts to Meet the Expectations of College-Bound Students. Noel-Levitz, Coralville (2010). https://files.eric.ed.gov/fulltext/ED541562.pdf. Accessed 3 Apr 2019
15. Saleh, A.A.: Exploring the use and the impacts of social media on teaching and learning science in Saudi. Procedia Soc. Behav. Sci. **182**, 213–224 (2015). https://doi.org/10.1016/j.sbspro.2015.04.758
16. Sayabek, Z., Madiyarova, A., Ulan, T., Gulvira, A., Aizhan, K., Zhanar, T.: Role of leaders in developing expertise in teaching and their influence on teachers in Kazakhstan. Acad. Strat. Manag. J. **17**(3), 1–13 (2018)
17. Mahadi, S.R.S., Jamaludin, N.N., Johari, R., Fuad, I.N.F.M.: The impact of social media among undergraduate students: attitude. Procedia Soc. Behav. Sci. **219**, 472–479 (2016)
18. Solis, B.: The Essential Guide to Social Media (2008). https://www.briansolis.com/2008/06/essential-guide-to-social-media-free/. Accessed 3 Apr 2019

Modelling of Software Producer and Customer Interaction: Nash Equilibrium

T. Czegledy[1], R. V. Fedorenko[2(✉)], and N. A. Zaichikova[2]

[1] University of Sopron, Sopron, Hungary
[2] Samara State University of Economics, Samara, Russia
fedorenko083@yandex.ru

Abstract. The article aims to simulate the process of interaction between software provider and his client. The authors consider the problem of information support of foreign economic activity. For successful integration of regions into the system of world economic relations, it is necessary to provide exporters and importers with modern software. The problem is the mismatch between the strategies of the software provider and the client. An additional complication is the volatility of the parties' behaviour. Both the provider and the user of the software must constantly review their strategy of behavior. To find the optimal variant of interaction, the authors calculated the Nash equilibrium.

Keywords: Game theory · Software · Foreign economic activities · Nash equilibrium · Math modelling

1 Introduction

The use of modern software products is common in the analysis of the processes of international and interregional trade. For example, in Halkos et al. article authors explored the possibilities of using the MATHEMATICA program, which pictures the multilateral trade connections between regions [6].

Modern software products are of great importance for companies engaged in foreign trade operations. Researchers use programs to analyze a variety of data, as well as to facilitate procedures for negotiating contracts. Dietz noted that relational contracts and reputational networks are nowadays far more effective due to developments in the field of information and communication technology [4].

The success of regional integration depends on the ability of regional companies to compete. In conditions of rapidly developing technologies, exporters and importers have to pay serious attention to the issues of information support of their own activities. Organizations are forced to adapt their Enterprise Resource Planning (ERP) systems. It is important to perform upgrades in order to react to rapidly changing business environments, technological enhancements and rising pressure of competition [2]. Researchers have noted serious difficulties in determining key criteria for choosing an information system. Particularly difficult is the selection of software for small and medium-sized companies. They need a more flexible information support system.

Many researchers turned to the game theory toolkit to study the problems of interaction between individual enterprises. Yakhneeva et al. investigated the possibility

S. I. Ashmarina et al. (Eds.): ISCDTE 2019, LNNS 84, pp. 298–307, 2020.
https://doi.org/10.1007/978-3-030-27015-5_36

of applying game theory methods to modeling the interaction between a client and a software provider [14]. A number of articles were devoted to problems arising from participants in foreign economic activity. Fedorenko et al. [5] presented the results of mathematical modeling of the interaction between participants in foreign economic activity and customs service providers. Nagurney et al. c used Nash equilibrium to solve interoperability problem of freight service providers and shippers [9].

2 Methodology

The authors have developed a model of bimatrix game, the parties of which are a software provider and a participant in foreign economic activity (client). In order to find the optimal strategy for the interaction of players, the construction of Nash equilibrium was carried out, the possibilities of maximizing the gain for each of the parties were determined. The units of analysis are:

1. Developer and/or software provider.
2. A client facing a business automation task.

We modeled the interaction situation using a bimatrix game, a non-zero-sum game in which each player has two strategies.

Player A - developer and/or software provider implementing strategies A1, A2, payment matrix - A.
Player B is a small and medium client who is a second player and implements strategies B1, B2, the payment matrix is B.

To achieve their goals, the provider (player A) can choose two strategies:

A1 - to offer standardized software products with the sale of a full license to the client and the possibility of subsequent maintenance and renewal by subscription;
A2 - to develop a cloud service, offering the client access to the necessary software modules for a renewable subscription.
The client can continue to cooperate (B1) or leave (B2).

Conventional numbers and parameters included in the matrix A:

R – revenue from selling standard boxed software,
v – number of purchases of boxed versions,
t – subsequent service and subscription update revenue,
u – the number of requests for subsequent updates and maintenance,
s – development and promotion costs of standard boxed software versions,
r – cloud subscription service revenue,
U – number of requests for service through the cloud service,
S – cloud development costs.
Restrictions: $R > r$; $S > s$; $U > u$; $U > v$.

The matrix of the provider's winnings depending on the client's different strategies is presented in the Table 1.

Table 1. Provider win matrix

Player A	B_1	B_2
A_1	$a_{11} = Rv + tu - s$	$a_{12} = R - s$
A_2	$a_{21} = rU - S$	$a_{22} = r - S$

Source: calculated and worked out by the authors.

Similar client payoff matrix has the following parameters:

-R – purchase costs for a standard boxed software product,
v – number of purchases of boxed versions,
-r – cloud subscription service costs,
t – subsequent maintenance costs and subscription upgrades when buying a boxed version,
u – number of requests for subsequent updates and maintenance,
U – number of requests for service through the cloud service,
W – benefit that a customer has when buying software that best meets the needs of the customer's business and its subsequent update,
w – benefit that a customer has when buying software and then leaving for another toolkit using free software analogues or unlicensed software,
J – benefit when using a cloud service, due to the possibility of abandoning its own information infrastructure and the ability to connect the necessary modules on demand,
j – benefit of avoiding the provider due to the possibility of using the management tools studied as a result of working with the purchased software with the use of free software analogues or unlicensed software.

Restrictions:

$$R > r;\ W > w;\ U > u;\ U > v;\ J > j \tag{1}$$

The matrix of client's winnings depending on the various strategies of the provider is presented in the Table 2.

Table 2. Client winnings matrix

Player B	B_1	B_2
A_1	$b_{11} = -Rv - tu + W$	$b_{12} = -R + w$
A_2	$b_{21} = J - rU$	$b_{22} = j - r$

Source: calculated and worked out by the authors.

3 Results

Solving a bimatrix game in a noncooperative version in pure strategies comes down to finding the maximin strategies of the players, i.e., strategies that provide players with Therefore, to begin the analysis, we will find *MaxiMin* for player *A*, choosing the minimum values in the lines first, and then determining the largest of them. The maximum guaranteed winnings for player *A* will be equal to $\max_i \min_j a_{ij} = a_{12} = R - s$, and *MaxiMin* strategy is *A1*. Player *B*'s maximum guaranteed winnings will be equal to $\max_j \min_i b_{ij} = b_{22} = j - r$, and client's *MaxiMin* strategy is B_2. With this distribution of winnings, there is no equilibrium Nash situation in pure strategies.

Then we will look for a Nash equilibrium in mixed strategies. According to the basic theorem of the existence of a Nash equilibrium, in any bimatrix game there is at least one such equilibrium.

We find the Nash equilibrium strategies of the players:

$$\begin{cases} \begin{cases} M_A(x^*,y^*) \geq M_A(1,y^*), \\ M_A(x^*,y^*) \geq M_A(0,y^*), \end{cases} & (2) \\ \begin{cases} M_B(x^*,y^*) \geq M_B(x^*,1), \\ M_B(x^*,y^*) \geq M_B(x^*,0), \end{cases} & (3) \end{cases}$$

In this case, expected value is equal to:

$$M_A(x, y) = \bar{X} \cdot \left(A \cdot \overline{Y}^T\right),$$

$$M_B(x, y) = (\bar{X} \cdot B) \cdot \overline{Y}^T.$$

The first player's win function of the for arbitrary strategies *(x, y)* is:

$$M_A(x, y) = x_1(a_{11}y_1 + a_{12}y_2) + x_2(a_{21}y_1 - a_{22}y_2).$$

Since $x_2 = 1 - x_1$, $y_2 = 1 - y_1$, then, substituting, we get:

$$M_A(x_1, y_1) = x_1(y_1(a_{11} + a_{22} - a_{12} - a_{21}) + a_{12} - a_{22}) + y_1(a_{21} - a_{22}) + a_{22}.$$

The second player's win function for arbitrary strategies *(x, y)* is:

$$M_B(x, y) = y_1(b_{11}x_1 + b_{21}x_2) + y_2(b_{12}x_1 + b_{22}x_2).$$

Since $x_2 = 1 - x_1, y_2 = 1 - y_1$, then the condition is:

$$M_B(x_1, y_1) = y_1(x_1(b_{11} + b_{22} - b_{12} - b_{21}) + b_{21} - b_{22}) + x_1(b_{12} - b_{22}) + b_{22}.$$

We define the winning function of the first player when the second player applies the Nash equilibrium strategy y^*, and the first player applies an arbitrary mixed strategy x (in the sense of defining the system (2), (3)):

$$M_A(x, y^*) = x(y^*(a_{11} + a_{22} - a_{12} - a_{21}) + a_{12} - a_{22}) + y^*(a_{21} - a_{22}) + a_{22}.$$

According to Nash's definition, for all mixed strategies x the condition is:

$$M_A(x^*, y^*) \geq M_A(x, y^*).$$

Then, substituting into the last inequality the expressions obtained for the gain functions, we get:

$$(x^* - x)(y^*(a_{11} + a_{22} - a_{12} - a_{21}) + a_{12} - a_{22}) \geq 0, \forall x \in [0; 1].$$

If the Nash equilibrium strategy does not coincide with any pure one, that is, $0 < x < 1$, then the previous inequality holds if and only if the conditions are:

$$y^*(a_{11} + a_{22} - a_{12} - a_{21}) + a_{12} - a_{22} = 0. \quad (4)$$

From here we obtain the value y^*, for which there exists a mixed strategy x^*, that does not coincide with the pure strategies of the first player.

Similarly, for all mixed strategies of the second player y, the inequality is:

$$M_B(x^*, y^*) \geq M_B(x^*, y).$$

And the condition under which the second player has an equilibrium mixed strategy y^*, that does not coincide with his pure strategies is:

$$x_1(b_{11} + b_{22} - b_{12} - b_{21}) + b_{21} - b_{22} = 0. \quad (5)$$

According to corollary (4) of the basic Nash inequality:

$$y_1^* = \frac{a_{22} - a_{12}}{a_{11} + a_{22} - a_{12} - a_{21}} = \frac{r - S - R + s}{R(v - 1) + tu - r(U - 1)}.$$

Taking into account the natural restrictions imposed on the variables of the problem and inequalities (1), for y_1^* we obtain the fulfillment of the conditions for the frequencies: $0 \leq y_1^* < 1$. According to the normalization condition $y_2^* = 1 - y_1^*$, then:

$$y_2^* = \frac{Rv + tu - rU + S - s}{R(v - 1) + tu - r(U - 1)}.$$

For x_1^* also, using corollary (5) of the main Nash inequality, we get:

$$x_1^* = \frac{b_{22} - b_{21}}{b_{11} + b_{22} - b_{12} - b_{21}} = \frac{j - J + r(U - 1)}{W + j + r(U - 1) - tu - R(v - 1) - J - w}.$$

Given the constraints imposed on the variable tasks (0), for x_1^* we also obtain the fulfillment of the conditions for the frequencies $x_2^* = 1 - x_1^*$, then:

$$x_2^* = \frac{U + p(n - 1)}{W + U - u + p(n - 1)}.$$

We find the average winning of the player *A* or the function of winning the first player $M_A(x^*, y^*)$ from the definition of expected value:

$$M_A(x^*, y^*) = x_1^*(a_{11}y_1^* + a_{12}y_2^*) + x_2^*(a_{21}y_1^* + a_{22}y_2^*).$$

Performing a test taking into account the natural constraints imposed on the variables of the problem and inequalities (1), we obtain $M_A(x^*, y^*) > 0$.

We find the average winning of the player *B* or the function of winning the first player $M_B(x^*, y^*)$ from the definition of expected value:

$$M_B(x^*, y^*) = y_1^*(b_{11}x_1^* + b_{21}x_2^*) + y_2^*(b_{12}x_1^* + b_{22}x_2^*),$$

Performing a test taking into account the natural constraints imposed on the variables of the problem and inequalities (1), we obtain $M_B(x^*, y^*) > 0$.

As a result of the search for solutions of systems of inequalities (2), (3), by substituting the available values of the gains, we obtain a set of solutions. For subsystem (2) it consists:

(1) of all situations *(0, y*)*, where $0 \leq y^* \leq \frac{r - S - R + s}{R(v-1) + tu - r(U-1)}$;
(2) of all situations $\left(x^*, \frac{r - S - R + s}{R(v-1) + tu - r(U-1)}\right)$, where $0 < x^* < 1$;
(3) of all situations *(1, y*)*, where $\frac{r - S - R + s}{R(v-1) + tu - r(U-1)} \leq y^* \leq 1$.

The set of solutions of subsystem (3) consists:

(1) of all situations *(x*, 0)*, where $\frac{j - J + r(U-1)}{W + j + r(U-1) - tu - R(v-1) - J - w} \leq x^* \leq 1$;
(2) of all situations $\left(\frac{j - J + r(U-1)}{W + j + r(U-1) - tu - R(v-1) - J - w}, y^*\right)$, where $0 < y^* < 1$;
(3) of all situations *(x*, 1)*, where $0 \leq x^* \leq \frac{j - J + r(U-1)}{W + j + r(U-1) - tu - R(v-1) - J - w}$.

The solution of the game as a whole is the values of $\mathbf{x}^*$ и $\mathbf{y}^*$, common to the considered sets. The image of the sets in the form of graphs (Fig. 1, for the subsystem (2) – bold line, for the subsystem (3) – dash line) indicates that there is one common point C located at the intersection of the broken lines. This allows us to conclude that there is a single Nash equilibrium in the game under study.

Thus, the solution in mixed strategies is the following player behaviors:

- for player *A* it is optimal to apply the A_1 strategy (to offer standardized software products with the sale of a full license to the client and the possibility of subsequent maintenance and updating by subscription) with a frequency $x^* = \frac{j - J + r(U-1)}{W + j + r(U-1) - tu - R(v-1) - J - w}$ and apply the A_2 strategy (developing a cloud service, offering the client access to the necessary software modules for a renewable

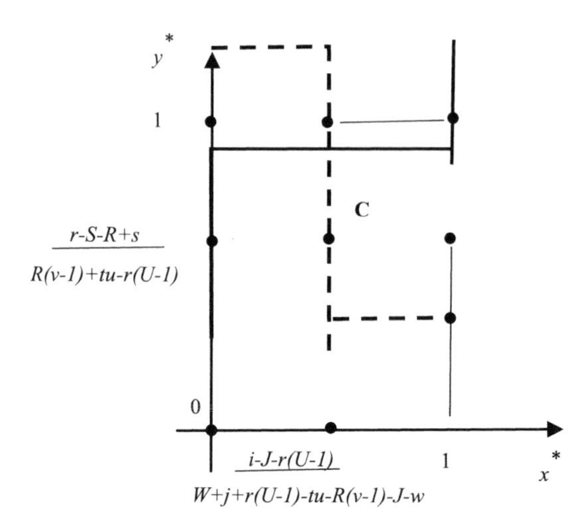

Fig. 1. Graphic solution of the problem of finding the equilibrium state (Source: worked out by the authors).

subscription) with the frequency $1 - x^* = \frac{-tu - R(v-1)}{W + j + r(U-1) - tu - R(v-1) - J - w}$. In this case, the average gain will be $M_A(x^*, y^*) \geq M_A(x, y^*)$.

- for player B it is optimal to apply the B_1 strategy (continue cooperation) with the frequency $y^* = \frac{r - S - R + s}{R(v-1) + tu - r(U-1)}$ and apply the B_2 (leave) with frequency $1 - y^* = \frac{Rv + tu - rU + S - s}{R(v-1) + tu - r(U-1)}$. In this case, the average gain will be $M_B(x^*, y^*) \geq M_B(x^*, y)$. Moreover, the equilibrium winnings of players A and B exceed their winnings with *MaxiMin* strategies.

4 Discussion

Successful work of participants of foreign economic activity is determined by their ability to effectively handle large volumes of information. The most advanced method of data processing can be considered the use of specialized software, namely computer systems for data analysis (SAS, SPSS, STATISTICA, Stata, etc.) [10].

The modern level of technology development opens up many opportunities to simplify the process of providing information to companies. Cloud technologies are actively developing. Cloud technologies allow traders to reduce their own costs and provide access to the necessary tools. For small and medium enterprises, this may be the most effective way to provide information. Cloud computing is widely used because there is more demand for flexibility in obtaining and cost-effectively releasing computing resources [8]. Cloud technologies provide the best price/quality ratio.

However, there are certain problems with the use of cloud technologies. Researchers highlight the problem of determining ownership of an information system. The cloud is made up of hardware and software, which is property of the service provider. On the other side, there is a definite need to ensure that users retain control over their data [3]. In addition, the cloud provider is not sure that customers will stay with him forever. In order to distinguish themselves from their competitors, cloud providers should offer superior services that meet customers' expectations [15].

The implementation of the development process and the subsequent maintenance of the software product is associated with a number of difficulties. Many researchers note the absence of a unified system of work organization and the need to use separate approaches in each specific case. Estimating methods assume an effort/duration trade-off relationship based mostly on theory or expert judgment [12].

A difficult problem is to determine the required level of quality of the developed software. This problem rises in the work of Barney, et al. Software quality is complex with over investment, under investment and the interplay between aspects often being overlooked as many researchers aim to advance individual aspects of software quality [1].

Users of cloud services also experience difficulties of effective interaction. Somu et al. in his work offers Hypergraph based Computational Model and Minimum Distance-Helly Property algorithm for ranking the cloud service providers [11]. The big problem is the level of customer confidence in the provider and its software products. So, despite the fact that many solutions can be automated, not all managers completely trust the recommendations of the machine. Empirical evidence suggests that managers habitually judgementally adjust the output of such solutions, such as replenishment orders or re-order levels [13].

Game theory toolkit can be used as a solution to the problems of interaction between software providers and their clients. The traditional assumption in the theory of games is that players are opaque to each other - if player changes strategies, this change in strategies does not affect the choice of strategies of other players. Halper et al. notes that translucent players may achieve significantly more efficient outcomes than opaque ones [7]. One of the effective methods for finding a mutually beneficial cooperation strategy is the Nash equilibrium model proposed in the article.

5 Conclusion

Improving the competitiveness of regional companies in the foreign market is impossible without the use of modern software. However, for many small and medium participants of foreign economic activity, the need for specialized software arises from time to time. In these conditions, it becomes important to find the optimal strategy of interaction with the software provider.

The authors used economic-mathematical modeling to substantiate the choice of the strategy of interaction between the software provider and his client. Accounting for the second party behavior was important for the model construction. The provider can get the maximum income by implementing the boxed versions of its software without investing in their constant updating. In the presence of high demand, this approach will provide maximum income. However, exporters and importers often have needs for specific software that is not available in a standard boxed product. The inability to connect the necessary modules as needed will lead to the gradual departure of customers. At the same time, there are quite a large number of customers who do not need cloud service. Thus, we can conclude that a mixed strategy would be optimal. The use of economic and mathematical tools of game theory, analyzed in the article, allows you to find the optimal strategy of interaction between the software provider and its clients.

Acknowledgments. The reported study was funded by RFBR and FRLC according to the research project №. 19-510-23001.

References

1. Barney, S., Petersen, K., Svahnberg, M., Aurum, A., Barney, H.: Software quality trade-offs: a systematic map. Inf. Softw. Technol. **54**(7), 651–662 (2012). https://doi.org/10.1016/j.infsof.2012.01.008
2. Barth, C., Koch, S.: Critical success factors in ERP upgrade projects. Ind. Manag. Data Syst. **119**(3), 656–675 (2019). https://doi.org/10.1108/IMDS-01-2018-0016
3. Bartolini, C., Santos, C., Ullrich, C.: Property and the cloud. Comput. Law Secur. Rev. **34** (2), 358–390 (2018). https://doi.org/10.1016/j.clsr.2017.10.005
4. Dietz, T.: Contract law, relational contracts, and reputational networks in international trade: an empirical investigation into cross-border contracts in the software industry. Law Soc. Inq.-J. Am. Bar Found. **37**(1), 25–57 (2012). https://doi.org/10.1111/j.1747-4469.2011.01281.x
5. Fedorenko, R., Zaychikova, N., Abramov, D., Vlasova, O.: Nash equilibrium design in the interaction model of entities in the customs service system. Math. Educ. Res. J. **11**(7), 2732–2744 (2016)
6. Halkos, G., Tsilika, K.: Trading structures for regional economies in CAS software. Comput. Econ. **48**(3), 523–533 (2016). https://doi.org/10.1007/s10614-015-9515-6
7. Halpern, J., Pass, R.: Game theory with translucent players. Int. J. Game Theory **47**(3), 949–976 (2018). https://doi.org/10.1007/s00182-018-0626-x
8. Meghana, H., Asish, O., Lewlyn, L.: Prioritizing the factors affecting cloud ERP adoption – an analytic hierarchy process approach. Int. J. Emerg. Markets **13**(6), 1559–1577 (2018). https://doi.org/10.1108/IJoEM-10-2017-0404
9. Nagurney, A., Shukla, S., Nagurney, L., Saberi, S.: A game theory model for freight service provision security investments for high-value cargo. Econ. Transp. **16**, 21–28 (2018). https://doi.org/10.1016/j.ecotra.2018.09.002
10. Skorodumov, P., Badanin, D.: Construction of a virtual environment of economic modeling. Quest. Territorial Dev. **4**(24), 1–12 (2015)
11. Somu, N., Kirthivasan, K., Sriram, V.: A computational model for ranking cloud service providers using hypergraph based techniques. Future Gener. Comput. Syst. Int. J. Escience **68**, 14–30 (2017). https://doi.org/10.1016/j.future.2016.08.014
12. Symons, C.: Exploring software project effort versus duration trade-offs. IEEE Softw. **29**(4), 67–74 (2012). https://doi.org/10.1109/MS.2011.126
13. Syntetos, A.A., Kholidasari, I., Naim, M.M.: The effects of integrating management judgement into OUT levels: in or out of context? Eur. J. Oper. Res. **249**(3), 853–863 (2016). https://doi.org/10.1016/j.ejor.2015.07.021

14. Yakhneeva, I.V., Agafonova, A.N., Fedorenko, R.V., Shvetsova, E.V., Filatova, D.V.: On collaborations between software producer and customer: a kind of two-player strategic game. In: Ashmarina, S., Mesquita, A., Vochozka, M. (eds.) Digital Transformation of the Economy: Challenges, Trends and New Opportunities. Advances in Intelligent Systems and Computing, vol. 908, pp. 570–580. Springer, Cham (2020). https://doi.org/10.1007/978-3-030-11367-4_56
15. Zheng, X., Martin, P., Brohman, K., Xu, L.: CLOUDQUAL: a quality model for cloud services. IEEE Trans. Ind. Inform. **10**(2), 1527–1536 (2014). https://doi.org/10.1109/TII.2014.2306329

The Readiness of the Economy for Digitalization: Basic Methodological Approaches

E. L. Sidorenko[1(✉)] and Z. I. Khisamova[2]

[1] Moscow State Institute of International Relations (University) of the Ministry of Foreign Affairs, Moscow, Russia
12011979@list.ru

[2] Krasnodar University of the Ministry of Internal Affairs of Russia, Krasnodar, Russia

Abstract. The study represents a comprehensive analysis of modern approaches to the assessment of digitalization. In the context of the qualitative transformation of industrial infrastructure and transition to digital tools for economic development, it is urgent to develop the most understandable and methodologically consistent methodology for assessing the efficiency of digitalization. In addition, such a methodology must meet both purely political and opportunistic objectives - to ensure a uniform approach to assessment and to allow predicting digitalization trends for the short and medium term. The authors review the existing models assessing the digital economy, propose the author's classification of models and test it when considering current foreign approaches. Special attention is paid to advantages and disadvantages of the models used and an attempt is made to harmonize them with each other.

Keywords: Assessment · Business · Digitalization · Digital technologies · Digital transformation · Index · Index structure

1 Introduction

The rapid development of the digital economy has become the main in the global development agenda. The course on digital transformation is proclaimed by the leading world powers and transnational corporations. Such a desire for a universal digital transformation of production processes is due to the potential of digital technologies for business development. They break down barriers and lead to the restructuring of entire industries [15], opening up new horizons for business.

The simultaneous progress in the areas of the Internet of Things (IoT), Big Data analysis, cloud computing and artificial intelligence (AI) makes giant innovations possible and will fundamentally transform economic activity, government and society into the next two decades. To use these benefits, countries need to create conditions conducive to the deployment of infrastructure networks and services. The adoption of development programs (concepts) that promote experimentation and innovation, while at the same time moderating implications of possible risks to information security, confidentiality and employment [10] also becomes relevant.

S. I. Ashmarina et al. (Eds.): ISCDTE 2019, LNNS 84, pp. 308–316, 2020.
https://doi.org/10.1007/978-3-030-27015-5_37

Undoubtedly, the prospects for the introduction of new technologies are also obvious for Russian companies whose business activities are carried out under conditions of a strict sanctions regime, which significantly reduces opportunities for new long-term investments.

As noted in the KPMG study [12], for the overwhelming share of large Russian companies, digital transformation is a real opportunity to reduce company costs, increase productivity, optimize customer interaction processes, and develop business. In 2017–2018 the most advanced Russian companies implemented pilot projects.

However, despite this, most large Russian corporations have not adopted comprehensive digitalization programs. Digital activity is limited, as a rule, to single pilot projects for individual digital solutions. The state is interested in a fundamental simplification of administrative procedures, streamlining the process of regulating business activities and, as a result, in the growth of the small and medium business segment.

The target indicators of the national program "Digital Economy of the Russian Federation"[1] illustrate the successful operation of a number of competitive companies - operators of ecosystems and organizations implementing projects in priority areas of international scientific and technical cooperation in the field of digital economy, as well as the creation of a number of enterprises for creating digital technologies and platforms and providing digital services in global markets by 2024. However, the achievement of these indicators is impossible without assessing the current level of digitalization of Russian business, identifying key technologies that have been successfully tested.

The influence of digitalization on national economic and social life depends on the place of the state in the international arena. To determine (assess) the integration of digital transformation processes into the economic and social life of a country, we use indicators quantifying these parameters.

Of particular importance is the assessment of digital solutions, as well as the analysis of digital projects in the country's economy in light of the growth in government spending on the national program. Separate attention should be paid to the assessment of the perception of digitalization by business, since without understanding of digital transformation mechanisms by the corporate management, it is impossible to create large-scale digital transformation programs.

2 Methodology

The methodological basis of the study is a systematic approach to the assessment of the digital economy. In the course of the study, the main practical and doctrinal approaches determining the efficiency of digital infrastructure were classified and analyzed. Such traditional scientific methods as dialectical, logical, method of scientific generalizations, content analysis, comparative analysis, synthesis, source study, etc. were used in processing factual material. Their use made it possible to ensure the validity of the

[1] National program "Digital Economy of the Russian Federation" (approved by the Presidium of the Presidential Council of the RF for Strategic Development and National Projects of 24.12.2018 № 16).

analysis, theoretical and practical conclusions. The authors also used a set of statistical methods: summaries and groupings, averages and relative values, index, analysis of time series, selective, and others.

3 Results

The overall purpose of this study is to assess current processes and then determine the areas of business for companies that require actions that will enable companies to get practical guidance for their own digital transformation.

Each company has its own way in the digital economy, so you need to start with an analysis of the current situation and goals of each company. The current situation of the company will help assess the following questions: What are its strategic goals for the next few years? Does the company plan to implement measures for the digital economy? What technologies and systems are already implemented and how do they function within the company?

The answers to these questions can be used to determine which characteristics of the company still need to be acquired for the successful implementation of the digitalization concept. The index will assess technological, economic, and organizational aspects of the company with a focus on business processes. The structure of the company for assessment is proposed to be divided into four areas, characterized in 4 sub-indexes, combining 44 indicators.

I. Sub-index of the social component of the project:
Social audit; Creation of new jobs; Staff restructuring; Wage level and working conditions of staff; Staff welfare; Increasing digital and financial literacy of staff; Creation of new types of jobs in areas related to the provision of digital services; Capacity building to strengthen activity centers; Personnel health and safety; Compliance with the requirements of national labor laws; Compliance with the requirements of the International Labor Organization; Mitigation of staff reduction.

II. Sub-index of the economic efficiency of the project:
Net present value; Internal rate of return; Yield index; Simple payback period; Payback period, taking into account discounting; Discount rate; Average rate of return; Profitability index; Investment index; Fixed investment.

III. Sub-index of innovation (information equipment of the project, compliance with the criteria of the National Project of the digital economy):
Digital competencies through sensors and activators; Automated data acquisition; Decentralized (preliminary) data processing; Communication efficiency; Task-based interface; Specialized user interface; Data analysis; Contextual data provision; Data management; Resiliency of IT infrastructures; Situational data warehouse; New - digital businesses; Network infrastructure.

IV. Sub-index of company/project management culture assessment:
Management style; Confidence in processes and information technology; Recognition of the use of errors; Openness to innovation; Data-based learning and decision making; Continuous professional growth; Increased transparency

and security of client-company interaction; Environmental friendliness (losses for ecology and ensuring environmental safety).

To calculate the index, the values of indicators are taken from the financial and other statements of the company (sub-index 1, 2). Determining the integration should be based on characteristics for the digital economy (sub-index 3). Characteristics are assessed using a questionnaire, which assess the characteristics for each process, as well as the absolute values presented in open sources and company reports (sub-index 4). The values of indicators are reduced to a single unit of assessment and are normalized. The value of each sub-index is calculated as the arithmetic average of the values of the corresponding group of indicators. If necessary, the values of sub-indexes are weighed in a specific research situation. In the general case, the weights of sub-indexes are equal to each other – 0.25. The index value is calculated as the sum of weighted values of sub-indices.

4 Discussion

Today, there are a number of generally accepted methods for assessing the digital economy, which are widely reflected in the media and official sources. The purpose of this study is to assess the efficiency and conformity with the current realities of the existing methods for assessing economic digitization in general, and business, in particular. Our study allowed us to divide the existing methods of assessing digitalization into several subgroups depending on the subject of the study:

Indices Assessing the Level of ICT Development

This group includes studies aimed at determining the level of development of information and telecommunication technologies, in particular, assessing the spread of broadband access to the Internet, the presence of telephone communications.

The undoubted advantage of such indices is the ability to assess the state in terms of their readiness for digital transformation - the presence of a basis in the form of a developed information infrastructure, reflecting the penetration of digitalization into various areas of government and society.

The key index for assessing the development of modern technologies is the Information and Communication Technologies Development Index (ICT Development Index, IDI), as measured by the International Telecommunication Union [10]. The index consists of 11 statistical indicators reflecting the availability and use of ICT. The index has 3 sub-indices:

"Access", which characterizes the parameters of providing mobile communications and broadband Internet access;

"Use", which characterizes the degree of public involvement in the use of ICT;

"Skills", which reflect the intellectual and educational level of the population.

Despite the generally recognized authority of this index, it should still be noted that the index criteria currently do not reflect all the technologies available and actively used within the digital economy, such as the Internet of Things, Big Data, cloud computing, blockchain, CRM, robotization. In our opinion, the lack of measurements about the implementation of these technologies significantly reduces the representative value of the Index in assessing the development of information and communication technologies.

Among the indices assessing the degree of development and commercialization in the country of advanced technologies, the Global Connectivity Index (GCI-C) developed by Huawei [9] is worth mentioning.

The index consists of 16 indicators in two dimensions: current engagement and growth momentum. The first indicator assesses the current state of affairs in each country and determines its position in the race to the ubiquitous broadband Internet connection, while the last indicator assesses the speed of development towards the same goal. The theoretical basis of the GCI-C coefficient is the traditional theory of price supply and demand (availability of goods), with a superstructure as an experience category. Supply and demand correlate positively – the growth of supply increases the availability of goods and develops the user experience, which stimulates demand. In turn, this raises the level of expectations of new experience and warms up the offer. The impulse to growth is shown by what could be called "additional demand" – the growth of key factors stimulating demand, taking into account investments and the favor of state bodies. For each of the 16 indicators characterizing the elements of infrastructure, investment, availability of broadband Internet and the availability of demand, the values are ranked in order of their assessment. For example, speaking about the indicator of mobile phone connectivity to the Internet, states are distributed in order of the prevalence of this technology in 2013, while the percentile formula is applied to the 20th, 40th, 60th and 80th dividing points on the rating scale. Then quantile estimates are applied to 5 groups, distributing from 1 (worst implementation) to 5 (best implementation).

This group should also include the Networked Readiness Index (NRI), conducted by the World Economic Forum, and characterizing the level of development of information and communication technologies (ICT) in different countries of the world according to 53 parameters. The indicators are grouped into 3 groups: availability of conditions for ICT development; willingness of citizens, business and government to use ICT; level of ICT use in the public, commercial and public sectors. The results of the calculation are published annually in the framework of the report of the World Economic Forum Global Information Technology Report [2, 7]. However, it is worth noting that the last report was published in 2016, according to which the Russian Federation ranked 41 among 143 countries on network readiness. The New Economy Index is similar, aimed at a comprehensive assessment of the broadband at home and the average speed of the Internet connection [1].

Indices Assessing the Level of Acceptance by Society, Authorities or Various Digitalization Groups

In this group, several indexes should be noted. Thus, the World Bank annually calculates the Digital Acceptance Index (DAI) [18] - a composite index that measures the extent to which digital technologies are distributed within and between countries. Among the advantages of the index, the authors note the degree of availability and introduction of digital technologies in all key economic actors: people, enterprises (firms) and governments, thereby providing a complete picture of technology distribution, as well as building calculations based on their own measurements from the

World Bank databases, thereby providing a certain level of confidence. The index is based on three sectoral indicators: business, people, authorities - where each indicator has the same weight.

- DAI (Business) is a group of four average indicators: the percentage of enterprises that have websites, the number of information security servers, download speed, state coverage with a 3G connection.
- DAI (People) is a group of 2 average indicators from the Gallup's public opinion questionnaire: access to mobile communication in the house and access to the Internet in the house.
- DAI (Government) is a group of 3 average indicators: the core of the management system, public Internet services, digital identification.

In our opinion, it is worth highlighting the maturity index of the digital economy, developed by the Scottish government. The peculiarity of the considered index is the update of indicators, based on modern realities. Thus, the "use of joint consumption" and "cyber resistance" indicators have undergone significant adjustments. The updated index includes 5 positions: adoption, use of joint consumption, benefits, skills and cyber resistance, combining 15 indicators. Each indicator is given a score based on its relative importance in the sense of digital maturity. The maximum score can be 100 points:

Indices Assessing the Cost of Digitization
The Global Innovation Index GII is notable in this group [6]. It characterizes the potential of innovation activity of states and the results obtained. From 2016, the Global Innovation Index is calculated as the average value of two sub-indices: the sub-index of innovation costs (innovation resources), which characterizes the costs of creating institutions, conducting R & D and human capital, creating infrastructure, developing market and business; as well as a sub-index of innovative results, reflecting the actual achievements in these areas: technological and creative achievements. According to the results of the assessment in 2018, Russia in the Global Innovation Index occupied 77th place, significantly worsening its performance in comparison with the previous year.

Indices Assessing Digitalization of Individual Economic Sectors
In 2019, the Institute for Statistical Studies and Economics of Knowledge (ISSEK) has computed a business digitalization index that characterizes the speed of adaptation to digital transformation, the level of use of broadband Internet, cloud services, RFID technologies, ERP systems and the involvement of business organizations in electronic commerce. Russia in the specified index takes the 28th place, dividing it with Hungary [11].

Despite the authors' appeal to the speed of digital transformation, it is impossible to recognize the results of the assessment, which have sufficient representativeness. Thus, the index reflects only calculations for the countries of Europe, Russia, South Korea, Turkey and Japan, while the index claims to be global. The calculation method is also unclear. The research of the Boston Consulting Group [4], McKinsey Global Institute [13] and OECD [14] also assess the level of economic digitalization in Russia and in the world.

Indices Assessing Digitalization of Public Administration
The most authoritative in this group is the United Nations (UN) United Nations Global E-Government Development Index, which assesses the readiness and ability of public authorities to use ICT and provide services in digital form. The advantage of the index is its representativeness in terms of ranking countries by the level of e-government development. The index is formed on the basis of three sub-indices: the degree of coverage and quality of Internet services, the level of development of the ICT infrastructure, human capital.

Composite Indexes
Composite indices assess various areas of digitalization. Among the most common are the Digital Economy and Society Index (DESI), developed by the European Commission [8], calculated on the basis of 5 groups of indicators that combine 31 criteria, each of which characterizes a certain digitalization trend and gives a characteristic of the EU country. To calculate the total coefficient for each group of indicators, experts from the European Commission determined a certain number of points. Each indicator from the "Connectivity" group (reflecting the degree of access to communication systems based on the digital presentation of information, including access to broadband Internet) gets 25% of the total ratio. The indicators of the Human Capital/Digital Skills group have a similar weight in the overall ratio.

The indicators from the Integration of Digital Technology by businesses group are estimated at 20%, since the use of ICT in business is the main engine of progress. Finally, "Online activities or use of the Internet" (Use of Internet by citizens) and "Digital Public Services" are estimated at 15%.

To compare the level of digitalization of the European Union countries with the whole world, the International Digital Economy and Society Index (I-DESI) is used, which contains several extended criteria of the DESI Index.

Similar criteria are also the basis of another composite index Digital Evolution Index 2017 [5]. However, in this index, unlike DESI, emphasis was on the innovation climate (investment in R & D and digital start-ups).

In Russia, ROSATOM has developed the National Digital Economy Development Index, which is part of the World Bank's Digital Economy Country Assessment Initiative, DECA [16]. According to the results of the pilot rating of 32 countries in 2018, the degree of readiness, use and impact of digital technologies on socio-economic development was determined for certain factors that create conditions for digital transformation: human capital, R & D and innovation, business environment, government policy and regulation, information security, digital infrastructure and the digital sector.

Composite indices also differ by territory. Thus, global indices, such as I-DESI, Digital Evolution Index 2017, regional DESI and in-country indices, ranking the subjects of the individual state, such as the Digital Economy Volume Index in US GDP, are calculated by the US Bureau of Economic Analysis [3].

In relation to the Russian Federation, one of the regional indices should be noted in the Rating of Russian Regions by the level of development of information society, calculated by the Ministry of Digital Development, Communications and Mass Communications of Russia. The rating methodology includes 120 indicators and 17 sub-indexes.

The Digital Russia Index of the subjects of the Russian Federation, developed by the Moscow School of Management of the Skolkovo Innovation Center, is also devoted to regional digitalization of constituent entities of the Russian Federation. The index consists of 7 sub-indices and their weights, which allow, using an additive model, obtaining the resulting weighted average of the digitalization index for each subject of the Russian Federation [17].

A comprehensive study of digitalization indices suggests that digitalization, as a world trend, is being actively measured today. At the same time, the subject areas and the composition of indicators of existing approaches do not reflect an integrated approach in assessing the integration of digital projects in the country's economy.

In our opinion, in the conditions of digital transformation, the provision of means for establishing companies' current readiness for Industry 4.0 and identifying concrete measures that will help them reach a higher maturity stage and extract maximum economic benefits from digitalization seems more relevant than ever.

5 Conclusion

Summarizing the conducted study, it should be concluded that today the scientific and expert community has an active interest in assessing various spheres of activity of the state and society in which digitalization processes take place. At the same time, the issues of involvement of business structures, especially small and medium-sized businesses have not been given special attention today. It determines the weak awareness of companies (small and medium managers) about new business models and advantages opened by digitalization. We believe that the index developed by the authors will allow creating an objective picture about the degree of interest of SMEs in the digital transformation and to determine further development.

Acknowledgements. The research was conducted by the authors with the financial support of the Russian Foundation for Basic Research, Project no. 19-011-20091.

References

1. Atkinson, R.D., Wu, J.J.: The 2017 New State Index Benchmarking Economic Transformation in the States (2017). https://ssrn.com/abstract=3066923 or http://dx.doi.org/10.2139/ssrn.3066923. Accessed 3 June 2019
2. Baller, S., Dutta, S., Lanvin, B. (eds.): The global information technology report 2016. Innovating in the digital economy. World economic forum (2016). http://www3.weforum.org/docs/GITR2016/WEF_GITR_Full_Report.pdf. Accessed 19 June 2019
3. Barefoot, K., Curtis, D., Jolliff, W., Nicholson, J.R., Omohundro, R.: Defining and measuring the digital economy. Bureau of economic analysis (2018). https://www.bea.gov/system/files/papers/WP2018-4.pdf. Accessed 19 June 2019
4. Boston Consulting Group: Russia online? Catch up impossible to fall behind. The Boston Consulting Group (2016). http://image-src.bcg.com/Images/Russia-Online-ENG_tcm26-152058.pdf. Accessed 19 June 2019

5. Chakravorti, B., Chaturvedi, R.Sh.: Digital planet 2017. How competitiveness and trust in digital economies vary across the world. The Fletcher School, Tufts University (2017). https://sites.tufts.edu/digitalplanet/files/2017/05/Digital_Planet_2017_FINAL.pdf. Accessed 19 June 2019
6. Dutta, S., Lanvin, B., Wunsch-Vincent, S. (eds.): The global innovation index 2018: Energizing the world with innovation. Cornell University, INSEAD, and WIPO (2018). https://www.wipo.int/edocs/pubdocs/en/wipo_pub_gii_2018.pdf. Accessed 19 June 2019
7. Dutta, S., Geiger, T., Lanvin, B.: The global information technology report 2015. ICTs for inclusive growth. World economic forum (2015). http://www3.weforum.org/docs/WEF_GITR2015.pdf. Accessed 19 June 2019
8. European Commission: DESI Report 2018 – Telecoms Chapters (2018). https://ec.europa.eu/digital-single-market/en/news/desi-report-2018-telecoms-chapters. Accessed 19 June 2019
9. Huawei: Global Connectivity Index (2017). www.huawei.com/ilink/en/download/HW_367221. Accessed 19 June 2019
10. ITU: Report "Measuring the Information Society" (2017). https://www.itu.int/en/ITU-D/Statistics/Documents/publications/misr2017/MISR2017_Volume1.pdf. Accessed 26 Dec 2018
11. Kevesh, M.A., Filatova, D.A.: Business digitalization index. Institute for Statistical Studies and Economics of Knowledge. High School of Economics, Moscow (2019). https://issek.hse.ru/data/2019/02/27/1193920132/NTI_N_121_27022019.pdf. Accessed 19 June 2019. (in Russian)
12. KPMG: Growing pains: 2018 Global CEO Outlook (2018). https://assets.kpmg/content/dam/kpmg/sk/pdf/2018/2018-CEO-Outlook-report.pdf. Accessed 2 June 2019
13. McKinsey & Company: Digital Russia. New reality (2018). https://www.mckinsey.com/ru/our-work/mckinsey-digital. Accessed 14 Feb 2019. (in Russian)
14. OECD: Measuring the digital transformation: a roadmap for the future. OECD Publishing (2019). https://www.oecd.org/publications/measuring-the-digital-transformation-9789264311992-en.htm. Accessed 19 June 2019
15. PwC: Future of the industry: breaking down barriers. PwC Publication Series on Development Prospects (2017). https://www.pwc.ru/ru/publications/the-future-of-industries.html. Accessed 11 Apr 2019. (in Russian)
16. Rosatom State Corporation: National Digital Economy Development Index: Pilot implementation (2018). https://rosatom.ru/journalist/news/tsentr-kompetentsiy-rosatoma-predstavil-na-zasedanii-nablyudatelnogo-soveta-ano-tsifrovaya-ekonomika/. Accessed 19 June 2019. (in Russian)
17. Skolkovo: Methodology for calculating the Digital Russia Index of the constituent subjects of the Russian Federation. Skolkovo, Moscow (2018). (in Russian)
18. World Bank: World development report. Digital Adoption Index (DAI): Measuring the Global Spread of Digital Technologies. Background note. Digital Dividends. WDR (2016). http://pubdocs.worldbank.org/en/587221475074960682/WDR16-BP-DAI-methodology.pdf. Accessed 19 June 2019

Influence of Digitalization on Motivation Techniques in Organizations

E. P. Troshina and V. V. Mantulenko(✉)

Samara State University of Economics, Samara, Russia
mantoulenko@mail.ru

Abstract. The article is aimed at analyzing goals and objectives of staff motivation in the context of globalization of information processes, as well as some ways of providing material and non-material motivation of the staff taking into account actual trends formed in the modern world. The authors identified basic technological systems used in modern organizations to improve competitiveness and new ways of material and non-material motivation of personnel. Within the framework of this topic, the analysis of opinions was carried out and the correlation of different motivational approaches in the conditions of digitalization was revealed. In addition, a significant result of this research is the systematization of approaches to staff motivation in the context of the economy digitalization, taking into account the existing fundamental theories of motivation.

Keywords: Motivation · Digitalization of the economy · Potential · Competitiveness · Management decisions

1 Introduction

The purpose of this article is to determine the most effective methods of motivation of employees in the globalized world of processes and information technologies that affect the change of the economy. The authors also tried to identify the impact of these methods on the behavioral strategies of employees and the development of their professional competencies.

According to the Program “Digital Economy of the Russian Federation” approved by the order of the Government of the Russian Federation of July 28, 2017 N 1632-p. [7], digitalization is a change in the personnel qualification management system, which should become the basis for a qualitative update of human resources in the conditions of changing technological foundations in the future [2].

Many organizations are introducing information technology to improve their competitiveness. The introduction of new information technologies and computer programs changes the attitude to human resources in the organizations. The companies begin improving work and production processes, among them there are changes in organizational behavior (e.g., the division of responsibilities between employees of the company), improvement of technology used by the organization, redistribution of knowledge, introduction of new information methods in the process of information transfer (development of collective responsibility of employees), etc. [1, 6].

S. I. Ashmarina et al. (Eds.): ISCDTE 2019, LNNS 84, pp. 317–323, 2020.
https://doi.org/10.1007/978-3-030-27015-5_38

2 Methods

In this research, we used methods of analysis, synthesis and synthesis of information, the method of survey (questioning). The survey was conducted among 500 entrepreneurs from traditional industries (representative sample) from 40 regions of Russia and 120 high-tech companies (residents of SKOLKOVO).

3 Results

In the context of digitalization, one of the approaches to company management, namely management from the position of optimizing the use of human resources, has become increasingly widespread. This approach means the fullest use of each employee' opportunities and the motivation of employees. The need of organizations for low-skilled personnel is decreasing, which led to the formation of new methods of material and non-material motivation of employees.

However, before moving directly to the motivation of staff in the conditions of digitalization, it was interesting to know the opinion of university graduates and employees of various business areas of the Samara region on this topic to identify barriers to the development of the digital economy. Our survey has shown that the most significant obstacle in the way of digitalization (in respondents' opinion) is the lack of understanding of the nature and significance of the economy digitalization as a whole, as well as the assumption of the inconsistency of their own qualifications to the required one and, as a consequence, the fear of losing or not getting a job. In order to minimize barriers, it is necessary to introduce digital technologies step by step, thoughtfully carrying out explanatory work with employees of enterprises and increasing their level of confidence to digitalization. In addition, you should provide employees with the development opportunities and stimulating them for training and retraining in accordance with the innovative changes.

Motivation of employees has the following characteristics: explanation of human behavior, its orientation and activity, the employees desire to meet their needs through the work, stimulation of someone to activities aimed at achieving the goals of the organization. In general, motivation is the main engine of human behavior and activity, and especially it is necessary in the process of formation of a future professional.

Staff motivation depends on a number of factors, such as individual needs of each person, self-actualization, self-assessment, etc. Methods of material motivation of employees have recently become very widespread not only in foreign companies, but also at domestic enterprises. This is due to the importance to meet physiological needs and to meet the spiritual needs of employees, which require funding. Among the material motivation means, we can identify the following types of motivation (Table 1).

The employer can apply separate types of material motivation or develop the own motivation system in the organization, by means of a combination of various methods available in the real practice, thus certain types of material motivation can be effective and actual in one organization, and useless or even harmful in the conditions of other enterprises [5].

Table 1. Types of material motivation of personnel in the modern world

№	Type of material (financial) motivation	Characteristics
1.	Loans	The salary itself is not always one of the most effective tools to motivate staff, because its increase or decrease involves the use of a large number of procedures
2.	Bonus payment (cash awards)	It is the most common way of material motivation of employees. It can be represented by the "13th salary", separate remuneration which can be assigned according to indicators of employees' work efficiency or by any other methods
3.	Gifts	They can be presented in the form of direct products or services of the organization or in the format of various other goods or other values important to employees. Examples of gifts may include vouchers and tickets to cultural events
4.	Fines	Directly, fines may be used for violation of labor discipline. Employees may be fined by the forfeiture of a bonus if they have some disciplinary sanctions, and in some state structures it may be in the form of demotion with a corresponding decrease in wages
5.	Benefits and compensation	Providing additional material assistance to employees, compensation for rental housing, payment for the use of mobile communication means or public transport

Source: compiled by the authors.

In contrast to the material, non-material (intangible) motivation of the organization's staff can be much more effective as a tool of labor stimulation in the development of management strategies. It is connected primarily with the fact that, for example, the increase in payments motivates employees to work more actively only for a short period, while the intangible methods of staff motivation help to maintain employees' interest in the work result, stimulating creative activity and professional competences development. The types of intangible motivation are presented in Table 2.

Due to the intangible motivation means presented above, holistic employee incentive systems can be developed that will function effectively, taking into account legal, economic, political and social external conditions in which the organization operates. At the same time, each organization is able to create, apply and combine its own methods of non-financial motivation for the staff. The set of motivation tools will depend on the nature of the staff of a particular organization, the specifics of the work and the chosen management style [3].

In addition, if we take as a basis Maslow's hierarchy of needs, then in modern conditions, employees are motivated by the upper levels, including self-actualization, esteem and belongingness needs (f.e., respect from the working team and the company's management, social needs) [4]. Ways to meet these needs in the context of digitalization of the economy are presented in Table 3.

Table 2. Typology of intangible motivation of employees in the context of digitalization of the economy

№ п/п	Type of material (intangible) motivation	Comments
1.	Social	It involves increasing the employee's sense of self-importance by engaging in management decision-making, work in a team, delegation of authority
2.	Psychological	Creating the most favorable environment for employees, establishing relationships in the working team, taking into account the interests of all employees
3.	Moral	The use of motivation methods that allow to realize the need of each employee for respect by the working team and the management of the organization
4.	Organizational	It is implemented by taking care of employees of the company, the organization of work environment and rest facilities during breaks at work

Source: compiled by the authors.

Table 3. Ways to improve the staff motivation in conditions of digitalization in accordance with Maslow's hierarchy of needs

№	Needs	Ways of satisfaction
1.	Self-esteem	- Realization of own potential - Employee training - Delegation of authority - Creative approach to problems solving
2.	Respect from the staff and management of the organization	- The right to make important management decisions - Possibility of training other employees and representation of the organization at exhibitions and forums - Independence in solving some problems - The possession of unique knowledge
3.	Social needs	- Establishing contacts with colleagues - Improvement of corporate culture of the organization
4.	Safety needs	- Stability of wages - The presence of a social package
5.	Physiological needs	- Provision of benefits for the purchase of housing - Free transport for employees

Source: compiled by the authors.

Thus, in the conditions of digitalization of the economy and the active development of management activities in organizations, the efficiency of companies can be improved using the knowledge and skills of each employee of the enterprise.

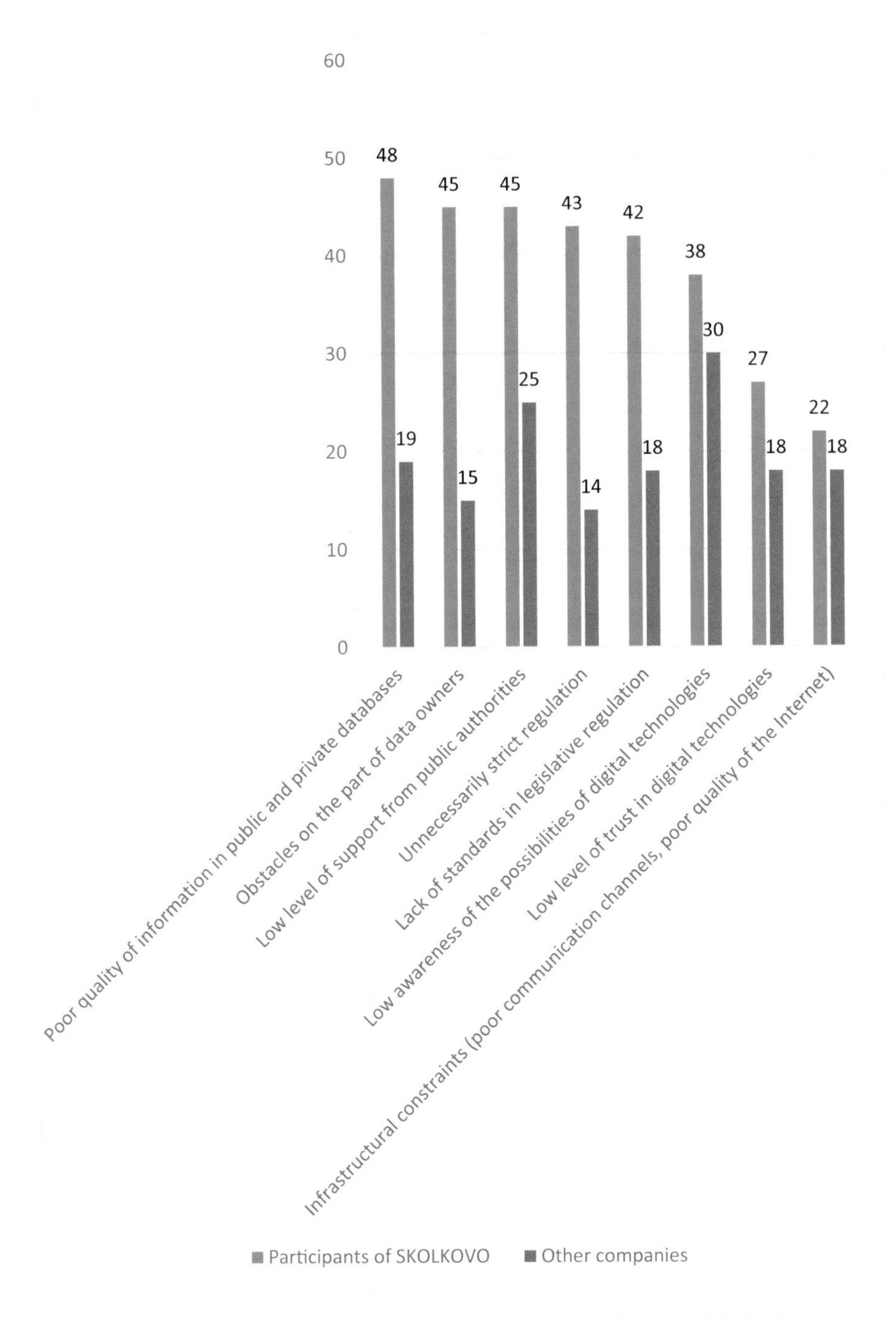

Fig. 1. Barriers to the transition of enterprises in the digital business model, % of respondents (Source: compiled by the authors).

However, on the way to the digital economy, there are various difficulties, the so-called barriers of digitalization. The determination of the degree of their impact on the motivation of personnel requires the identification of the main types of these barriers in the Russian Federation. To solve this task, a survey was conducted among entrepreneurs from traditional industries (representative sample) from different Russian regions and some high-tech companies (residents of SKOLKOVO). According to the survey presented in the form of a histogram (Fig. 1), it can be concluded that SKOLKOVO participants consider as the most significant barriers for the development of the digital business model: the low quality of information in public and private databases, obstacles from data owners and low level of support from public authorities. At the same time, other companies in the market consider a low level of awareness of the possibilities of digitalization as the most significant barrier in this sphere.

Based on the respondents' responses, we see that one of the barriers to the digitalization of the economy is the low level of staff confidence in digital technologies. We suppose that this barrier exists because of the lack of the required level of qualification by employees. In this regard, the respondents were asked about the existence of measures for retraining and/or advanced training of the organization's personnel at their enterprises, the results of which are presented in Fig. 2.

According to the survey results, it becomes obvious that less than half of the respondents have a centralized program that helps employees adapt to the transition to digital technologies, and their organizations have got a practice of sponsoring educational events and trainings for employees of the organization.

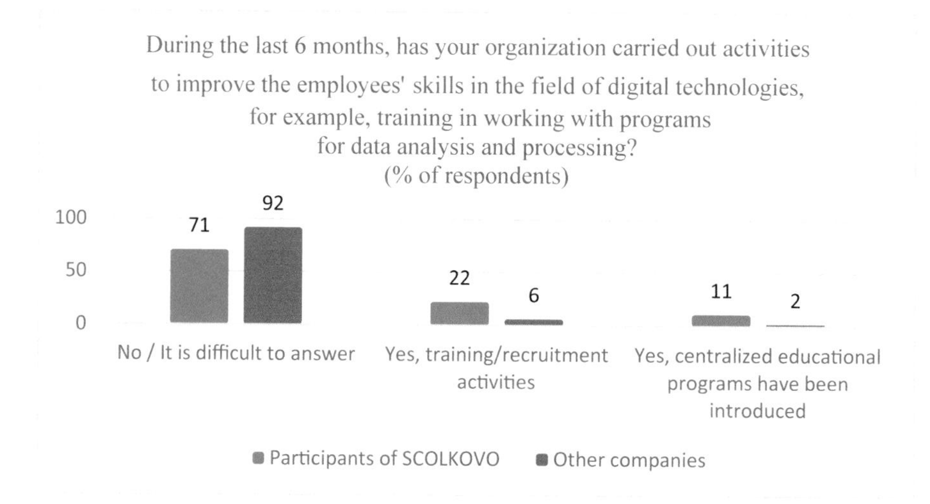

Fig. 2. The presence of courses for professional development/retraining in the organizations (Source: compiled by the authors).

4 Conclusion

Based on the above, it can be concluded that one of the main obstacles to the development of the digital economy is the lack of confidence in the new technology among the employees of organizations. It is often so because of their low qualification and the lack of measures to improve it in most enterprises. In this regard, the level of motivation at these enterprises will be reduced, which indicates the need for the development of non-financial ways of employees' motivation, because only these tools can stimulate the efficiency of employees in the long term. However, the current system of staff motivation is largely outdated and can harm the organization and lead to the loss of significant personnel reserves. The management of the companies needs to follow the current trends in the development of motivational systems and adapt to the rapidly changing conditions of management.

Thus, within this research work, the methods of personnel motivation in the conditions of digitalization of the economy and development of modern organizations were analyzed and systematized. Also, one of the results of the study was the analysis of the survey results among university graduates and employees of organizations of the Samara region to identify barriers to building a digital model of the economy. The authors proposed some options for their minimization. The conclusions and recommendations made in this article relate to the issues of staff motivation in the context of digitalization of the economy and are applicable in real business practice of many Russian companies.

References

1. Ashmarina, S.I., Mantulenko, V.V., Troshina, E.P.: Readiness to changes as one entrepreneurial value of the innovation-oriented economy. In: Ashmarina, S., Vochozka, M. (eds.) Sustainable Growth and Development of Economic Systems. Contributions to Economics, pp. 157–166. Springer, Cham (2019). https://doi.org/10.1007/978-3-030-11754-2_12
2. Development of HR processes and use of digital tools in Russian companies. Headhunter research service, July 2017. https://hhcdn.ru/file/16480569.pdf. Accessed 30 Mar 2019. (in Russian)
3. Evplova, E.V.: To the question of material and non-material motivation. Perspect. Sci. Educ. **2**, 104–108 (2013). https://cyberleninka.ru/article/n/k-voprosu-o-materialnoy-i-nematerialnoy-motivatsii. Accessed 17 June 2019. (in Russian)
4. Kotlyarov, I.D.: Theories of motivation Abraham Maslow and Frederick Herzberg: experience of mathematical formulas. Bull. St. Petersburg Univ. Ser. 12: Sociol. **1**(1), 220–225 (2009). https://cyberleninka.ru/article/n/teorii-motivatsii-abrahama-maslou-i-frederika-gertsberga-opyt-matematicheskoy-formalizatsii. Accessed 17 June 2019. (in Russian)
5. Salihov, A.Ah., Abdrahimova, R.G.: Motivation. Symbol Sci. **5–1**, 188–190 (2016). https://cyberleninka.ru/article/n/motivatsiya-personala-4. Accessed 30 Mar 2019. (in Russian)
6. Shitova, T.F.: Use of ERP-systems in enterprise management. Manag. Issues **5**(48) (2017). http://vestnik.uapa.ru/ru/issue/2017/05/21/. Accessed 17 June 2019. (in Russian)
7. The order of the Government of the Russian Federation of July 28, 2017 N 1632-p.: About the approval of the program "Digital Economy of the Russian Federation". http://base.garant.ru/71734878/. Accessed 30 Mar 2009. (in Russian)

Digitalization of Business and Management: Realities and Prospects

Digital Talents: Realities and Prospects

E. P. Barinova, E. N. Sheremetyeva, and A. S. Zotova(✉)

Samara State University of Economics, Samara, Russia
azotova@mail.ru

Abstract. The article analyzes the problems and key aspects of digital transformation. It is noted that not only the economy, but also the social sphere and education is undergoing digital transformation. The research emphasizes the importance of digital talents, new generations and the changing worldview of management. Russian companies are showing interest in talent management systems. The supply of qualified personnel, and especially managers, is rapidly lagging the demand that grows together with the market. Digital economy transformation demands the talent development program (Talent Management) to be aimed at creating a system of attracting, hiring, developing and using talented employees who can achieve exceptional business results. The main tasks are the coordination of talent management considering the global strategy, building a differentiated talent management architecture and differentiating strategic positions involving digital talents. Learning Agility is becoming the key competence in planning the development of future digital leaders. Learning Agility includes mental agility, self-awareness, people agility и change agility. That is why it is suggested to target 3 main perspectives: new knowledge acquiring, new skills development and changing the mentality.

Keywords: Digital transformation · Digital economy · Digital talents · Digital jobs · Innovative educational trends

1 Introduction

Structural, technological, geopolitical changes of the global economy are associated with revolutionary transformations of business, digital connection, the development of intelligent (smart) components of technology, technologies of augmented and virtual reality; transfer, storage and processing of data. Technologies of artificial intelligence, joint digital perception, computer recognition of emotions and many others change the business landscape and management system and have a significant impact on business models and companies' strategies. To ensure the country's competitiveness at the international level, digital transformation of the educational space in management and technological processes and models, as well as increased interaction with enterprises at the meso-, macro- and mega-levels is required. This will allow to fully realize the modern potential of the digital economy by integrating digital technologies into the system of cross-sectoral and cross-country interaction of universities and smart enterprises.

S. I. Ashmarina et al. (Eds.): ISCDTE 2019, LNNS 84, pp. 327–334, 2020.
https://doi.org/10.1007/978-3-030-27015-5_39

2 Methodology

The digital economy generates new effects that are subject to rapid changes and are not sufficiently studied in economics and interdisciplinary research. The proposed theoretical models are often adjusted significantly after a short period of time. The constant transformation of the conceptual tools, methodological and terminological borrowings, the lack of differentiation of concepts and their limited understanding complicates the functioning of key institutions which are generally in charge of creating the conditions for the development of digital economies (personnel and education, the formation of competencies and technological models, normative regulation). Changes are also subject to traditional economic categories and terminological tools.

The basic component of a modern economic organization is to work with data and use information and communication systems in the management process. Researchers identify four criteria for analyzing the digital economy: economic, technological, spatial and demographic, associated with the sphere of employment. Other additional criteria are used to determine the researchers' own views. However, most definitions are based on the conviction that effective development of areas of activity in the digital economy is possible only if there are developed platforms, technologies, institutional and infrastructural environment [1]. To measure the development of the digital economy, OECD countries have developed a system of indicators. They characterize not only the development of the high-tech sector of the economy, but also the cost of education, additional retraining; mobility and international cooperation in the field of science and innovations, international knowledge flows, etc. [7, 10].

The study of the processes of preparing digital talents causes the use of a neo-institutional approach that makes it possible to analyze the basic incentive to enhance the interaction of economic agents under the conditions of digitalization of socio-economic processes. It is necessary to increase the efficiency of formal institutions functioning, which will allow to use the unrealized potential of the socio-economic system and start the process of its self-development. The use of systems analysis methods will allow to structure the relationships between elements and factors of development, as well as to identify "bottlenecks" in the development of the economic ecosystem.

3 Results

Using the UK experience, it is necessary to differentiate the definitions used to characterize digital talents: the digital sector, the digital professions, the digital economy. The digital economy includes both digital and non-digital professions. Digital professions involve working in digital classes, but they are not part of the digital sector. The digital economy is understood as all working capital involved in digital sectors, as well as those who are engaged in digital occupations, outside the digital sectors [13].

The tendencies of management decentralization, the rejection of rigid hierarchical relations, granting of expanded powers to employees determine the nature of modern organizations. One of the key problems is the training of digital talents, and, in this connection, the fundamental transformation of the education system, since the training of such specialists involves not only the training of technological innovations, but also the features of the transformation of modern business processes.

For the continued development of the organization, there is a need in digital leaders capable to create, lead and develop teams, maintain connections between people and their involvement, and develop a culture of innovation, resistance to risk, and continuous improvement. Organizations face the difficult task of creating new, more productive ways to train specialists, since previous approaches of succession planning do not work. The innovative models in this area include rethinking the general concept of leadership, identifying probable digital leaders in the organization, and developing a talent management system.

Key competencies that a future digital leader should have is the ability to retrain and self-learn, possessing cross-cultural and intergenerational interaction skills. A digital leader should be able to distribute responsibility and use innovative management methods related to empowerment and knowledge sharing.

Requirements of business talents to the workplace, motives and methods of activity, as well as worldview, are significantly different from employees of other abilities and ages. The need to attract them forces companies to rethink their attitude not only to recruitment and management methods, but also to organizational management systems. Adaptive principles (agile), the concept of self-learning organizations and Laloux's concept of teal organizations are becoming more common. Business talents do not actually separate work from their lifestyle and want to receive a set of additional social services, which is also an additional difficulty for their involvement in the organization.

Modern trends in the development of education are determined by changes in the models and scale of education, the increasing role of project and problem-oriented learning, and its personalization. Innovative educational trends are provided by the development of digital technologies that override the old approaches and bring new opportunities. The transformation of traditional education requires the correlation of all learning formats, and not just the introduction of new technologies.

Under the conditions of digitalization there is a transformation of the processes carried out in the framework of the activities of enterprises and organizations to achieve their goals. The real economy is distinguished by a more complex system of interconnections and a greater number of processes involved in the creation of the final product. The changes that must occur in the system of higher education are due to transformations both in the areas of activity of university graduates and the macroeconomic challenges of modernity [2]. The changes of the educational process happen due to innovative trends describing ways of perceiving information, as well as globalization directly related to education (the graduate's labor market is now not a single region, but the whole world). The formation of a new educational space is possible now thanks to the development of technologies that allow to transform existing approaches and use newly emerging or just emerging opportunities. Consequently, the greatest learning efficiency can be achieved using a blended learning approach – building a training program from both synchronous e-learning and asynchronous elements.

World universities use a variety of methodological approaches to learning, allowing to prepare students for a professional career. The experience of European universities is an example of building practice-oriented learning models and the use of case studies - stories showing the business problems faced by enterprises in different sectors of the economy [10]. The use of such models stimulates students' debates about work methods for finding solutions and forces them to apply knowledge obtained in off-line as well as on-line-formats to real-life scenarios. The problem of interaction between enterprises and higher education institutions is urgent due to the increasing speed of digitization of enterprises in the real sector of the economy. Effective examples of such cooperation can be found in Russian Federation experience namely in Samara region.

Since 2015, the Unified system of measures is being implemented in the Samara region, aimed at identifying creative and talented young people aged 14 to 30 years. The unified organizational and methodological e-platform serves as the basis of systematic work carried out to include talented young people in the activities of leading regional research organizations and enterprises.

In 2016, the first level of the Unified System of Measures was implemented and the scientific and educational program was called "Vzlyot". The program is aimed at attracting schoolchildren to research activities and to solving regionally significant problems.

In 2017, schoolchildren successfully completed 900 research projects, which exceeds the value of the previous year 5 times. The "Vzlyot" program is becoming one of the effective tools for attracting highly motivated school graduates to enroll in universities in the Samara region.

In the academic year of 2017, the Unified System of Measures covered the university and postgraduates (the so-called scientific and educational program "Flight" ("Polyot"). The peculiarity of the "Flight" program is the formation of different-age youth research teams, including students of all ages, and carrying out research and innovation tutored by experts from leading enterprises and organizations in the region.

According to the results of these two programs, in the 2016/17 academic year, 900 schoolchildren and 650 students were included in the Governor's register of creatively talented youth of the Samara region in the field of science, technology and technology. The registry was created in order to identify, develop and support talented children and youth in the Samara region, to involve it in solving the tasks of the socio-economic, scientific, technical and innovative development of the region.

As part of the "Vzlyot" scientific and educational program, schoolchildren carry out their first research projects with distance counseling of leading scientists from the region (which is especially important for schoolchildren living in distant local places). The results of the projects are submitted to the regional competition "Rise", according to which the authors of the best projects are credited to the Governor's register of creative and talented young people in the field of science and technology and use various incentives and support from the regional budget.

Since 2017, the winners and awardees of the final stage of the "Rise" competition have been recipients of social support measures in the form of full or partial educational grant at universities located in the Samara Region. At present, more than 5,000 schoolchildren, more than 700 students and young scientists of higher educational institutions, 350 scientific consultants, about 2,000 teachers and university professors, and 20 leading enterprises of the region are involved in these programs. From November 2018 to March 2019 alone, about 1,800 new schoolchildren registered to express their desire to participate in the "Vzlyot" program. The total number of participants has already exceeded 17600. The capacity of a software product - an information system at the first stage is about 20 thousand participants. If this rate continues, in a year the system will not be able to process all the necessary number of applications of young people. Therefore, there is no other choice but to develop the system further.

Learning Agility includes the ability and willingness to learn, adapt and apply knowledge and experience in a constantly changing environment. Researchers consider it as the main indicator and the most important factor in determining talents with high potential [3, 5, 8, 13]. It includes teaching new skills for working in new conditions, rather than previously formed skills. Learning ability is an early indicator of leadership competence. In the field of talent management, it is widely recognized as vital for long-term leadership success [4–6]. Agility training model is positioned as a multidimensional concept and includes:

- Self-Awareness – The degree to which an individual has personal insight, clearly understands his or her strengths and weaknesses, is free of blind spots, and uses this knowledge to perform effectively;
- Mental Agility – The extent to which an individual embraces complexity, examines problems in unique and unusual ways, is inquisitive, and can make fresh connections between different concepts;
- People Agility – The degree to which one is open-minded toward others, enjoys interacting with a diversity of people, understands their unique strengths, interests, and limitations, and uses them effectively to accomplish organizational goals;
- Change Agility – The extent to which individual likes change, continuously explores new options and solutions, and is interested in leading organizational change efforts;
- Results Agility – The degree to which an individual is motivated by challenge and can deliver results in first-time and/or tough situations through resourcefulness and by inspiring others [9].

The Learning Agility model promotes informed and proved decisions regarding the availability of potential for development. The LAT test allows to use this concept in the practice of selection for training, for enrollment into the personnel reserve for further top management positions, as well as for solving some specific problems. In the modern world, the ability to retrain is one of the most important characteristics of employees, the importance of which in the XXI century will only grow. In this regard, it is necessary to reload 3 directions of modern education: learning new knowledge, learning new skills, mentality changing.

4 Discussion

The 2018th rating of countries economy digitalization, based on the E-intensity index, places Russia on the 39th position, and the Internet penetration rate is about 85% (in 2016 - 70.4%). The views of researchers worldwide are broadly presented on the topics of predictions of digital economy development, types of technological strategies and ways of business ecosystems evolution. Russian authors focus on the highly specialized issues of digital transformation. Many questions are debatable and require further development. Since the 2000s, the organizational forms of innovative firms have been divided into two groups: Japanese-type organizations (J-form) and a wide range of adhocratic organizations [1]. Organizations of these types focus on systematic innovation. The difference in them is that the organization of the J-form type relies on the knowledge embedded in operational processes, group interaction and a strong corporate culture. Training and knowledge accumulation occur at the lowest levels of the organization and is determined by corporate cross-functional interaction. Career stability is ensured by preferential recruitment from internal sources. New knowledge is formed through the synthesis, synthesis and combination of existing knowledge. However, in a digital economy, organizations like J-forms are not as effective as they are not intended to search for fundamentally new knowledge and solutions [16]. Educational organizations are more suitable for breakthrough areas, but their weak point is relatively little accumulation of knowledge.

Digital projects are carried out by qualified personnel, to define which the term "digital talents" is used. In most cases, the transformation team includes a small but super-qualified "core" of digital talents and a wider group of personnel who do not have unique knowledge and competencies. At the same time, both categories of personnel are equally important for ultimate success. Digital talent is the ability to deal with numbers and various kinds of precise symbols. This definition limits the understanding of a specialist in the conditions of digitalization. In our opinion, digital talent is a universal analytical skill in a certain field of activity, which allows us to achieve high results based on the adoption of innovative decisions. Talent management is, above all, the company's activities aimed at finding, attracting highly efficient employees and retaining them in the company. The use of resources of "digital talents" provides companies with undoubted competitiveness, since their advantage is non-standard thinking and high labor productivity [9].

Digital talent is a new type of employee that understands its value to employers. They are entrepreneurial, decision-oriented based on data analysis, focused on flexible forms of employment and have experience working in multidisciplinary international teams. Digital talents are unevenly distributed across regions of the world. This is the conclusion of a new Digital Economy and Society Index developed by the European Commission and released recently. Data shows that the picture of how digital countries are varies across the EU and that borders remain an obstacle to a fully-fledged Digital Single Market [11–13].

It should be noted that economists predict a shortage of highly qualified employees who are able to adapt, implement and optimize processes based on new digital technologies. According to a Gartner study, 30% of technical jobs will be vacant due to lack of digital talent. The BCG study [14] notes that this is a major issue. The use of traditional human resource management strategies is proved to be inefficient in order to attract professionals who have digital competencies and are able to adapt to the digital environment quickly. The traditional professions are receding into the past, new ones that will carry out digital transformation as a supportive process of business are emerging [15].

Digital transformation of an enterprise means not only the introduction of individual technologies and the redesign of processes, but also a change in the culture of the organization, a change in the mentality of employees, which inevitably changes the role of the manager and manager in managing human resources. Reducing the need for management, its decentralization, the abandonment of rigid hierarchical relationships, as well as numerous technological innovations demand management system modification.

Statistics show an increase in the share of the working-age population employed in the services sector. For example, in Western Europe, USA, Japan, their share reaches 70% or more. Moreover, most of them are connected in one way or another with the processing of certain data, which convincingly proves the accelerated development of the digital economy. One of the existing problems is the lack of the methodology for counting workers in the digital economy. For example, you can associate the growth of the digital economy with an increase in the number of computer technology specialists, telecommunications companies, and analysts whose main task is data processing. However, at the same time, there was an increase in workers of "non-digital" professions or having a rather weak connection with it - spheres of commerce, lawyers, etc.

5 Conclusion

In the context of global markets, the need to accelerate the creation of innovations that realize new emerging opportunities determines the interaction of many independent actors - companies, universities, research centers, consumers, suppliers and other economic partners that form business ecosystems. World practice shows that the formation of ecosystems requires a change in the mental attitudes of managers and the expansion of the horizon of vision.

The Russian education system has a high potential for training specialists in the digital economy, which is especially important because in a digital economy, people will focus mainly on the realization of new opportunities and the systemic organization of interaction in the ecosystem of people and machines. The introduction of original organizational and technological solutions to create an effective infrastructure of the digital economy will allow for the integration and development of specific cases based on modern principles of the digital economy.

Universities need to train specialists who are competent in solving urgent problems faced by enterprises in the real sector of the economy and possess information about the requirements for knowledge, skills and abilities of graduates. Coordination of competencies, communications and skills between the university and the enterprise is necessary to develop actual practical economic tasks and cases that meets the requirements of practice-oriented learning.

The management of an innovative university is based on a transformation towards a professional entrepreneurial organization. Such a transformation necessitates the development of innovative training courses and the modernization of the classical courses of the innovation component, considering contemporary environmental challenges. The problems of developing technological entrepreneurship based on platform technologies, the formation of business ecosystems, the strategic management of high-tech companies are inextricably linked with the conduct of applied research in the digital economy and the transformation of the educational infrastructure of digital platforms.

Acknowledgment. The research is done in the frame of the state task of the Ministry of Education and Science of the Russian Federation № 26.9402017/PC "Change management in high education system on the basis of sustainable development and interest agreement".

References

1. Arenkov, I.A., Smirnov, S.A., Sharafutdinov, D.R., Yaburova, D.V.: Transformation of the enterprise management system in the transition to the digital economy. Russ. J. Enterp. **19** (5), 1711–1722 (2018). https://doi.org/10.18334/rp.19.5.39115. (in Russian)
2. Ashmarina, S.I., Kandrashina, E.A., Izmailov, A.M., Mirzayev, N.G.: Gaps in the system of higher education in Russia in terms of digitalization. In: Ashmarina, S., Mesquita, A., Vochozka, M. (eds.) Digital Transformation of the Economy: Challenges, Trends and New Opportunities. Advances in Intelligent Systems and Computing, vol. 908, pp. 437–443. Springer, Cham (2019). https://doi.org/10.1007/978-3-030-11367-4_43
3. Bersin, J.: The disruption of digital learning: ten things we have learned (2017). https://joshbersin.com/2017/03/the-disruption-of-digital-learning-ten-things-we-have-learned/. Accessed 08 Apr 2019
4. De Meuse, K.P., Dai, G., Hallenbeck, G.S.: Learning agility: a construct whose time has come. Consult. Psychol. J. **62**(2), 119–130 (2010). https://doi.org/10.1037/a0019988
5. De Meuse, K.P., Dai, G., Swisher, V.V., Eichinger, R.W., Lombardo, M.M.: Leadership development: exploring, clarifying, and expanding our understanding of learning agility. Ind. Organ. Psychol. **5**(3), 280–286 (2012). https://doi.org/10.1111/j.1754-9434.2012.01445.x
6. Dries, N., Vantilborgh, T., Pepermans, R.: The role of learning agility and career variety in the identification and development of high potential employees. Pers. Rev. **41**(3), 340–358 (2012). https://doi.org/10.1108/00483481211212977
7. Global Talent Monitor - SHL Russia (2017). https://www.shl.ru/uploads/file/ceb-gtm-infographic.pdf. Accessed 08 Apr 2019
8. Eichinger, R.W., Lombardo, M.M.: Learning agility as a prime indicator of potential. Hum. Resour. Plann. **27**(4), 12–16 (2004)
9. Kuznetsova, S.A., Markova, V.D.: Digital economy: new facets of research and teaching in management. Innovation **7**(225), 20–25 (2017). (in Russian)
10. Kupriyanovsky, V.P., Sukhomlin, V.A., Dobrynin, A.P., Raikov, A.N., Shkurov, F.V., Drozhzhinov, V.I., Fedorova, N.O., Namiot, D.E.: Skills in the digital economy and the challenges of the education system. Int. J. Open Inf. Technol. **1**, 19–25 (2017). (in Russian)
11. Press Release of European Commission: How digital is your country? (2015). https://ec.europa.eu/digital-single-market/en/desi. Accessed 15 Apr 2019
12. Silzer, R., Church, A.H.: The pearls and perils of identifying potential. Ind. Organ. Psychol. **2**(4), 377–412 (2009). https://doi.org/10.1111/j.1754-9434.2009.01163.x
13. Sokolov, I.A., Kupriyanovsky, V.P., Namiot, D.E., Drozhzhinov, V.I., Bykov, A.Y.U., Sinyagov, S.A., Karasev, O.I., Dobrynin, A.P.: State, innovation, science and talents in measuring the digital economy (UK Case Study). Int. J. Open Inf. Technol. **5**(6), 33–48 (2017). (in Russian)
14. Strack, R., Dyrchs, S., Kotsis, A., Mingardon, S.: How to gain and develop digital talent and skills. Bcg.com (2017). https://www.bcg.com/de-de/publications/2017/people-organization-technology-how-gain-develop-digital-talent-skills.aspx. Accessed 08 Apr 2019
15. Wagner, T., Herrmann, S., Thiede, S.: Industry 4.0 impacts on lean production systems. Procedia CIRP **63**, 125–131 (2017)
16. Whitley, R.: The institutional structuring of organizational capabilities: the role of authority sharing and organizational careers. Organ. Stud. **24**(5), 667–695 (2003). https://doi.org/10.1177/0170840603024005001

Harmonization of Financial and Credit Resources of Commercial Organizations in the Digital Economy

E. A. Serper, O. A. Khvostenko(✉), and M. A. Pershin

Samara State University of Economics, Samara, Russia
olegkhvostenko@yandex.ru

Abstract. The purpose of the study is to justify the need for short-term forecasting of financial and credit resources and the budgeting of daily cash flow to improve the organization solvency and strengthen financial stability in the digital economy. The scientific novelty is that the authors have presented the mathematical description of the model for determination of financial and credit resources requirements in a commercial organization at any time in the digital economy. The digital economy places increased demands on the composition and source of the financial and credit resources of commercial organizations. These requirements have been analyzed by the authors in this paper.

Keywords: Budgeting, cash · Commercial organizations · Digital economy · Financial and credit resources · Monetary assets

1 Introduction

Macroeconomic situation in Russia and in the digital economy is characterized by increasing various risks in commercial organizations' activities. Consequently, it requires a careful approach to the development of financial forecasts, plans and budgets. The implementation of integrated budgeting improves the efficiency of financial and credit resources management, optimizes cash flows and allows to make grounded decisions in financial management.

The main document reflecting the results of forecasting and planning of financial and credit resources is the cash flow budget that determines the following criteria:

- financial and credit resources amount, which is necessary and sufficient for the organization's operational activities,
- reasons of shortage or excess of funds in Bank accounts,
- sources of maintaining the optimal amount of temporarily free funds,
- need for borrowed funds, the amount and terms of financial borrowing,
- most profitable short-term investment areas of temporarily released financial and credit resources.

In the digital economy, the cash flow budget of a modern commercial organization is compiled for a year and includes quarter, month and decade sections for operating, investment and financial activities of organizations. If necessary, it can be formed on a

S. I. Ashmarina et al. (Eds.): ISCDTE 2019, LNNS 84, pp. 335–341, 2020.
https://doi.org/10.1007/978-3-030-27015-5_40

daily basis, which allows to assess quickly and efficiently the state of cash reserves and flows. This research was conducted because there are many problems with the imperfection of budgeting methodological approaches and the lack of clear regulation in the specific series in Russian commercial organizations.

2 Theoretical Background

The analysis of macroeconomic indicators allows us to conclude that the operating conditions of commercial organizations are not favorable enough at the present time. There is a decreasing return on investment in the financial market against a lack of liquid assets and a shortage of payment instruments, which is aggravated by the introduction of economic sanctions against Russia. Despite the fact that over the past few years the share of loss-making organizations decreased, profit and profitability in 2017 did not increase, as shown in Table 1:

Table 1. Dynamics of organizations' financial results (without small businesses) for the Russian Federation in 2015–2017

No	Indicator description	2015	2016	2017
1.	Balance for loss and gain, bln rbl	7,503	12,801	10,320
2.	Comparison with the corresponding period of the preceding year (as %)	173.6	157.0	91.5
3.	Value of returns	12,654	15,823	12,276
4.	Relative share of profit-making organizations (as %)	67.4	70.5	73.7
5.	Amount of loss	5,151	3,022	1,956
6.	Relative share of loss-making organizations (as %)	32.6	29.5	26.3
7.	Return on total assets (as %)	3.7	5.9	5.3
8.	Profitability of goods sold, products, works done and services (as %)	8.1	7.6	7.5

Source: compiled by the authors according to the Federal State Statistics Service [9]

The situation had a negative impact on the payment discipline. On 1 January 2018, the organizations' credit debts amounted to 44,481 bln rbl, with 2,616 bln rbl or 5.9% overdue credit debts; 1,961 bln rbl or 75% debts to suppliers and contractors for goods, works and services; 4.3% debts for budget payments and extra-budgetary funds; and 20.7% other overdue debts [9].

Preliminary figures indicate that on 1 November 2018, the amount of overdue credit debts amounted to 3366.5 bln rbl. The share of overdue payments in the total amount of credit debts increased by 0.2% in October 2018, and it amounted to 7.1% on 1 November 2018 [9]. The amount of overdue debts, according to the economic activities, is shown in Table 2.

To solve the problem of reducing non-payment at the micro level in the digital economy, it is necessary to increase the level of financial planning in every commercial organization. This will allow to identify, at the appropriate times, the need for the formation of additional financial and credit resources and to determine the funds raising sources. When comparing national and foreign practices, it becomes obvious that there is a fundamental difference in forecasting and planning approaches. In developed countries the key issue is to determine the optimal size of the cash balancen which can be invested (since they do not generate income and remain on the current account), but in Russia, the main task is to find money to finance operating activities, pay fixed-term and overdue liabilities, and repay debts.

Table 2. Size and structure of overdue credit debts of organizations according to the economic activities

No	Economic activities	On 1 November 2018, bln rbl	% of total
1.	Agriculture and forestry	21.9	0.7
2.	Mining	265.1	7.9
3.	Manufacturing	1273.9	37.8
4.	Electric energy, gas and steam provision	581.7	17.3
5.	Water supply, water discharge and waste recycling	36.6	1.1
6.	Construction	124.8	3.7
7.	Wholesale and retail distribution	628.9	18.7
8.	Shipment and storage of cargoes	60.8	1.8
9.	Information and communications	24.0	0.7
10.	Other economic activities and services	348.8	10.3
	Total	3366.5	100.0

Source: compiled by the authors according to the Federal State Statistics Service [9]

There are two most well-known models of money asset management, which are used in the world practice: the model by Miller-Orr and Baumol-Tobin. The model of optimal cash balance calculation, proposed by Miller and Orr [5], is based on the stochastic process in which the receipt and expenditure of funds in each period of time are independent random events with equal probabilities.

The essence of the model is as follows. A commercial organization establishes a minimum cash reserve (LCL), which includes, for example, an insurance reserve and a compensation balance.

The amount of funds in the account continuously changes between the lower limit (LCL) and the upper limit (UCL). If the balance reaches the upper limit, the organization invests funds in various financial facilities in order to bring the balance to the optimal level (return point – RP). If the cash reserve reaches the lower limit of the LCL, the organization implements its investments or obtains a loan and thus, replenishes the cash reserve to the optimal level of RP.

Since the management purpose in the model by Miller-Orr is to minimize total costs, the formula for calculating the optimal cash reserve is as follows:

$$\mathrm{RP} = \sqrt[3]{\frac{3 \times b \times \sigma^2}{4 \times i}} + LCL \tag{1}$$

where:

RP is a target cash balance between maximum and minimum values,
b – maintenance costs of one transaction with financial facilities,
σ – variance (mean-square deviation) of daily cash turnover,
i – the average daily level of alternative income losses on cash assets keeping, which is calculated as the average daily interest rate on short-term financial transactions,
LCL – minimum cash balance that must be left in the account to make current payments.

In the model by Miller-Orr, the account balance is considered as a value that can change randomly and take values that cannot be predicted and planned in advance. When the organization's cash expenditures are stable and predictable over equal periods of time, the Baumol-Tobin model is applied [5].

According to this model, the optimal cash balance depends on two factors: the cost of a replenishment transaction and the opportunity costs of replenishment maintaining. Thus, the optimal cash balance (or OCB) is calculated by the formula:

$$OCB = \sqrt{\frac{2 \times F \times T}{k}} \tag{2}$$

where:

OCB is the optimal cash balance,
F – fixed costs of buying and selling securities or loan servicing,
T – transaction costs of cash balance replenishment,
k – opportunity costs of cash balance maintaining (interest rate on liquid securities).

The presented models can be used only under ideal conditions in a developed market economy if there is a regular resources flow, and income is credited to the accounts of stable functioning organizations. Currently, in our country, a different situation is being created. The main task is not to determine the amount of excess funds for investment, but to forecast and minimize the risks of the lack in financial and credit resources. It requires a different approach to the development of the cash flow budget.

To determine the need for financial and credit resources of a commercial organization at a given time (more precisely, on the morning working day), we propose to divide all planned receipts and payments into four groups:

1. Compulsive receipts and payments, when the date and amount are known or can be calculated with high probability; for example, loan repayment and tax payments.

2. Payments and receipts, when the total amount is known but they may be delivered, in full or by installments, at any time within a specified period including contract payments to supply or sale produts.
3. Amounts that cannot be determined precisely in advance but can be calculated approximately on the basis of factual information from the previous period or, taking into account the seasonality factor of production and sales, on the basis of factual information for the relevant periods of previous years.
4. Cash that will be raised or placed only under certain conditions, i.e. investments in securities and other investments.

In this case, the balance at the end of the day will be calculated by the formula:

$$\begin{aligned} ML_d = MB_d + SR1_d - PC1_d + \frac{SR2}{WD} - \frac{PC2}{WD} + \frac{SR3 * KSR}{WD} \\ - \frac{PC3 * KPC}{WD} + DI_d - CI_d + DF_d - CF_d \end{aligned} \quad (3)$$

where:

ML_d is minimum account balance to be maintained at the end of the day,
MB_d – cash account balances at the beginning of the day,
$SR1_d$ – first group's receipts expected today,
$PC1_d$ – first group's receipts needed to be delivered today,
SR2 – second group's receipts expected for the planned period,
PC2 – second group's receipts needed to be delivered for the planned period,
WD –number of working days for the planned period,
SR3 – third group's receipts expected for the planned period,
KSR – correction factor of the receipts probability in the third group,
PC3 – third group's payments which are planned to be delivered for the scheduled period,
KPC – correction factor of the payment amount in the third group (including the adjustment for expected inflation),
DI_d – receipts from investment operations expected today,
CI_d – payments for investment activities scheduled for today,
DF_d – receipts from financial transactions expected today,
CF_d – payments for financial transactions scheduled for today.

If the equality in the formula (3) is observed with positive values of all indicators including MB_d и ML_d, then the commercial organization is expected to have sufficient financial and credit resources not only during the entire period of budgeting but, in particular, for each planned day. If the value of ML_d is predicted to be less than zero, it is necessary to promptly obtain additional financial and credit resources or, accordingly, reduce costs. Then, the formulas for determining the sufficient amount of financial and credit resources for the planned period of time ("enough money", or EM), as well as on the morning working day (EM_d), will be as follows:

$$EM = ML - MB - SR1 + PC1 - SR2 + PC2 - SR3 * KSR + PC3 * KRC - DI + CI - DF + CF \quad (4)$$

$$EM_d = ML_d - MB_d - SR1_d + PC1_d - \frac{SR2}{WD} + \frac{PC2}{WD} - \frac{SR3 * KSR}{WD} + \frac{PC3 * KPC}{WD} - DI_d + CI_d - DF_d + CF_d \quad (5)$$

3 Discussion

The works of many national and foreign scientists are devoted to the formation of the necessary fundings and possible types of use of financial and credit resources in accordance with the current needs of commercial organizations. These problems are only increasing under current conditions of digitalization in the economy.

The fundamental premises in the basic concept of financial management, which provides for an alternative choice of sources for financial and credit resources as the organizations' costs grow, are presented in the works by Myers and Majluf [6]. This concept was further developed in the works by Krasker [3], Narayanan [7] and a number of other researchers.

In the Russian economic science, the issues of improving the monetary assets management are reflected, in particular, in the works by Limitovsky [4], Rogova and Blinova [8]. Planning, budgeting and optimization of organizations' income and expenses are considered in detail by Bykova, Stoyanova [2] and others. Amir, Auzair and Amiruddin affirm that cost management is an important resource to develop organization's competitiveness [1].

The research of various directions in cash flows optimization in the budgeting system will allow to solve the problem of their synchronization, and will help strengthen financial discipline and solve the non-payments problem under the current conditions of digitalization in the economy.

4 Conclusion

The abovementioned formulas can be used to form the cash flow budget and the income and expenses budget, when dealing with budget modelling in commercial organizations. This will lead to the daily harmonization of cash flows, which in turn will help reduce the risk of non-payment and strengthen the financial stability of commercial organizations in the digital economy.

References

1. Amir, A., Auzair, S.Md., Amiruddin, R.: Cost management, entrepreneurship and competitiveness of strategic priorities for small and medium enterprises. Procedia Soc. Behav. Sci. **219**, 84–90 (2016). https://doi.org/10.1016/j.sbspro.2016.04.046
2. Bykova, E.V., Stoyanova, E.S.: Financial Art of Commerce: Monograph. Perspectiva, Moscow (1995). (in Russian)
3. Krasker, W.: Stock price movements in response to stock issues under asymmetric information. J. Financ. **41**(1), 93–105 (1986)
4. Liimitovsky, M.A.: Influence model of the capital structure on the company's value based on realistic conjectures. Corp. Financ. Manag. **1**, 36–48 (2006). (in Russian)
5. Miller, M., Orr, D.: A model of the demand for money by firms. Q. J. Econ. **80**(3), 413–435 (1966)
6. Myers, S.C., Majluf, N.S.: Corporate financing and investment decisions when firms have information that investors do not have. Financ. Econ. **13**(2), 187–221 (1984). https://doi.org/10.1016/0304-405X(84)90023-0
7. Narayanan, M.P.: Debt versus equity under asymmetric information. J. Financ. Qant. Anal. **23**(1), 39–51 (1988)
8. Rogova, E., Blinova, A.: The technical efficiency of Russian retail companies: an empirical analysis. Zeszyty Naukowe Uniwersytetu Ekonomicznego w Krakowie. Cracow Rev. Econ. Manag. **5**(977), 171–185 (2018)
9. Finance section. Official statistics. Federal State Statistics Service. http://www.gks.ru/wps/wcm/connect/rosstat_main/rosstat/ru/statistics/finance/. Accessed 22 Apr 2019

Improving Russian Agribusiness Competitiveness Within the Digital Transformation Framework

N. V. Molotkova[1], M. N. Makeeva[2], M. A. Blium[3](✉), B. I. Gerasimov[3], and E. B. Gerasimova[2]

[1] Tambov State Technical University, Tambov, Russia
[2] Financial University under the Government of the Russian Federation, Moscow, Russia
[3] FSUE «STANDARTINFORM», Moscow, Russia
blyumarina@gmail.com

Abstract. The development of innovation is of great importance for any state. Innovations expand the horizons for the development of the economy through the creation of a new competitive product or method of its production, the emergence of market segments, etc. In the Tambov region, as in the agro-industrial region, great attention is paid to the development of agriculture. Currently, a promising area for innovation in the agricultural sector is the development and implementation of innovative technological solutions, such as satellite tracking and aeromonitoring with the provision of analytical data based on them, precision farming, automation of production processes. Currently, in the city of Tambov, work is underway to open the first in the region Technopark in the field of information technology "Mielta", which will combine the existing innovation infrastructure and IT-enterprises in a single mechanism. The purpose of Technopark is the development of innovative business forms in the region, maintaining and developing the synergy effect from the optimal interaction of resident enterprises of the Technopark, their joint work, the exchange of knowledge and technology.

Keywords: Competitiveness · Agribusiness · Digital transformation

1 Introduction

Today, the main global trend is a course for innovative development. Innovation is not only scientific and technical progress, but also a powerful engine for the economy of any state.

The concept of "innovation" is closely related to the so-called fourth industrial revolution or Industry 4.0. The term itself appeared in Germany in 2011. It means the massive introduction of information technology in the production and maintenance of human needs. It covers all spheres of human activity. Industry 4.0 is not just new technologies, it is a fundamentally new approach to defining products and methods for their production and consumption.

S. I. Ashmarina et al. (Eds.): ISCDTE 2019, LNNS 84, pp. 342–350, 2020.
https://doi.org/10.1007/978-3-030-27015-5_41

The innovative technologies include the Internet of Things (IoT, Internet of things), cloud services, big data analytics, robotics, artificial intelligence and much more. But, despite the rapid spread of high technologies and innovations around the world and the favorable attitude towards them of mankind, in general, enterprises that are engaged in innovative activities still have difficulties in promoting innovative products. This is due to the fact that in a rapidly changing economy, specialists associated with the promotion and sales of innovative products have not yet had enough experience, skills and knowledge in this area. The second reason is that innovative products, as a rule, are complex and multi-component, therefore, they are difficult to sell and difficult to explain to consumers the principle of action and the benefits they will receive from using the product. The third reason is the lack of readiness for innovation of the target audience for which the product is designed. Despite the fact that innovative products are very different and to develop uniform requirements for their promotion is problematic, these factors make it necessary to develop a strategy for promoting innovative products.

2 Methods

The study was carried out by the following methods: methods of systematic approach and structural-functional analysis, expert assessments and forecasting, content analysis of scientific literature and other economic research methods.

Combined methodological approach allowed the authors to structure the process of study and to identify its basic stages. The first stage involves content analysis of scientific literature and gathering the materials for analysis. The second stage, practice-oriented, was focused on studying the experience of leading enterprises in the sphere of digitalization of Russian agribusiness.

Methods of promoting innovation have been developed and studied since the mid-twentieth century, when the rapid development of technologies began, and they began to actively appear in the lives of ordinary people. After the Second World War, humanity was actively transforming new technologies, which were invented initially for military needs, into a civilian field of activity. Thus, innovations were studied by Schumpeter [8]. From modern authors it is worth highlighting the works of Khotyasheva [5]. However, most research identifies an innovative product as a new product or service in already existing industries and describes the launch of a new product on an already existing market. At the same time, there are practically no works on the promotion of innovative products in the market segment, which is only at the stage of formation, and therefore is innovative in itself.

3 Results

The development of innovation is of great importance for any state. Innovations expand the horizons for the development of the economy through the creation of a new competitive product or method of its production, the emergence of market segments, etc.

Innovation (English “innovation” - novelty, alteration, novation) refers to the use of innovations (novations) in the form of new inventions, discoveries, new technologies, new forms of organization of production and labor, new methods of meeting social needs.

The concepts of “innovation”, “novelty”, “novation” are often identified, although there are differences between them. Innovation is a new order, a new method, an invention, a new phenomenon. The phrase “novation” literally means the process of using innovations [4].

Once a novation is accepted for distribution, it becomes an innovation. Innovation is a materialized result obtained from investing in new equipment or technology, in new forms of organizing the production of labor, maintenance, management, etc. [3].

4 Discussion

It should be noted that it is the market that gives an economic assessment to the idea of the practical use of new knowledge and determines its subsequent fate - the rapid introduction into production or oblivion for many years [6].

There is no direct relationship between the emergence of the idea, its materialization and commercialization, as confirmed by various studies. An enterprise is not obliged to carry out a full cycle of innovative entrepreneurial activity from the R&D stage to sales [2]. Therefore, for the successful implementation of the development and its commercialization, the cooperation of various enterprises plays an important role, due to which a synergy effect arises.

There are factors contributing to the acceleration or inhibition of the innovation process. The development of innovation is influenced by such factors as the availability of scientific and technical infrastructure and advanced technologies, state support of innovation, flexibility of the organizational structure of the enterprise.

Many factors impede the development of innovation. First of all, it is the high cost of spending on innovation, the acquisition of the necessary equipment and materials, the lack of own funds of enterprises, the complexity of commercialization of development, marketing problems, insufficient qualifications of personnel, difficulties with licensing and obtaining patents. Also the braking factors are the rigidity in planning, difficulties in inter-branch and inter-organizational interactions, orientation to the established markets and short-term payback of products.

At present, innovations in the global economy are the engine of growth and can become a lever that will help transform short-term economic growth into long-term one. A key factor in industrial growth is consumer demand and the ability to maximize the consumption of goods and services, and this can be achieved through the use of innovation. Therefore, the main global trend is connected with the fourth industrial revolution or Industry 4.0. The material world connects with the virtual, creating a single digital ecosystem. The idea quickly spread throughout the world, most countries today are heading for digitalization.

The main trends of Industry 4.0 are as follows:

(a) Decentralization of the production of products and resources, as well as a much more flexible management of the scale of production in order to reduce costs.
(b) Total imparting to all things the functions of artificial intelligence, the transformation of each thing into a consumer and source of information. The active participation of "smart" things in their own design, creation and repair.
(c) Automation of services through the massive use of artificial intelligence - the gradual transformation of the entire service industry into an industry controlled by the interaction of client and service artificial intelligence with the active use of "big data" as a source of information for prediction and planning.
(d) The rapid reduction of human participation in the interactions between things.
(e) The widespread creation of augmented reality institutions and infrastructure and protocols for communicating with "smart" things and devices.
(f) The rapid expansion of the "passive entrepreneurship" of the population through the development of electronic trading systems and the use of various resources of households and residents.
(g) Total expansion of the blockchain technology and the like.
(h) Development of alternative networks like the Internet, and their integration into the infrastructure of augmented reality [9].

The principles of the construction of Industry 4.0 were also formulated, following which, with the introduction of technology, enterprises will be able to achieve better results.

The principle of compatibility means the ability of machines, devices and people to interact and communicate with each other. In fact, it declares the Internet of Things (IoT) to be one of the essential technologies.

The principle of transparency follows from the principle of compatibility. In the virtual world, a digital copy of real objects is created, which exactly repeats what happens to a physical object. As a result, all the processes that take place with the object accumulate the most complete information that must be collected and transmitted using sensors.

The third principle is technical support. The principle means the use of decision support systems through technology collection, analysis, data visualization and forecasting. It can also mean replacing humans with robots when performing dangerous and routine operations.

The principle of decentralized management decisions is to maximize the automation of business processes to ensure effective work without the intervention of people. People in this case play the role of supervisors.

As mentioned earlier, Industry 4.0 is inextricably linked with the concept of the Internet of Things. One of the researchers at the Massachusetts Institute of Technology K. Ashton in 1999 gave such a definition to the term "Internet of things" (Internet of things, IoT) is the concept of a computer network of physical objects ("things") equipped with embedded technologies for interacting with each other or with external environment, considering the organization of such networks as a phenomenon that can restructure economic and social processes, excluding the need for human participation from some of the actions and operations [1].

Modern Russia needs an impulse to launch innovative development and restructure the economy. This does not mean that so far we have not had innovations. Researchers estimate that 25–30% of economic growth in Russia over the past 20 years has accounted for the growth of total factor productivity. With the active participation of the state in Russia over the past years, many elements of the national innovation ecosystem were created and transformations carried out in many areas. Nevertheless, innovations in the country were not the key driver of economic growth - this role was given to the prices of natural resources and the connection of unused capacity and labor. Now innovations should come to the fore, as the possibilities of other drivers have been exhausted [7].

For many years of market economy, the state has attempted to plan and forecast innovative development of the country. So, in 2005, “The Main directions of the policy of the Russian Federation in the field of development of the innovation system for the period up to 2010” were adopted, in 2006 – “The Strategy for the development of science and innovations in the Russian Federation until 2015”. In 2011, “The Strategy for the Innovative Development of the Russian Federation for the period up to 2020 (CIR-2020)” was adopted [12].

The purpose of SIR-2020 is to transfer the Russian economy by 2020 to an innovative path of development. In order to achieve this goal, the following tasks were set:

- development of human resources in the field of science, education, technology and innovation;
- increasing the innovative activity of the business and accelerating the emergence of new innovative companies;
- the widest possible introduction of modern technologies in the activities of government bodies;
- the formation of a balanced and sustainably developing research and development sector;
- ensuring the openness of the national innovation system and economy, as well as the integration of Russia into the global processes of creating and using innovations;
- activization of the implementation of the innovation policy carried out by state authorities of the constituent entities of the Russian Federation and municipalities [10].

State innovation policy is aimed at increasing the innovative activity of business, creating favorable conditions for the emergence, formation and development of entrepreneurship with the aim of developing and producing new types of products and technologies, as well as the interaction of innovative forms of business with each other. An important role is played by government support.

The Russian Federation supports innovative projects in:

- promoting the development of research in promising areas through the provision of financial support, for example, the allocation of grants, investments through funds;
- granting privileges at the legislative level, for example, lowering the property tax rate, profit tax, tax rates;
- creation of mechanisms for state regulation of innovation activities;

- promoting the development and implementation of programs to promote innovative products;
- formation of state orders for innovation;
- improvement of the legal framework of innovation;
- staffing innovation;
- creation of innovation infrastructure.

Over the past six years since the adoption of SIR-2020, the efforts of the Russian state in the field of innovative development have yielded obvious positive results recorded by the recognized world rankings, primarily the World Bank's "Doing Business", the INSEAD Business School's Global Innovation Index, and Global Competitiveness Index "Davos Economic Forum" [7].

In the Doing Business rating, which is a target indicator of two key state initiatives - the National Entrepreneurial Initiative and the Rating of Innovative Development of the Regions of the Russian Federation - Russia moved up from 123rd place in 2011 to 51st in 2016. This suggests that our state has reduced the gap to the leading countries by more than a half.

According to the Global Innovation Index (GII) rating, in six years Russia has improved its result from 64th to 43rd place, that is, approximately one and a half times. An almost identical picture is also observed in the Global Competitiveness Index (GCI) rating, in which Russia grew from 63rd to 43rd place in the same period [7]. When studying the indicators of the "GII" and "GCI" ratings, which are responsible for the dynamics and degree of innovative development, the picture becomes less optimistic.

Much of the growth in the GII rankings occurred in the Innovation Input Subindex component, reflecting the country's investment in innovation development, including efforts to create a favorable innovation environment. For this component, Russia's place has risen in six years from 82nd to 44th, that is, the country has improved its result about two times. For the second component, Innovation Output Subindex, which reflects innovation activity, Russia only slightly improved its position, rising from 51st to 47th place.

In the GCI rating, Russia has not been able to improve its position since 2010. GCI distributes countries into three groups depending on the main driving force of their economic development at this stage:

- countries driven by production factors;
- countries driven by production efficiency;
- countries driven by innovative production.

In the first group are the poorest countries that survive by offering cheap labor or natural resources. The countries of the second group compete at the expense of efficiency - that is, they produce something that is already widely available, but they do it better than the countries of the first group. And only countries of the third group offer innovative solutions to the world, that is, they compete by creating new products and technologies [7]. For several years, Russia in the GCI rating is in the second group.

Judging by the dynamics in the ratings, the country has not been able to make an "innovative breakthrough" since the adoption of CIR-2020. World rankings point out the weakness of the Russian innovative economy.

The Tambov region in the Rating of the Subjects of the Russian Federation, by the value of the Russian Regional Innovation Index, is in the second group of regions and occupies 25th place in the rating, having improved it since 2015. The key to success and breakthroughs from 2015 - improving the quality of innovation policy. It can be concluded that the most important factor is the innovative development of the region is government support.

In general, the territorial distribution of regions with different levels of innovative development can be characterized as stably uneven. However, the regions with the highest rating of innovation development are concentrated in the following federal districts: Central, Siberian and Volga.

In October 2017, an agreement was signed between the administration of the Tambov region and the Skolkovo Foundation, under which it is planned to develop the innovation infrastructure of the region [11]. In the Tambov region, priority is given to digitalization and the development of the IT industry. These areas are key to the adopted strategy for the development of the Tambov region until 2035. The Tambov region has become a pilot region for the implementation of IT projects and IoT technologies in the agro-industrial complex.

For the successful implementation of these programs, it is necessary to solve a number of tasks related to the creation of new high-tech industries and innovative companies, the commercialization and use of their scientific and technological developments, including in agricultural enterprises. In general, in Russia, only 15% of enterprises are engaged in the creation of innovative products. In the Tambov region in 2016, 10.6% of enterprises carried out innovative activities. In total, they shipped innovative products in the amount of 8.7 billion rubles. The cost of technological innovation increased from 2.3 billion rubles. In 2014 to 6.8 billion rubles. In 2016, of which own funds of enterprises - 3.8 billion rubles. According to the administration of the Tambov region, by 2025 it is planned that the shipment of innovative products in monetary terms will increase to 13 billion rubles, and the cost of innovation will increase to 11 billion rubles.

Despite the fact that the number of innovative enterprises is increasing (for comparison, in 2014 in the Tambov region 25 enterprises carried out scientific developments, and in 2016 - 32), the share of innovative goods and services generally decreases: from 6.3 in 2014 to 4, 5 in 2016. Enterprises of the Tambov region, working in the field of innovations, spend 14% of funds for research and development of new products, and 1.7% for the acquisition of new technologies.

If we touch upon the introduction of innovations in small enterprises, we can say the following. By their very nature, small businesses have to be more active in the market, using their flexibility and ability to quickly reorient. Therefore, it is often these enterprises that are the pioneers of new products and new technologies in various industries [10].

In the Tambov region, as in the agro-industrial region, great attention is paid to the development of agriculture. Currently, a promising area for innovation in the agricultural sector is the development and implementation of innovative technological solutions, such as satellite tracking and aeromonitoring with the provision of analytical data based on them, precision farming, automation of production processes.

Currently, in the city of Tambov, work is underway to open the first in the region Technopark in the field of information technology "Mielta", which will combine the existing innovation infrastructure and IT-enterprises in a single mechanism. The purpose of Technopark is the development of innovative business forms in the region, maintaining and developing the synergy effect from the optimal interaction of resident enterprises of the Technopark, their joint work, the exchange of knowledge and technology.

The project is scheduled for the period from 2017 to 2022 and is aimed at creating a specialized infrastructure in the Tambov region that can create the conditions for the development of innovative business. The project involves the creation of a complex that includes all types of infrastructure elements necessary for the commercialization: production areas, laboratories, separate offices and a coworking zone, conference rooms, meeting rooms, a consulting center, service companies and representative offices of large companies - potential customers for innovation.

Potential residents of Technopark are innovative enterprises, software developers, equipment manufacturers operating in one of the areas of specialization of Technopark. They will be able to implement several projects at the same time in the Technopark, while at the same time cooperating in the work on the implementation of one project.

The main areas of specialization for Technopark residents are:

- IoT (Internet of Things) technologies;
- Electronic instrumentation, primarily the development and production of equipment in the field of navigation and satellite monitoring, precision farming, unmanned systems.
- Information and telecommunication technologies, including the creation of software and software solutions.
- GLONASS/GPS technology.

5 Conclusion

The main function of Technopark in the field of information technology will be to provide support for the innovation projects of residents from the moment of their occurrence to the moment of introduction. A prerequisite for the successful implementation of the project at this stage is the development of the technological base, the creation of conditions for innovative companies to conduct development work (R & D), and a reduction in time and financial costs.

The approach to the technological equipment of Technopark with specialized equipment for R & D, the production of prototypes and small series of products is based on the principle of placing the most demanded equipment on its site. To meet the needs of highly specialized technical facilities, it is planned to develop Technopark's network interaction with the nearest scientific, educational and production centers. This will significantly improve the efficiency and reduce the cost of the project.

In order to achieve greater efficiency, the management of the production complex will be concentrated in one place, providing a single point of scientific and technological services in the immediate vicinity of future residents, as well as the necessary retrofitting equipment. Such integration of Technopark with scientific and educational institutions will allow optimal use of the existing technological potential for solving the problems of residents.

On the territory of Technopark it is planned to open the Electronics Prototyping Center - a platform where small innovative companies will be able to create a prototype of their product. To create prototypes companies are provided with specialized equipment, software and workplace. For experimental and small-scale production in the Technopark, a Center for experimental production will be created.

Thus, Technopark in the field of high technologies "Mielta" will contribute to the interaction of innovative enterprises to develop the potential of the region in the areas of IT technologies and the technologies of the Internet of things, turning them into one of the main factors of economic growth in the Tambov region.

References

1. Ashton, K.: That 'Internet of Things' Thing. In the real world, things matter more than ideas. RFID J. (2009). https://www.rfidjournal.com/articles/view?4986. Accessed 17 Jan 2019
2. Bogomolov, I.S.: Innovative and Project Management. SFU Publishing House, Rostov-on-Don (2014). (in Russian)
3. Deryagin, A.V.: Science and innovation economy in Russia. Innovations **5**(82), 15–27 (2005). (in Russian)
4. Gruzinov, V.P.: Enterprise Economics. "Finance and Statistics", Moscow (2000). (in Russian)
5. Khotyasheva, O.M.: Innovative Management. YURAYT, Moscow (2016). (in Russian)
6. Minko, L.V.: Business Planning in Innovation Management. Publishing House of TSTU, Tambov (2013). (in Russian)
7. National Innovation Report (2016). http://www.rvc.ru/upload/RVK_innovation_2016_v.pdf. Accessed 15 May 2018. (in Russian)
8. Schumpeter, J.A.: Theory of Economic Development. Capitalism, Socialism and Democracy. EKSMO, Moscow (2007). (in Russian)
9. Shpurov, I.: New industrial order. Expert, **40**(1002) (2018). http://expert.ru/expert/2016/40/industriya-4_0/. Accessed 8 June 2018. (in Russian)
10. The Strategy for the Innovative Development of the Russian Federation for the period up to 2020 (CIR-2020). Analytical Center for the Government of the Russian Federation (2011). http://ac.gov.ru/projects/public-projects/04840.html. Accessed 15 Mar 2019. (in Russian)
11. Tambov Region Administration (2018). https://www.tambov.gov.ru. Accessed 15 Mar 2019. (in Russian)
12. The main directions of the policy of the Russian federation in the field of development of innovation system for the period until 2010 (2005). https://elementy.ru/Library9/r2473.htm/. Accessed 17 Jan 2019. (in Russian)

The Process of Production Digital Transformation at the Industrial Enterprise

A. A. Chudaeva(✉), I. A. Svetkina, and A. S. Zotova

Samara State University of Economics, Samara, Russia
chudaeva@inbox.ru

Abstract. The research work studies the problems of production process digital transformation at the industrial enterprises. Stage by stage process of introduction of digital transformation in production is analyzed. The issues of financing of such transformation at the enterprise are studied. Currently, in order to ensure the competitive advantages of industrial enterprises, digital transformation processes are necessary in all areas of their activities. However, the share of industrial enterprises implementing projects on digitalization of production is low, as the process is associated with several problems. The solution of the problems that impede the digital transformation of the production of an industrial enterprise must be comprehensive. The plan for the introduction of digital transformation process (DTP) in five stages, with the definition of the cost of implementing each of the stages and the role of man in digital production, is proposed. The authors investigated the positive consequences for the company from the process of transformation of production and issues of the DTP project financing. The conditions under which the Industry Development Fund co-finances projects aimed at introducing digital and technological solutions are described. These solutions are designed to optimize production processes at the enterprise and aimed at digitizing existing production facilities within the framework of the Russian Industry Digitalization program.

Keywords: Digital transformation process · Digital transformation of production · Industrial enterprises · Digital production · Production process · Economic security

1 Introduction

A traditional company is turning into a company with "digital thinking", following the path of Digital Transformation to provide a competitive advantage in all areas of its business: manufacturing, business processes, marketing and customer interaction [3]. According to Rosstat the share of industrial enterprises that used special software to control automated production and/or individual technical means and technological processes in the period from 2003 to 2017, is small, and from 2011 to 2017 there is a decrease in this value. The share of enterprises that used CRM, ERP, SCM systems is also small, but the dynamics of change according to this criterion is positive. However, it is not possible to understand how much of this are the ERP-systems with the transformation of production processes considering only these data [5, 6].

S. I. Ashmarina et al. (Eds.): ISCDTE 2019, LNNS 84, pp. 351–358, 2020.
https://doi.org/10.1007/978-3-030-27015-5_42

The low interest of enterprises in production digitalization is also confirmed by the data from a joint study of the company "Tsifra" and the Ministry of Industry and Trade of the Russian Federation [1], the results of which are reflected in Table 1.

Table 1. The costs of industrial enterprises for digitalization and IT infrastructure development

The share of industrial enterprises	The costs of industrial enterprises for digitalization and IT infrastructure development, the share of enterprise budget
55%	Not more than 1%
6%	More than 5%

Source: composed by the authors on the basis of data from the resource [1]

According to data from industrial enterprises of the developed countries, the costs for digitalization and development of information technology infrastructure are like the costs of Russian industrial enterprises, and rarely exceed 5% of the budget [1].

The basic condition for digitalization, according to researchers, is the equipment of an industrial enterprise with software numerical control equipment (CNC). In the Russian Federation, only fourteen percent of the factories have more than 50% of such equipment. The ranking of industries of the Russian Federation by the number of CNC machines is shown in Table 2.

Table 2. The ranking of Russian Federation industries depending on the number of CNC machines

#	Type of industry	Number of CNC machines
1.	Aircraft industry	30%
2.	Instrument engineering	20%
3.	Machinery	A little bit more than 10%
4.	Car industry	Less than 10%
5.	Heavy machinery	Less than 10%

Source: composed by the authors on the basis of data from the resource [1]

According to the study [1], about eighty percent of the surveyed enterprises intend to purchase additional CNC machines within three years. A prerequisite for the implementation of the digital transformation process (DTP) is the presence of an automated planning and accounting system (ERP systems) at the enterprise. According to the study [1], in which at the beginning of 2018, 200 medium and large enterprises were surveyed, most of which are engaged in machine-tool construction and heavy engineering, 20% of respondents lack automated planning and accounting systems. At the same time, the majority of respondents used the ERP system from 1C (46%), other domestic companies (another 4%) or their own development (9%). The systems from Microsoft are reported to be used by 7% of respondents, systems from SAP −5%.

What is the reason for such an inactive position of Russian enterprises with regard to the use of special software to control automated production and/or individual technical means and technological processes?

2 Methodology

Digitalization of production involves building the architecture and hierarchy of the system, introducing software, specialized application software designed to solve problems of synchronization, coordination, analysis and optimization of production within any production MES-system (Manufacturing Execution System). According to experts, for large companies the costs can be hundreds of millions of dollars. Each stage of digitalization of production generates new costs. Thus, the installation of a conditional sensor leads to the need for implementation and a device that allows data to be transferred from the sensor. At the same time, the next device must receive data - and so on throughout the chain of interacting components of the production process. Consequently, digital production requires both powerful stations and data warehouses, as well as a multitude of mobile devices, special interfaces, etc., which leads to huge costs associated with the implementation of the digitalization process of production, and its subsequent operation.

The goal of an industrial enterprise is to make a profit, the increase of which is possible due to an increase in the volume of output, a reduction in the cost of its production and sale, a reduction in labor intensity, and an improvement in the quality and competitiveness of products. Determining the effectiveness of digital production transformation is possible by measuring the costs of implementing this process and increasing the profit (result) due to the introduction of digital tools. The determination of costs and benefits is associated with the complexity of the digital transformation process and the problems it generates.

The authors used such research methods as analysis, synthesis, description and comparison. Their application is determined by the theoretical nature of the study which included the following stages: formulation of the problem, analysis of information on this topic, comparison and description of different scientific views on the studied issues, synthesis of different approaches to the problem.

3 Results

The digital transformation process (DTP) of an industrial enterprise is associated with such problems as:

- appreciation of diverse business processes;
- lack of understanding among the structural subdivisions of the DTP;
- lack of common understanding of DTP;
- the gap between the existing regulatory documentation describing business processes and the actual use of information technology;
- lack of confidence in IT tools;

- lack of understanding of the results of applying IT tools;
- lack of regulatory documentation in terms of the formation and implementation of DTP;
- creation of an innovative platform for DTP;
- formation of a portfolio of digital projects and a system for their monitoring and updating;
- formation of digital interaction with business partners;
- building a culture of digital solutions in the direction of business strategy.

In addition to the above, you can specify additional negative factors [2, 9].

The solution of the problems that impede the digital transformation of the production of an industrial enterprise must be comprehensive. In our opinion, it is necessary to develop a phased plan for the introduction of DTP with the definition of the cost for each stage implementation.

At the first stage, an enterprise needs to conduct a comprehensive study of already automated production processes, a kind of "production inventory" in order to determine its own level of digitalization. You need to know the current capacity and capabilities, the results of work that require changes, as well as the target state of the enterprise, from which the desired results can be achieved.

At the second stage, it is necessary to conduct a survey of the company's employees in order to form a common understanding of the DTP, to design changes in the enterprise behavior patterns and business partners.

In the third stage, it is necessary to identify gaps between standards, norms and technologies of DTP to determine the depth of penetration of digitalization into production and management processes. The gap between the current capabilities and the target state should be turned into a strategic plan, the implementation of which should be based on the priorities of the enterprise.

The fourth stage should be aimed at filling the gaps in the formation of an innovative platform for the DTP, a portfolio of digital projects, the system for monitoring and updating the projects, and digital interaction with business partners.

At the fifth and final stage, a culture of digital solutions should be introduced for each DTP, a business strategy should be adopted, and special attention should be paid to information and economic security.

It should be noted that there is a widespread opinion about the exclusion of a person from production processes during the implementation of the DTP. However, recent studies [4] show that artificial intelligence can be used to automate certain processes, but much greater effect can be achieved with the addition and expansion of human capabilities through digitalization. It can free the employees of the enterprise from the monotonous routine actions. Robotic (automated, with software) production systems perform repetitive monotonous tasks in accordance with a human-defined algorithm.

Analysis of publications and research in the field of digitization of industrial enterprises shows that the DTP of industrial enterprise production leads to such positive results as: optimization of production and logistics operations; real-time production line monitoring; optimization of logistics routes and prioritization of shipments; reducing equipment downtime and repair costs; increase equipment load; increase equipment performance; rapid prototyping and quality control; analysis of large amounts of data in the development and improvement of products; increasing the efficiency of R & D and product development; reduction of electricity and fuel consumption; reduction of production losses of raw materials; reduction of resource consumption and production losses [8, 10].

Thus, the implementation of DTP projects is an objective reality and necessity in the conditions of scientific and technological development and competition. Production digital transformation will lead to consolidation of responsibility for the business processes of managers. And transmission of preliminary results of the analysis and methodological apparatus should lead to stabilization of the digital circuit and, accordingly, strengthen the system of economic security. Since DTP is included in the monitoring of the economic security system to determine the current state and the desired state of the life cycle of an industrial enterprise.

This process allows you to combine digital transformation with the overall strategy of the enterprise, as it relates to key business indicators. It also helps to substantiate current and capital costs in the context of digital transformation, to evaluate any activity in the long-term goals of the "digitized" organization.

4 Discussion

Digital production is estimated not only by the totality of production assets and personnel, but also by the level of intangible assets used: development strategies, production methodologies, intellectual property, information, competencies, skills and abilities, ability to cope with uncertainty, etc.

Projects aimed at digitization of production can be financed from different sources. These include both internal funds (profit, depreciation, funds from the sale of surplus assets) and external funds in relation to the company: borrowed (bank loans, issuance of corporate bonds, budget funds of various levels, provided on a returnable basis, investment leasing) or attracted (issue of shares, contributions of investors in the authorized capital). The ratio of different sources depends on the economic situation in the country and in the world.

The initiative of the Ministry of Industry and Trade of the Russian Federation was established in 2014, it concerns the transforming the Russian Fund for Technological Development into the Industrial Development Fund. It was founded to modernize Russian industry, organize new productions and provide import substitution. It offers preferential conditions for co-financing projects aimed at digitizing existing industries under the Digitalization of Industry program designed to co-finance projects aimed at introducing digital and technological solutions designed to optimize the production processes in the enterprise the conditions for financing projects under this program are shown in Table 3.

The introduction of digital production should lead to the following changes: (1) an increase in labor productivity; (2) the maximum quality of products; (3) the complexity of the product; (4) increased requirements to the staff; (5) production automation at all stages of product manufacturing, including its development.

At present, the consumer is a participant in the interaction with the enterprise and, therefore, an element of the digital production and digital enterprise being created. This

Table 3. The conditions for financing projects under the program "Industry digitalization" of Industrial Development Fund

Basic conditions	
Loan duration	Up to 60 months
Loan sum	20–500 mln.rub
Extra conditions	
Interest rate	1% with Russian software (included in the list of Ministry of communication (https://reestr.minsvyaz.ru) and/or elaborated in the frame of the projects participating in National technological initiative (https://asi.ru/nti/)) or system integrator of the Russian Federation (should be included in the latest rankings of IT-companies (РБК+, CNews, Tadviser)) and is not a subsidiary of the non-residential company of Russia; 5% other cases
Co-financing	$\geq$ 20% of the project budget, including own assets, private investors resources, bank financing
Productivity growth per 1 employee	$\geq$ 5% annually, starting from the 2 year of getting the grant
Total project budget	From 25 mln.rub

Source: compiled by the authors based on data of [7].

leads to the need to include the consumer in the DTP chains. To do this, the company should not only digitize production, but also the processes associated with it: product development, technology definition and its implementation, supply and marketing, production planning and sales system.

DTP should solve two main tasks: to minimize the cost of production and increase the net revenue, while maintaining product quality at a consistently high level. This means that all stages of the production process must be fully manageable and transparent. To do this, a single information space is created at the enterprise, where high-tech equipment, analytical and management IT systems exchange data non-stop: (1) at the technological level, it is represented by engineering infrastructure; (2) at the level of production these are monitoring systems and analytical tools that process the data obtained from the equipment and help in a timely manner to influence the main factors of production in a timely manner; (3) at the management level it is the all business processes synchronization towards the fulfillment of a single goal.

Production management is associated with the organization of the collection, transmission and processing of various information, which serves as the basis for the development and adoption of appropriate management decisions. The complex of interrelated organizational and technical elements forms the production and management framework, regulates and organizes all information flows. And further, in the process of digital transformation of production, information flows are formed, which are accompanied by a system of economic security of a manufacturing enterprise. In relation to the digital transformation process, internal and external flows are distinguished. Internal transformational digital flow adjusts production, management, administrative and other business processes within the enterprise. External transformational digital flow regulates the security of processes and forms of interaction with customers, suppliers, regulators and other business partners.

5 Conclusion

Thus, the digitization of the production of enterprises in the real sector of the Russian economy is possible by attracting the required amount of investment. Enterprises do not have their own funds for such projects; it is necessary to attract credit resources that are currently limited in the Russian Federation due to sanctions. As a result, large industrial enterprises attract foreign investment and claim to state support for the development of breakthrough technologies, investment projects, long-term projects, and projects for technical re-equipment and modernization. Medium-sized industrial enterprises need R&D support, development of applied technologies, cheap loans. Small industrial enterprises are concerned with providing venture financing, providing conditions for creating new products, providing infrastructure, providing funds for investment projects, financial support for expansion projects.

For support, you can contact both the authorities of the Russian Federation and state development institutions. Any industrial enterprise developing in the field of digital production may apply for subsidizing interest rates on loans, subsidizing the cost of replenishing working capital, subsidizing and supporting research and development, subsidizing capital expenditures, budgetary investments, government guarantees, targeted tools under state programs, preferences in public procurement, special investment contracts, support for foreign economic activity, the provision of direct financing (grant, venture or debt). The choice of state support instruments depends on the scale of the project and the stage of the product life cycle.

References

1. Balenko, E.: Industry scrimps digitalization. Technol. Media **117**(2841)(0307) (2018). https://www.rbc.ru/newspaper/2018/07/03/5b3a26a89a794785abc9f304. Accessed 10 Apr 2019. (in Russian)
2. Belzer, M.: Industry digitalization: fashionable trend or necessary condition for keeping competitiveness? (2018). https://promdevelop.ru/tsifrovizatsiya-promyshlennosti-modnyj-trend-ili-neobhodimoe-uslovie-dlya-sohraneniya-konkurentosposobnosti/. Accessed 10 Apr 2019. (in Russian)
3. Chudaeva, A.A., Mantulenko, V.V., Zhelev, P., Vanickova, R.: Impact of digitalization on the industrial enterprises activities. In: SHS Web Conference, vol. 62, article 03003 (2019). https://doi.org/10.1051/shsconf/20196203003
4. Daugherty, P.R., Wilson, J.H.: Human + Machine: Reimagining Work in the Age of AI. Harvard Business Review Press, Boston (2018)
5. Federal State Statistics Service (Rosstat): Share of organizations that used personal computers, by constituent entities of the Russian Federation for 27 July 2018 (2019). http://www.gks.ru/free_doc/new_site/business/it/it7.xls. Accessed 10 Apr 2019. (in Russian)
6. Fin-book. What is digital economy in simple words? (2019). https://fin-book.ru/chto-takoe-tsifrovaya-ekonomika-prostymi-slovami/. Accessed 10 Apr 2019. (in Russian)

7. Industry development funds. The program "Industry digitalization" (2019). http://frprf.ru/zaymy/tsifrovizatsiya-promyshlennosti/. Accessed 10 Apr 2019. (in Russian)
8. LTD "Mckinsey and Company CIS". Digital Russia: new reality (2017). https://www.mckinsey.com/~/media/mckinsey/locations/europe. Accessed 10 Apr 2019. (in Russian)
9. TAdviser. Digital economy of Russia. State Business IT (2019). http://www.tadviser.ru/index.php/%D0%A1%D1%82%D0%B0%D1%82%D1%8C%D1%8F:%D0%A6%D0%B8%D1%84%D1%80%D0%BE%D0%B2%D0%B0%D1%8F_%D1%8D%D0%BA%D0%BE%D0%BD%D0%BE%D0%BC%D0%B8%D0%BA%D0%B0_%D0%A0%D0%BE%D1%81%D1%81%D0%B8%D0%B8. Accessed 3 June 2019. (in Russian)
10. TAdviser. Digital Russia: NEW REALITY. State Business IT (2017). http://www.tadviser.ru/images/c/c2/Digital-Russia-report.pdf. Accessed 10 Apr 2019

Digitization of the Agricultural Sector of Economy as an Element of Innovative Development in Russia

O. V. Mamai[1], I. N. Mamai[1], and M. V. Kitaeva[2](✉)

[1] Samara State Agricultural Academy, Samara, Russia
[2] Samara State University of Economics, Samara, Russia
kmv_1965@mail.ru

Abstract. The authors consider development of agricultural sector in terms of implementing the concept of digitalization, reasons for a slow introduction of innovations in the agricultural sector, possibilities of using information and communication technologies (ICT) in agricultural production, and principles of digitalization of the agricultural sector. The aim of the contribution is to identify opportunities for the development of the agricultural sector based on the application of the principles of digital economy. The methodological core research is a systematic and institutional approach to the digital platform for the effective development of the agricultural sector of the economy. The authors used methods of comparison and analogy, analysis and synthesis. As a methodological base, regulatory documents were used for sustainable development of the agro-industrial complex in the Russian Federation. Modern information technologies contribute to improving the efficiency of production processes in the agricultural sector of the economy. However, domestic farmers use elements of the digital economy in limited quantities due to lack of financial resources. To speed up this process, it is necessary to create programs for enhancing innovation processes, as well as managerial personnel who will be able to introduce new digital technologies, expand the sphere of application of the digital economy through public investment, and the formation of various cooperative associations.

Keywords: Agricultural sector · Development · Digital economy · Efficiency · Innovation · Information and communication technology

1 Introduction

The world has already entered the era of digital globalization, defined by data streams that contain information, ideas and innovations. Smart devices are becoming smaller, faster, cheaper, more powerful and will be the key to solve problems. Today is the time when intelligent digital solutions should help the agricultural sector of the economy to cope with the problems of increasing productivity and sustainable development. Innovative development of the agricultural sector is impossible without its digitalization. New information and communication technologies (ICT) have become modern tools of progress. Digitalization is the use of digital technology to change the business

S. I. Ashmarina et al. (Eds.): ISCDTE 2019, LNNS 84, pp. 359–365, 2020.
https://doi.org/10.1007/978-3-030-27015-5_43

model and provide new opportunities for generating income and creating value. This is the process of transition to digital business [1]. W. Brian Arthur writes about the emergence of a second economy - digital, which in its volume will catch up with the traditional economy in ten years, and in 20–30 and overtake it. The authors forecast the increase in the productivity of the world economy by about 2.4% per year [2]. Thus, the purpose of the contribution is to identify the opportunities of the agricultural sector of the economy based on the application of the principles of the digital economy.

2 Methodology

The methodological core research is a systematic and institutional approach to the formation of a digital platform for the effective development of the agricultural sector of the economy. Research methods include economic analysis and analysis of statistical data. The contribution is based on data from the Federal State Statistics Service, the Ministry of Agriculture of the Russian Federation. The methodological basis of the contribution was dialectic principles, methods of system analysis of economic phenomena: analysis and synthesis, a method of scientific abstraction, an economic-statistical method, a method of expert evaluation, methods of comparison and analogies, etc.

3 Results

The study showed that Russia has set ambitious goals for digitization of the agricultural sector. In general, the digitalization program should include the following issues:

- Create a single information management space of the agricultural sector and increase its transparency, ensure food security;
- Improve the quality and efficiency of management decisions by agricultural producers;
- Increase the reliability of the agricultural census results;
- Expand the range of services provided in electronic form, introduce automatic reporting;
- Optimize costs for the development and maintenance of the System of State Information Support in the field of agriculture.

However, the risks of introducing digital technologies in the agricultural sector should be noted:

- Insufficient level of knowledge in the field of ICT;
- Lack of the required number of IT specialists in the agricultural sector (the number of people employed in agriculture is 4,706 thousand people (6.5%), of which about 113 thousand people in IT, therefore, there is a shortage of IT specialists in the industry at least 90 thousand person);
- Transformation of professions (about 40% of professions may disappear by 2030);

- Lack of the required number of devices and sensors made in Russia (high import dependence);
- Cyber attacks.

On the one hand, there is a low level of ICT development for the agricultural sector. On the other hand, there is a threat of increasing technological backwardness, and, consequently, the problem of the country's economic security [5, 6, 10–14]. If this level is not raised in the near future and the number of specialists with ICT does not increase, there is no need to talk about digitalization of the agricultural sector of the economy.

4 Discussion

Today, the transition to the digital economy is one of the key priorities of Russia's development, because it is the level of digitalization that will determine the country's competitiveness in the new technological order. Accordingly, for the Russian Federation to reach a new level of economic development, social sectors, it is necessary to have own advanced developments and scientific solutions. The analysis of statistical data shows that Russia is striving for leading indicators of digitalization, but so far has average values.

According to the IDI Development Index, Russia is in 45th place out of 176 countries surveyed, in E-Government Development Index - 35th out of 193 countries, in Global Cybersecurity Index - 10th place and is in the top 20. In general, according to the International Index of the Digital Economy and Society in 2016, Russia almost reached the national average. Its value was 0.47, with an average for the countries of 0.54. However, it should be noted that domestic costs of research and development in the priority area "Information and Telecommunication Systems" increase annually in Russia (Fig. 1).

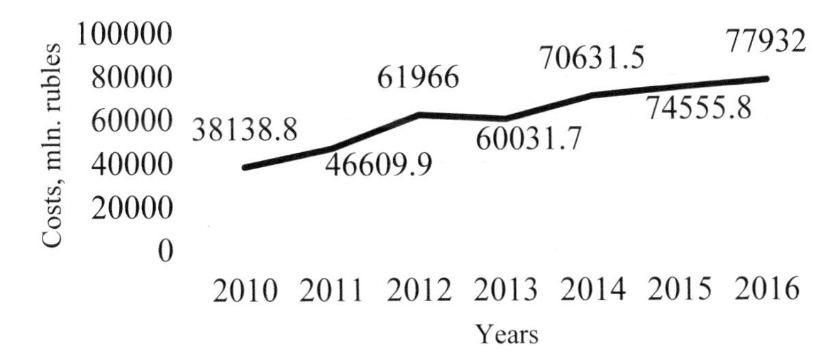

Fig. 1. Internal costs of research and development in the priority area "Information and Telecommunication Systems", in current prices (Source: compiled by the authors based on [3])

The activity of the agricultural sector of the economy is aimed at providing the population with food and obtaining raw materials for a number of industries. The industry is one of the most important, represented in almost all countries. According to Rosstat, in 2017 the amount of investment in ICT amounted to 3.6 billion rubles or 0.5% of total investments in fixed assets. This is the lowest indicator by industry, which indicates low digitalization of the domestic agricultural sector of the economy. At the same time, almost all scientists and practitioners [4–7] recognize the importance of digitalization for the agricultural sector and see the following opportunities in it:

- Risk reduction;
- Track climate change;
- Increase the yield of agricultural structures and productivity of animals;
- Timely planning of field work, reduce production costs based on the efficient use of resources and scientifically based approaches;
- Reduce transaction costs for the purchase and sale and simplify the supply chain of products from the field to the consumer;
- Reduce the shortage of skilled labor;
- Timely provision of critical information to rural producers, etc.

The study showed that the main objectives for digitalization of the agricultural sector of Russian economy are:

- Improve management efficiency;
- Dispatch and aggregate of data flows to create end-to-end chains from agricultural production to consumption with deep integration into related sectors of the digital economy;
- Introduce new professions in agricultural production;
- Increase the incomes of agricultural producers and the quality of life in the countryside;
- Create conditions for subsidizing the installation, processing and transmission of data from cyber-physical systems (Internet of Things platforms) with deep integration into the digital agriculture platform, as a principle encouraging the introduction of digital solutions;
- Improve the efficiency of interaction of participants and the state with the transition to digital format of data exchange, reduce types of reporting;
- Create conditions for the industry transition to a through-production cycle with minimization of intermediaries and trading margin;
- Develop a generally accessible structured bank of knowledge and technology in the context of agricultural sub-sectors and regions.

At the same time, the following tasks of the digital transformation of agriculture are recognized as important:

- Transition to digital agriculture, precision farming and active use of digital technologies to increase productivity;
- Integrate objective data flows of agricultural producers and government data into the digital agriculture platform to ensure global planning in the industry and provide accurate recommendations to market participants, including using ICT;

- Integrate functionality of the digital agriculture platform, which provides agricultural producers with access to state, banking and insurance products;
- Develop digital mechanisms and measures of support in the agricultural sector of the economy;
- Ensure the traceability of agricultural products (tags, chips, identifiers, technologies, devices, systems);
- Stimulate domestic development and access to various digital open platforms (digital field, herd, control equipment, greenhouses, etc.);
- Provide a package of personal (matrix) technological solutions for market participants;
- Introduce online trading platforms and systems to promote agricultural products;
- Predetermine proposals for the adjustment of legal acts and regulatory and technical requirements for the transition to the digital agriculture;
- Develop educational and methodological complexes (standards, methods, programs) training;
- Ensure the compatibility of processes and standards of production with the worldwide in order for Russia to take the leading position as an exporter of agricultural products.

Analysis of information and communication technologies for the agricultural sector of the economy showed their diversity. Many innovations can be successfully applied in Russia.

Insects, diseases, and weeds—all of these terrible farmer nightmare cause severe damage to crops and can reduce the yield. Modern plant protection can help the farmer overcome these problems and create fairly safe and affordable conditions for growing crops. In most cases, stress factors affecting plants are detected only when significant damage has already been done. At the moment, farmers often have little choice and they use plant protection products to cure plants that they can save.

Innovative digital farming technology can help solve this problem. For example, satellites and unmanned aerial vehicles are able to perform detailed field observations, and remote sensors can regularly take photographs and measure radiation in the long wavelength range. These invisible signals reveal a great deal of information about the state of crops, such as their general life force. Thus, stressors that threaten plants can be detected long before they become visible to the human eye.

By integrating the data, the farmer can anticipate the spread of pests and diseases or identify them at an early stage. This data is used to create detailed real-time field maps that allow you to apply crop protection and other valuable materials. For example, Bayer Field Manager [8] provides farmers with instant access to strategies that apply to specific conditions in order to make the most effective use of crop protection.

Developments in the field of agricultural production management systems are also very interesting, for example, the agricultural production management system of Agro Network Technologies (ANT) [9]. It allows you to ensure the full cycle of management of agricultural production of agricultural holding:

- Provide a single database of reliable information with the integration of adjacent systems;
- Automated and unified paperwork and reports;

- Improved efficiency of land use;
- Improved accuracy of resource planning and consumables;
- Increased personal responsibility.

Therefore, this ANT system makes it possible to turn an avalanche of reports into a controlled flow, visualize everything that is happening in the fields, and provide a relevant and understandable basis for making management decisions.

Thus, the capabilities of information and communication technologies make it possible to improve the agricultural sector of the economy and bring it to a new level.

5 Conclusion

An important direction in the development of the agricultural sector of the economy is the introduction of the principles of the digital economy, which will create an institutional environment that meets modern requirements and realities, reduces transaction costs and generally improves the efficiency of the agricultural sector of the economy. Innovative development that is adequate to the upcoming scale and objectives of the scientific and technical transformation of agricultural production is possible if there is a properly organized and well-functioning innovation system of the agricultural sector of the economy, which is a set of interacting organizations that are participants in the process of creating and developing innovations with a comprehensive innovation process [15]. This also applies to digitalization of the agricultural sector. Modernization of the infrastructure of the agricultural sector and the creation of its new elements require coordinated actions of all its participants, both at the intergovernmental and state levels, as well as at the level of regions and individual organizations. A reasonable combination of various forms and types of interstate support for them and technological solutions is required [16].

Summarizing the above, we can conclude that in recent years the IT structure has been expanding and improving in Russia, many interactions of participants in innovation processes are being disseminated into the Internet environment, new disciplines related to economy digitalization appear in the learning processes. There are ambitious plans to create a high-tech agricultural sector. All these factors continue to improve and improve. At the same time, all projects require huge financial support from the state and new specialists.

References

1. Abdrakhmanova, G.I., Vishnevsky, K.O., Volkova, G.L., Gokhberg, L.M., et al.: Digital Economy Indicators in the Russian Federation: 2018: Data Book. National Research University Higher School of Economics, HSE, Moscow (2018)
2. Ablaev, I.M.: On the question of the economic content of innovations. Questions Econ. Law **55**, 88–91 (2013)
3. Agro Network Technologies agricultural production management system. https://ant.services/website/sections/7. Accessed 10 Nov 2018
4. Arthur, W.B.: The second economy. McKinsey Q. **4**, 90–99 (2011)

5. Dokin, D.B.: Universal intellectual system for designing and analyzing complex structures of the agro-industrial complex. Bull. Novosibirsk State Agrarian Univ. **2**(23–2), 83–88 (2012)
6. Babkin, A.V. (ed.): Formation of a New Economy and Cluster Initiatives: Theory and Practice (Monograph). Polytechnic Publishing House, St. Petersburg (2016)
7. Gray, J., Rumpe, B.: Models for digitalization. Softw. Syst. Model. **14**(4), 1319–1320 (2015)
8. Babkin, A.V. (ed.): Innovations and Import Substitution in industry: Economics, Theory and Practice (Monograph). Cult-Inform-Press, St. Petersburg (2015)
9. Kapelyuk, Z.A.: From the information and communication revolution to the modernization of the agro-industrial complex. In: Digital Transformation of the Economy and Industry: Problems and Prospects (Monograph). Publishing House of St. Petersburg Polytechnic. University of Peter the Great, St. Petersburg (2017)
10. Koptelov, A.: Information technologies in agriculture. Agribus. Comput. Sci. Equip. Technol. **12**, 60–64 (2010)
11. Kurcheeva, G.I.: Benchmarking analysis of the sites of enterprises of the dairy industry. Pract. Mark. **6**(208), 26–32 (2014)
12. Mamai, O., Mamai, I.: The agricultural sector. In: Raupelienė, A. (ed.) Proceedings of the 8th International Scientific Conference "Rural Development 2017: Bioeconomy Challenges", pp. 1167–1173. Aleksandras Stulginskis University, Lithuania (2017)
13. Mamai, O.B.: Modern tools for managing the innovative development of the agricultural sector of the region. In: Modern Economy: Ensuring Food Security: Collection of Scientific Papers of the IV International Scientific and Practical Conference, pp. 91–96. Kinel Publisher: Samara State Agricultural Academy (2017)
14. Mod Field Manager FieldSell v 1.0. http://farming-simulator15.ru/2016/11/16/mod-menedzher-polej-fieldsell-v-1-0-rus-farming-simulator-17/. Accessed 10 Nov 2018
15. Nechaev, V.I.: State support of the agricultural technology platform of the EAEU. Econ. Agric. Russia **7**, 81–85 (2017)
16. Sandu, I.P., Afonina, E.B.: Innovations: the path from the beginnings to the present. Science **6**(19) (2013). https://naukovedenie.ru/PDF/01EFTA613.pdf/. Accessed 10 Nov 2018

Competitiveness of Project-Oriented Recreational Organizations in the Context of Marketing Technologies Transformation

E. N. Sheremetyeva[1(✉)], E. P. Barinova[1], and N. V. Mitropolskaya-Rodionova[2]

[1] Samara State University of Economics, Samara, Russia
lena_scher@mail.ru
[2] Odintsovo Campus of MGIMO University, Odintsovo, Russia

Abstract. The article considers the need to use marketing technologies for recreational organizations with a project-oriented approach. The main advantages of marketing technology and project management in enhancing the competitiveness of a recreational organization from fundamentally distinctive positions are presented. Competitive advantages of marketing technologies are formulated, which have novelty for a given application and help to reveal the competitive advantages of a recreational organization in the constant struggle for the consumer (recreant) and the use of various tools and methods to increase competitiveness. Attention is paid to aspects of the project-oriented approach and its components. The advantages of the participants from the professional management of the project-oriented activity of the recreational organization are revealed. Determined that the effectiveness of marketing technologies and a project-oriented approach contribute to the development of project-oriented recreational organizations, the formation of a fundamentally different work culture in general and the management of a recreational organization in particular, and various categories of participants in project-oriented recreational activities will benefit from professional management. Of particular importance is the strategic focus of the recreational organization, including the methodology and tools of effective marketing technologies, which are an integral part of the work in carrying out project activities, as well as creating conditions for the emergence of a "project culture" in a recreational organization means knowledge of modern science of project management. Accordingly, the use of marketing technologies at the present stage is an important factor in the development of recreational services in project-oriented recreational organizations.

Keywords: Competitiveness · Digitalization · Marketing technology · Project-oriented · Recreational organization · Recreational services

1 Introduction

Today's marketing is a business philosophy and allows modern recreational organizations (RO) to anticipate the desires of consumers (tourists) and to meet their needs. The use of marketing technologies helps not only in the organization and management

S. I. Ashmarina et al. (Eds.): ISCDTE 2019, LNNS 84, pp. 366–372, 2020.
https://doi.org/10.1007/978-3-030-27015-5_44

of RO, but and allows a creative approach in the management process. In Russia at the present time the problem of assessing the effectiveness of marketing activities is especially important for RO, which has to survive not only in conditions of instability of the external environment, due to political and economic factors, but also a serious competition and foreign direct RO – competitors [3, 5, 8].

In the increasingly competitive market of recreational services (RS) successful RO is determined by the efficiency of marketing and marketing techniques, and knowledge of current marketing tools and ability to apply them in practice gives a competitive advantage RHO, helps to achieve the desired results in the market of recreational services [13, 15].

2 Methods

Content analysis, informative-target analysis and patent search were chosen as the main research methods. The methodological foundations are used to define the goals and strategies of marketing technologies, identify interpretations and adequacy of the image perception of a recreational enterprise, which allows obtaining more accurate information on specified criteria and finding the most reliable and relevant data on indicators of the level and trends of the organization competitiveness in the field of recreational services.

3 Results

In the works of domestic and foreign authors on the theme of organization and use of techniques and methods of marketing, the term "marketing technology" involves various methodological schemes aimed at structuring and rationalization of marketing activities. However, specific studies on the development of content marketing technologies in project-oriented recreational organizations, determination of their specificity, the main ways of development and implementation are not enough.

The purpose of this article is determined in the development of adequate approaches to achieving competitive advantage, based on the use of advanced marketing technologies, taking into account the specific characteristics of the industry recreational services. The objective is to develop a theoretical guidance for the application of marketing technologies in the process of formation of competitive strategies aimed at expanding the potential of RO with a design-oriented approach [7].

4 Discussion

The task of increasing the competitiveness of domestic producers of RS closely linked with the formation of strategic decisions based on marketing techniques. Marketing technology is a system of actions and operations, aimed at bringing administrative activities of the entity in accordance with the needs and expectations of the consumer [4]. Marketing technology of RO are components of the marketing process, which can effectively accommodate and implement RU on the regional market. There are advertising campaigns with a purpose to talk about new RS and attraction of more potential visitors, and making available to them information about discounts and new qualitative characteristics of the RS to make it more competitive.

A key element is the empowerment of RS such distinctive features that are perceived by consumers (tourists) as the best ways to meet existing needs. Not many RO, forming the package offers, clearly understand that only the perception of tourists is able in the last instance to identify them as "best", "quality", "irreplaceable". Even fewer of those who exactly understand what the basis is of any assessment of consumer perception [14].

Of particular relevance to this issue is also due to the increase in the share of RS in the composition of the recreational offerings of RO operating in the regional market. The perception of RS due to their impalpability, heterogeneity, one-time provision and consumption differs significantly from the perceptions of the physical goods.

In the constant struggle for the consumer (recreant) RO apply various tools and techniques to enhance competitiveness. Marketing techniques are one of the methods aimed at the development and successful activity of RO. We can identify five key technologies of marketing of RO [2]:

1. Segmentation - dividing the market of RS on the segments with the aim of the study of demand (buyers) recreants, their relationship to RS;
2. Target (from English "target" - aim - allows you to select from a number of consumers (recreants) the target audience for which it is intended or that the advertising RS;
3. Positioning is more advantageous position of RS in the market with relation to RS - competitors;
4. The analysis includes RU market research, demand for different RS, study of changes in the pricing policy on the market, consumer attitudes to data of RS and others, with the aim of increasing demand for its RS;
5. Forecasting is an important step in marketing research. It includes the assessment of the prospects of development of the market or segment, trends in prices and other market conditions in the forecast period, which can be short-, medium - and long-term.

New marketing techniques, new marketing communications and even new marketing strategies, they are all united in a unique "marketing mix", with fresh ideas and unusual resources.

In today's rapidly changing world, RO, which in practice apply the methods of project management, has a competitive advantage over those RO that operate instinctively because this increases their agility and maneuverability.

Project-oriented approach allows you to create a new corporate culture management - project management in RO. Project-oriented approach provides a systemic approach to the implementation of the planned actions, whether there are recreational and entertainment activities, conclusion on the regional market of new RS or development of the region. In any case it requires a clearly defined goal, the desired results of the project, timelines, resources, total cost, and the risks (environmental impact and interaction of the participants themselves), and the creation of a system of planning, monitoring and control of progress, form a team and to coordinate the efforts of all performers. The advantages of such a clearly structured and purposeful system of creating RS and presenting these services to consumers are obvious [10].

Abroad project management approach is taught by many educational institutions and professional communities, for example AACEI - Association for the Advancement of Cost Engineering International with a half-century history. Its ideology is shared by General Electric, Intel Corporation and other companies with a worldwide reputation. This Association along with its own, in the training uses standards PMBOK (Project Management Body of Knowledge), ICB (Interactive Cell Broadcast), PMI (Project Management Institute). However, they do not differ from each other, as well as their Russian counterparts, one of which NTC. These standards are the benchmark for almost all of the schools in our country [12].

Until recently, in domestic practice with the concept of "project" is usually connected with the idea of the set of project-estimate documentation for construction of buildings, structures or technical devices. Professional project management is connected with the concept of the project communicates the process of implementing a series of interventions for the creation of a new product or service within the established budget, time and quality.

Project approach contributes to the formation of a fundamentally different culture of work in General and management of RO in particular. It allows you to define "immediate goal", not cancelling previously defined strategic objectives, as a tool for the development of specific program goals in RO [11].

Project management improves organizational skills of the staff of RA, they organize everything around workflows, clearly imagining what resources are needed, what is the budget and the time required to achieve results, what are the responsibilities of each member of RHO. In short, project management is the future of RO, new ROUX and new competitive advantages. The benefits can also be attributed, and that reduces the time required to produce jobs for the employees begin to understand each other without words, to think in the same categories. For example, if it says: you need to build a hierarchical structure of the project, you do not need long and tedious to explain the essence of this job. In a project-oriented organization clearly defines the duty and responsibility of each member of the project team, for a specific period of time, which is painted in the form of a graph or network and associated with the overall objective of the project [6].

Different categories of project-oriented activities RO will receive the following benefits from professional management [9]:
Investors:

(1) Transparency of public and private projects;
(2) Reduction and control of risks;
(3) Expand the range of investors and investment opportunities;
(4) Save the investment of resources by improving the efficiency of use of project funds;
(5) Increase return on investment.

Project managers and owners:

(1) Improve competitiveness;
(2) Improve return on invested capital;
(3) Additional income;
4) Improve manageability.

State:

(1) the validity and clarity of the planning and implementation of projects and programs;
(2) control over expenditure of funds, resources and deadlines;
(3) reduce risks, costs, time and resources;
(4) reduce costs to the budgets of all levels and other

Society:

(1) increase the efficiency of the economy as a whole;
(2) increase investment in the social sector from the budgets of all levels and guide RO;
(3) improve the quality and standard of living.

All tasks and proposals need to be worked out to the level of the project, as the project is a card on which you want to move forward to achieve results. You need to use the whole complex project management - from project initiation to achievement of results and objectives of the project. All of this will be project-oriented approach.

However, manufacturers of RS, mastering marketing techniques in conjunction with the project, often do not have a theoretically sound method of preparation and decision-making in this specific field, which requires special strategic competence in marketing, project management, competition [1].

Competitive potential of RO in the application of marketing technologies from the point of view of project-oriented approach can be most efficiently realized using the principles of strategic marketing and management.

Creating conditions for the emergence of a "project culture" in RO denotes the knowledge of the modern science of project management. The most effective means obtaining such knowledge as hiring a professional project managers and consultants for the production of project management. If you act in two directions simultaneously, the result will surpass expectations. You also need to train the managers of the marketing department of RO design technologies and approaches in order to achieve mutual understanding in action of project and operational managers. RO, whose marketing is a large unit, with a large number of employees and projects, might think about organizing project office, which could be involved in the evaluation, selection and control of projects, which is very important today.

5 Conclusion

The problem of implementing marketing technologies in the management of RO quite acute for many RO. Not all RO can afford the organization of the marketing department or hiring chartered marketers and project managers, while managers RO do not possess the necessary knowledge. However, the application of marketing technologies at this stage is an important factor in the development of RS in project-oriented RO. Further investigation RO-oriented project management should be aimed at developing teaching skills in the application of marketing technologies, taking into account the specifics of the RS.

References

1. Antontseva, M.S.: Stimulating entrepreneurial activities based on the implementation of projects. In: Shatalov, M.A. (ed.) Fundamental and Applied Research: From Theory to Practice, Materials of the II International Scientific and Practical Conference, vol. 3, pp. 69–74. AMiSta, Voronezh (2018)
2. Baboyan, E.S., Sidorova, A.M., Alekseeva, E.V.: To the question of the estimation of efficiency of marketing activity and marketing research. In: Kamalieva, I.R. (ed.) Social and Humanitarian Trends: History and Modernity. Collection of Articles of the International Scientific and Practical Conference, pp. 120–127. AntroVita, Moscow (2018). http://soc-is.ru/wp-content/uploads/. Accessed 04 Apr 2019
3. Bagozzi, R.P., Rosa, J.A., Celly, K.S., Coronel, F.: Marketing-Management. Walter de Gruyter GmbH & Co KG, Oldenbourg (2018)
4. Beck, A.A.: Role of marketing activities in the life of the enterprise. Alley Sci. **6**(22), 182–185 (2018). https://alley-science.ru/domains_data/files/75June2018/ROL%20MARKETINGOVOY%20DEYaTELNOSTI%20V%20ZhIZNI%20PREDPRIYaTIYa.pdf. Accessed 04 Apr 2019
5. Chernev, A.: Strategic Marketing Management. Cerebellum Press, Chicago (2018)
6. Keegan, A., Ringhofer, C., Huemann, M.: Human resource management and project based organizing: fertile ground, missed opportunities and prospects for closer connections. Int. J. Project Manag. **36**(1), 121–133 (2018). https://doi.org/10.1016/j.ijproman.2017.06.003
7. Kerzner, H.: Project Management Best Practices: Achieving Global Excellence. Wiley, Hoboken (2018)
8. Martinsuo, M., Hoverfält, P.: Change program management: toward a capability for managing value-oriented, integrated multi-project change in its context. Int. J. Project Manag. **36**(1), 134–146 (2018). https://doi.org/10.1016/j.ijproman.2017.04.018
9. Minaev, D.V.: Role of project management in ensuring sustainable development. In: Shamakhov, V.A. (ed.) The Collection: State and Business. Modern Problems of Economics. Materials of the X International Scientific Practical Conference, vol. 2, pp. 65–73. North-West Institute of Management of the RANEPA Under the President of the Russian Federation, St. Petersburg (2018)
10. Mujanovic, E., Damnjanovic, A.: Marketing management and social entrepreneurship. In: Hammes, K., Klopotan, I., Nestorovic, M. (eds.) Economic and Social Development (Book of Proceedings), 30th International Scientific Conference on Economic and Social Development, pp. 77–88. Varazdin Development and Entrepreneurship Agency (VADEA), Belgrade (2018)

11. Prajogo, D., Toy, J., Bhattacharya, A., Oke, A., Cheng, T.C.E.: The relationships between information management, process management and operational performance: internal and external contexts. Int. J. Prod. Econ. **199**, 95–103 (2018). https://doi.org/10.1016/j.ijpe.2018.02.019
12. Project Management Institute: A Guide to the Project Management Body of Knowledge (PMBOK® Guide)-(SIMPLIFIED CHINESE) (2018). https://b-ok.org/book/2825190/ce5880. Accessed 04 Apr 2019
13. Usmanova, T.Kh.: Mechanisms of project and program implementation of social and economic development strategies. In: Gerasimov, V.I. (ed.) Russia: Trends and Development Prospects. Yearbook, pp. 126–129. Institute of Scientific Information on Social Sciences of the Russian Academy of Sciences, Moscow (2018)
14. Sheremetyeva, E.N., Mitropolskaya-Rodionova, N.V.: Marketing in international projects for project-oriented companies. In: Sheremetyeva, E.N. (ed.) Actual Problems of Science, Economy and Education of the XXI Century. Materials of the III International Scientific-Practical Conference, 5th March–26th September 2014, pp. 300–312. Samara Institute (Branch) of Plekhanov Russian University of Economics, Samara (2014)
15. Zybinskaya, R.R.: Marketing research in the field of youth tourism development. Alley Sci. **6**(22), 169–173 (2018). https://alley-science.ru/domains_data/files/63June2018/MARKETINGOVYE%20ISSLEDOVANIYa%20V%20OBLASTI%20RAZVITIYa%20MOLODEZhNOGO%20TURIZMA.pdf. Accessed 04 Apr 2019

XBRL Reporting in the Conditions of Digital Business Transformation

O. V. Astafeva[1(✉)], E. V. Astafyev[2], E. A. Khalikova[2], T. B. Leybert[2], and I. A. Osipova[1]

[1] Financial University under the Government of the Russian Federation, Moscow, Russia
astafeva86@mail.ru
[2] Ufa State Petroleum Technological University, Ufa, Russia

Abstract. The concept implementation of business digitalization and completeness of information disclosure in accordance with international financial reporting standards presupposes the transition of Russian companies to reporting in XBRL format. The purpose of the study is limited by its conceptual problem and its definitions for the purposes of integrated reporting. The authors of the article also suggested the structure of the integrated reporting, including topics and the approximate composition of financial and non-financial indicators for the Russian companies. The basic conceptual advantages of XBRL format and its development prospects in the Russian practice have been formulated in the article.

Keywords: XBRL · Digital transformation · Business model · Digital technology · Integrated reporting · Business reporting

1 Introduction

Active development and the implementation of digital technologies substantially alter the business environment. Digital business transformation involves a new approach to working with data due to the fact that companies using digital technology are able to find new profit sources and ways to create value. Data analytics allows us not just to accumulate data, but even monetize the data because nowadays companies can offer their clients something new, through collecting data from certain devices, analyzing them and identifying patterns that will allow clients to find more accurate solutions. Possibilities emerge for companies to predict certain events and situations before problems arise. At the same time, the requirements for the submission change, as well as for the processing and the use of the output information that is contained in the financial statements for the subsequent interpretation of the indicators of financial activity of the company by the various interested parties.

The recently observed in both Russian and foreign practice growing interest of the controlling bodies to the reporting format pushes the reporting companies to the activate the standardization process of electronic reporting taxonomy in disclosing corporate information provided to external users. In this regard, the companies are interested in implementing software products that are capable of enhancing agility in

S. I. Ashmarina et al. (Eds.): ISCDTE 2019, LNNS 84, pp. 373–381, 2020.
https://doi.org/10.1007/978-3-030-27015-5_45

business management. XBRL format that is used by most G20, allows you to improve the quality of data management, adopt modern unified reporting data exchange format for market participants with all the stakeholders, which leads to the active application of the IT products for further internal analysis of the company's own information.

In the context of the transition to the digital economy, the real sector understands the urgent need for the transformation of the business model focused on assessing the feasibility of strategic goals and ultimately on the increase of the value of companies. The transition from the old model to the new business model is only available provided the introduction of breakthrough digital technologies both in the core business processes and servicing.

Today the real sector of the economy, presented by the major oil companies in Russia, is actively engaged in the digital transformation of the business. The digital transformation is understood as "the deep and extensive changes in the productive and social processes related to the universal replacement of analog network systems to digital systems and as a result extensive use of digital technologies" [1].

Scientists Shishkin, Timashev, Solovykh, Volkov and Kolonskih believe the main difference between digital transformation and automation is that digital technologies are implemented not just in individual production systems but in the business model of the company as a whole, and this technique allows to obtain reporting information on the state of business models in real time mode [2].

Thus, PAO (PJSC) ANK «Bashneft» implements today a strategic project "digital oilfield" in accordance with which the following digital technologies are implemented; these technologies are presented in Fig. 1 (Appendix A).

Of course, business transition to digitalization involves the need for reliable and real-time reporting information for all groups of interested parties, the so-called stakeholders, who make strategic and operational decisions [2]. Earlier in their scientific works, the group of authors [3, 4] drew up a matrix of the interested user groups of reporting financial information (Fig. 2, Appendix B).

Today, however, the stakeholders require not only the disclosure of information about the property and financial status of the companies, but also about the results of the activities in the context of all business areas, including investment and innovation, environmental, industrial security, and social. In addition, the company must be informationally open in terms of corporate governance, assessing the impact of uncertainties and risks on its activities and much more.

In current international accounting practice, such a form of accounting information openness is presented in XBRL (eXtensible Business Reporting Language) reporting format. In the professional community of accountants extensively discussed is the transition of the most important large companies in Russia to such a presentation of information with a view to improving the quality, transparency and disclosure for stakeholders. However, there are still open questions in relation to conceptual framework, composition of indicators, qualitative and quantitative content of reporting and the development of a road map for the transition from the traditional format of financial statements to XBRL format.

2 Methodology

In the context of digitalization of Russian business, the main trend of transformation of the analytical processes is the structuring of financial reporting by expanding the analyticity of its information.

There are two possible ways to achieve the requisite financial information structure, namely the specification and expansion of the analytical reporting forms or providing the necessary explanations and transcription to them. Global accounting community chose the second path – reporting in electronic form with the complete explanations of the reporting.

Today, throughout the world, the adopted uniform standard of business reporting is standard XBRL. There is a group of countries where this format is already mandatory for application; and there are countries where this format is still under development (Table 1).

Table 1. International practice of the XBRL format application

Stages of implementation	Countries where XBRL format is used
Mandatory	Australia, Belgium, Bermuda, Brazil, United Kingdom, Germany, Denmark, Israel, India, Indonesia, Ireland, Spain, Italy, Canada, China, Luxembourg, Netherlands, Poland, Saudi Arabia, U.A.E., Singapore, United States, Turkey, France, South Korea, Japan
In development	Argentina, Hong Kong, European Union, South Africa, Malaysia, Mexico, Russia, Romania, Thailand, Taiwan, Switzerland, Sweden

Source: compiled by the authors

We shall carry out research in the field of conceptual framework "XBRL Reporting" and we will try to find out what principal opportunities such a reporting format presents. A large number of scientific publications of both Russian and foreign scientists is available on this topic, the results of the conceptual framework are presented in Table 2.

Table 2. Conceptual framework analysis of "Reporting in XBRL format"

Author/source	Brief description
Kernan [5]	This is the language, which is the extension of the commonly used markup language (XML) in the area of financial reporting
School of Accountancy Shidler, College of Business University of Hawai'i at Mānoa [6]	This is the business reporting language, based on XML, for presenting the results of the activity of the company in accordance with the metadata. Over the years, this format has been the reporting format for financial statement analysis
Piechocki, Felden, Gräning, Debreceny [7]	This is the de facto XML-based standard for the distribution of metadata associated with the business reporting information. The standard allows the parties to create reporting solutions for a wide range of information value creation chains

(*continued*)

Table 2. (*continued*)

Author/source	Brief description
Kaspin [8]	This is the electronic language of business reports, that is the new method of "labelling" of financial information in reporting or the so-called "tags" as a way to communicate information to users of integrated reporting
Fomina, Fomin [9]	This is the extensible business reporting language, a common computer language for electronic transmission of business and financial data
Grushko, Grushko [10]	It is an open standard for presentation of the financial statements in electronic format based on XML
Reporting rules in XBRL format and its submission to the Bank of Russia [11]	Extensible business reporting language is the transfer format of regulatory, financial and other reporting

Source: compiled by the authors

Studies have shown that the XBRL format has been recognized by the widespread global accounting community as the modern digital format for reporting of financial and non-financial information. Moreover, studies have shown that the majority of Russian scientists believe that XBRL format supports information that eventually is revealed in the integrated reporting of the companies.

The main advantages of the XBRL format in the digital business transformation include:

(1) presentation of any information, including non-financial specific for industries and companies;
(2) switching from costly manual processes of collecting, processing and data analysis to automatic formation of data, relying on software of the XBRL format;
(3) the possibility of extending quality of the reported information of companies, which can be adjusted in line with the strategic objectives and business models;
(4) removal of replicated or backup data of accounting functions and reporting data to build a unified corporate accounting system for collecting and processing information;
(5) reducing accounting risks and unreliability of reporting in the company.

3 Results

As the results of scientific research in the area of companies' reporting, the authors proposed the structure and enlarged composition of integrated reporting, which is presented in Table 3.

Table 3. Enlarged composition of XBRL format supported integrated reporting indicators for Russian companies

Sections	Group	Composition of indicators
Financial reporting	General information	Company name, industry, ownership and operating structure, structure of management, corporate governance, number of employees, competitive environment, market position, etc.
	Reporting of financial results and the company's value	Quantitative performance indicators, revenue structure, factors influencing the change in financial results, distribution of resources, a portfolio of assets and ownership structure, the size and structure of CAPEX, the size and structure of OPEX, the amount and structure of accounts payable and receivable, tax and other deductions in the context of recipients, the amount of financial assistance to authorities, dividend policy, management remuneration scheme
Non-financial reporting	Risk-reporting	Risk management system of the company, information on risk factors affecting the value of the company, the quantification of risk-cost factors
	Report on business prospects	Indicators of the company's strategic goals in business segments, key factors of external and internal environment, sources of funding
	Environmental reporting	The amount of emissions and discharges of pollutants into the environment and its compliance with the regulatory level of acceptable environmental impact, relative indicator of pollutant emissions into the atmosphere and its compliance with the level of the world's leading companies, the relative proportion of emissions of pollutants in waste water and its compliance with the level of the world's leading companies, the volume of waste of 3rd, 4th, and 5th class of danger placed on environmentally sound objects of the sufficient capacity
	Social reporting	The system of remuneration of staff, funding level of social package of privileges and guarantees for employees, the level of expenses for the realization of social programs of the company per worker in terms of expenses, percentage of staff receiving medical examination, average number of employees, the average annual number of training hours per worker, funds allocated by the company for training and retraining of personnel, etc.

Source: compiled by the authors

The set of indicators presented in Table 3, is of a recommendatory nature and is not limited to a composition, which is offered by the authors. In our view, this is the most comprehensive set of indicators that should be reflected in the reporting of a public company as well as maintained by XBRL format.

4 Discussion

Most Russian academics and financial analysts [1, 2, 9–13] agree that XBRL reporting is an integral part of the formation of the integrated reporting of all public companies that are on the way to digital transformation of their business processes. In the long run, the chief financial regulator – the Central Bank of Russia marked the scope and financial institutions that will have to switch to the XBRL format by the year 2021 [11]. The transition relates to the depositaries, investment funds, pension funds, insurance companies, credit institutions, pawnshops, microfinance institutions of consumer cooperatives.

Given the high cost of the XBRL format introduction project, large companies are also ready to invest in the development of the IT platform on the formation of integrated reporting, which will meet the requirements of international standards.

5 Conclusion

Thus, XBRL format reporting can be seen as a new promising trend in the Russian practice of information and analytical support to corporate management. Its focus lies in technology, efficiency, quality and analytical provision of information to all interested parties with a view to meeting the interests of owners and other interested stakeholders, as well as improving its efficiency and growth of business value.

Appendix A

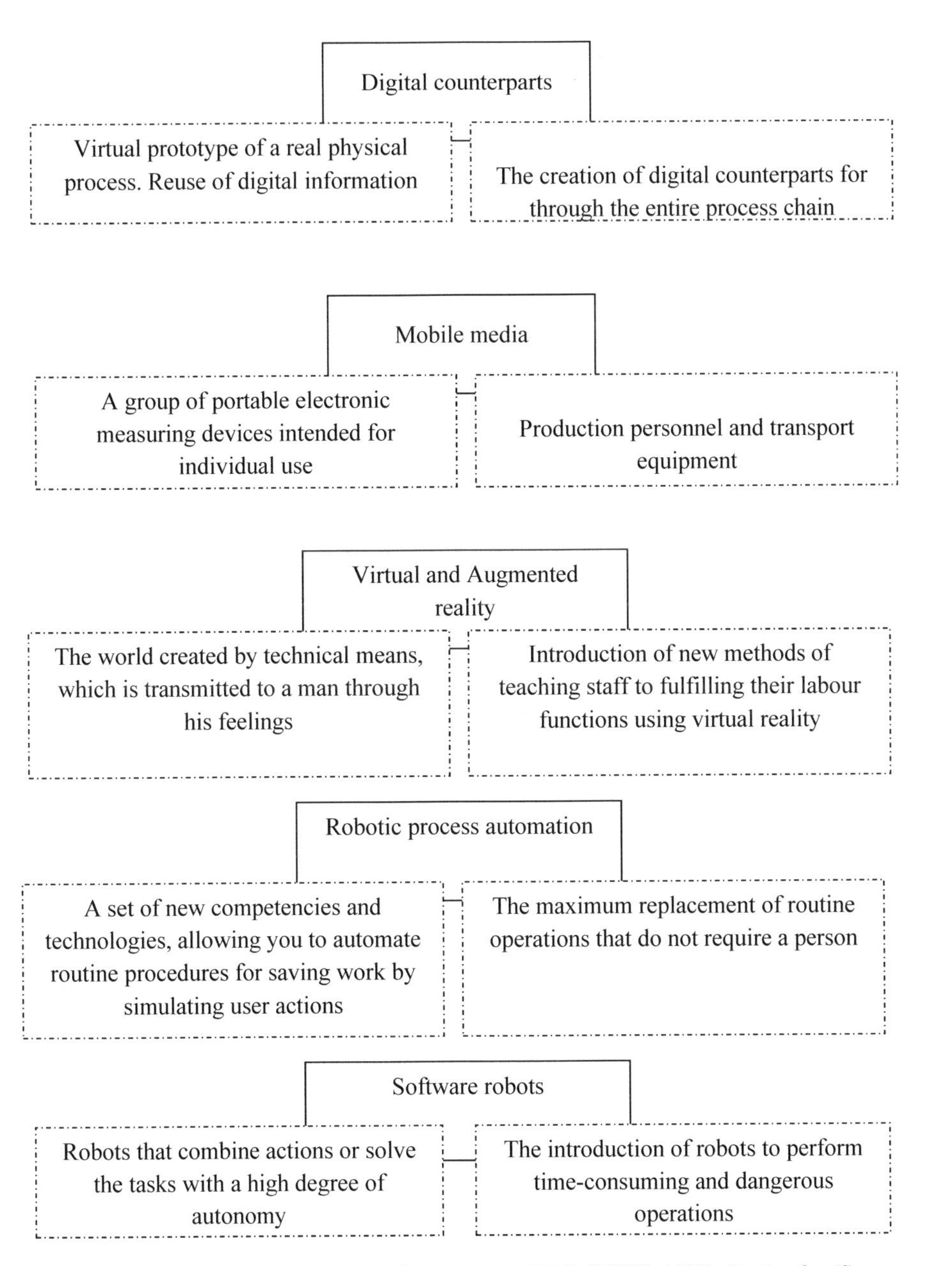

Fig. 1. Digital technologies in the Russian company PAO (PJSC) ANK «Bashneft» (Source: compiled by the authors based on [2])

Appendix B

Legality of the transaction	Going concern	Profitability of business	Structure of capital, incomes, expenses of business segments, net assets	Socio-economic development	Solvency	Property status, settlement status	Financial sustainability
16	16	15	11	14	13	17	17
15	14	9	8		12	8	7
10	5	3	5		7	7	6
8		2	3		6	1	4
5		1	1		4		3

1	Statistical bodies
2	Shareholders
3	Investors
4	Creditors
5	MEDIA representatives
6	Banks
7	Financial markets
8	Manufacturing Companies/Competitors
9	Tax authorities
10	Arbitration courts
11	Management
12	Suppliers
13	Personnel
14	Society
15	Auditors
16	Trade-unions
17	Arbitration managers

Fig. 2. Matrix of stakeholders – users of reporting financial information (Source: compiled by the authors)

References

1. The G20 Initiative for the development and cooperation in the field of the digital economy. http://kremlin.ru/supplement/5111. Accessed 25 Apr 2019. (in Russian)
2. Shishkin, A.N., Timashev, E.O., Solovyh, V.I., Volkov, M.G., Kolonskih, A.V.: Digital transformation of PJSC ANK “Bashneft” from concept to realization. Oil Ind. **3**, 7–12 (2019). (in Russian)
3. Astafeva, O.V., Astafiev, E.V.: The study of approaches to sustainable development of the national economy. World Sci. Discov. **3–9**(63), 3904–3917 (2015). (in Russian)
4. Khalikova, E.A., Shajhetdinova, A.R.: The classification of accounting and reporting information to user groups. Bull. Econ. Manag. **1**, 54–57 (2015). (in Russian)
5. Kernan, K.: XBRL around the world. J. Account. **206**(4), 62–66 (2008)
6. Debreceny, R., Felden, C., Ochocki, B., Piechocki, M., Piechocki, M.: XBRL for Interactive Data: Engineering the Information Value Chain, pp. 1–214 (2009). https://doi.org/10.1007/978-3-642-01437-6
7. Piechocki, M., Felden, C., Gräning, A., Debreceny, R.: Design and standardisation of XBRL solutions for governance and transparency. Int. J. Discl. Gov. **6**(3), 224–240 (2009)
8. Kaspin, L.E.: The possibilities of using XBRL during the formation of integrated reporting. Innovative Dev. Econ. **1**, 148–149 (2013). (in Russian)
9. Fomina, O.B., Fomin, M.V.: Modern trends in corporate reporting. Bull. TvGU Econ. Manag. Ser. **23**, 148–159 (2014). (in Russian)
10. Grushko, A.N., Grushko, E.S.: Credit organizations: towards the integrated reporting. Bull. TvGU Econ. Manag. Ser. **3**, 150–158 (2015). (in Russian)
11. Reporting rules in XBRL format and its submission to the Bank of Russia. https://www.cbr.ru/Content/Document/File/33582/rules_XBRL.pdf. Accessed 25 Apr 2019. (in Russian)
12. Druzhilovskaya, T.Yu., Druzhilovskaya, E.S.: Improving the financial reporting of organizations in the digital economy. Account. Anal. Auditing **6**(1), 50–61 (2019). https://doi.org/10.26794/2408-9303-2019-6-1-50-61
13. Pustylnick, I., Temchenko, O., Gubarkov, S.: Analysis of use of XBRL based accounting data in financial research. WSEAS Trans. Bus. Econ. **14**, 471–483 (2017)

Digital Transformation in the Management of Contemporary Organizations

O. V. Astafeva[1], E. P. Pecherskaya[2(✉)], T. M. Tarasova[2], and E. V. Korobejnikova[3]

[1] Financial University under the Government of the Russian Federation, Moscow, Russia
[2] Samara State University of Economics, Samara, Russia
pecherskaya@sseu.ru
[3] Samara State Technical University, Samara, Russia

Abstract. Globalization processes and the active development of innovative technologies in the present period of time are the most significant factors that affect the activities of modern organizations and form new challenges to maintain their competitiveness. Modern managers are very important to understand that the formation of the digital economy leads to the transformation of different sectors of the economy, and therefore requires the construction of new business models and approaches to the management of organizations, resulting in new challenges and requirements for the work of managers. Thus, the inevitable digitalization of the economy changes working conditions and requires new competencies among company employees. In the conditions of high competition, they need to adapt to the ongoing changes in the external environment.

Keywords: Digital economy · Digitalization · Skilled worker · Digital competence · Educational organization

1 Introduction

Digitalization processes of globalization will have an increasing influence on the activities of organizations, and data in digital form will act as the most important factor ensuring the success of the development of the organization. Consequently, Industry 4.0 is an objective phenomenon that provides modern organizations with many tools to identify new growth points in development. Therefore, the inaction of individual participants in the business environment regarding the formation of the digitalization of society can be attributed to the manifestation of the modern form of "Luddism", since in the next few years, due to the active use of information technologies, most sectors of the economy will inevitably change.

In the technological development of society, researchers identify three key stages with fairly conventional boundaries: automation, informatization, digitalization. In the present period of time, when the majority of the world's population is actively using the Internet and mobile devices (Fig. 1), thanks to modern technology, the hierarchy is

S. I. Ashmarina et al. (Eds.): ISCDTE 2019, LNNS 84, pp. 382–389, 2020.
https://doi.org/10.1007/978-3-030-27015-5_46

shrinking, organizations are becoming increasingly networked. The observed transformation of organizational structures from rigid bureaucratic to more flat and networked, entailing a reduction in the number of levels in the management hierarchy of large organizations, is a generally important change that helps to obtain the necessary information to make an informed timely decision.

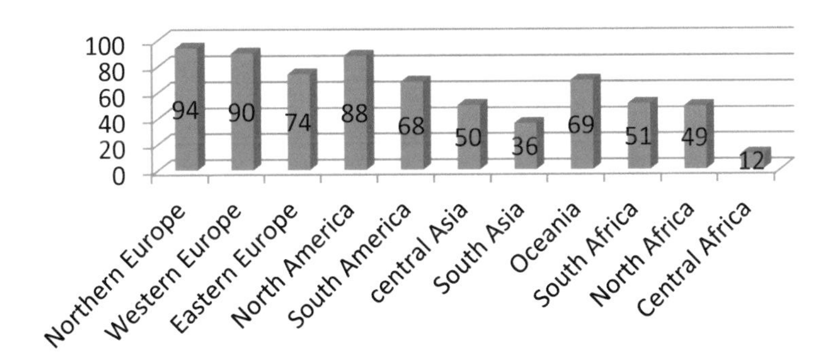

Fig. 1. Internet access by region (Source: compiled by the authors based on [10])

Today, when the majority of the world's population is actively using the Internet and mobile devices (Fig. 1), the hierarchy in the management of large organizations is shrinking, organizations are becoming more networked, and organizational structures are changing from rigid bureaucratic to more flat and networked. This helps to obtain the necessary information to make an informed timely decision; it can be considered the completion of the informatization stage and the beginning of the digitalization of society.

Despite the relatively small number of Internet users in the countries of South Asia, these countries have the highest rates of growth of users of social networks (Fig. 2).

Digitalization involves the use of a complex of information technologies for providing services based on digital platforms and analyzing big data. Of the 7.6 billion people in the world, 5.135 billion people are mobile phone users. The active growth of mobile telephony users is also due to the increase in more affordable smartphones, which opens up tremendous opportunities for companies to develop the market, especially in developing countries, where mobile technologies are literally the only channel for product promotion (Fig. 3). In this regard, the role of local government, local execution and regional differences is increasing. Therefore, in the conditions of increasing influence of globalization and digitalization, in order to achieve greater success, not only on a global scale, but also at the local level, it is necessary to explore the possibilities and demands of regional markets. According to the Institute of Growth

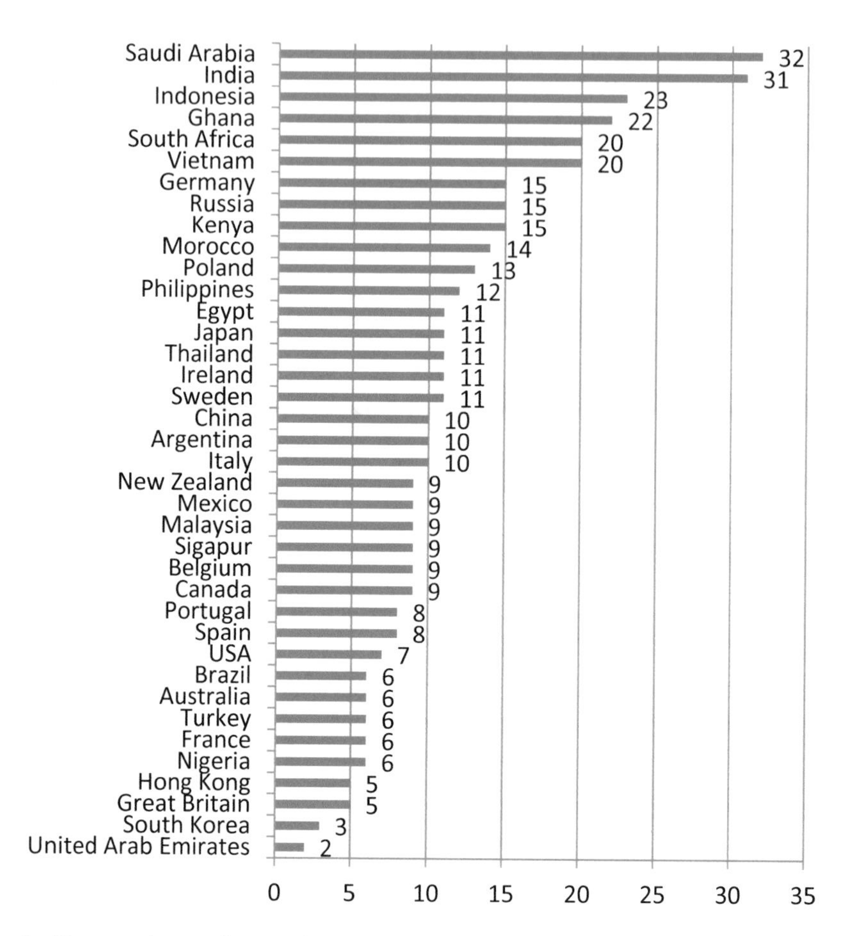

Fig. 2. The growth rate of users of social networks (Source: compiled by the authors based on [10])

Economics named after P.A. Stolypin the contribution of the digital sector to the domestic economy is small compared with developed countries [9] (Fig. 4).

The aim of the research is to analyze the transformational changes in the management of organizations and the challenges that the participants of the digital economy need to pay attention to. The subject of this research is the changes that occur in the management of organizations in connection with the formation of the digital economy.

The research question is to define the role of educational organizations in the process of preparing competitive workers for the digital economy. The research used analytical methods, methods of comparative analysis, theoretical modeling, induction, generalization.

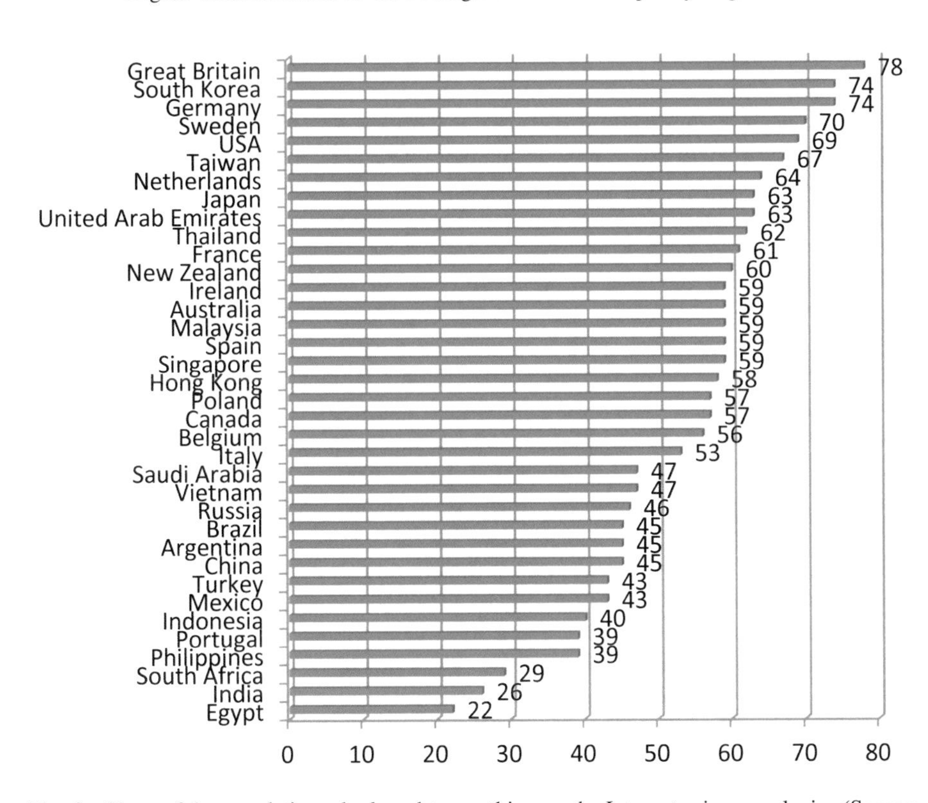

Fig. 3. Share of the population who bought something on the Internet using any device (Source: compiled by the authors based on [10])

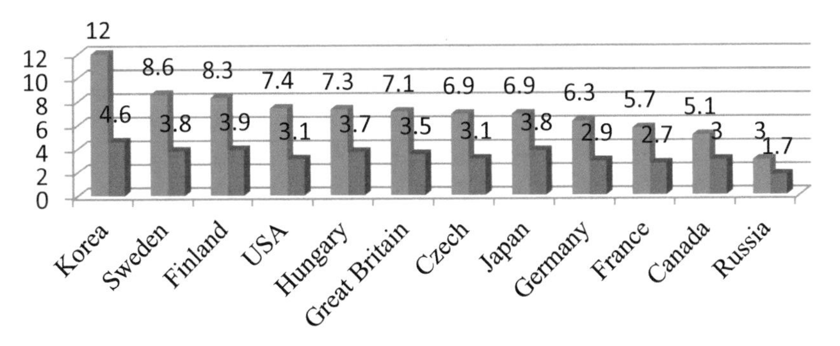

Fig. 4. Domain of the digital sector in the internal gross product and the share of employed in the IT sector from the total number of employees, 2017, % (Source: compiled by the authors based on data of OECD, HSE, Institute for Economic Growth)

2 Methodology

The key element that forms the company's competitiveness is a person interested in the introduction and active use of new technologies and tools. Therefore, active work with people is becoming a more significant task in the activities of managers. In the conditions of the development of the digitalization of increasing the impact of factors such as data analysis, robotization, machine training, there is a technocracy of working with staff, conjugated to both the positive moments in terms of development methods between employees, recruiting, training, monitoring and monitoring of the results of employees of workers, and with a number of dangers for the development of the organization. The risks of the digitalization of the human resources management process are due to the fact that the role of analysis of large data and formal indicators increases when working with people and decision-making and ignoring the characteristics of employees who are not subject to digitizing such as emotional intellect, leadership, charisma, responsibility, diligence, ethnicity, which reduces the effectiveness of managing individual vivid personalities that can perform key employees of the organization forming its competitive advantage. At the same time, against the background of the increased deficit of highly qualified personnel, the humanitarian process of the humanization is observed, expressed in improving working conditions, contributing to the personal and professional development of the employee, which protrudes the key strategic resource of the organization when toughening competition in the labor market [2–4, 7, 8, 11]. Subsequent generations of workers entering the labor market are adapted to work in multitasking, tend to dictate their conditions in labor relations and often change jobs with unsatisfying working conditions, which contributes to the growth of labor mobility and strengthening the power of workers in transforming market relations worker-employer. This circumstance contributes to the formation of a new organization culture, in which every employee feels responsibility and is able to take the initiative and participate in the introduction and implementation of changes, is the most time-consuming and labor-intensive process that requires much more effort from the manager and resources than introducing digital technologies into process. However, it is the transformation of the culture and system of thinking of workers that will allow moving to a new model of moral contract, under which each employee accepts responsibility for the most effective functioning of the part of the company in which he works, and also shows willingness to participate in the process of continuous learning necessary ensuring efficiency in the face of constant change. In exchange, the company provides not just a guarantee of employment. This is achieved by providing each employee with opportunities for continuous training to protect and expand the possibilities of its use not only in the company itself, but also in the labor market. At the same time, the company creates such a creative and stimulating working environment that not only encourages employees to use their skills and abilities to increase the competitiveness of the company, but also motivates them to remain in the company, even if they have the opportunity of alternative employment.

A new moral contract is a fundamental change in the philosophy of management, its transformation: from treating people as a liability of a corporation, a source of value appropriation, to considering them as a source of responsibility and a resource that adds value. With this approach, the driving force behind the functioning of the market is not the all-powerful mind of the top management of companies, but the initiative, constructiveness and skills of all employees. Companies bear moral responsibility for the long-term guarantee of workers' safety and well-being and for assisting them in achieving the full realization of their capabilities in the chosen field.

3 Results

The general dissemination of digitalization and the penetration of the main areas of society's activity will inevitably affect changes in the structure of the workforce and will contribute to the emergence of new professions (in particular, the Ministry of Labor and Social Protection of the Russian Federation approved the professional standard "Consultant in the field of digital literacy development (digital curator)", October 31, 2018, № 682n, contributing to the emergence of a new profession for the development of the digital economy), the disappearance of a number of old professions and the blurring of boundaries between related and unrelated professions.

In this regard, educational organizations face new challenges to ensure the training of qualified workers who meet new requirements in terms of the formation of individual digital competencies, which are proposed to understand the skills in relevant professional activities based on an understanding of the priority directions of development of the technosphere and confident use of information and communication technologies [1]. Also, educational organizations should provide employees with opportunities to acquire new qualifications in a short period of time, lifelong learning and a portfolio of qualifications that will allow an employee to be in demand in different areas of the company, not afraid to change jobs and, as a result, feel confident (Fig. 5).

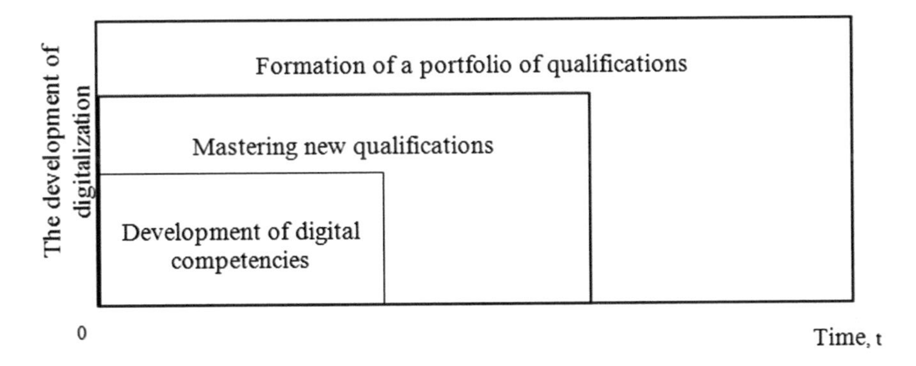

Fig. 5. The educational process throughout life (Source: compiled by the authors)

Consequently, the development of distance learning in educational organizations, allowing working people to change qualifications, and closer cooperation with employers to provide trainees with the opportunity to become familiar with the features of work at work and in real business will be quite an important task. At the same time, educational organizations will have to compete for students with the companies themselves, as many leading enterprises create conditions for retraining workers in the workplace, turning the production site into a kind of "educational institution".

At present, the possibility of acquiring knowledge and developing working skills in highly specialized areas in mass open online courses is undervalued. The results of the research of online courses [5] indicate that now an average of 15% of students receive a certificate of completion of the course, but when the employer provides financial support to employees taking online courses, the proportion of those who completed the training increases to 58%. At the same time, only 20% of respondents who underwent on-the-job training received financial assistance from the employer or free time for training in online courses. Such a low figure is explained by the fact that employers are currently not very willing to invest in the training and development of employees, preferring to attract qualified personnel from outside, or fearing that after training, employees will leave the organization. However, after managers realize that an online course can help to maintain a learning culture in the organization and achieve high returns with low financial costs and minor disruptions in employee work schedules, online courses will be a serious alternative to traditional retraining courses in educational institutions that slowly adapt programs to changing market requirements.

Another important challenge that educational organizations will have to face in the near future is finding a balance in the training of skilled workers who, on the one hand, should be able to solve complex tasks in the current conditions of increasing uncertainty and be focused on obtaining a specific end result, on the other hand, a number of top managers, employees first need to be able to find the necessary information and apply it in practice, and not act as "man - computer" with a large array of knowledge in the head [6]. In this regard, it is possible that a change in the education system will ultimately lead to the fact that the country will have a limited number of universities aimed at training "thinkers" who are able to create innovations for graduates, and other educational organizations will be engaged in training qualified specialists who can quickly adapt to the changing conditions of the digital space thanks to the advanced soft skills.

4 Conclusion

In order for companies to not only adapt to changing conditions, but to act as a driver for the development of individual sectors of the economy and markets, management should aim at taking a proactive stance in the use of new technologies and working methods. The passive behavior of the company can lead to its full displacement by competitors, including international participants. Thus, the inevitable processes of digitalization require from all participants in the digital economy readiness for changes, continuous learning and development to maintain their competitiveness and relevance in the modern conditions of digitalization.

References

1. Astafieva, O.V.: Problems of management training in the digital economy. Probl. Manag. Theory Pract. **5**, 134–141 (2018)
2. Averina, L.V., Pecherskaya, E.P., Astafeva, O.V.: Effective staff training for the contract system in the conditions of the digital economy: opportunities and limitations. In: Ashmarina, S., Mesquita, A., Vochozka, M. (eds.) Digital Transformation of the Economy: Challenges, Trends and New Opportunities. Advances in Intelligent Systems and Computing, vol. 908, pp. 510–517. Springer, Cham (2020). https://doi.org/10.1007/978-3-030-11367-4_51
3. Averina, L.V., Pecherskaya, E.P., Rakhmatullina, A.R.: Readiness to changes as one of educational values of innovation-oriented procurement. In: Ashmarina, S., Mesquita, A., Vochozka, M. (eds.) Digital Transformation of the Economy: Challenges, Trends and New Opportunities. Advances in Intelligent Systems and Computing, vol. 908, pp. 429–436. Springer, Cham (2020). https://doi.org/10.1007/978-3-030-11367-4_42
4. Chudaeva, A.A., Mantulenko, V.V., Zhelev, P., Vanickova, R.: Impact of digitalization on the industrial enterprises activities. In: Mantulenko, V.V. (ed.) 2018 SHS Web Conferences 17th International Scientific Conference "Problems of Enterprise Development: Theory and Practice", vol. 62, p. 03003 (2019)
5. Hamori, M.: Eternal students go online. Harvard Bus. Rev. (2019). https://hbr-russia.ru/management/upravlenie-personalom/792633. Accessed 22 Apr 2019
6. Mordashov, A.: How Industry 4.0 changes management and production. Harvard Bus. Rev. (2018). https://hbr-russia.ru/liderstvo/lidery/a24981. Accessed 22 Apr 2019
7. Pecherskaya, E.P., Averina, L.V.: FIP as an innovative mechanism in the era of the digital economy. In: Khasaev, G.R., Ashmarina, S.I. (eds.) Modernization of Accounting, Control and Analytical Processes in the Digital Economy. Collection of Scientific Articles of the I All-Russian Scientific and Practical Conference, pp. 70–75. SSEU, Samara (2018)
8. Pecherskaya, E.P., Averina, L.V., Kozhevnikova, S.A.: ERP implementation challenges: case-study of the Russian Federation Astra Salvensis, Special issue, pp. 411–423 (2018)
9. Stolypin, P.A.: Russia: From digitalization to the digital economy. Materials of the Institute of Economics of Growth, September 2018. http://stolypin.institute/wp-content/uploads/2018/09/issledovanie_tsifrovaya-ekonomika-14-09-18-1.pdf. Accessed 22 Apr 2019
10. Sergeeva, Yu.: Internet 2017–2018 in the world and in Russia: Statistics and trends (2018). https://www.web-canape.ru/business/internet-2017-2018-v-mire-i-v-rossii-statistika-i-trendy/. Accessed 22 Apr 2019
11. Sokolov, D.: Human resource management strategies in the context of modern trends in labor relations. Probl. Manag. Theory Pract. **9**, 124–130 (2018)

Digital Reality and Perspective of the Management of Educational Organizations

E. A. Mitrofanova[1], I. V. Bogatyreva[2(✉)], and V. V. Tarasenko[3]

[1] State University of Management, Moscow, Russia
[2] Samara State University of Economics, Samara, Russia
scorpiony70@mail.ru
[3] Orenburg State Pedagogical University, Orenburg, Russia

Abstract. The study considers the reality and prospect of the digital transformation of the economy and society using the example of the digital transformation of the management of educational organizations. The authors of the study clarify the essential features and possibilities of the digital transformation of the management of educational organizations, and identify the natural links between the reality of the digital transformation of the management of educational organizations and the formation of digital competences of managerial staff. Recommendations to improve the training of managerial staff for the digital transformation of the management of educational organizations are given.

The methodological basis of the study is systemic and competence-based approaches; methodological basis - analysis and synthesis, statistical methods, self-assessment and expert assessment, questioning; the empirical basis is the data of the federal statistical observation of the activities of Russian educational organizations, the results of the qualitative assessment of managerial staff of educational organizations (N = 300).

Keywords: Educational organization · Development of managerial staff · Digitalization · Digital transformation of the management · Managerial staff

1 Introduction

The current stage of development of society is characterized by the increasing role of knowledge, information, digital and information and communication technologies [1, 2, 4, 5, 10]. Responding to contemporary challenges (trends) of global informatization and digitalization, Russia has developed a number of strategic policy documents that reinforce the leading role of education when developing information society, forming the national digital economy, maintaining national interests and national priorities (Strategy of the Information Society Development in the Russian Federation for 2017–2030, the Program “Digital Economy of the Russian Federation”, the State Program of the Russian Federation “Development of Education” for 2018–2025, the National Project “Education” for 2019–2024 and corresponding federal projects in the field of education (“Modern School”, “Digital Educational Environment”, “Teacher of the Future”), etc. In the submitted documents, the objectives for the development of

S. I. Ashmarina et al. (Eds.): ISCDTE 2019, LNNS 84, pp. 390–397, 2020.
https://doi.org/10.1007/978-3-030-27015-5_47

education as a social institution are determined by digitalization of the educational process, the educational result, educational conditions and management of the educational organization [2].

This study is devoted to the analysis of the reality and prospect of the digital transformation of the management of educational organizations in the context of the global digital transformation of the economy and society. The purposes of the study are the following:

1. Clarify the essential features and possibilities of the digital transformation of the management of educational organizations.
2. Identify the natural links between the reality of the digital transformation of the management of the educational organization and the development of digital competences of managerial staff.
3. Develop recommendations to improve the training of managerial staff for the digital transformation of the management of educational organizations.

2 Methodology

The methodological basis of the study is systemic and competence-based approaches; methodological basis - analysis and synthesis, statistical methods, self-assessment and expert assessment, questioning; the empirical basis is the data of the federal statistical observation of the activities of Russian educational organizations, the results of the qualitative assessment of managerial staff of educational organizations of the Orenburg region (N = 300).

3 Results

The term "digitalization" as a system-forming concept of the reality and prospect of the digital transformation of the management of educational organizations does not have an established definition. In a broad sense, the term "digitalization" is defined as a transition from the analog form of information to the digital one; a transition to digital communication, recording and transmission of data using digital devices. In turn, a number of authors rightly clarify the concept of "digitalization", which is not only the transition of information into digital form, but also a comprehensive, infrastructural, managerial, behavioral and cultural solution [6]. The decisive role in making such decisions is played by managerial staff, their intellect, experience, and intuition, and digital technologies that allow optimizing the decision-making process and increasing its efficiency.

McKinsey's research suggests that digital technology is changing the organizations' operating model, increasing cost-effectiveness and revealing new opportunities in the market. For the services sector, the benefits of the digital transformation is a multiple cost reduction (by 40–60%) and a significant acceleration of new products launch into the market, creating partnerships with organizations in related fields for integrating resource capabilities and disseminating experience [12]. The studies of the

Higher School of Economics also confirm the positive effect from digital technologies, which is reflected in simplification and acceleration of organizational processes (business processes), as well as in improving the accuracy and quality of work [3].

The Programs "The Digital Economy of the Russian Federation" and "Development of Education" implemented in Russia are designed to promote a digital ecosystem in which educational organizations and their managerial staff play a leading role in improving: (1) the quality of training for the digital economy and (2) the effectiveness of the management of the educational organization as a social institution. At the same time, educational organizations, unlike the business environment, often demonstrate a conservative approach to modern technologies, including digital technologies. For example, 13.5% of Russian students use e-learning; 2.5% of students use distance learning technologies; 1.8% of students use the network form (Fig. 1).

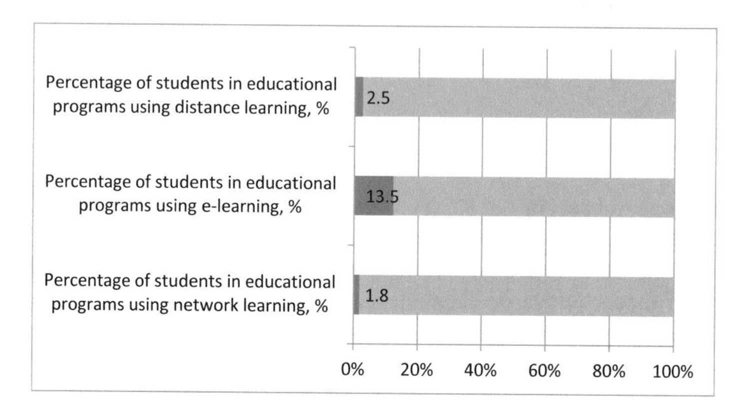

Fig. 1. The share of students in educational programs using modern technology, in % of the total number of students (Source: compiled by the authors)

Figure 2 presents data confirming the conservative approach of most Russian educational organizations to digital control technologies. So, special software for solving management tasks is used only in 31% of Russian educational organizations, electronic document management systems – in 45% of educational organizations, and computer testing programs for students and staff – in 46% of educational organizations (Fig. 2).

According to the authors of the study, the data presented in Figs. 1 and 2 may indicate a low level of readiness of the overwhelming majority of managerial staff in Russian educational organizations to provide education digitalization, including the digital transformation of the management of educational organizations. This conclusion was confirmed during the study of the digital competence development of managerial staff of Russian educational organizations (N = 300).

The term "digital competence", "information and communication competence" and "digital literacy" do not have an established definition. The International ICT Literacy Panel uses the term "information and communication competence", defined as the willingness to use digital technologies, communication tools and/or networks to access,

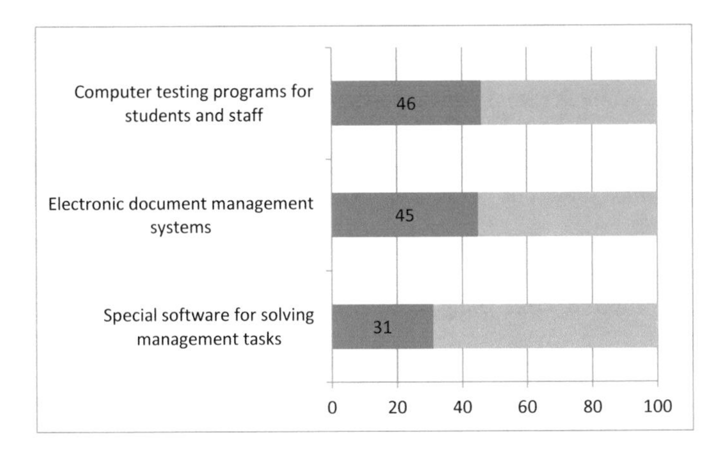

Fig. 2. The share of educational organizations using digital management technologies, in % of the total number of educational organizations (Source: compiled by the authors)

manage, integrate, assess, create and transmit information in compliance with ethical and legal norms in order to successfully live and work in knowledge society. In the scientific works of P. Gilster concerning the problem of digitalization, the term "digital literacy" is used and characterized as the ability to critically use and understand information that we receive through computers from a wide range of sources [4]. Specifying the definition proposed by Gilster, Martin considers digital literacy as the ability, attitudes and awareness of individuals to use digital means and tools for managing, identifying, accessing, evaluating, integrating and synthesizing digital resources in a certain way; communicating with other people, as well as building systems of new knowledge with the aim of constructive social actions in the context of specific life situations [8]. The Russian scientific literature often uses the term "digital competence", which refers to the individual's readiness to confidently, effectively, critically and safely select and apply digital technologies in various activities, as well as his readiness for such activities [11]; a dynamic quality system of the subject in the availability of knowledge, skills and abilities, beliefs and values (responsibility) associated with the use of digital devices and technologies [7]. The theoretical analysis of the scientific literature, concerning the issue of digitalization, allowed the authors to clarify the concept of "digital competence of managerial staff of the educational organization". It is an integrative personal quality, expressed in the psychological and professional readiness of managerial staff to provide digitalization of education including designing and applying digital systems and technologies to achieve digital development goals of education and educational organization.

A study of psychological attitudes of managerial staff made it possible to conclude that a significant proportion of respondents (64%) are skeptical about the digital transformation of the management of the educational organization. This attitude of respondents is due to risks in ensuring the safety of storage and distribution of information and, as a consequence, the need for duplication of information in print and electronic forms.

The study of the professional readiness of managerial staff to digitalize education (understanding digitalization of education, knowledge of features and capabilities of modern digital technologies in management) led to the conclusion that the overwhelming proportion of respondents (73%) determines digitalization of education truncated, limited to indicators of annual statistical observation of Russian organizations implementing programs of general education:

- Increase the number of personal computers/laptop computers (laptops, tablets) in the educational organization;
- Increase the speed of connecting personal computers/laptops (laptops, tablets) to the Internet;
- Ensure the functioning of the official site on the Internet;
- Introduce and use an electronic diary, an electronic journal of academic achievement and an electronic library;
- Introduce and apply distance learning technologies and e-learning in the educational process [9].

Managerial staff has trouble enumerating modern digital educational management technologies. For example, 68% of respondents do not understand the essence and capabilities of electronic accounting, which is explained by the lack of demand for this management tool in the context of the current practice of centralized accounting organized by municipal education authorities. The consequence of the limited understanding of digitalization of education by managerial staff is the corresponding degree of its readiness to develop and introduce digital technologies in the educational process and the educational management system. At the same time, the other 26% of respondents are aware of the availability and capabilities of integrated management systems of the educational organization (1C Program: Educational Institution and its analogues), but experience difficulties in their implementation (legal, human, financial, material and technical).

Having studied the psychological and professional readiness of respondents to digitalize education and manage educational organizations, we have led to the conclusion that it is necessary to improve the training of managerial staff. The lack of a systematic approach to the development of digital competences of managerial staff of the educational organization has updated the development and implementation of the educational module "Digital technologies for managing the educational organization" (Table 1).

At this stage of the study, the approbation of the educational module "Digital technologies for managing the educational organization" is being carried out within the framework of the regional model of advanced training for managerial staff of educational organizations of the Orenburg region. Following the results of testing the educational module, the effectiveness of digital competences of managerial staff of educational organizations will be assessed and additional recommendations will be developed to improve this process.

Table 1. The structure of the educational module "Digital technologies for managing the educational organization"

Sections of the educational module	Academic hours
Section 1. The place and role of the educational organization in the digital transformation of the economy and society	2
Section 2. Digital technologies support strategic planning activities of the educational organization	4
Section 3. Digital technologies in the management of the educational process	4
Section 4. Digital technologies for assessing and monitoring the quality of education	4
Section 5. Digital technologies in the management of staff and resources of the organization	6
Section 6. Digital technologies in information management and Public Relations	4
Section 7. Development and testing of the project introducing digital technologies into the management system of the educational organization	10
Final certification (defense of the project)	2
Total:	**36**

Source: compiled by the authors.

4 Discussion

The analysis of the results shows that the education system in Russia is an active participant in the digital transformation of the economy and society. At the state level, programs are developed and implemented to promote digitization of the education system as a social institution. First of all, the emphasis is on digitalization of the educational process, the conditions and results of its implementation, and to a lesser extent on the digital transformation of the management of educational organizations. Annual statistical observation of Russian educational organizations also does not contain indicators of the digital transformation of the management of educational organizations. These factors limit the need of managerial staff for a targeted digital transformation of the management of educational organizations. The situation is becoming worse by the lack of a systematic approach to the development of digital competences of managerial staff of educational organizations, which is observed in the insufficient psychological and professional readiness of managerial staff to digitalize the management. According to the authors of the study, the developed educational module "Digital technologies for managing the educational organization" cannot be considered a universal solution to the problems identified, but its implementation is designed to update the psychological and professional readiness of managerial staff to digitalize the management of educational organizations. This conclusion is confirmed by Naisbitt's conclusion that modern digital technologies help to manage information society only to the extent that its members are able to use them [10].

5 Conclusion

The study of the reality and prospect of the digital transformation of the economy and society using the example of the digital transformation of the management of educational organizations allowed the authors of the study:

(1) Refining the essential features and possibilities of the digital transformation of the management of educational organizations, allowing optimizing the decision-making process, increasing their efficiency, providing the comprehensive, infrastructural, managerial, behavioral and cultural solution;
(2) Identifying the logical links between the reality of the digital transformation of the management of the educational organization and the insufficient level of development of digital competences of managerial staff;
(3) Developing recommendations on the inclusion of the educational module "Digital technologies for managing the educational organization" in the training of managerial staff for the digital transformation of the management, which will allow updating the psychological and professional readiness of managerial staff to digitalize the management of educational organizations.

The next stage of the study involves the results of testing the educational module "Digital technologies for managing the educational organization", the results of which will assess the efficiency of digital competences of managerial staff and develop additional recommendations for improving the digital transformation of the management of educational organizations.

References

1. Abdrakhmanova, G.I., Plaksin, S.M., Kovaleva, G.G.: Internet economics in Russia: approaches to definition and assessment. Forsyth **11**(1), 55–65 (2017). (in Russian)
2. Castells, M.: Information Age: Economy, Culture, Society. HSE, Moscow (2000). (in Russian)
3. Digital economy: global trends and practices of Russian business. https://imi.hse.ru/pr2017_1. Accessed 16 Apr 2019. (in Russian)
4. Gilster, P.: Digital Literacy. Wiley Computer Publishing, New York (1997)
5. Inozemtsev, V.L.: Modern post-industrial society: nature, contradictions, perspectives: studies. Manual for University Students. Moscow: Logos (2000). (in Russian)
6. Keshelava, A.V., Budanov, V.G., Rumyantsev, V.Yu., et al.: Introduction to the Digital Economy. VNII Geosystems, Moscow (2017). (in Russian)
7. Lomasko, P.S., Simonova, A.L.: Fundamental principles of professional ICT competence of teachers in the context of smart education. Bull. Tomsk State Pedagogical Univ. **7**(160), 78–84 (2015). (in Russian)
8. Martin, A., Madigan, D.: Digital Literacies for Learning. Facet Publishing, London (2006)
9. Mitrofanova, E.A., Simonova, M.V., Tarasenko, V.V.: Potential of the education system in Russia in training staff for the digital economy. In: Ashmarina, S., Mesquita, A., Vochozka, M. (eds.) Digital Transformation of the Economy: Challenges, Trends and New Opportunities. Advances in Intelligent Systems and Computing, vol. 908, pp. 463–472. Springer, Cham (2020)

10. Naisbitt, J.: Megatrends. AST, Moscow (2003)
11. Soldatova, G.U., Nestik, T.A., Rasskazova, E.I., Zotova, E.Yu.: Digital competence of adolescents and parents. Results of the All-Russian study. Foundation for Internet Development, Moscow (2013). (in Russian)
12. The Digital Russia: a new reality. The McKinsey Report. www.tadviser.ru/images/c/c2/Digital-Russia-report.pdf. Accessed 16 Apr 2019. (in Russian)

Strategic Purchasing Control of the Industrial Enterprise: Digitalization and Logistics Approach

I. A. Toymentseva(✉), N. P. Karpova, and T. E. Evtodieva

Samara State University of Economics, Samara, Russia
tia67@rambler.ru

Abstract. Economic instability of our time especially touches industrial enterprises, which results in decreasing of product competition and flexibility of the logistics system response of industrial enterprises for environmental changes. That is why business managers are increasingly paying attention to modern methods and ways of doing business, based on the principles of logistics and digitalization. Within the existing enterprise management system, it is impossible to study the running business processes. In addition, there is no integrated information system that can take into account the changing needs of end customers. There is still a breach of delivery terms and a high percentage of spoilage in enterprises, which lead to the enterprise efficiency decrease. To solve these problems, we need a clearly formulated strategy that will allow us to make right management decisions and develop logistics measures. Digital transformation is a way to improve the product competitiveness. Russian heavy engineering industry needs to use digital tools both in production and business models, which will allow to maintain its presence in the world market and not lose share in the national market.

Keywords: Digitalization · Information system · Logistics strategy · Suppliers · Supply

1 Introduction

The relevance of the study is not in doubt because industry is of strategic importance for Russia. National heavy industry is just beginning the digitalization process, while as western companies are already profiting from innovations. Digitalization allows to lead to a new level such processes as design, production and enterprise management. The human labor proportion is declining and it is possible to move on to autonomous digital production cycles.

Supply plays a huge role in achieving the enterprise strategic goals, as it is necessary to constantly improve the product quality and customer service. Therefore, the main supple chains include the coordination of input and output flows in the enterprise, development of a constantly updated database of suppliers, minimizing costs in the supply situation and providing high quality customer service.

To achieve these strategic goals, it is necessary to build partnerships with all participants of the logistics system of industrial enterprises [8, 15]. A well-managed

S. I. Ashmarina et al. (Eds.): ISCDTE 2019, LNNS 84, pp. 398–407, 2020.
https://doi.org/10.1007/978-3-030-27015-5_48

supply chain develops strategic and operational objectives, comparing the participants' actions at each level [10, 12].

It is the use of logistics in strategic management of the industrial enterprise that leads to the optimization of both quantitative and qualitative parameters of logistics processes. Thus, to increase a competitive position in the industrial products market, it is necessary to apply logistics to control resource flows in the supply chain [1, 9]. Therefore, it is important to adopt the strategic management concept, which is the main activity of leading companies.

The authors of the paper studied the procuring activity of OAO "Volgotsemmash". They defined strategic guidelines in the supply management process and interaction with material and technical resources suppliers, identified strategic objectives, offered the economic indicators system and gave an integrated assessment of functional subsystems of supply logistics in industrial enterprises, based on the mathematical modeling.

2 Methodology

To identify the sources of industrial enterprise efficiency in the supply chain, the following experimental and theoretical methods were used: functional and cost method, integrated assessment, statistical method, economic and mathematical method and others.

To determine strategic guidelines in the supply management process of OAO "Volgotsemmash", strategic management tool was used - Balanced Scorecard, BSC, developed by Kaplan and Norton [4–6].

To form the method of evaluation of suppliers in OAO "Volgotsemmash", the expert method with the involvement of specialists from the Center of competitive procurement arrangements was used. The tendencies of heavy engineering enterprises providing for conjuncture-forming factors, which influence the development, were revealed based on SWOT-analysis.

The experimental base of the study to identify the sources of industrial enterprise efficiency in the supply chain was OAO "Volgotsemmash".

The study included several phases. At the first phase, the authors identified the problem of the research. The second phase involved the analysis of scientific literature on the research problem, as well as the choice of research methods. The third phase was directly related to the study: the definition of strategic guidelines in the supply management process and interaction with material and technical resources suppliers, identification of strategic goals in the industrial enterprise development and elaboration of the functional strategic planning for purchasing department in the industrial enterprise.

3 Results

There are many ways of classifying purchases in the scientific researches on supply activities in the enterprise. The authors' classification of purchases in industrial enterprises most fully corresponds to the supply objectives (Fig. 1).

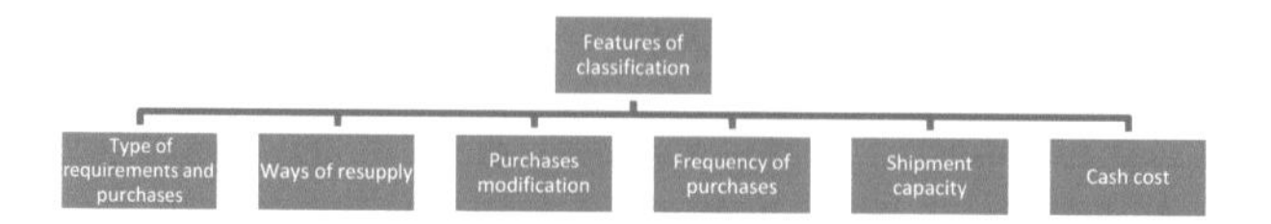

Fig. 1. Features in classification of purchasing (Source: compiled by the authors)

1. *Type of requirements and purchases.* Purchases are divided into following types: raw materials, components, energy resources, additional services and goods for sale.
2. *Ways of resupply.* They include one-time purchases, repeat purchases and urgent orders, which can lead to the disbalance in working capital.
3. *Purchases modification.* These are new purchases related to the emerging needs for the new product manufacturing, or the purchase of material resources from a substitute supplier.
4. *Frequency of purchases.* Basically, purchases are carried out frequently, that is to say repeatedly, and the supplier offers a large discount. On the other hand, there are also one-time purchases.
5. *Shipment capacity.* They are small shipments and large quantities of products transported over long distances by the best carriers who were identified in the course of their rating.
6. *Cash cost.* This type of classification is based on Pareto curve, or ABC analysis.

This extended classification of purchasing presented by the authors will contribute to the efficient choice of raw materials and components according to their type, transportation volume, orders frequency and minimum cost, which will ensure the best rationalization and coordination of supply activities in the industrial enterprise.

The studies have helped to identify many factors that have a serious impact on the work of national industrial enterprises. They are low frequency of purchases, high degree of equipment deterioration, low technological level of production capacity, insufficient investment, inequality in the territorial distribution of production capacity and low life cycle of equipment.

The industrial enterprise efficiency depends on how efficiently planning, management and control of the supply process is carried out, and this also affects the financial stability, strong competitive positions in the industrial equipment market and the enterprise profits received from the sale of manufactured products.

The acquisition process for each industrial enterprise is individual and depends on the product specifics, industry and production capacity. Therefore, a stepwise algorithm on supplier selection in industrial enterprises was developed (Fig. 2).

According to the presented algorithm, at the stage " Identifying intra-firm needs and (or) nomenclature of material resources", the need for material resources, raw materials and components in all enterprise departments is revealed, taking into account the production schedule and minimum prices.

At the stage "Identifying and evaluating requirements imposed on suppliers", it is necessary to develop a list of evaluation criteria: weights, sizes, frequency and volumes of deliveries, as well as service and warranty obligations at the hands of both suppliers and manufacturers.

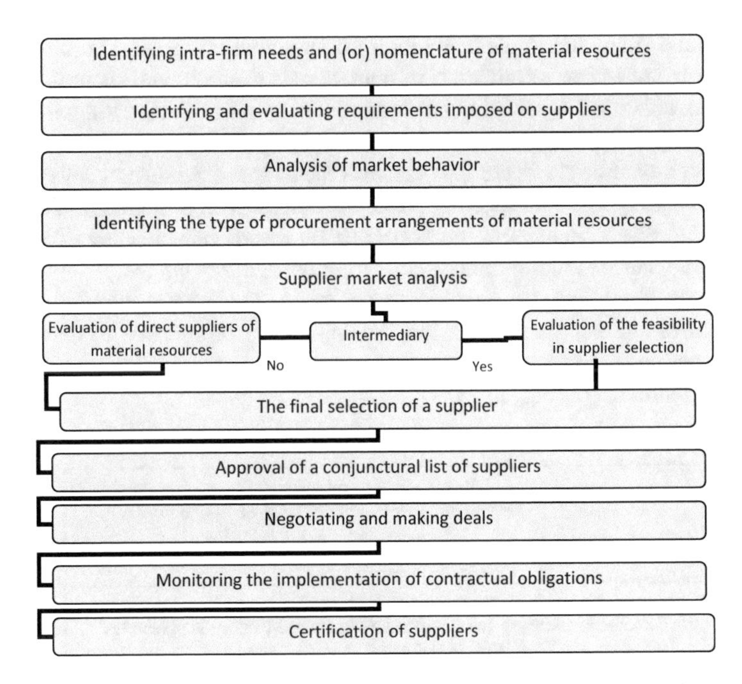

Fig. 2. Algorithm of supplier selection (Source: compiled by the authors)

At the stage "Analysis of market behavior", it is important to conduct a full review of raw materials markets, determine the existing market opportunities and demand, evaluate the offers from existing suppliers and new suppliers, as well as substitutes suppliers. It is also essential to solve the problem of organizing your own production or external procurement, taking into account the ratio of service level and costs.

At the stage "Identifying the type of procurement arrangements of material resources", based on characteristics and complexity of products as well as a production type, and according to the authors' classification of purchasing, the sort of purchases is established. If suppliers or parameters of bought material resources change, the type of "modified purchases" is used, while as the type of "new purchases" is caused by the needs of a new internal user. A portfolio of orders is made.

At the stage "Supplier market analysis", it is necessary to make a preliminary assessment of possible sources of purchases. At the same time, it is important to compare the material resources quality and their price, in accordance with intra-firm needs. Such an assessment should be carried out with the assistance of independent experts. At this stage, intermediaries offers should be considered and an economic assessment of their performance should be made.

At the stage "Final selection of a supplier", an integrated assessment of price, quality, reliability and delivery terms is carried out.

At the stage "Negotiating and making deals", all the details of upcoming deals are discussed, and the final delivery terms and prices are adjusted and approved. After that, the contract is signed.

At the final stage, the implementation of contractual obligations is monitored, and the issue of extension or termination of contractual relations with them is resolved.

The main condition in selecting the resources supplier in OAO "Volgotsemmash" is supplies quality [11]. The level of delivery reliability in the company can be characterized by such parameters as compliance with the terms schedule and delivery volume, spoilage reduction and warranty service of the delivered material and technical resources. Supplier's loyalty can be estimated by means of efficiency of response on claims, taken measures and elimination of the revealed defects [3].

In order to determine the strategic guidelines in OAO "Volgotsemmash" in purchasing control and supply chain management, it is recommended to define strategic objectives, activities and key indicators of supply efficiency on the basis of Balanced Scorecard methodology (Fig. 3) [4, 7].

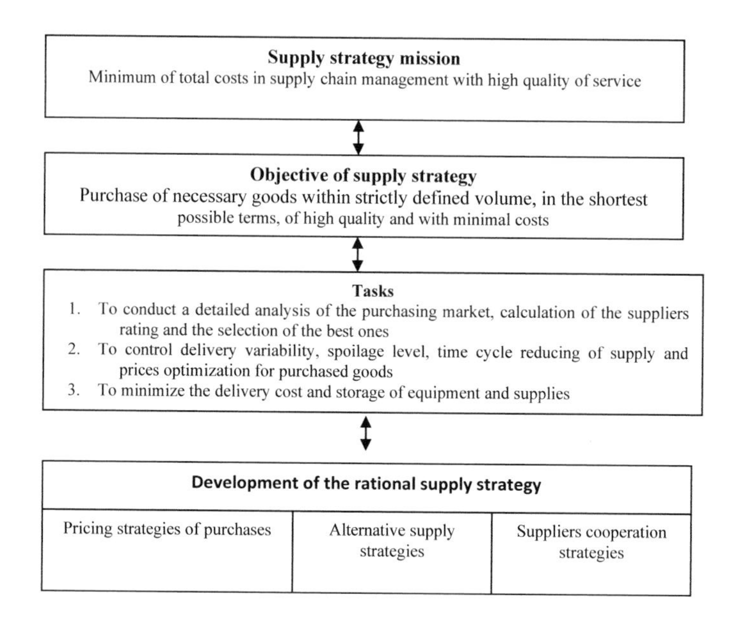

Fig. 3. The algorithm of supply strategy development in purchasing (Source: compiled by the authors)

Figure 3 shows the algorithm of the rational strategy development in supply from among the alternatives.

The best results are achieved by those companies that base their activities on logistics management methods. It contributes to strengthen interaction between the parts of the logistics system "purchases - production – sales". Therefore, integrated logistics provides end-to-end management of resources flow in the enterprise, bringing together all its structural units, which will improve the efficiency of the entire supply chain [2].

SWOT analysis of supply activities defined that the situation in the studied enterprises is almost the same. According to experts, the strengths of their activities are a flexible payment system, developed network of suppliers, qualification level of staff involved in the supply sector and the availability of intermediaries in this system. The weaknesses of the enterprises include the long delivery terms of equipment and supplies, lack of computerized inventory control system and high level of the fleet deterioration. The threats for enterprises include price increase of raw materials and competition from foreign manufacturers. Among the favorable factors in the supply process, we can identify the availability of modern information technology, reduction of customs duties and state support of national producers.

Considering supply as a functional area of logistics, the following strategic areas of economy management (SAEM) are identified:

- SAEM 1 – purchases,
- SAEM 2 – shipment,
- SAEM 3 - warehousing and storage.

The integrated (complex) index for each SAEM is calculated by the following formula (compiled by the authors):

$$S_i = k_i \sum_{i=1}^{n} S_{11}^{ij} + k_i \sum_{i=1}^{n} S_{21}^{ij} + k_i \sum_{i=1}^{n} S_{31}^{ij} + k_i \sum_{i=1}^{n} S_{41}^{ij}$$

Where

S_{11}^{ij} - is total estimated figures of the supply activity level,
S_{21}^{ij} - total estimated figures of delivery variability,
S_{31}^{ij} - total estimated figures of shipment quality,
S_{41}^{ij} - total estimated figures of supply efficiency,
k_i - significance ratios of i-group (SAEM)

Table 1 presents the calculations of integrated indicators for each strategic area of economy management.

Table 1. Integrated assessment according the strategic areas of economy management in OAO "Volgotsemmash"

SAEM			Purchases	Shipment	Warehousing and storage
			k_1	k_2	k_3
Item		Rank	**0.5**	**0.17**	**0.33**
$\sum S_{11}^{ij}$	Total estimated figures of the supply activity level	**3**	1.5	0.51	0.99
$\sum S_{21}^{ij}$	Total estimated figures of delivery variability	**2**	1	0.38	0.25
$\sum S_{31}^{ij}$	Total estimated figures of shipment quality	**2**	1	0.34	0.66
$\sum S_{41}^{ij}$	Total estimated figures of supply efficiency	**1**	0.5	0.17	0.33
Integrated assessment			4	1.4	2.23

Source: compiled by the authors.

Based on a strategic matrix, a map of functional strategies in the main logistics chains, such as purchases, shipment, warehousing and storage, has been developed (Table 2). The map represents the matrix, where the vertical line is the strategic areas of economy management (SAEM) and the horizontal line is scores.

Table 2. Strategy map (compiled by the authors)

Score	Purchases	Shipment	Warehousing and storage
4–5	Long-term partnership strategy	Strategy based on the environment protection	Strategy focused on the consumer
3–4	Strategy of forming alliances with suppliers and customers	Strategy of diversifying services	Strategy of improved productivity
2–3	Price discount strategy	Strategy based on time parameters	Strategy based on time parameters
1–2	Strategy of minimizing overall logistics costs	Best value strategy (or economic cost strategy)	Strategy of minimizing overall logistics costs
0–1	Active search strategy	Logistics outsourcing strategy	Strategy of minimizing investment in logistics infrastructure

Source: compiled by the authors.

This strategy map allows you to choose the optimal strategy for each SAEM. The implementation of the developed functional strategies is associated with evaluation of group of strategic and operational actions in the supply chain in the industrial enterprise. The optimal strategy depends on the square of the matrix, in which the calculated value of the integrated index for a particular SAEM falls.

Thus, according to the developed strategy map and the calculations of integrated assessment for each strategic area of economy management, the following optimal strategies for OAO "Volgotsemmash" are determined:

- SAEM 1 (purchases) - long-term partnership strategy,
- SAEM 2 (shipment) - best value strategy,
- SAEM 3 (warehousing and storage) - strategy based on time parameters.

4 Discussion

The main purpose of the paper is to determine the sources of efficiency in purchasing in industrial enterprises in course of their interaction with material and technical resources suppliers and end consumers. According to scientists, operation of enterprises should not be isolated. It is necessary to unite all participants' efforts in the supply chain for the best results. This reaffirms that industrial enterprise management should use integrated digitization. It helps to specify the readiness degree of individual parts, availability of materials for product manufacture and employment rate of workers at each stage of a production process. It also facilitates to obtain information about product cost, planned and actual costs. All this allows managers to show a more flexible approach to enterprise management. Therefore, we consider it reasonable to extrapolate the conclusions obtained in the course of the study of the activities in OAO "Volgotsemmash" to the entire market of metallurgical products.

Developing the theory of strategic interaction with key suppliers of material and technical resources, the authors proposed a method for developing evaluation of suppliers in OAO "Volgotsemmash". The authors developed a stepwise algorithm to select suppliers for industrial enterprises, which will contribute to the achievement of high quality supplies.

According to analysts, the production of competitive industrial equipment needs the development of rational management decisions (strategies) in all functional areas of the enterprise. The study showed that ensuring the effective operation of the industrial equipment manufacturer is largely determined by purchasing activities including variability, reliability and consistency in procurement management and control, which have a significant impact on the financial stability, production efficiency and competitiveness in the enterprise. According to the strategy map, developed by the authors, and results of the integrated assessment for each strategic area of economy management (SAEM), the optimal development strategies were determined.

Thus, only a competent logistics approach and digitalization will eliminate the identified problems in operation of the industrial enterprise in the supply chain. Logistics in purchasing material resources for production needs helps improve relationships with suppliers and the purchased resources quality, reduce costs and time for their shipment and increase the efficiency of logistics processes, as well as the production of high-quality industrial equipment at competitive prices.

The use of digital technologies gives companies a new impetus to develop and change the production paradigm. The key objective today is not risk assessment, but the adoption to the principles of digitalization, due to the revision of basic management algorithms, optimization of production processes and the development of new technological regulations [13, 14].

5 Conclusion

The successful operation of any industrial enterprise depends on the availability of raw materials, goods and services supplied by other organizations. Supply activities to arrange and control purchases are aimed to provide the enterprise with necessary quantity and quality of raw materials, goods and services at the right time, in the right place, from a reliable supplier, with good service and at a reasonable price. Purchasing (supply) is a main functional area of logistics and is of paramount importance in every organization.

The study allowed to determine a group of actions to improve the efficiency of purchasing activities in the supply chain of industrial enterprises. For this purpose, the authors developed an extended classification and typology of purchasing of the industrial equipment manufacturer, which most fully corresponds to the objectives in the sphere of the activity under question. The paper also presents the main stages of purchasing in industrial enterprises and their role in the logistics system.

In order to improve purchasing activities, the authors propose to comply with the following recommendations:

- constantly search for new sources of supply of raw materials in the context of financial appeal and product quality,
- to adopt the principles of "Industry 4.0", which means the transfer to digital format of both vertical processes in the enterprise and horizontal relations of the manufacturer with customers, contractors, partners, shipping companies, etc.,
- to determine the best purchasing option by forecasting, developing the most rational schemes and conditions for the resources delivery to the enterprise,
- to identify the compliance degree of the bought resources with the strategic objectives of the purchasing activities in the industrial enterprise, and to carry out the purchasing modeling based on the economic and mathematical methods.

In order to identify problems and sources of efficiency in the activities of industrial enterprises, a map of the functional strategies for purchasing department in the industrial enterprise was developed, which allows you to choose the optimal strategy for each SAEM.

References

1. Banchuen, P., Sadler, I., Shee, H.: Supply chain collaboration aligns order-winning strategy with business outcomes. IIMB Manag. Rev. **29**(2), 109–121 (2017). https://doi.org/10.1016/j.iimb.2017.05.001
2. Biggemann, S., Kowalkowski, C., Maley, J., Brege, S.: Development and implementation of customer solutions: a study of process dynamics and market shaping. Ind. Mark. Manag. **42** (7), 1083–1092 (2013). https://doi.org/10.1016/j.indmarman.2013.07.026
3. Bowersox, D.J., Closs, D.J., Cooper, M.B., Bowersox, J.C.: Supply Chain Logistics Management, 4th edn. McGraw-Hill Education, New York (2002)
4. Kaplan, R.S.: Conceptual foundations of the balanced scorecard. In: Handbooks of Management Accounting Research, vol. 3, pp. 1253–1269 (2009). https://doi.org/10.1016/s1751-3243(07)03003-9
5. Kaplan, R.S., Norton, D.P.: The Balanced Scorecard: Translating Strategy into Action. Harvard Business School Press, Boston (1996)
6. Kaplan, R.S., Norton, D.P.: Balanced Scorecard: Translating Strategy into Action. Olimp-Biznes, Moscow (2017). (in Russian)
7. Karpova, N.P.: Strategic Logistics of Supply: Monograph. Creative Economy, Moscow (2011). (in Russian)
8. Kumar, D., Rahman, Z.: Sustainability adoption through buyer supplier relationship across supply chain: a literature review and conceptual framework. Int. Strateg. Manag. Rev. **3**(1–2), 110–127 (2015). https://doi.org/10.1016/j.ism.2015.04.002
9. Kumar, G., Banerjee, R.N., Meena, P.L., Ganguly, K.K.: Joint planning and problem solving roles in supply chain collaboration. IIMB Manag. Rev. **29**(1), 45–57 (2017). https://doi.org/10.1016/j.iimb.2017.03.001

10. Lee, T., Nam, H.: An empirical study on the impact of individual and organizational supply chain orientation on supply chain management. Asian J. Shipp. Logist. **32**(4), 249–255 (2016). https://doi.org/10.1016/j.ajsl.2016.12.009
11. Linders, M.R., Dzhonson, F., Flinn, A.Ye., Firon, G.: Procurement and supply management. Yuniti-Dana, Moscow (2014). (in Russian)
12. Saak, A.E.: Traceability and reputation in supply chains. Int. J. Prod. Econ. **177**, 149–162 (2016). https://doi.org/10.1016/j.ijpe.2016.04.008
13. Sousa, M.J., Rocha, A.: Digital learning: developing skills for digital transformation of organizations. Future Gener. Comput. Syst. **91**, 327–334 (2019). https://doi.org/10.1016/j.future.2018.08.048
14. Toimentseva, I.A.: Marketing information systems in management decision-making. Vestnik Samara State Univ. Econ. **6**(104), 119–123 (2013). (in Russian)
15. Waters, D., Rinsler, S.: Global Logistics: New Directions in Supply Chain Management, 7th edn. GB & USA by Kogan Page (2014)

Digital Transformation in Business

S. Ziyadin[1(✉)], S. Suieubayeva[2], and A. Utegenova[1]

[1] Al-Farabi Kazakh National University, Almaty, Kazakhstan
sayabekz@gmail.com
[2] D. Serikbayev East Kazakhstan State Technical University, Almaty, Kazakhstan

Abstract. The aim of the research is to identify the clear definition of digital transformation and to present a structured framework with digital transformational stages, activities and results. This research contains a literature review which is giving insight into the fundamental comprehension of digital transformation. Findings indicate that even though digital transformation is a well-known idea a method for the organized digital transformation of business models is missing. The research gives an explicit meaning of digital transformation of business models and phases for the digital transformation of business models. In addition, the paper offers the roadmap for digital transformation.

Keywords: Business model · Digital transformation · Digitalization · Innovation

1 Introduction

Digital transformation affects all sectors of society, in particular economies. Companies now are given an opportunity to radically change their business models by new digital technologies like social networks, mobile, big data, Internet of things, other innovations like blockchain. This mostly involves changes of the core business operations and modifies products and processes, as well as organizational structures, as companies ought to set up management practices to conduct these complex transformations [13]. Consequently, society overall is facing a radical change due to the development of digital technologies and their extensive implementations of all markets [4]. To add to the expanded interest from clients, organizations are facing even harder competition because of globalization [24] and putting strain to go digital before others do, looking to survive and accomplish competitive benefits [2].

The digital transformation phenomenon has been explored widely in different academic domains, resulting in a crude overview of the field. What is still unaccounted for is an unmistakable definition for the digital transformation of business models, a methodology how to digitize business models, which stages and instruments ought to be considered and what models and empowering influences exist. In this paper, there is a framework presented for digital transformation of business models by analysing the existing literature and practical examples.

S. I. Ashmarina et al. (Eds.): ISCDTE 2019, LNNS 84, pp. 408–415, 2020.
https://doi.org/10.1007/978-3-030-27015-5_49

2 Theoretical Background

2.1 Defining Digital Transformation

Multiple definitions of digital transformation can be found in literature. The concept of digital transformation is formed by the merger of personal and corporate IT environments and encapsulate the transformational effect of new digital technologies such as social, mobile, analytical, cloud technologies and the Internet of Things (SMACIT) [26]. In a broader sense, digital transformation is presented as the integration of digital technologies and business processes in a digital economy [12]. A comparative boundless view sees it as the utilization of innovation to radially enhance the execution or reach of ventures [23]. More accurate recognition under the influence of digital transformations implies three organizational aspects: from the outside, with the improvement of the client's experience and the change of his entire life cycle; on the inside, the impact on business objectives, basic leadership and hierarchical structures; and in general, when all business sections and opportunities are influenced, usually leading to completely new business models [6].

The depicted idea of digital transformation shows that its multifaceted nature exceeded the level of past transformations endowed with IT capabilities, new transformations. This is upheld by the fact that DT is viewed as one of the real difficulties in all industries lately, without exception, and even in spite of the fact that organizations perceive its fundamental significance, they still confront numerous obstacles that repress them from starting, not to mention profiting by, digital transformation [17]. They battle to get business profits by new digital advancements, as contending needs lead the rundown of normal hindrances [7]. This may be because of an absence of lucidity about the distinctive accessible choices and components that directors need to consider in their transformation approach [6]. Fitzgerald et al. (2014) recommend that a noteworthy minority of organizations have prevailing with regards to building up the privilege administrative and mechanical aptitudes to pick up transformational impacts from new digital innovations [5]. They additionally recommend that extra authority and institutional difficulties are likewise looked by businesses today. Authority challenges incorporate an absence of criticalness, vision and bearing, though institutional ones are identified with the mentalities of more seasoned specialists, inheritance innovation, development weariness and legislative issues. Institutional difficulties can best be clarified by the way that most innovation empowered transformations include a specific level of protection from change showed in the conduct of specific individuals, who decline to acknowledge the new situation [11]. This social hindrance is frequently thought little of and for the most part not perceived by organizations [20].

To conclude, we infer that digital transformation is a more mind boggling kind of technology enabled business transformation, which needs to address the vital jobs of new digital advances and capacities for effective digital development in the digital world [27]. This paper characterize it as the process through which organizations meet numerous new digital innovations, upgraded with universal network, with the expectation of achieving predominant execution and managed upper hand, by changing various business measurements, including the business model, the client encounter (involving digitally empowered items and administrations) and tasks (containing processes and basic leadership), and all the while affecting individuals (counting abilities ability and culture) and systems (counting the whole esteem framework).

2.2 Research Questions and Research Design

In the introduction, the brief literature review provided with fundamentals of digital transformation. To fill in as a base for a Roadmap, I examined six digital transformation strategy frameworks encountered in the literature and shed light on their nature [6, 8, 15, 18, 20, 25].

Despite the fact that these methodologies make esteemed commitments, they do not totally cover digital transformation of business models and don't determine the digital transformation's application. In my contribution, I initiate the digital transformation of business models and build up a Roadmap including a few stages.

> Based on the problem described and current understanding, we will answer the following research questions:
> How do businesses define and execute their digital business transformation strategies?
> What does an organized methodology for digital transformation of business models resemble?

For our literature review, we applied case study strategies since this procedure is attractive for portraying and analysing applicable cases in which grounded-hypothesis can be produced. The authors analysed existing definitions, methodologies and examples for digital transformation of business models. We conducted the literature review to gain insight into current research in digital transformation and developed a Roadmap for the digital transformation of business models.

3 Discussion

Looked with various digital transformation challenges, organizations have perceived the need to oversee this perplexing undertaking by defining and executing an unmistakable technique to keep pace with the new digital reality [13]. This is bolstered by different partners inside associations: administrators see the capability of developing digital advances, but they are indistinct about how to accomplish their transformation objectives. Specialists in the business world are all in understanding that the capacity to digitally rehash the business isn't just about the advances being received, yet rather about a radical vital and social change inside the association [20], and corporate representatives similarly believe in a core role that strategy plays for effectively implementing new technologies [5].

The two levels, individual and hierarchical, are along these lines encouraged to fathom this key basic behind any digital mix and transformation endeavors [8]. Regardless of the vital significance of planning a committed technique that coordinates all the prioritization, coordination instruments and execution ventures of digital transformation, the scholarly community still neglects to give a reasonable rule that locations a all inclusive transformation system [6]. "We require key systems that are gone for intentionally tackling the remarkable capacities of digital innovation that are inserted into items to increase upper hand" [27]. This "developing feeling of criticalness about the need to make effective methodologies for the digital commercial center" [10] makes it evident that the idea of an organization level digital transformation appears to depend on a methodology point of view, or, in other words the writing, and additionally by and by. Three Major Strategy Levels are the most commonly met in the world of business:

Corporate Strategy: What set of business should we be in?
Business Strategy: How should we compete in this business?
How important is a functional strategy and how does it form a competitive advantage?

How important the strategy is and how useful it is for each situation should pay attention to the fact that the choice can be limited at the expense of the participants involved in each level [14].

The plan of a digital transformation strategy expects organizations to make appropriate key choices in a few key areas and this segment incorporates the applicable ones experienced in the writing, as summarised below.

From a business perspective, an underlying business case must be made, in which the longterm destinations should be clear and furthermore overweighing the simple quest for brisk additions. Firmly identified with this are choices about the potential requirement for a general change readiness evaluation with the end goal to understand the current condition of an organization's execution, and to distinguish potential issues, vulnerabilities, openings and the related dangers [8, 11].

Innovative technological choices are urgent while bringing new rising advances into organizations. Among territories being investigated here is the job that advancements play in the firm to accomplish vital objectives. This can be isolated into an empowering job for new business openings or a simply steady one to satisfy current business necessities. Nearly identified with that is an organization's disposition towards new advances and its capacity to misuse them for future business objectives. The relationship with investments allows us to predetermine the possibilities of the impact of digital technologies on the business [7] and also suggests special change programs to guarantee the organization to simultaneously develop with technologies [22].

Various vital choices must be taken in connection to changes in the communication with clients. Organizations are urged to research conceivably new advantages made they would say through digitally improved changes to the client venture [19]. This can be accomplished by investigating all the client contact focuses and incorporating the organizations' communications crosswise over different digital, and additionally physical, stages [1]. Changing the client experience can likewise be refined through the presentation of digitally upgraded items and administrations [6]. Moreover, interests in R&D can additionally enable associations to create digitized answers for foresee client needs instead of just reacting to existing ones [18].

Managerial decisions area have a monetary component, which develops around picking how to back the digital transformation try, subsequent to evaluating the monetary weight on the present business [13]. The emphasis on cultivating advancement is additionally a basic component to be examined [11], in which managers are urged to see digital advancement as an indispensable piece of their strategy [8]. Nimble and new adaptable working, alongside a thought of a base up advancement processes, should drive a progressing digital transformation [25]. Supervisors need to additionally be mindful of the principal significance that different capacities play, to be specific, authoritative, innovation based, item related and digital capacities [3]. Looking at their organizations' vital resources and abilities through a digital focal point may enable directors to pinpoint which existing resources can be utilized, which capacities can be utilized in new ways and whether or not new abilities are should have been brought into the organization [16]. If the digital divide is overcome, many developing countries would be able to distribute their products, increase their client base and form trade partnerships [21].

Different organizational decisions are likewise distinguished in the writing, in which organizations are encouraged to investigate the workers, culture, ability and range of abilities, and authority. Organizations should survey the requirement for building up a collective work condition and guarantee that the transformation venture is staffed effectively [7]. Workers are regularly surveyed from a development perspective, in which their jobs, aptitude and capacities are examined [20]. This empowers organizations to arrange themselves into a classification of digital development and furthermore causes them to explore their transformation in an organized way. Further contemplations are required with respect to the vital changes in the organization's way of life, which are gone for adjusting it to work with new advancements as opposed to forcing these advances on representatives.

The authors conduct research and focus on the six pillars of digital transformation often found in the world literature [6, 8, 15, 18, 20, 25]. They direct most of the attention to the previously discussed content of the strategy, where everything is gradually projected. This echoes the idea that transformation is a process that goes through stages, leaning on each other, rather than on individual events [9]. Having identified one of the main distinctive stages defined for all structures, as well as for equipping managers with a structured approach during the transition from one stage to the next, we form a new reincarnation strategy.

Phases distinguished in Digital Transformation Frameworks:
Initiation: Understanding digitalization openings, dangers and effect;
Ideation: Imagining transformation measurements as alternatives for the business;
Assessment: Evaluating digital preparation levels and distinguishing holes;
Commitment: Communicating the vision and incorporating the fundamental individuals;
Implementation: Proceeding with the activity plan in different areas;
Sustainability: Validating and streamlining the activity plan persistently;

A Roadmap is given here dependent on the introduced ways to deal with digital transformation and dependent on existing speculations about business model advancement, the Roadmap for digital transformation of business models is clarified as pursues:

Digital reality: In this stage, Digital reality, the organization's current business model is outlined alongside an esteem added examination identified with partners and a study of client prerequisites. This gives a comprehension of the Digital Reality for this organization in various territories.

Digital aspiration: Based on the Digital reality, targets with respect to digital transformation are characterized. These goals identify with time, accounts, space and quality. Digital desire hypothesizes which destinations ought to be considered for the business model and its components. Therefore, goals and business model measurements are organized.

Digital potential: Within this Digital potential stage, best practices and empowering agents for the digital transformation are gathered. This fills in as a beginning stage regarding Digital potential and the outline of a future digital business model. For this reason, diverse alternatives are inferred for every business model component and consistently consolidated.

Digital fit: The Digital Fit stage takes a gander at choices for the plan of the digital business model, which are assessed to decide Digital Fit with the current business model. This guarantees one satisfies client necessities and that business goals are accomplished. The assessed mixes are then organized.

Digital integration: Digital usage incorporates conclusion and usage of the digital business model. The different mix alternatives are additionally tightened inside a digital execution system. The Digital usage likewise incorporates the plan of a Digital Customer Experience and the digital esteem creation organize that portrays combination with accomplices. Moreover, assets and capacities are likewise recognized in this stage.

4 Conclusion

The aim of this paper has been to demonstrate the current state of academic research in digital business transformation strategies and give a broaden understanding oft he definition of digital transformation. It uncovers the options how organizations digitally change and outlines organizations' inside inspirations, and additionally their outer triggers. At long last, we recognize numerous measurements with respect to the substance of methodology plan in the digital time, and furthermore remark on the technique process.

References

1. Berman, S.J.: Digital transformation: opportunities to create new business models. Strategy Leadersh. **40**(2), 16–24 (2012). https://doi.org/10.1108/10878571211209314
2. Bharadwaj, A.S.: A resource-based perspective on information technology capability and firm performance: an empirical investigation. MIS Q.: Manag. Inf. Syst. **24**(1), 169–193 (2000)
3. Bonnet, J., Subsoontorn, P., Endy, D.: Rewritable digital data storage in live cells via engineered control of recombination directionality. Proc. Natl. Acad. Sci. U.S.A. **109**(23), 8884–8889 (2012). https://doi.org/10.1073/pnas.1202344109
4. Ebert, C., Duarte, C.H.C.: Requirements engineering for the digital transformation: industry panel. In: Proceedings of 2016 IEEE 24th International Requirements Engineering Conference, RE 2016, pp. 4–5 (2016). https://doi.org/10.1109/re.2016.21

5. Fitzgerald, M., Kruschwitz, N., Bonnet, D., Welch, M.: Embracing digital technology: a new strategic imperative. MIT Sloan Manag. Rev. (2013). https://sloanreview.mit.edu/projects/embracing-digital-technology/. Accessed 28 Mar 2019
6. Hess, T., Benlian, A., Matt, C., Wiesböck, F.: Options for formulating a digital transformation strategy. MIS Q. Exec. **15**(2), 123–139 (2016)
7. Kane, G.C., Palmer, D., Phillips, A.N., Kiron, D., Buckley, N.: Strategy, not technology, drives digital transformation. MIT Sloan Manag. Rev. Deloitte Univ. Press (2015). https://sloanreview.mit.edu/projects/strategy-drives-digital-transformation/. Accessed 28 Mar 2019
8. Kaufman, I., Horton, C.: Digital transformation: leveraging digital technology with core values to achieve sustainable business goals. Eur. Financ. Rev. (December–January), 63–67 (2015)
9. Kotter, J.P.: Leading change: why transformation efforts fail. IEEE Eng. Manag. Rev. **37**(3), 42–48 (2009)
10. Kulatilaka, N., Venkatraman, N.: Strategic options in the digital era. Bus. Strategy Rev. **12**(4), 7–15 (2001). https://doi.org/10.1111/1467-8616.00187
11. Kumar Basu, K.: The leader's role in managing change: five cases of technology-enabled business transformation. Global Bus. Organ. Excell. **34**(3), 28–42 (2015). https://doi.org/10.1002/joe.21602
12. Liu, D.Y., Chen, S.W., Chou, T.C.: Resource fit in digital transformation: lessons learned from the CBC Bank global e-banking project. Manag. Decis. **49**(10), 1728–1742 (2011). https://doi.org/10.1108/00251741111183852
13. Matt, C., Hess, T., Benlian, A.: Digital transformation strategies. Bus. Inf. Syst. Eng. **57**(5), 339–343 (2015). https://doi.org/10.1007/s12599-015-0401-5
14. Mills, J., Platts, K., Gregory, M.: A framework for the design of manufacturing strategy processes: a contingency approach. Int. J. Oper. Prod. Manag. **15**(4), 17–49 (1995). https://doi.org/10.1108/01443579510083596
15. Parviainen, P., Tihinen, M., Kääriäinen, J., Teppola, S.: Tackling the digitalization challenge: how to benefit from digitalization in practice. Int. J. Inf. Syst. Proj. Manag. **5**(1), 63–77 (2017). https://doi.org/10.12821/ijispm050104
16. Ross, J.W., Sebastian, I.M., Beath, C.M.: How to develop a great digital strategy. MIT Sloan Manag. Rev. **58**(2), 7–9 (2017)
17. Schuchmann, D., Seufert, S.: Corporate learning in times of digital transformation: a conceptual framework and service portfolio for the learning function in banking organisations. Int. J. Adv. Corp. Learn. (iJAC) **8**(1), 31–39 (2015)
18. Sebastian, I.M., Moloney, K.G., Ross, J.W., Fonstad, N.O., Beath, C., Mocker, M.: How big old companies navigate digital transformation. MIS Q. Exec. **16**(3), 197–213 (2017)
19. Valdez-de-Leon, O.: A digital maturity model for telecommunications service providers. Technol. Innov. Manag. Rev. **6**(8), 19–32 (2016)
20. von Leipzig, T., Gamp, M., Manz, D., Schöttle, K., Ohlhausen, P., Oosthuizen, G., Palm, D., von Leipzig, K.: Initialising customer-orientated digital transformation in enterprises. Procedia Manuf. **8**, 517–524 (2017). https://doi.org/10.1016/j.promfg.2017.02.066
21. Watkins, M., Ziyadin, S., Imatayeva, A., Kurmangalieva, A., Blembayeva, A.: Digital tourism as a key factor in the development of the economy. Econ. Ann.-XXI **169**(1–2), 40–45 (2018). https://doi.org/10.21003/ea.V169-08
22. Webb, N.: Vodafone puts mobility at the heart of business strategy: transformation improves performance of employees and organization as a whole. Hum. Resour. Manag. Int. Digest **21**(1), 5–8 (2013). https://doi.org/10.1108/09670731311296410
23. Westerman, G., Bonnet, D., McAfee, A.: Leading Digital: Turning Technology into Business Transformation. Harvard Business Review Press, Boston (2014)

24. Westerman, G., Calméjane, C., Bonnet, D., Ferraris, P., McAfee, A.: Digital transformation: a roadmap for billion-dollar organizations. MIT Center for Digital Business and Capgemini Consulting (2011). https://www.capgemini.com/wp-content/uploads/2017/07/Digital_Transformation__A_Road-Map_for_Billion-Dollar_Organizations.pdf. Accessed 28 Mar 2019
25. Westerman, G., Tannou, M., Bonnet, D., Ferraris, P., McAfee, A.: The digital advantage: how digital leaders outperform their peers in every industry. MIT Sloan Management and Capgemini Consulting, MA (2012). https://www.capgemini.com/wp-content/uploads/2017/07/The_Digital_Advantage__How_Digital_Leaders_Outperform_their_Peers_in_Every_Industry.pdf. Accessed 28 Mar 2019
26. White, H.C.: Identity and Control: How Social Formations Emerge, 2nd edn. Princeton University Press, Princeton (2008)
27. Yoo, Y., Lyytinen, K.: The new organizing logic of digital innovation: an agenda for information systems research. Inf. Syst. Res. **21**(4), 724–735 (2011). https://doi.org/10.1287/isre.1100.0322

Advantages and Disadvantages of Automated Control Systems (ACS)

M. Vochozka(✉), J. Horák, and T. Krulický

School of Expertness and Valuation, Institute of Technology and Business, Okružní 517/10, 37001 České Budějovice, Czech Republic
vochozka@mail.vstecb.cz

Abstract. ACS is finding more application in our world today. Applicability of this technology ranges from households to large industries. In today's industries, there is a high demand for production maximization and automation in everyday life. This has led to a literary review, researching the advantages and disadvantages of the technology. Automated systems have advanced from mere application in buildings or vehicles to performing smart task in aircrafts, spaceships, factories, and ships, etc. This review uses a multidisciplinary lens to examine the subject from a broader perspective. Research papers available in the web of science and Scopus in recent decade is examined to help the researcher pinpoint the benefits and consequences related to the use of ACS. The critical review of literatures led to the conclusion that ACS make life easier for humans, enhance economic growth and can be applied in almost all fields. On the other hand, ACS leads to unemployment and can subdue rather than to serve humans in the near future.

Keywords: Automation · Control systems · Technologies · Industries

1 Introduction

1.1 Background

Automation technology requires minimal human effort. The use of various control systems to operate equipment is referred to as automatic control. Automation in other words can simply be defined as the use of electrical or electronics, mechanical and/or computerized solutions to control or operate systems in vehicles, manufacturing, construction, sea ports as in intelligent automated crane systems, etc. Applicability of this technology ranges from households to large industries. The term automatic was not widely use up to the 1947's until when Ford Motor company created an automation department. Though the term can be traced earlier back in 762 B.C. Technological revolutions in the 21st century have led to great development of research in automation and robotic science. This development in return has advantages and disadvantages on it end users. Human error crashes accounted for 93% of total vehicle crashes in the United States [7].

S. I. Ashmarina et al. (Eds.): ISCDTE 2019, LNNS 84, pp. 416–421, 2020.
https://doi.org/10.1007/978-3-030-27015-5_50

1.2 Current Trend

In today's industries, there is a high demand for production maximization and automation in everyday life. That increasing demand has led to the employment of robots instead of humans to carryout complex and even simple but repeated task. ACS are therefore now considered very important in industries because of their accuracy, reliability, and efficiency and can perform same task repeatedly. Modern aircraft manufacturing companies, car manufacturing factories and ships now uses automated control systems (ACS) to perform different functions.

1.3 Scope of the Review

The scope of this review is to pinpoint the advantages and disadvantages of ACS. This work is limited only to published works from 2009–2019 in the web of science and Scopus data bases. More than 24,562 open access related journals, articles, and conference papers were found using the key words. The review can be a good resource to raise awareness about the lapses with automation in addition to unravelling the merits and demerits of ACS. Also, this review can be a motivator to senior researchers to conduct detail primary research on the topic, and will also serve as a referencing material when carrying out similar work for other professionals.

2 Review of Related Literature

2.1 Application of ACS

Some researchers confirm that ACS can be used in controlling streetlights [3, 6, 12]. They designed different systems, and in so doing were able to achieve their research aim in the sense that they reduce time consuming manual switching of streetlights by a light dependent resistor and infrared-sensors. Their findings further proofed that ACS are not only use in factories, but also can be used to turn on, dim, turn to high state of streetlights, while during day-time the streetlights will remain OFF. Also, the system can skip DIM conditions at night and are only turn ON based on the detection of objects. Nalamwar et al. [7] proposed that using intelligent ACS in solar energy sector will enable the full utilization of photovoltaic (PV) system and will improve durability. In their research they developed a smart automated monitoring & controlling system for the solar. They believed that the system will be smart enough to alert responsible persons in case there is a malfunction, error display, and even when the system is due for maintenance through emails or sms. The system is believed to improve performance of solar panel system and data from it can be easily monitored and analyzed.

Naujoks et al. [8] are totally against drivers being merely occupants or passengers in automated vehicles. They believed that drivers should be required to concentrate and supervised the operations of the automated system. They proposed a need for drivers to be fully grounded and to participate in relevant system when driving. A proposal for design recommendations from their empirical research on which they derive an initial set of principles and criteria to guide the development and design of automated vehicle was put forward. Yoon and Ji [12] examined the same topic and investigated the influence of non-driving-related task (NDRT) on takeover performance in a highly automated driving (HAD) context and the effect of workload on driver's takeover performance. A driving simulator was used in their research to evaluate how well a driver resumes control of a vehicle after being in a HAD situation during which they performed a NDRT. They found a significant difference in visual performance and takeover capability based on the task carried out. The concluded that reaction times when reaching for the steering wheel did not differ among the tasks. Seppelt and Lee in their study evaluated the advantages and value associated with providing drivers continuous feedback on the limits and behavior of imperfect vehicle control automation system [10]. They found out that ACS remove drivers from active control, relying on timely interventions of the system when disaster occur. Distinct warnings, as a form of comments to accurately inform drivers about system behavior, fail to keep drivers alert about its contiguity to running limits.

Oborski [9] focused his research on the integration of machine operators with information flow in manufacturing process. He developed a special IT system connecting together machine operators, machine control, process and machine monitoring with companywide IT systems. He fully automated information flow between management and manufacturing process. Automated system performed two functions at the same time, which for humans might be impossible; whilst information at the management level about particular orders are taken, on-line information about manufacturing process and manufactured parts are also given. Vidros et al. [11] reveal that automated systems are, used by human resources managers in the fashion industry to complete recruitment of new employees and believed that it makes the process efficient, cost-efficient, and more immediate. However, the consequences of online exposure of such traditional business procedures are harmful to the employees and hiring organizations due to the fact that the system can be hacked and secret information made public.

Harahap, Adyatma and Fahmi [5] vehemently support that ACS plays an important role in most industries if not all. ACS usage can be encountered in almost all fields of industry, not only in the manufacturing world but also on many other things such as elevators in office buildings, logistics operations, hospitals, sea ports, etc. Zhang et al. [13] on the other hand explored factors affecting customers' acceptance of automated vehicles (AV). Shockingly, their discoveries support that underlying trust offers some other and presumably progressively indispensable pathway for different variables to impact clients' adoption of systems with uncertainty. In practice, their findings provide guidance for designing interventions that will improve overall public acceptance of ACS. They utilized a self-regulated poll to gather experimental information for the research and the questionnaire comprised of three segments; the principal segment estimated demographic qualities, the second segment got some information about respondents' driving related data and the third segment began with a concise meaning of AV with a question of whether they respondents have ever known about AV before the study.

Cost and manpower management and the integration of other systems in clinics enhance successful implementation of ACS. Delaney et al. [2] believe that automation is applicable in the health system in treatment planning. The revealed that Rapid-PlanTMPT (Varian Medical Systems, Palo Alto, California, USA) is an automated knowledge-based planning solution, which could reduce variability and increase efficiency. They explained that the system uses a library of previous IMPT treatment plans to generate a model which can predict organ-at-risk dose for new patients, and guide IMPT optimization. In conclusion the believed that the system demonstrated efficiency and consistency as compared with manual plans. They asserted that automated planning solution could also assist in clinical trial quality assurance.

2.2 Impact of ACS on Human Behavior

The application of ACS is not only limited in industries. In recent years' research have revealed that automation has influenced human behavior in various ways; hence the field of politics is not an exception. Gingrich [4] reveals that early responses to automation have shaped how those affected negatively by technological change responded politically. She found out that those exposed to technological change are both more likely to vote for the mainstream left and right populists. Differences in compensation have a limited direct or indirect effect. Where spending and labor market regulation does matter, it heightens both left and right-populist voting among affected groups.

2.3 Challenges and Opportunities of ACS

Acemoglu and Resptrepo [1] examined some of the consequences and opportunities of automation and pinpointed that new technologies will render labor redundant in the sense that tasks previously performed by humans can be automated and new job potentials can be created, and humans will have comparative advantage as well. They asserted that ACS reduces employment and the labor share, and may even reduce wages, but the creation of new job opportunities will have positive impact. The balance of the system is an outcome of the reality that ACS decreases production cost and consequently discourages similarly automation and further encourages the creation of new function. Also, inequality will increase during transitions because of faster automation and the introduction of different function.

3 Findings

In vehicles, automated control system is used to carry out different functions: First, ACS reduces error crashes caused by humans during driving. Second, ACS in automated vehicles reduces road congestion and fuel emissions by causing the stop-and-move waves so that traffic doesn't move as it does when humans are driving on well planned routes. Third, ACS enables drivers to be more relax and be at leisure free from driving activities and can be involved in other activities. Fourth, ACS enables the disabled and aged people whom without automated control systems in vehicles will not be fit to drive to comfortably drive. Also, ACS in nuclear science reduces time in data entry and increase productivity. In industries, ACS as compare to humans do not fear operations, don't get tired and lack work place conflict of interest. In mining automated on-board systems are used in operations to capture data and maintenance recording in machines [14]. Automation technology manage increasing workload demands and enhance clinical laboratory performance. Automated control robots to be specific can make minimal mistakes when carrying out job task and can work 24/7 with higher productivity. In smart homes, they are comfortable, secure, private, economical, safe and are flexible in usage. However, ACS in cars shifts driver's attention from being active to passive controllers from the demands of the roadway and even from the system. As a result, they misjudge the entire system due to over-relying on the system, thereby leading to serious consequences when system failure occurs especially when there is an influence on changeover of control. In general, it was found out that the rapid takeover of automation in industries has led to a reduction in human employment and remuneration. Safety risk due to system or equipment failure is worrisome as well as privacy risk, bearing in mind that personal data could be hacked or even transmitted to third parties.

4 Conclusion

The advantages and disadvantages of ACS have not been too clear, research in the field, has been generalized and limited. With that in mind, the literature reviewed in this research reveals the overall benefits and consequences of the application of ACS. Therefore, it can be emphasized that ACS help manage increasing workload demands and limited labor force. Primary research is needed to specifically examined the advantages and disadvantages of using ACS in individual industries or field of application, and to innovate effective training programs to address the challenges for future workforce.

References

1. Acemoglu, D., Restrepo, P.: The race between man and machine: implications of technology for growth, factor shares, and employment. Am. Econ. Rev. **108**(6), 1488–1542 (2018)
2. Delaney, A.R., Verbakel, W.F., Lindberg, J., Koponen, T.K., Slotman, B.J., Dahele, M.: Evaluation of an automated proton planning solution. Cureus **10**(12), e3696 (2018). https://doi.org/10.7759/cureus.3696
3. Fagnant, D.J., Kockelman, K.: Preparing a nation for autonomous vehicles: opportunities, barriers and policy recommendations. Transp. Res. Part A: Policy Pract. **77**, 167–181 (2015)
4. Gingrich, J.: Did state responses to automation matter for voters? Res. Politics **6**(1) (2019). https://doi.org/10.1177/2053168019832745
5. Harahap, R., Adyatma, A.F., Fahmi, F.: Automatic control model of water filling system with Allen Bradley Micrologix 1400 PLC. In: IOP Conference Series-Materials Science and Engineering, Talenta – Conference on Engineering, Science and Technology. IOP Publishing, Bristol (2018). https://doi.org/10.1088/1757-899x/309/1/012082
6. Mumtaz, Z., Ullah, S., Ilyas, Z., Aslam, N., Iqbal, S., Liu, S., Madni, H.: An automation system for controlling streetlights and monitoring objects using Arduino. Sensors **18**(10) (2018)

7. Nalamwar, H.S., Ivanov, M.A., Baidali, S.A.: Automated intelligent monitoring and the controlling software system for solar panels. In: International Conference on Information Technologies in Business and Industry 2016 IOP Publishing IOP Conference Series: Journal of Physics: Conference Series, vol. 803, no. 2017, p. 012107 (2017). https://doi.org/10.1088/1742-6596/803/1/012107
8. Naujoks, F., Wiedemann, K., Schömig, N., Hergeth, S., Keinath, A.: Towards guidelines and verification methods for automated vehicle HMIs. Transp. Res. Part F: Traffic Psychol. Behav. **60**, 121–136 (2019)
9. Oborski, P.: Integration of machine operators with shop floor control system for Industry 4.0. Manag. Prod. Eng. Rev. **9**(4), 48–55 (2018)
10. Seppelt, B.D., Lee, J.D.: Keeping the driver in the loop: dynamic feedback to support appropriate use of imperfect vehicle control automation. Int. J. Hum.-Comput. Stud. **125**, 66–80 (2019)
11. Vidros, S., Kolias, C., Kambourakis, G., Akoglu, L.: Automatic detection of online recruitment frauds: characteristics, methods, and a public dataset. Future Internet **9**(1) (2017)
12. Yoon, S.H., Ji, Y.G.: Non-driving-related tasks, workload, and takeover performance in highly automated driving contexts. Transp. Res. Part F: Traffic Psychol. Behav. **60**, 620–631 (2019)
13. Zhang, T., Tao, D., Qu, X., Zhang, X., Lin, R., Zhang, W.: The roles of initial trust and perceived risk in public's acceptance of automated vehicles. Transp. Res. Part C: Emerg. Technol. **98**, 207–220 (2019)
14. Zvonarev, I.E., Shishlyannikov, D.I.: Information and diagnostic tools of objective control as means to improve performance of mining machines. In: Zykova, A., Martyushev, N. (eds.) IOP Conference Series: Materials Science and Engineering: International Conference on Mechanical Engineering, Automation and Control Systems. IOP Publishing, Bristol (2017)

Using Artificial Intelligence in Company Management

J. Vrbka(✉) and Z. Rowland

Institute of Technology and Business, School of Expertness and Valuation, Okružní 517/10, 37001 České Budějovice, Czech Republic
vrbka@mail.vstecb.cz

Abstract. Thanks to technological progress, artificial intelligence is currently used in different areas of our lives. The use of artificial intelligence in business and finance has a promising future. Artificial intelligence is inspired by the behavior of biological patterns, having also the ability to learn and then capture these strongly non-linear dependencies. The advantage of artificial neural networks consists in their capability of working with big data, in the precision of their results or easier use of the obtained neural network. The objective of this contribution is to carry out systematic literary research of the most renowned scientific resources and find out whether it is possible to use artificial intelligence in practice, in company management. After a clearly defined process of selecting the appropriate scientific outcomes, these studies are explored and conclusions are made. A total of 31 publications fulfilled the criteria. The publications more or less agree on the practical applicability of artificial intelligence in company management.

Keywords: Artificial intelligence · Artificial neural networks · Company management

1 Introduction

Artificial intelligence aims to complete and even take over practically all the tasks currently performed by people. Due to the influence of external forces and technological progress, artificial intelligence is being used in various fields. A promising future is especially using artificial intelligence in finance [1, 33].

Simply said, the main objective of artificial intelligence is to understand human intelligence and design computer systems that would be able to imitate the patterns of human behaviour and generate knowledge applicable for problem-solving. Artificial intelligence shall thus be able to learn and understand new concepts, shall be able to learn from experience ("on its own"), to make reflections and conclusions, to imply importance, and to interpret symbols in context. Due to such abilities, artificial intelligence has been successfully implemented in various fields, such as gaming, semantic modelling, human performance modelling, robotics, machine learning, data mining, neural networks, genetic algorithms, and expert systems [14, 19].

S. I. Ashmarina et al. (Eds.): ISCDTE 2019, LNNS 84, pp. 422–429, 2020.
https://doi.org/10.1007/978-3-030-27015-5_51

2 Data and Methods

This review is based on the procedure of systematic literary research recommended by Fink [8]. It includes selecting a research question, bibliographic database or database of articles, selecting search terms, applying practical criteria for screening, applying criteria for methodological screening, reviewing and results synthesis. For the purpose of the review, a research question will be formulated, and two most renowned scientific databases (Scopus, Web of Science) will be searched. Using more additional documents, books, papers, etc. would result in greater variability of a peer-review process and limitation of their accessibility [15].

In this stage, terms and combinations of terms (artificial neural networks, business management, artificial intelligence, business value growth, effective management, business processes, risk management, etc.) will be used for finding suitable publications, from which only those written in professional English will be chosen. All available publications or all those published from the year 2000 up to present will be reviewed, with the preference of newer and more valuable publications. The following stage will include the selection of suitable publications that are relevant to the topic and not duplicated. The final sample of publications will be reviewed thoroughly; inappropriate publications will be excluded from the list. The remaining quality and the most important publications will be used for the purposes of this contribution.

3 Systematic Literary Research

In order to achieve the objective of the contribution, a research question is formulated: "Is it really possible to use artificial intelligence for business management in practice?" Continuous business processes optimization remains a challenge for enterprises. In times of digital transformation, faster changes in internal and external framework conditions, and new customers′ demands for the fastest and highest quality goods and many other, enterprises should make their internal process as good as possible [24].

Artificial intelligence surpasses people both in data processing and computing power. It also makes an advance at strategic thinking, creativity, and social interaction skills with almost human cognitive skills. How long will it take to have digital managers running corporations? Will management become a commodity created by electronic brains [7]?

Research related to the field of economics often arises on the basis of artificial intelligence, by means of artificial neural networks. Artificial neural networks try to copy the processes in the 2012 human brain and neural system using computer devices. The term "artificial neural network" appeared first in biology and psychology. According to Klieštik [17], artificial neural networks are computing models inspired by biological neural networks, specifically by the behavior of neurons. Vochozka and Machová [30] claim that this is the reason for using them in modeling very complex strategic decisions. Beiranvand et al. [5] state that the results of neural networks are very promising and their performance and accuracy in dealing with key indicators of an enterprise is significantly higher than when using conventional statistical methods. The advantage of artificial neural networks consists in their capability of working with big data, in the precision of their results or easier use of the obtained neural network [26]. Artificial neural networks represent intelligent technologies for data analysis that differ from other traditional technologies [13]. According to Santin [25], neural networks have many advantages compared to common technologies. They are capable of fast and high-precision analysis of complex patterns and are very flexible in their use. Horák [12] sees the disadvantage of neural networks mainly in the fact that they require high-reaquality data and architecture definition.

Artificial intelligence reappears in the mainstream of business technologies especially in relation to trading systems that provide competitive advantages in all industry sectors, including electronics, manufacturing, marketing, engineering, and communication. Today´s artificial intelligence technology, which has been designed rather to enable better use of human abilities than to replace people, provides an extraordinary range of applications that create new connections between people, computers, knowledge, and the physical world [29]. Some of the applications include distribution and retrieval of information, database mining, product design, manufacturing, inspection, training, user support, resource planning, and complex resource management. Artificial intelligence technologies help enterprises reduce latency in business decisions, minimize frauds, and increase revenue [3].

Most of the existing artificial intelligence studies in financial literature deal only with predicting market movements. Vella and Ng [28] focus on the methods of applying artificial intelligence to risk management decisions. The authors propose to use an innovative fuzzy logic model. By applying a hybrid method combined with a popular model of neural network trends prediction, the authors′ results show a significant improvement of performance compared to the standard neural network approaches.

In the corporate world, communication is a dynamic process including all types of information exchange. It does not refer only to communicating ideas, but the question "Can communication be better? "The main objective of Arputhamalar and Kannan´s contribution [2] is to introduce a managerial decision-making model based on artificial neural networks. Managerial decision-making models based on neural networks behave like a "black box". They usually work in two phases. The first phase is typically focused on learning. An artificial neural network processes data and on the basis of topology, algorithms, and functions, it acquires a context. There are many alternatives to learning methods for various applications. In the second phase, the artificial neural network is seen as an expert that produces an output based on the learning and knowledge acquired in the first phase. In this phase, the importance of its quality is growing. Managers need highly accurate information presented in a way that does not require spending more time than necessary for decoding the information. Therefore, dashboard is a great choice.

Artificial intelligence in general and artificial neural networks provides a huge amount of knowledge for improving managerial decision making. Moreover, artificial neural networks and artificial intelligence serve as a knowledge repository and distribution scheme for the organizations that facilitate managerial responsibilities. In his article, Walczak [31] explores how different applications of artificial neural networks and other artificial intelligence applications can be tailored to facilitate management, improve manager´s performance and in some cases perform management activities.

Integrated Management Systems (IMS) can be seen as one unified management system dealing with various requirements of other management systems. Their ability to show and describe complex data, context and knowledge in real time is invaluable and becomes an integral part of modern and efficient management. Artificial neural networks are one of the modern trends in evaluating financial and non-financial health of enterprises. Artificial neural networks are suitable for modeling and exploring complex, often irreversible strategic managerial decisions. The results of data processing can be presented in the form of a dashboard [18].

The objective of the paper written by Šustrová [27] is to verify the possibility of using artificial neural networks in the business management processes, especially in supply chain management. The author designed several neural networks models with diverse architecture for optimizing the level of an enterprise inventory. The survey results show that neural networks can be used for order cycle management and lead to a reduction of the amount of the goods purchased and storage costs. Optimal neural networks show results applicable for predicting the number of items to be ordered, as well as for reducing inventory purchases and costs.

Supply Chain Risk Management (SCRM) includes a wide range of strategies aimed at identifying, evaluating, reducing and monitoring unexpected events or conditions that could affect any part of the supply chain. SCRM strategy often depends on quick and adaptive decision making based on relatively large, multidimensional data sources. These characteristics make SCRM suitable area for the application of artificial intelligence technologies. The objective of the contribution written by Baryannis et al. [4] is to provide a comprehensive review of supply chain literature dealing with SCRM-related approaches falling within artificial intelligence. For this purpose, various definitions and classifications of supply chain risks and related terms, such as uncertainty, are examined. Subsequently, the authors conducted mapping studies in order to categorize the existing literature according to the artificial intelligence methodology used, from mathematical programming to machine learning, big data, and the specific role of SCRM.

In the current fast-changing world economy, business managers strive to reduce costs and focus on key competencies that have forced many of them to outsource some or all their productions. Choy et al. [6] found that integration of Customer Relationship Management (CRM) and Supplier Relationship Management (SRM) in order to facilitate the supply chain management in supplier selection through artificial intelligence has become a promising solution for manufacturers to identify suitable suppliers and business partners and at the same time to create a network on which all products, services, and distribution will depend.

Artificial intelligence is a crosscutting and marginal discipline involving computer science, cybernetics, informatics, psychology, philosophy, and decision-making. Gao et al. [10] attempt to use a combination of artificial intelligence and system engineering to design artificial intelligence planning and decision-making problem of innovative products in experiments. As an indispensable part of management and decision-making system, artificial intelligence deciding is controlled by computer planning - combining empiric coefficient with setting and planning the direction of the research, intelligent problem detection and intelligent decision-making.

The main objective of the research conducted by Khalyasmaa et al. [16] is to develop a methodology for the statistical survey of effective management of Russian power plants production facilities, which is based on artificial intelligence methods to create a system of analyzing information on analyzing new generation technological risks. In the experimental part, the authors used mathematical modeling. The main research method is a method of adaptive neural-fuzzy inference and based on the training of artificial neural networks, characteristic sub-functions are created. As a modeling tool, Matlab software was used. The result of the research is the mathematical model of supporting decision making and a set of evaluation criteria for the influence of various administrative and technological risks and methods of their predicting. The model can be used as an independent tool of the automated system of integral risk assessment or can be used as a module for modern enterprise systems of managing power companies´ production assets.

Since the late 1970s, artificial intelligence has shown a great contribution to improving human decision-making processes and subsequent productivity in various business activities due to its ability to recognize business models, learn business phenomena, search for information, and intelligently analyze data. Despite widespread acceptance of the model as a decision support tool, artificial intelligence´s application was only limited in supply chain management. To fully exploit the potential benefits of artificial intelligence for supply chain management, Min [22] explores various sub-areas of artificial intelligence that are the most suitable for solving practical problems related to supply chain management.

A well-timed and adequate response to market changes is a key issue for each enterprise, whose aim is to be competitive and to reduce risk. To remain on the market, an enterprise must be able to analyze the situation on the market and make the right decisions [11]. The main objective of the contribution written by Nenortaite and Butleris [23] is to present a decision model that could bridge the gap between business rules management and artificial intelligence approaches. The proposed model is based on the application of artificial neural networks algorithms and SWO. In the proposed decision model, artificial neural networks are applied in order to analyze data and calculate decisions. Subsequently, the SWO algorithm is applied. The main idea of this algorithm application is to choose the best artificial neural networks for decision making.

Business Process Management (BPM) is a key element of today´s organizations. Despite the fact that over the years the main focus on supporting the processes in highly controlled domains, nowadays, many areas of BPM community interests are characterized by constantly changing the unpredictable environment and growing amount of data influencing the execution of process instances. Under such dynamic conditions, BPM systems must increase the level of automation to ensure reactivity and flexibility necessary for processes management. On the other hand, the artificial intelligence community focused on exploring dynamic domains that include active control of computing entities and physical devices (e.g. robots, software agents, etc.). In this context, automated planning, which is one of the oldest areas in artificial intelligence, is conceived as a model approach to automation of autonomous behavior of the model [20].

The development of tourism is currently an important part of the national economy and its growth and development. To ensure growth, managers are looking for new efficient tools to optimize decision making. The article written by Gallo et al. [9] deals with the issue of dashboards based on neural networks and their use in management decision-making processes. The research result is a proposal of balanced order and prediction model using financial and non-financial indicators using artificial intelligence, which enables to achieve a high level of efficiency and accuracy when evaluating financial and non-financial health of enterprises in the hospitality industry. The proposed model also brings a new managerial and scientific insight into the in-depth performance analysis of these devices.

Artificial intelligence extends the limits of current performance in data processing and analysis. Since it is a significant improvement in the management of public data, the conceptual study by Wirtz and Müller [32] deals with using artificial intelligence in public management structures in relation to their risks and side effects.

Adaptive logistic control can be achieved through artificial intelligence and optimization processes. Mičieta et al. [21] deal with the management of adaptive logistic with the aim to enable direct communication between the systems of logistics management and logistics facilities, machines, robots, and mobile robotic systems using artificial intelligence and adaptive behavior in logistics processes.

4 Conclusion

The criteria described in the chapter dealing with the data and methods were met by 31 publications that had been analyzed in detail, read, and processed in the aforementioned part of the work. The publications processed were especially the most valuable articles from impacted journals and articles from conference proceedings. None of the publications was published before 2000.

Based on the aforementioned analyses of publications it can be stated that the use of artificial intelligence in business management is a relatively demanding, yet very necessary tool applicable in every enterprise. There are countless methods for managing a business, each of which is outstanding in a certain way. The method of artificial neural networks is one of the best ones, ensuring a quality business management.

However, currently, it is not sufficiently used in business practice, although there are a large number of scientific texts dealing with the issue, some of them being more valuable. The potential of this topic is thus high. The research question, whether it is possible to use practically artificial intelligence in business management, is verified by extensive analysis of available scientific literature. The authors conducted a systematic literature review of the most renowned scientific resources, drew conclusions from the findings, and answered the research question. The objective of the contribution was achieved.

References

1. Antonescu, M.: Are business leaders prepared to handle the upcoming revolution in business artificial intelligence? Qual. Access Success **19**(53), 15–19 (2018)
2. Arputhamalar, A., Kannan, S.P.: Written correspondence – the foremost channel of information transfer in organisations. Qual. Access Success **17**(151), 111–114 (2016)
3. Bai, S.A.: Artificial intelligence technologies in business and engineering. In: IET Conference Publications, International Conference on Sustainable Energy and Intelligent Systems, pp. 856–859 (2011)
4. Baryannis, G., Validi, S., Dani, S., Antoniou, G.: Supply chain risk management and artificial intelligence: state of the art and future research directions. Int. J. Prod. Res. **57**(7), 2179–2202 (2018)
5. Beiranvand, V., Abu Bakar, A., Othman, Z.: A comparative survey of three AI techniques (NN, PSO, and GA) in financial domain. In: IEE Proceedings of the 7th International Conference on Computing and Convergence Technology, pp. 332–337 (2012)
6. Choy, K.L., Lee, W.B., Lo, V.: An intelligent supplier relationship management system for selecting and benchmarking suppliers. Int. J. Technol. Manag. **26**(7), 717–742 (2003)
7. Ferràs-Hernández, X.: The future of management in a world of electronic brains. J. Manag. Inq. **27**(2), 260–263 (2017)
8. Fink, A.: Conducting Research Literature Reviews, 3rd edn. Sage, Los Angeles (2010)
9. Gallo, P., Gallo, P.J., Timková, V., Šenková, A., Karahuta, M.: Use of dashboards in predicting the development of the company using neural networks in hotel management. Geojournal Tourism Geosites **22**(2), 307–316 (2018)
10. Gao, W., Qi, Q., Dong, L., Liu, C.: Application of artificial intelligence in innovation experiment management system engineering. In: Jing, W., Ning, X., Huiyu, Z. (eds.) Proceedings of the 8th International Conference on Management and Computer Science, pp. 171–175. Atlantis Press, Paris (2018)
11. Gressley, S., Horák, J., Kováčová, M., Valašková, K., Poliak, M.: Consumer attitudes and behaviors in the technology-driven sharing economy: motivations for participating in collaborative consumption. J. Self-Gov. Manag. Econ. **7**(1), 25–30 (2019)
12. Horák, J.: Using artificial intelligence to analyse businesses in agriculture industry. In: Horák, J. (ed.) SHS Web of Conferences: Innovative Economic Symposium 2018 – Milestones and Trends of World Economy, p. 01005. EDP Sciences, France (2019)
13. Horák, J., Krulický, T.: Comparison of exponential time series alignment and time series alignment using artificial neural networks by example of prediction of future development of stock prices of a specific company. In: Horák, J. (ed.) SHS Web of Conferences: Innovative Economic Symposium 2018 – Milestones and Trends of World Economy, p. 01006. EDP Sciences, France (2019)
14. Jia, Q., Guo, Y., Li, R., Li, Y., Chen, Y.: A conceptual artificial intelligence application framework in human resource management. In: Chang, F.K., Li, E.Y., Li, E.Y. (eds.) Proceedings of the International Conference on Electronic Business, pp. 106–114. International Consortium for Electronic Business (2018)
15. Jones, M.V., Coviello, N., Tang, Y.K.: International entrepreneurship research (1989–2009): a domain ontology and thematic analysis. J. Bus. Ventur. **26**(6), 632–659 (2011)
16. Khalyasmaa, A.I., Dmitriev, S.A., Valiev, R.T.: Grid company risk management system based on adaptive neuro-fuzzy inference. In: IEEE Proceedings of 2017 XX International Conference on Soft Computing and Measurements, pp. 892–895. IEEE, New York (2017)

17. Kliestik, T.: Models of autoregression conditional heteroskedasticity garch and arch as a tool for modeling the volatility of financial time series. Ekonomicko-manažerské spectrum **7**(1), 2–10 (2013)
18. Kopia, J., Kompalla, A., Ceausu, I.: Theory and practice of integrating management systems with high level structure. Qual.-Access Success **17**(155), 52-29 (2016)
19. Lawrynowicz, A.: Production planning and control with outsourcing using artificial intelligence. Int. J. Serv. Oper. Manag. **3**(2), 193–209 (2018)
20. Marrella, A.: What automated planning can do for business process management. In: Teniente, E., Weidlich, M. (eds.) Business Process Management Workshops, pp. 7–19. Springer, Berlin (2018)
21. Mičieta, B., Staszewska, J., Biňasová, V., Herčko, J.: Adaptive logistics management and optimization through artificial intelligence. Commun. - Sci. Lett. Univ. Žilina **19**(2A), 10–14 (2017)
22. Min, H.: Artificial intelligence in supply chain management: theory and applications. Int. J. Logistics Res. Appl. **13**(1), 13–39 (2010)
23. Nenortaite, J., Butleris, R.: Business rules management improvement through the application of particle swarm optimization algorithm and artificial neural networks. In: Targamadze, A., Butleris, R., Rutkiene, R. (eds.) Information Technologies' 2008, Proceedings, pp. 84–90. Kaunas University of Technology Press, Kaunas (2008)
24. Paschek, D., Luminosu, C.T., Draghici, A.: Automated business process management – in times of digital transformation using machine learning or artificial intelligence. In: Bondrea, I., Inta, M., Simion, C. (eds.) MATEC Web of Conferences, vol. 121, p. 04007. EDP Science, France (2017)
25. Santin, D.: On the approximation of production functions: a comparison of artificial neural networks frontiers and efficiency techniques. Appl. Econ. Lett. **15**(8), 597–600 (2008)
26. Šuleř, P.: Using Kohonen´s neural networks to identify the bankruptcy of enterprises: Case study based on construction companies in South Bohemian region. In: Dvouletý, O., Lukeš, M., Mísař, J. (eds.) Proceedings of the 5th International Conference Innovation Management, Entrepreneurship and Sustainability, pp. 985–995. Oeconomica Publishing House, Prague (2017)
27. Šustrová, T.: An artificial neural network model for a wholesale company's order-cycle management. Int. J. Eng. Bus. Manag. **8**, 1–6 (2016)
28. Vella, V., Ng, W.L.: A dynamic fuzzy money management approach for controlling the intraday risk-adjusted performance of AI trading algorithms. Intell. Syst. Acc. Financ. Manag. **22**(2), 153–178 (2015)
29. Vochozka, M., Horák, J.: Comparison of neural networks and regression time series when estimating the copper price development. In: Ashmarina, S., Vochozka, M. (eds.) Contributions to Economics, pp. 169–181. Springer, Heidelberg (2019)
30. Vochozka, M., Machová, V.: Determination of value drivers for transport companies in the Czech Republic. Nase More **65**(4), 197–201 (2018)
31. Walczak, S.: Artificial neural networks and other AI applications for business management decision support. In: I. Management Association (ed.) Intelligent Systems: Concepts, Methodologies, Tools, and Applications, pp. 2047–2071. IGI Global, Hershey (2018). https://doi.org/10.4018/978-1-5225-5643-5.ch091
32. Wirtz, B.W., Müller, W.M.: An integrated artificial intelligence framework for public management. Public Manag. Rev. **21**(7), 1076–1100 (2018)
33. Zhang, X., Chen, Y.: An artificial intelligence application in portfolio management. In: Zheng, X. (ed.) Proceedings of the International Conference on Transformations and Innovations in Management, pp. 37, 86–104. Atlantis Press, Paris (2017)

Digital Infrastructure of the Economy: Tools, Platforms and Mechanisms

Model of Sustainable Development of Regional Ecological and Economic Systems

N. V. Lazareva[1], G. S. Rosenberg[2], and O. A. Sapova[1](✉)

[1] Samara State University of Economics, Samara, Russia
loli_air@mail.ru
[2] Institute of Ecology of the Volga Basin, Russian Academy of Sciences, Tolyatti, Russia

Abstract. The main feature of mathematics, which is very essential for a scientific ideology, is its ability to transform the solution of sophisticated problems into standardized logic circuits. The article discusses the problems of creating geo-and eco-information models of large territories with the involvement of the expert system REGION. Geoinformatics operates with huge amounts of data, which are divided into extensive and intensive indicators. In this sense, geoinformatics is closely connected with ecoinformatics and digitalization (especially when predicting the structure and dynamics of complex socio-ecological-economic systems [SEES]). Some algorithms for the synthesis of complex (intensive) indicators are discussed. An example is given of using the REGION expert system for the comprehensive assessment of SEES of the Volga basin.

Keywords: Geo-information and eco-information systems · Database · Socio-ecological-economic system

1 Introduction

The ideological basis of technological civilization is scientism, which is based on the belief that there is a small number of precisely formulated nature laws, on the basis of which everything in nature is predictable and manageable. Nature is considered to be as a giant machine that can be controlled if the principle of its functioning is known. This scientific ideology often plays the role of the "religion of technological civilization".

2 Theoretical Justification

The basic dogma of scientism is confidence in mathematization. It (the dogma) states that everything (or, at least, everything essential) in nature can be measured, turned into numbers or other mathematical objects, and that by performing various mathematical manipulations on them, all natural and social phenomena can be predicted and subjugated. This confidence is already included in the call of Galileo Galileo (Galileo Galilei): "Measure everything that is measurable, and make measurable what is immeasurable." E. Kant (Immanuel Kant) said that each area of consciousness is a science as much as it contains mathematics. A. Poincare (Jules Henri Poincaré) wrote

S. I. Ashmarina et al. (Eds.): ISCDTE 2019, LNNS 84, pp. 433–440, 2020.
https://doi.org/10.1007/978-3-030-27015-5_52

that the final, ideal phase of scientific concept development is its mathematization. In a certain sense, we can say that we live in a mathematical civilization and, perhaps, die with it [15].

The main feature of mathematics, which is very essential for scientific ideology, is its ability to transform the solution of sophisticated problems into standardized logic circuits. The issue of world cognition can be found in N. Bourbaki (Nicolas Bourbaki; assumed name of the group of French mathematicians), this is the possibility of a compact record of observable phenomena, as a compact record is exactly what gives us the opportunity to predict and manage. It is curious that a compact record of observable phenomena in science is considered as a theory even when no theorizing is associated with it. The theory is, in fact, a logical construction that allows us to describe the phenomenon significantly shorter than it is possible to do with direct observation [9]. The example is the periodic system of D.I. Mendeleev, that was a compact record of a vast variety of phenomena in inorganic chemistry, it immediately began to be regarded as some very significant contribution to the theory of chemistry, although when Mendeleev's table appeared, no theorization was associated with it at all. Thus, the increase in the mass of non-generalized facts leads to the fact that they gradually turn into ballast.

One of the ways of processing large data arrays that arise in the simulation of socio-ecological-economic systems (SEES) of large territories is the creation of geo-or eco-information systems. The tendency to combine the concepts of applied cartography and geoinformatics resulted in the emergence of the thesis of "mathematical certainty" of geographical (digitized) maps [1, 3, 6]. Since there is no clear definition of geoinformatics, the combination of computer and telecommunication technologies (models) of data processing will be understood to solve the problems of analyzing the SEES.

On this path (the convergence of the concepts of cartography and geoinformatics) there are many pitfalls. First of all, there are maps and geo (eco) information models of very different style. For example, a cartographer, using the "language of the map", seeks primarily to visualize information to make it readable, without thinking about some of its "mathematical certainty." The main product of geo (eco) information technology is the generation of new information by algorithmically targeted "chewing" and "digesting" an existing data array.

3 Discussion

More than 50 years ago, the development of geographic information systems (GIS) began. Going through the stages of creating simplified maps and coarse imitations of paper atlases, modern software and hardware systems consistently summarized the experience and aesthetics of traditional mapping and learned how to make map products of the highest quality. Electronic maps which were obtained using such GIS-industry products, such as Arcview, MapInfo, etc., have become more accurate than ordinary hand geometrically, more varied in color, bar, halftone, and bright design. Simultaneously with the assimilation of traditional achievements, geographic information mapping gradually reached a new level. Today, geo-informatics cartographers are increasingly thinking about creating panoramic works of art that are completely different from traditional maps and atlases. For example, three-dimensional digital modeling allows you to build three-dimensional images, and animations give the cards a dynamic aspect they need.

Geoinformatics impresses by incredible datasets, it operates with them, therefore the results are unique and reproducible. However, the generation of information inherent in GIS technologies is meaningfully interesting only when someone from outside, a representative of a different field of knowledge, or other science has invested a specific task in the "mouth of geoinformatics". In this sense, geoinformatics is closely intertwined with ecoinformatics and digitalization (especially when predicting the structure and dynamics of complex SEES).

When comparing any data in scientific areas characterizing a phenomenon or process in time and space, indices which mean relative statistical values were widely applied. These indices show how much the level of a phenomenon studied under given conditions differs from the level of the same phenomenon in other conditions. They personify the attempt to calculate easily and match complex objects or systems consisting of disparate elements. Calculated indicators obtained on the basis of the index method can be used in more complex mathematical models to characterize the development of the analyzed processes over time or across territory, to identify the structure, interrelationships, and the role of individual factors in the dynamics of complex systems.

We consider the methods of calculating so-called general indices, which are a vector of values of the resulting complex indicator, obtained as a result of information convolution (reduction) of a certain subset of individual indicators. At present time the common scheme for such data generalization in ecology and economics are methods based on the additivity hypothesis of individual contributions. The complex indicator obtained in this way is a vector of the same dimension as the basic one, each ***i-th*** component of which is calculated using one of the following formulas ("Summation" algorithm):

$$\text{Simple sum } X_i = \sum_{j=1}^{P} B_{ij}; \tag{1}$$

$$\text{Weighted sum } X_i = \sum_{j=1}^{p} K_j \cdot B_{ij}; \tag{2}$$

$$\text{simple means } X_i = \left(\sum_{j=1}^{p} B_{ij} \right) / p; \tag{3}$$

$$\text{weighted means } X_i = \left(\sum_{j=1}^{p} K_j \cdot B_{ij} \right) / \sum_{j=1}^{p} K_j, \tag{4}$$

where ***Bij*** are the components of the ***j-th*** vector generating the subsets of ***p*** initial indices, expressed in the normalized scale; ***Kj*** are weights reflecting the relative importance of the ***j-th*** indicator in the construction of the generalized indicator. The multiplier ***Kj*** is an arbitrary positive or negative number given by expert estimation methods. The composition of the generating subset can include both the original and previously synthesized generalized indicators. The formulas are mutually reducible: for example, if we take ***Kj*** = 1, then the complex indicator calculated by the formula "weighted sum" will be equal to the simple sum of the points of original indicators.

In some cases, a multiplicative model is used to obtain a complex indicator, for example, which is easily reduced to an additive model by logarithm of the initial variables.

However, the relevant question is: to what extent is the additivity hypothesis applicable to socio-ecological-economic indicators? By the nature of mapping the subject area of individual indicators can be attributed to two main types: extensive, or volumetric, and intensive, or relative.

Extensive indicators usually have a sense of stock or flow. The values of the stock type are recorded at a specific point in time and have elementary measurement units: instance, ton, joule, meter, dollar, etc. Examples include the accumulation of humus in soil, the unit cost of production, population size, and overall morbidity. The flow type values are determined only for a specific period of time and have the dimension "volume per unit time": production per day or during the vegetative period, the amount of incoming energy per hour, the amount of biological resources taken out of the ecosystem (for example, fish catch) or the total incidence etc.

The values of the stock and flow are rigidly interconnected:

$$\boldsymbol{S}_b[v] + \boldsymbol{P}_i\,[v/t]\boldsymbol{t} = \boldsymbol{S}_e[v] + \boldsymbol{P}_o\,[v/t]t,$$

where ***Sb*** and ***Se*** are stocks at the beginning and end of the period (***v*** is the unit of measurement), ***Pi*** and ***P0*** are flows to increase and decrease the stock (t - period). In particular, this ratio underlies the formation of tables of material and energy balance.

In our opinion, there is no reason to reject the additivity hypothesis of contributions for extensive indicators. Indeed, using of the simple sum of biomass of individual communities gives the total biomass of living organisms in the ecosystem, weighted by MPC, the amount of pollutant emissions into the atmosphere adequately assesses the overall level of its pollution, the organization's costs for some period quite adequately indicate the decrease in economic benefits as a result of disposals of assets, etc.

Intensive indicators are relations of extensive or intensive values. These indices may have different content, dimensions or be dimensionless, which is determined by the formula for their calculation. In the majority of cases, in order to obtain relative indicators, "division one into another" is applied: such intensive values do not have dimensions (that is, expressed in shares, percentages, parts per million, etc.). These include growth rates, spatial comparison coefficients, indicators of coenotic and territorial structure, gender inequality index (used in UN reports on human development), etc.

The main drawback of intensive indicators is their often non-additive nature (the impossibility of determining weighted average values; for example, for the condition characterized as an "ecological catastrophe", it is quite enough that only one of the analyzed components exceeds the lethal dangerous level of pollution, while all average parameters are within normal limits.

There can be initial data on the basis of which decisions are made, and the method of processing (mapping) of the initial data into a solution is called the model [14]. Thus, in the most general form, a model is a function that transfers the source data into a solution, and the specific method of transferring does not matter. In most cases, researchers and practitioners, as a rule, have little interest in the model formalism that was used in solution development. At the same time, it is obvious that the proposed solutions are represented under the conditions of incomplete information and assumptions of modeling methods, therefore some conclusions regarding the adequacy [7] and stability of obtained models to these allowable uncertainties are more important. The general scheme for assessing the sensitivity and stability of statistical procedures is described in detail in the monograph [14].

We illustrate these theoretical concepts with a specific example.

The expert system REGION has been developed at the Institute of Ecology of the Volga Basin of the Russian Academy of Sciences, which makes it possible to give a comprehensive assessment of the state of SEES of any territory [2, 5] with the availability of relevant databases [8].

In the database on the Volga basin [8], we select 11 medical-statistical indicators (total morbidity, carcinogenic neoplasms, diseases of the circulatory system, respiratory organs, digestion per 1000 people in 2017, etc.) and calculate a complex incidence rate indicator with three different algorithms, this indicator will present the summed up data by "one number".

- According to the first "Summation" algorithm, we can perform a simple summation of the scores of a standard normalized scale using the formula (1).
- In accordance with the second "Convolution" algorithm, we perform the reduction of 11 baselines to two main components, which in this particular case illustrate more than 64% of available statistical variation; we will calculate the complex indicators using the formula (5). The value of a complex indicator can be determined, for example, as the weighted distance from the shifted origin of the coordinates of two main components of factor analysis to each analyzed point:

$$x_{Pi} = \sqrt{[\lambda_1(f_{i1} - f_1^{\min})]^2 + [\lambda_2(f_{i2} - f_2^{\min})]^2}, \tag{5}$$

where ***fi1*** and ***fi2*** are the coordinates of ***i-th*** analyzed region in the space of two main components, $f_1^{\min}$ and $f_2^{\min}$ are the minimum values of corresponding factor scores; λ1 and λ2 are the values of matrix eigenvalues when determining the principal components.

- According to the third "Estimation" algorithm, the generalization of individual indicators is performed using the Pythagorean theorem (the position of each multidimensional point inside the "minimax cloud" is determined by its projection on the axis "minimum value - maximum value").

To compare the obtained results, we convert the calculated complex indices into a standard 6-point scale ([max - min]/6) and define the ranks for each territorial unit -

ordinal numbers in sorted lists ordered by increase of the resulting indicator for each version used (see Table 1).

Table 1. The values of complex indicators calculated on the basis of generalization of 11 medical and statistical features of three algorithms used (score – the value of the indicator in the standard normalized scale; SSD – the sum of deviations squares from the average score)

Region	Algorithm						SSD
	«Summation»		«Convolution»		«Assessment»		
	Score	Grade	Score	Grade	Score	Grade	
Bashkortostan	1	1	1	1	1	1	0
Kostroma Obl.	1	2	1	4	1	3	0
Tatarstan	1	3	1	2	2	8	0,67
Saratov Obl.	1	4	1	3	2	6	0,67
Astrakhan Obl.	2	5	2	6	2	5	0
Tula Obl.	2	6	3	10	3	9	0,67
Mordovia	2	7	2	5	1	4	0,67
Ryazan Obl.	2	8	3	12	1	2	2
Nizh. Novgorod Obl.	3	9	3	9	3	11	0
Moscow Obl.	3	10	4	14	3	12	0,67
Kirov Obl.	3	11	3	11	2	7	0,67
Ivanovo Obl.	**3**	**12**	**5**	**17**	**3**	**10**	**2,67**
Volgograd Obl.	4	13	4	13	5	20	0,67
Tver Obl.	4	14	4	15	4	14	0
Mari El	**4**	**15**	**2**	**7**	**5**	**17**	**4,67**
Kaluga Obl.	4	16	4	16	4	13	0
Chuvashia	**5**	**17**	**2**	**8**	**4**	**15**	**4,67**
Ulyanovsk Obl.	5	18	5	19	5	18	0
Penza Obl.	5	19	5	18	4	16	0,67
Yaroslavl Obl.	5	20	5	20	6	23	0,67
Udmurtia	6	21	6	21	6	21	0
Samara Obl.	6	22	6	22	5	19	0,67
Vladimir Obl.	6	23	6	23	6	24	0
Perm Obl.	6	24	6	24	6	22	0

Source: compiled by the authors.

4 Conclusion

The presented results indicate quite obvious solutions stability that depends little on the algorithm type. Based on Spearman correlation coefficient, rank sequences of territorial units which were formed by different methods have a high level of similarity: from 0.8 between algorithms 2 and 3 to 0.91 between algorithms 1 and 3. The null hypothesis formulated as "there is no correlation between samples "Deviates with a high level of significance. In 88% of cases, the calculated complex indicators either completely coincide, or there is a partial shift to the next gradation (a more or less significant shift takes place for the three territories in bold type).

The figure shows the map of generalized (intensive) indicator distribution of the state of the SEES on the territory of the Volga basin, which was determined by 13 extensive (first 8 parameters) and intensive indicators (Fig. 1):

(1) forest cover
(2) reforestation
(3) the share of protected areas
(4) population density
(5) the formation of toxic waste
(6) assessment of the cost of nature conservation,
(7) distribution of terrestrial vertebrate species,
(8) a variety of reptiles
(9) assessment of air pollution,
(10) assessment of water use
(11) generalized agricultural load
(12) assessment of the population incidence,
(13) complex indicator of anthropogenic load.

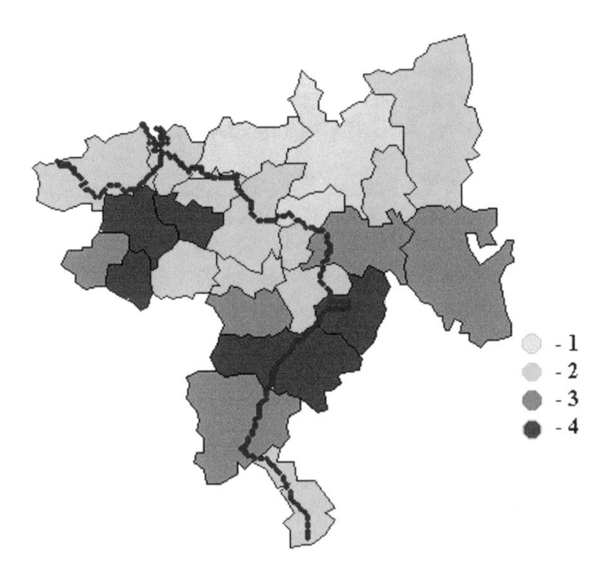

Fig. 1. The generalized indicator of ecological condition assessment of the territory of the Volga basin (1 – the best state; 4 – the worst state) (Source: compiled by the authors.)

The considered methodology for constructing synthetic computer maps (SEES models) and the EIS “REGION-VOLGABAS” developed the high quality of the comprehensive analysis of socio-ecological-economic systems of the territories of different scale - the Volga basin [5], Samara region [4, 11, 12], Ulyanovsk Region [13], Republic of Tatarstan [10], etc.

References

1. Belyaev, V.I., Ivakhnenko, A.G., Fleishman, B.S.: Imitation, self-organization and potential efficiency. Automation **6**, 9–17 (1979). (in Russian)
2. Brusilovsky, P.M., Rosenberg, G.S.: Testing the inadequacy of the simulation model of a dynamic system using MGUA algorithms. Avtomatika **6**, 43–48 (1981). (in Russian)
3. Chislenko, L.L.: On the need for statistical characteristics of taxons for modeling faunistic systems. J. Gen. Biol. **30**(4), 399–409 (1969). (in Russian)
4. Kostina, N.V.: Ecological information system of a large region as the basis of environmental monitoring. Regional environmental monitoring for the management of biological resources. IEVB RAS, Tolyatti (2003). (in Russian)
5. Kostina, N.V., Rosenberg, G.S., Shitikov, V.K.: The expert system of the ecological state of the basin of a large river. Izv. Samar. SC RAS **5**(2), 287–294 (2003). (in Russian)
6. Nalimov, V.V.: Theoretical biology? It is still not there. Knowl. Power **7**, 9–11 (1979). (in Russian)
7. Orlov, A.I.: Sustainability in Socio-Economic Models. Nauka, Moscow (1979). (in Russian)
8. Rosenberg, G.S.: Volga basin: On the way to sustainable development. IESB RAS, Cassandra, Tolyatti (2009). (in Russian)
9. Rosenberg, G.S., Krasnoshchekov, G.P.: Landscape in the interior (environmental problems of Tatarstan against the background of the Volga basin). Actual environmental problems of the Republic of Tatarstan, pp. 266–271. ANT, Kazan (1997). (in Russian)
10. Rosenberg, G.S., Krasnoshchekov, G.P., Krylov, Y.M.: Information for consideration (some data to the analysis of environmental safety and sustainable development of the Ulyanovsk region on the expert system REGION-VOLGABAS). IESB RAS, Tolyatti, Ulyanovsk (1997). (in Russian)
11. Rosenberg, G.S., Shitikov, V.K., Kostina, N.V., Kuznetsova, R.S., Lifirenko, N.G., Kostina, M.A., Kudinova, G.E., Rosenberg, A.G.: Expert information database of the state of socio-ecological-economic systems of different scales “REGION” (EIBD “REGION”). Certificate of State database registration No. 2015620402 dated 27 February 2015 (2015). (in Russian)
12. Rozenberg, G.S.: The environmental situation in the Samara region. IESB RAS, Tolyatti (1994)
13. Rozenberg, G.S., Lazareva, N.V., Simonov, Y.V., Lifirenko, N.G., Sarapultseva, L.A.: Ecology on the level of highly urbanized region. Int. J. Environ. Sci. Educ. **11**(15), 7668–7683 (2016)
14. Shitikov, V.K., Rosenberg, G.S., Kostina, N.V.: Methods of synthetic territory mapping (on the example of the VOLGABAS environmental information system). Quantitative methods of ecology and hydrobiology (collection of scientific papers dedicated to the memory of A.I. Bakanov), pp. 167–227. SamSC RAS, Tolyatti (2005). (in Russian)
15. Winberg, G.G.: Experience of using different biological information systems of water pollution in the USSR. Influence of pollutants on aquatic organisms and ecosystems of water bodies. Science, Leningrad (1979). (in Russian)

Digitalization of Education as a Basis for the Competence Approach

E. G. Repina(✉), O. V. Bakanach, and N. V. Proskurina

Samara State University of Economics, Samara, Russia
violet261181@mail.ru

Abstract. The authors analyze the electronic information and educational environment of the university as an innovative component of the competence approach under the digital transformation of the economy and defines its relationship with the formation of information and communication competence of students. The innovation of the proposed approaches is the transformation of the traditional system of education and its integration with electronic educational resources. The study presents the features of the electronic information and educational environment of the university, analyzes the theoretical foundations of the educational process in economic and mathematical courses using innovative technologies, and develops guidelines for the introduction of innovative technologies to train students in the modern information and educational environment.

Keywords: Electronic information and educational environment · Digitalization · Higher education · Innovations · Learning technologies

1 Introduction

Education and research competencies are among the basic directions of state policy for creating the necessary conditions for economy digitalization in Russia.

The key objectives of these areas are:

- Improving the education system which should provide the digital economy with competent personnel;
- Developing competencies in the digital economy.

The digital economy requires the education system not just to "digitize" individual processes. It needs an integrated approach that would set new objectives related to the introduction of innovative models that change the structure and content of the educational process in order to shape the key and professional competencies of information society.

The research objective is to develop and implement technology for students studying economic and mathematical courses in the information and educational environment, which creates conditions for ensuring a competence-based approach, interconnection of academic knowledge and practical skills.

S. I. Ashmarina et al. (Eds.): ISCDTE 2019, LNNS 84, pp. 441–447, 2020.
https://doi.org/10.1007/978-3-030-27015-5_53

To achieve the objective, the following tasks were solved:

- The features of the information and educational environment of the higher educational institution were studied (experimental site - Samara State University of Economics);
- The theoretical foundations of the educational process of the higher educational institution in economic and mathematical courses using innovative technologies in the modern information and educational environment were analyzed;
- The methodological recommendations to introduce innovative technologies for students studying the following courses: "Theory of Probability and Mathematical Statistics", "Econometrics", "Statistics" in the modern information and educational environment of the university were developed.

Having achieved the objectives, we will get methodological tools for students studying economic and mathematical courses in the information and educational environment, which will ensure the formation of subject and meta-disciplinary results at a high quality level fixed by federal state educational standards of higher education in the stated courses.

The innovation of the proposed approaches is the transformation of the traditional system of education and its integration with electronic educational resources.

2 Methodology

The methodological basis of the research is general scientific and special methods and tools of scientific knowledge, such as:

- Theoretical - the study of economic, mathematical, scientific, methodological, psychological and educational literature;
- Empirical - the study of the pedagogical experience of colleagues (observation, conversation, testing, the study of the products of students, the study of pedagogical documentation);
- Experimental - carrying out a transformative experiment when introducing innovative technologies in the information and educational environment of the university in such courses as "Theory of Probability and Mathematical Statistics", "Econometrics", "Statistics";
- Mathematical - registration, scaling and ranking of experiment outcomes.

3 Results

As a result of the experiment, we found out that within the development of information society, the basis of which is the exponential growth of information and convergence of information and educational resources, it becomes necessary to develop the electronic information and educational environment of the university. This is primarily due to the need to provide students with quick access to information and reference systems, databases, software, learning materials, scientific literature, training systems, knowledge control systems.

The combination of “MS Excel” and “Gretl” was recognized as the most optimal of applied computer data analysis packages, which allows forming the necessary competences in the field of economic and mathematical courses which the students of the University of Economics have to study at a high professional level.

Digitalization of the education sector entails the development of integration processes in the educational community, the rapid growth in the number of multidisciplinary research using information and communication technologies.

The process of teaching econometrics, for example, at the undergraduate level, is resource-provided with statistical information obtained as a result of studying Enterprise Economics. The integration of such educational components as a computer simulation model “Business Course Maximum” and econometrics workshop was held in order to provide students with practical skills in econometric modeling of firms’ performance indicators operating in a competitive environment.

The experimental work, carried out by the authors of this study when developing the educational environment of a modern university, using the example of economic and mathematical courses, is based on the trends in educational technology. At the present time the main trend in the development of higher education is the introduction of technological innovations aimed at maximizing the compliance of graduates with the requirements established by the federal state educational standard of higher education and the professional educational standard.

The efficiency of electronic educational resources has been proven experimentally in the formation of competencies, enshrined in state standards.

The authors selected 6 equal groups of bachelor students enrolled in 1 and 2 university courses of various training programs in 38.03.01 “Economics” (serial sample). The sample size of students was 150 people. Participation in the experiment of different age groups allowed leveling the various psycho-emotional development of the students’ personality, as well as different degrees of adaptation to the educational process at the university. For the representativeness of experiment outcomes, the authors tested students studying three courses (“Statistics” - 2 groups, “Probability Theory and Mathematical Statistics” - 2 groups, “Econometrics” - 2 groups). The basic student group in each course was trained in traditional pedagogical technology (classroom and extracurricular work omitting electronic educational courses), experimental - using electronic educational resources in the University Electronic Education System on the platform of Moodle.

The tested hypothesis proved the superior use of electronic educational resources in combination with the traditional learning methodology over the learning methodology only in the traditional form.

The outcomes of formed competencies fixed by state standards for each course were determined on the basis of a point-rating assessment system.

The point-rating assessment system has a clear structural framework in the form of course modules that are logically completed in terms of the subject and timeframes and have a functional load focused on learning outcomes.

The formation of competencies was assessed on a 100-point scale. The rating assessment was based on the outcomes of three forms of control: semester control; control of attendance, activity and independent (out-of-class) student work; final control. Online rating of students was carried out in the University Electronic

Education System. A student was successfully certified when he achieved at least 80 points on the basis of all forms of control. Learning outcomes of the experiment are given in Table 1.

Table 1. Learning outcomes of the experiment

Learning outcome	Learning technology		Total:
	Traditional	Traditional + electronic resources	
Certification (at least 80 points)	60	69	129
Unsatisfied (less than 80 points)	15	6	21
Total:	75	75	150

Source: compiled by the authors.

The relationship of dichotomous features was studied using applied statistical software Statistica (13.0). The hypothesis about the relationship of learning technology and the academic formation of competences in courses was tested using the Pearson criterion at 5% significance level $H_0 : p_{ij} = p_{i*} \cdot p_{*j} \, (i,j = \overline{1,2})$. The observation value was $\chi^2_{obs} = \frac{n_{**}(|n_{11}n_{22} - n_{12}n_{21}| - 0{,}5)}{n_{1*}n_{2*}n_{*1}n_{*2}} = 4.48$, the critical value was $\chi^2_{cr}(\alpha = 0.05; k = 1) = 3.8$. The occurrence of the observed criterion value in the critical region gives reason to believe that the tested hypothesis is valid. The tightness of the relationship between technology and learning was measured by Yule association: $Q = \left| \frac{n_{11}n_{22} - n_{21}n_{12}}{n_{11}n_{22} + n_{21}n_{12}} \right| = 0.71$. According to Cheddock scale, this relationship is classified as high. Learning outcomes of the experiment are presented in Fig. 1.

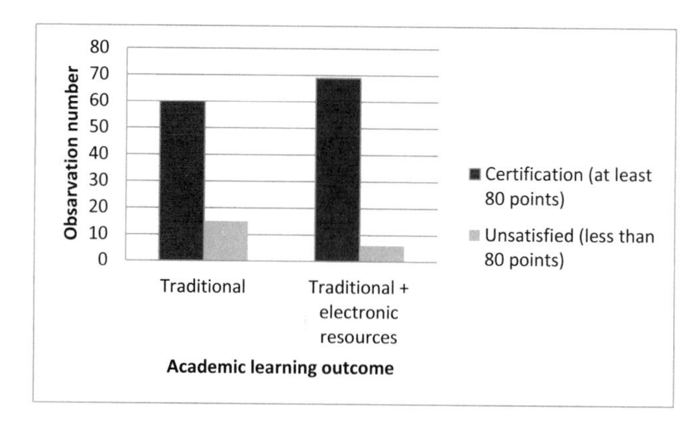

Fig. 1. Comparative analysis of the formation of competencies according to learning outcomes of the experiment (Source: compiled by the authors)

Thus, we have revealed that the use of the University Electronic Education System allows:

- Changing the paradigm of book-frontal education to student-centered education;
- Carrying out adequate and objective monitoring of learning outcomes of mastering the course through out its study;
- Implementing an interdisciplinary approach;
- Forming a qualitatively new level of information and communication competence.

Consequently, the effectiveness of the educational process with a high share of digital innovations has been experimentally proven in the educational process of the university.

4 Discussion

Modern society is becoming more "digitized" and associated with computers and algorithms that somehow mediate most of the daily activities of people [3–6]. Digital perception is an important approach to shaping the future of digital society [9]. As part of the implementation of the state program "Digital Economy of the Russian Federation" and "Strategies for the Development of Information Society in the Russian Federation for 2017–2030", the Government of the Russian Federation approved the project "Modern Digital Educational Environment" [14], aimed, in particular, at creating the electronic educational environment, the availability of online learning, self-education and the possibility of building individual educational learning routes [11].

The use of ICT in the education system changes the methods and forms of education, affects the technology of studying courses, thereby transforming the traditional educational environment into the qualitatively new one–electronic information and educational environment (EIEE). Is EIEE an innovation in the educational process, the outcome of which is a qualitatively new level of competences? Does the digitalization of the educational process correlate with the level of development of information and communication competence of future labor market participants?

Education under the functional approach was considered from the perspective of the system in [2, 12, 13]. The education system needs to develop the innovative educational environment and it is associated with the need to change the educational space, taking into account the diversity of all forms of educational strategies and technologies. A key factor in the implementation of innovative education is the competence approach [1, 7], developed in accordance with the main global educational trends, such as e-learning technologies, mobile education, including mobile learning platforms that ensure the development of communication, cooperation, creativity, critical thinking and needs for self-education to obtain multivariate solutions to professional problems [8, 10]. The main component of the innovative educational environment is information and communication technologies (ICT), integrated into the electronic information and educational environment of the educational institution.

Thus, the electronic information and educational environment contributes to the formation and implementation of new technological initiatives, becoming the basis for the growth of students' education mobility [15] and provides qualitatively new parameters of education, taking into account the strategic goals of economy digitalization.

5 Conclusion

A prerequisite for the breakthrough development of economy digitalization in Russia is the provision of competent labor resources in the field of information and communication technologies. To take up the challenge and respond to complexities of mastering relevant competencies in the digital age, innovative solutions are required. One of these solutions is based on the development and implementation of technology for training students in the information and educational environment of the university, which creates conditions for intensification of educational activities and qualitatively new educational results determined by the needs of information society.

References

1. Baksheeva, Z.K., Strogova, N.E.: Features of modern education systems of the different countries on the basis of competence-based approach. Humanitarian Res. **3**(20), 132–135 (2018). Russia
2. Barabanova, S.V., Kaybiyaynen, A.A., Kraysman, N.V.: Digitalization of engineering education in the global context (review of international conferences). Pedagogical Educ. Russ. **28**(1), 94–103 (2019). Russia
3. Cope, B., Kalantzis, M.: «Multiliteracies»: new literacies, new learning. Pedagogies: Int. J. **4** (3), 164–195 (2009). https://doi.org/10.1080/15544800903076044
4. Di Giulio, M., Vecchi, G.: Multilevel policy implementation and the where of learning: the case of the information system for school buildings in Italy. Policy Sci. **52**(1), 119–135 (2019)
5. Dufva, T., Dufva, M.: Grasping the future of the digital society. Futures **107**, 17–28 (2019). https://doi.org/10.1016/j.futures.2018.11.001
6. Gushchina, O.M.: Competence approach in creation of the information and educational environment of knowledge with using of electronic resources. Baltic Humanitarian J. **2**(11), 49–52 (2015). (in Russia
7. Hill, I.: Evolution of education for international mindedness. J. Res. Int. Educ. **11**(3), 245–261 (2012)
8. Klikunov, N.D.: The impact of network technologies on the transformation of Russian higher education. Pedagogical Educ. Russ. **3**, 78–85 (2017). (in Russian)
9. Mossberger, K., Tolbert, C.J., McNeal, R.S.: Excerpts from digital citizenship: the internet, society, and participation. First Monday **13**(2) (2008). MIT Press, Cambridge
10. Neustroyev, S.S., Simonov, A.V.: Innovative directions of development of electronic education. Hum. Educ. **3**(44), 9–15 (2015). (in Russia)
11. Nikulina, T.V., Starichenko, E.B.: Information and digital technologies in education: concepts, technologies, management. Pedagogical Educ. Russ. **8**, 107–113 (2018). (in Russia)
12. Repina, E.G.: Student movement as a search tool and pedagogical work with gifted youth: principles, characteristics, experience. Samara J. Sci. **6**(3(20)), 297–302 (2017). (in Russia)

13. Repina, E.G.: System and institutional approaches to the concept “education”. Differ. Equ. Adjacent Prob. **1**, 229–232 (2017). (in Russia)
14. Russian Federation Government: Modern digital educational environment in the Russian Federation. A priority project in the field of education (2016). http://neorusedu.ru. (in Russia)
15. Santha, K.: Teacher trainees beliefs concerning efficient teaching and learning-pedagogical spaces in focus. New Educ. Rev. **55**(1), 17–29 (2019)

Optimization of Higher Education in Economy Digitalization

N. V. Speshilova[1], V. N. Shepel[1], and M. V. Kitaeva[2](✉)

[1] Orenburg State University, Orenburg, Russia
[2] Samara State University of Economics, Samara, Russia
kmv_1965@mail.ru

Abstract. Changes currently taking place in the world, caused by digitalization processes, impose new requirements on all stages and levels of economic management. In the period of changing strategic goals of the effective development of modern complex socio-economic objects, it is impossible to do without qualified specialists. Under new economic conditions, intellectual potential is one of the key resources of the economy. After all, such an indicator as the human development index reflects the level of national development in the world economic community. The management of the education system at the modern level requires substantiation of decisions made based on the wide use of modeling of training processes, optimization of planning and organization of distribution of educational resources according to the criterion of cost minimization, as well as the use of modern information technologies. The contribution presents the results of studies of two of the four management functions - organization and control. It was established that combinatory and morphological analysis and synthesis should be used in optimal allocation of training time, taking into account curricula. Information technology control and self-learning material will improve feedback in the system, which will positively affect the quality of control. Next, you need to study the functions of planning and motivation.

Keywords: Digital economy · Economic and mathematical tools · Individualization of training · Organizational management

1 Introduction

One of the most current trends in the development of the economy and society is digitalization. The idea of digital transformation embraced the whole world.

A number of papers by domestic and foreign scientists describe various areas of the digital economy:

- Enterprise economics: forecasting the sustainable development of enterprises [4, 7, 12], assessing the financial condition of a business [8], planning the resources of the enterprise [18], etc.;
- Production processes: modeling the function of production [6], modeling of production risks [9], using IT systems that support business processes in the enterprise [21], and others.

S. I. Ashmarina et al. (Eds.): ISCDTE 2019, LNNS 84, pp. 448–457, 2020.
https://doi.org/10.1007/978-3-030-27015-5_54

Moreover, a key factor for successful professional growth in the modern digital world is a person's ability to continually learn, readiness to regularly get new knowledge about new emerging technologies. In this case, the development of self-organization, planning, and self-motivation skills can be considered necessary, but not sufficient. And this should be facilitated by a training process organized at each of its stages.

But, if we take into account that the development of digital technologies displaces primarily workers of "routine" labor, then mass education, which trains specialists in one program, becomes irrelevant. This means that the processes of economy digitalization naturally affect not only the production, but also its human resources.

The educational sphere of the knowledge-based economy [10] is one of the fundamental foundations of the knowledge economy, due to the fact that the education system of any country in the world is primarily a forge of personnel for all industries and the source of its intellectual elite. The quality of labor resources and their effective use are a prerequisite for economic success. This is becoming especially relevant today.

2 Materials and Methods

The task of optimal allocations of time provided by the curriculum for the study of each subject is based on the problem of resource allocation using the method of morphological synthesis [3]. This method is due to its successful application in other industries to solve the problem of resource allocation.

The methods of combinatory and morphological analysis and synthesis of optimal allocations are designed to find new solutions. It is based on the division of the system under consideration into subsystems and elements, the formation of subsets of alternative options for their implementation and the combination of various solution results, as well as the selection of the best ones. It is noted that "the goals of morphological analysis and synthesis of systems are: a systematic study of all conceivable solutions to the problem arising from the laws of the structure (morphology) of the object being improved, which allows us to take into account which, in a simple enumeration, could be overlooked by the researcher; implementation of a set of search operations on a morphological set of options for describing functional systems that meet the initial requirements, i.e. conditions of the problem [3].

The morphological set of options for describing functional systems is presented in the form of a morphological table. The total number of all variants N is defined as the Cartesian product of the sets of alternatives formed by each row of the morphological table. The generated variant of the system is a sample of alternatives one by one from each row of the morphological table and it differs from any other variant by at least one alternative.

Consider the task of resource allocation based on the method of morphological synthesis. As a resource, we will mean the total amount of hours (allowed for variation) devoted to the study of a number of subjects (according to the curriculum).

The distribution of time (the resource R_i - the total number of hours available according to the curriculum, which must be optimally distributed among the alternatives of the morphological matrix) can be formed by a finite number of alternatives – $A_{ij,}$ and for each *i-th* row the index *j* has its maximum value, which can take on different values for different disciplines. The alternatives of each row are assigned values of relative efficiency (Ψ_{ij}) and values required for the implementation of the resource (R_{Tij}).

To begin with, it is necessary to analyze the possibilities of mastering each *i-th* discipline by students within the selected groups. This can be done on the basis of the morphological table, in which groups of students will act as generalized functional subsystems (GFSS). Then the time series can be made up of the values obtained on the basis that the number of hours in this discipline of the corresponding training profile (specialty) for part-time education is chosen as the criterion determining the shortest possible time to master the material. This value is 25% of full-time form.

The values of Ψ_{ij} are given taking into account: the difficulty and intensity of the studied subject; specifics of the acquired specialty; memory property, the value of which increases with an increase in the study time according to the logistic curve (i.e. at a certain point in time there is no necessary skill growth, since there is a drop in interest due to "supersaturation"), etc.

And:

- We will distribute the resource (time) among all the combinations that include one alternative from each row of the morphological table, and enter the corresponding values of efficiency;
- We will generate the set of all possible combinations of alternatives, taking into account the inclusion in the combination of all *n* generalized functional subsystems by the method of complete enumeration;
- We will define the following results for each generated combination of alternatives:

$$R_T = \sum_{i=1}^{n} R_{Tij}; \tag{1}$$

$$\Psi = \sum_{i=1}^{n} \Psi_{ij}; \tag{2}$$

$$\Psi/R_T = \sum_{i=1}^{n} \Psi_{ij} / \sum_{i=1}^{n} R_{Tij}; \tag{3}$$

where R_T – is the total value of the required resource;
Ψ – is the total value of relative efficiency;
Ψ/R_T – is the relative efficiency per unit of required resource;
n – is the number of generalized subsystems included in the generated version of the system;
i – is the ordinal number of the generalized subsystem included in the generated version of the system;
j – is the sequence number of the alternative to the *i-th* generalized system;

Choose the desired combination of alternatives with the given target function and constraints:

$$\max \Psi/R_T = \max\left(\sum_{i=1}^{n} \Psi_{ij} / \sum_{i=1}^{n} R_{Tij}\right); \tag{4}$$

$$\max \Psi = \max \sum_{i=1}^{n} \Psi_{ij}; \tag{5}$$

$$\min\,(R_И - R_T); \tag{6}$$

$$R_T \leq R_I \tag{7}$$

$$R_I - R_T \leq C \tag{8}$$

where C –is the specified resource threshold (i.e., the number of hours that may remain unused, we have $C \to 0$).

Such a sequence of actions can be reproduced to distribute the time between the streams of students of different specialties within the same discipline. This approach can be limited to, if we are talking only about one subject. But if several disciplines are used, then it is possible to strengthen one of them through the use of hours, which were released in the other. Therefore, we can further consider a set of morphological tables according to the number of streams, each of which within the *i-th* GFSS can be filled on the basis of a preliminary analysis performed on the basis of the actions described above, and at the last stage not one but several of the desired combinations of alternatives are chosen that satisfy the selected earlier conditions.

3 Results

When organizing training, feedback is constantly in operation, informing the teacher about the state of knowledge of each student, about the difficulties that the student faces in the training process. It is established by various means of control (oral questioning, checking individual homework assignments, examinations, etc.). The problem of testing knowledge and skills has always been acute in relation to the competence, reliability and validity of the tools of its organization. Technical means of control, accelerating its activity, help the teacher to test knowledge.

The problem of creating and mastering an objective test of students' knowledge in education is especially relevant today. The active use of such systems helps to maintain the desired educational level of students, gives the teacher the opportunity to pay more attention to individual work with students. It should be emphasized that the control system does not exclude the teacher from the process of knowledge testing. Freeing him from many formal and time-consuming procedures, the system allows him to focus on individual problems of each student. Thus, the role of the teacher increases with the constant expansion of his capabilities.

With the help of a computer, continuous feedback is organized in the form of preliminary, current and boundary control, which contributes to the improvement of the training process and the quality of knowledge. The use of computer technologies that use the principles of the test approach, multi-point assessment scale and statistical methods of processing and analysis, is also justified by the possibility of organizing

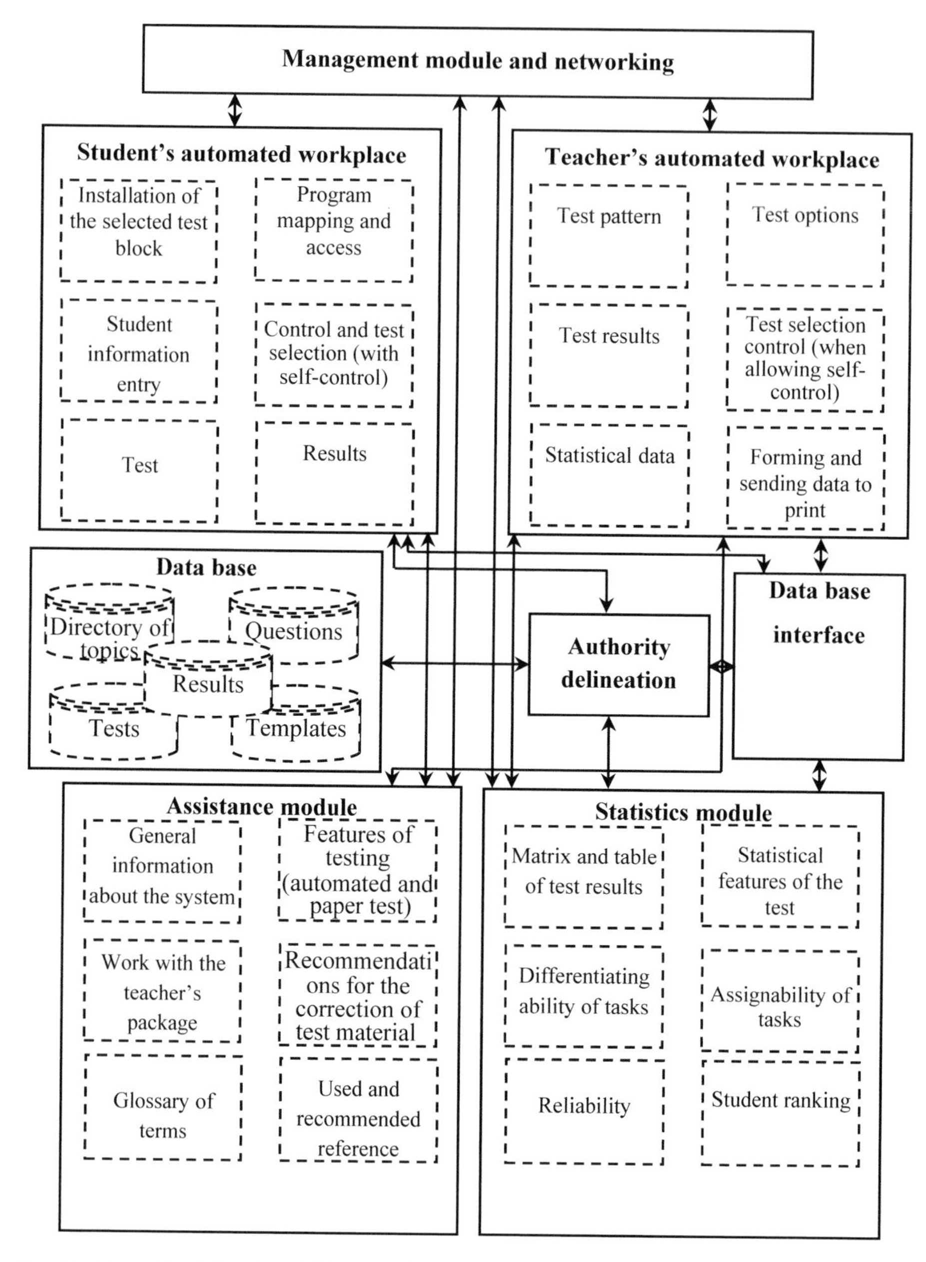

Fig. 1. Generalized functional diagram of monitoring and self-monitoring subsystem of student performance (Source: compiled by the authors).

self-learning and self-testing of acquired knowledge. This work requires the appropriate tools. One of such means can be a multifunctional testing system created by the authors (a generalized functional diagram which is shown in Fig. 1), which can be considered as one of the subsystems of the training control system.

This software product can function in the network and contributes to the objective knowledge monitoring. It uses an algorithm for generating test items with the assistance block developed on hypertext, and includes a modified set of indicators of scientific principles of test formation and methods for calculating them, taking into account the possible adjustment of the latter based on the accumulation of statistical information on test results. A module for the preparation, testing and adjustment of test tasks on a computer has been developed. In forming the coefficients for the statistical analysis of the test material, as well as for organizing assistance for testing tests, recommendations presented in [1, 2] were taken into account.

Since a pedagogical measuring (like any other) always contains measurement errors, it is inevitable to estimate errors and calculate the "true" value, which is also carried out in the "Statistics" module of this software product. There are several methods for assessing the reliability of pedagogical measures, one of which is implemented in the program in accordance with the formula Kuder-Richardson-8 [1]:

$$r_{kr-8} = \frac{\left(s_y\right)^2 - \sum_j p_j q_j}{2\left(s_y\right)^2} + \left[\left[\frac{\left(s_y\right)^2 - \sum_j p_j q_j}{2\left(s_y\right)^2}\right]^2 + \frac{\sum_j \left(r_{xjy}\right)^2 p_j q_j}{\left(s_y\right)^2}\right]^{\frac{1}{2}}, \tag{9}$$

where r_{kr-8} –is the test reliability factor;
$\sum p_j q_j$ –is the sum of variances of test tasks;
$(s_y)^2$ – is the value of the variance of test points;
$(r_{xjy})^2$ –is the value of squares of correlation coefficients of each task x_j with the sum of points y;
J –is numbers of tasks, starting with the first

From a practical point of view, "reliability is often understood as a measure of the sameness, repeatability or connectedness of two measurements of the same quality with the same test, or its parallel variants" [1].

The term "validity" as used in the software product means the measure of the suitability of test results for specific purposes. Validity depends on a huge number of factors, such as: the number and quality of tasks; the degree of completeness and depth of the content of the academic discipline; the method of selection of tasks in the test and their distribution by complexity, etc. There are several approaches to validating tests that differ depending on the criteria used [2], most of which are based on the opinion of experienced experts. Taking into account the composition of questions, an experienced specialist can assess the validity (suitability) of a test for testing knowledge. The value of validity can be determined by the correlation coefficient:

$$r_{xy} = \frac{\sum_{i=1}^{n} (x_i - m_x) \cdot (y_i - m_y)}{\sqrt{\sum_{i=1}^{n} (x_i - m_x)^2 \cdot \sum_{i=1}^{n} (y_i - m_y)^2}}, \tag{10}$$

where m_x, m_y –are arithmetic average indicators of the test and expert assessments;
n – is the number of students;
x_i – is the number of correct answers of the *i-th* student;
y_i – is the assessment of the *i-th* student by experts

Depending on the values of r_{xy} (the change in the range from -1 to $+1$), the degree of association is divided into:

$0 < |r_{xy}| < 0{,}4$ - weak, very low validity;
$0{,}4 \leq |r_{xy}| < 0{,}8$ - average, low validity;
$0{,}8 \leq |r_{xy}| \leq 1$ - strong, high validity.

In addition to the proposed and implemented in the program method for determining the reliability of the test noted above, it is possible to compare the indicators of responses to even and odd questions and calculate their rank correlation.

The reliability of the full test is assessed using the Spearman-Brown formula:

$$H = 2r_{xx}/(1 + r_{xx}), \tag{11}$$

where is the Spearman rank correlation coefficient:

$$r_{xx} = 1 - 6 \cdot \frac{\sum_{i=1}^{n} (x_i' - x_i'')^2}{(n^3 - n)}. \tag{12}$$

The reliability of the test at $H > 0.8$ is considered sufficient.

The program is implemented in the environment Inpise Delphi 5.0 using additional components from the library RXLib 2.75. The software includes: two automated workplaces - the student's automated workplace and the teacher's automated workplace, united by a single database; assistance module containing reference information; statistics module (Fig. 1).

4 Discussion

What and how should we teach a future graduate of a higher educational institution that he could find a place in a rapidly changing socio-economic space? How to manage the formation of knowledge and skills when training highly qualified specialists in demand in the labor market? How to organize the training process in the digitalized economy? Recently, various scientists have been trying to find answers to such questions:

Kupriyanovskiy et al. [5], Cornu [13], Frey and Osborne [14], Hembrooke and Gay [15], Hew [16], Kao and Tsai [17], Levine et al. [19], Sánchez et al. [22], Vavasseur and McGregor [23] and many others.

Digital technologies, coming into the education system, allow us to individualize the training process at various stages, including during the development of new material and during the control and self-control of knowledge. Digital technologies make it possible to use a wide range of tools to overcome the limitations of a classroom system with the same curriculum and the same time to master it.

Therefore, under current conditions digitalization of the education system cannot be limited to simply creating a digital copy of conventional textbooks, digitizing documents and providing all educational institutions with access to high-speed Internet. The approach should conceptually change, not only what and how to teach, but how to organize the management of the training process, including knowledge monitoring and allocation of training time, as a resource potential of a competent expert in the field of economics.

In order to choose the most optimal pathway of managing the training process, it is necessary to solve the problem of optimal allocation of time provided by the curriculum for the study of each subject. For this, it is necessary to select the most effective economic-mathematical method.

Procedures for the control and self-control of knowledge should be implemented through the use of valid and reliable tools based on IT-technologies. Search and formation of the latter should be given serious attention.

5 Conclusions

The optimal allocation of training time was carried out on the basis of adapting the resource allocation problem using the method of morphological synthesis. This approach can be successfully implemented to select the time to study the disciplines in forming the curriculum in the dean's offices at the faculties of higher educational institutions, as well as in determining the individual training path.

The testing system developed for monitoring performance and self-control of knowledge, which is a subsystem of the training management system [11, 20], contains interconnected modules and contributes to the creation of a test base at the scientific level. The developed software product provides the ability to:

- Comprehensive accumulation of statistical information, which among other things, can be used to adjust test tasks according to scientific principles;
- Check test tasks for reliability, validity, differentiating ability in an explicit form;
- Control and self-control with parameters specially set by the teacher;
- Formation of a rating assessment of the student in the group and the whole course;
- Statistical processing of the results obtained on paper.

References

1. Avanesov, V.S.: Composition of Test Tasks. Adept, Moscow (1998)
2. Avanesov, V.S.: Fundamentals of the Scientific Organization of Pedagogical Control in Higher Education. Mies, Moscow (1989)
3. Andreichikov, A.V., Andreichikova, O.N.: Analysis, Synthesis, Planning Decisions in the Economy. Finance and Statistics, Moscow (2000)
4. Bulgakova, I.N., Golovaneva, H.C.: Improving methods for predicting the sustainable development of an enterprise based on logit analysis. Mod. Econ.: Prob. Solutions **1**(25), 146–150 (2012)
5. Kupriyanovskiy, V.P., Sukhomlin, V.A., Dobrynin, A.P., Raikov, A.N., Shkurov, F.V., Drozhzhinov, V.I., Fedorova, N.O., Namiot, D.E.: Digital skills and challenges of the education system. Int. J. Open Inf. Technol. **5**(1), 19–25 (2017)
6. Matyukhin, B.B.: Model of the production function based on the law of diminishing productivity. Bull. Kamchatka State Tech. Univ. **13**, 30–34 (2010)
7. Orlov, A.I.: Sustainable Economic and Mathematical Methods and Models. Development and Development of Sustainable Economic and Mathematical Methods and Models for the Modernization of Enterprise Management. LAP (LAMBERT Academic Publishing), Saarbrücken (Germany) (2011)
8. Pelipenko, E.Y., Khalafyan, A.A.: Decision making support information system in sphere of small and medium business companies solvency. Sci. J. Kuban State Agrarian Univ. (Sci. J. KubSAU) **108**(4), 872–890 (2015). http://ej.kubagro.ru/2015/04/pdf/63.pdf
9. Potupchik, A.V., Pyshnograi, G.V., Chai, A.A.: Modeling of risk for output. Bull. Altai Acad. Econ. Law **1**, 87–90 (2012)
10. Shepel, V.N., Akimov, S.S.: Problems of knowledge extraction. In: Materials of the All-Russian Scientific and Methodical Conference "University Complex as a Regional Center of Education, Science and Culture", pp. 1562–1565. OGU, Orenburg (2015)
11. Shepel, V.N., Speshilova, N.V.: Multi-criteria deterministic decision-making problems of subsystems of higher educational institutions. Intellect. Innov. Investments **12**, 124–128 (2016)
12. Bansal, P.: Evolving sustainably: a longitudinal study of corporate sustainable development. Strateg. Manag. J. **26**(3), 197–218 (2005)
13. Cornu, B.: Digital natives: how do they learn? How to teach them? Policy brief. UNESCO Institute for Information Technologies in Education (2011). http://iite.unesco.org/pics/publications/en/files/3214698
14. Frey, C.B., Osborne, M.A.: The Future of Employment: How Susceptible are Jobs to Computerization? (2013). https://www.oxfordmartin.ox.ac.uk/downloads/academic/The_Future_of_Employment.pdf
15. Hembrooke, H., Gay, G.: The laptop and the lecture: the effects of multitasking in learning environments. J. Comput. High. Educ. **15**(1), 46–64 (2003). https://doi.org/10.1007/BF02940852
16. Hew, K.F.: Students' and teachers' use of facebook. Comput. Hum. Behav. **27**(2), 663–675 (2011)
17. Kao, C.-P., Tsai, C.-C.: Teachers' attitudes toward web-based professional development, with relation to Internet self-efficacy and beliefs about web based learning. Comput. Educ. **53**, 66–73 (2009)
18. Lepistö, L.: Taking information technology seriously: on the legitimating discourses of enterprise resource planning system adoption. J. Manag. Control **25**, 193–219 (2014)

19. Bowman, L.L., Levine, L.E., Waite, B.M., Gendron, M.: Can Students really multitask? Comput. Educ. **54**(4), 927–931 (2010)
20. Kitaeva, M.V., Speshilova, N.V., Shepel, V.N.: Mathematical models of multi-criteria optimization of subsystems of higher educational institutions. Int. Rev. Manag. Mark. **6**(S5), 249–254 (2016)
21. Mesjasz-Lech, A.: The use of IT systems supporting the realization of business processes in enterprises and supply chains in Poland. Pol. J. Manag. Stud. **10**(2), 94–103 (2014)
22. Sánchez, J., Salinas, A., Contreras, D., Meyer, E.: Does the new digital generation of learners exist? A qualitative study. British J. Educ. Technol. **42**(4), 543–556 (2011)
23. Vavasseur, C.B., McGregor, S.K.: Extending content-focused professional development through online communities of practice. J. Res. Technol. Educ. **40**(4), 517–536 (2008)

Socio-Technical Approach to a Research of Information Economy

O. V. Bakanach[1], O. F. Chistik[1(✉)], M. Y. Karyshev[2], and N. V. Proskurina[1]

[1] Samara State University of Economics, Samara, Russia
yurijchistik@yandex.ru
[2] Samara State University of Railway Engineering, Samara, Russia

Abstract. The article is devoted to the digital transformation of the economy and the necessity in existence of a profiled science-based statistical methodology in this area, due to the high relevance of the problems deal with quantitative assessment and analysis of the information and communication technologies effectiveness. The problems associated with the adaptation of the existing methodological "reserve" to the Russian statistical realities seem to be an urgent and promising direction of scientific research, leading both to the expansion of the theoretical basis and to obtaining practically significant results. The scientific novelty of the approach is to solve the fundamental problem associated with the scientific methodological development on quantitative evaluation of complex socio-technical systems interaction in the conditions of digital economy transformation and the formation of common information area. In this regard, the article presents the results of forming the information base and the system development of statistical indicators in social processes interaction within common information area; describes the integrated indicators, namely the Common index of infrastructure development in the information society and the Index of information economy.

Keywords: Composite index · Digital economy · Information and communication technology · Social engineering systems

1 Introduction

The problem to develop a complex scientifically based methodology for analyzing the interaction of complex socio-technical systems in the context of economy digitalization is extremely relevant. Passing to the information society, forming common information area is global in all spheres of modern society. Globalization is manifested in the complexity of social relations and affects, first of all, the spheres of economy, engineering and technologies, information, politics and law [11]. That is why, in our opinion, currently the interaction of different sciences in the research process of common information area has particular importance. This interaction allows expanding significantly the information data array, to consider the studied problems from different sides and, as a result, to improve the quality and reliability of the study. Sociology and statistics are sciences that play a significant role in meeting the needs of society and

S. I. Ashmarina et al. (Eds.): ISCDTE 2019, LNNS 84, pp. 458–465, 2020.
https://doi.org/10.1007/978-3-030-27015-5_55

managing relevant information [4]. The result their interaction is the process forming actual scientific knowledge, manifested in the combination of theoretical, methodological and empirical resources in sociology and statistics in the objective knowledge of multifaceted, dynamically changing phenomena [13, 18, 23].

The aim of the study is to develop a scientifically based methodology for analyzing the interaction of complex social engineering systems in the context of the economy digitalization.

To achieve this goal, the following tasks were solved:

- to form information research base and to develop a statistical indicators system of social engineering processes interaction in common information area;
- to develop integrated indicators of social engineering processes in common information area: the Composite index of infrastructure development in the information society and the Index of the information economy development.

2 Methods

The methodological procedures of our research are based on general scientific and special methods and techniques of scientific knowledge as a dialectical cognition method, logical and theoretical analysis, synthesis and analysis, generalization and systematic approach, revealing possible scientific research of socio-economic phenomena in the development, their relationship and interdependence. The statistical-sociological method is used as a method observing mass social phenomena, allowing establishing the repeatability of homogeneous phenomena in social processes and, based on the large numbers law and the theory of probability, to make statistical conclusions about these recurring phenomena stability. The spread spectrum of used methods has been expanded to include statistical and sociological tools such as online surveys, long-term online communities, including observations and other data collection procedures. To provide qualitative research methods, the survey instruments verification was carried out, testing the characteristics of quantitative research. A wide range of statistical methods are used as a reliable tool: the method of multivariate statistical grouping, the generalized indicators method, the "Pattern" method and the index method.

3 Results

The research in the field of information and communication technologies (ICT) is very relevant, since the effects of interaction between society and technology largely determine the comfort of modern human civilization. The research direction in this area allowed to generalize and systematize the scientific approach to the study of complex socio-technical systems, as well as to determine their place and role in the digital economic transformation. In the development process of theoretical and methodological foundations within statistical and sociological research, their interaction in common information area, there was formed the information base of the study and proposed a system of statistical indicators (indicators). On the basis of the international statistical methodology, the results composite indices of the information and communication technologies development in Russia have been developed. These include the Composite index of information society infrastructure development and the Index of information economy development. These indices make it possible to characterize the development level of the information society and they are analogous to international indicators. These indicators take into account the Russian specifics of the existing information base.

The composite development infrastructure index of information society (IIS) is represented by three sub-indices: ICT access, ICT usage and ICT skills. The first sub-index, which characterizes the population access token in Russian Federation to information and communication technologies, is calculated by 5 indicators of the system and allows evaluating the actual access results to telephones, terminals, the Internet, personal computers. The second sub-index is based on 3 indicators for employees of the whole services range. The last sub-index provides an assessment of ICT skills. For the purpose, the sub-index introduced 3 indicators characterizing the number of personal computers used in all educational institutions, organizing the educational process, an average of 1,000 trainee and 100 students.

The initial values of the partial indicators selected system are used in our analysis by rationing them on the basis of the "Pattern" method. In contrast to the calculation method adopted in international practice, in our case the range of index values fluctuation is limited by the interval [0;1]. The presented composite index can be used as an adequate representative assessment of the information society development. The calculated values of the proposed index allow a multidimensional grouping of Russian regions and give an appropriate interpretation of the obtained groups. The first two groups are designated as regressive, the third – proportional, the fourth and the fifth – progressive in relation to the process of forming information society from the standpoint of information and communication technologies development (Table 1).

Table 1. Multidimensional typological grouping of Russian regions by the Composite index values of IIS

The values of the Composite index	The number of regions	Group structure
<0,500	6	Republic of Tyva, Republic of Dagestan, Republic of Ingushetia, Republic of North Ossetia, The Chechen Republic., Zabaikalsky Territory
0,501–0,550	21	Bryansk Region, Vladimir Region, Ivanovo Region, Kursk Region, Lipetsk Region, Tambov Region, Tula Region, Republic of Adygea, Kabardino-Balkarskaya Republic, Republic of Kalmykia, Astrakhan Region, Republic of. Mari El, Saratov Region, Ulyanovsk Region, Republic of Altai, Republic of Buryatia, Altai Territory, Kemerovo Region, Omsk Region, Magadan Region, Jewish Autonomous Region

(continued)

Table 1. (*continued*)

The values of the Composite index	The number of regions	Group structure
0,551–0,600	23	Belgorod Region, Voronezh Region, Kaluga Region, Kostroma Region, Orel Region, Ryazan Region, Smolensk Region, Yaroslav Region, Vologda Region, Kaliningrad Region, Stavropol Region, Volgograd Region, Udmurtskaya Republic, Perm Region, Kirov Region, Orenburg Region, Samara Region, Sverdlovsk Region, Chelyabinsk Region, Irkutsk Region, Novosibirsk Region, Republic of Yakutia, Khabarovsk Territory
0,601–0,650	18	Tver Region, Komi Republic, Murmansk Region, Pskov Region, Krasnodar Region, Rostov Region, Republic of Bashkortostan, Republic of Mordovia, Republic of Tatarstan, The Chuvash Republic., Penza Region, Kurgan Region, Tyumen Region, Republic of Khakassia, Krasnoyarsk Region, Kamchatka Region, Amur Region, Sakhalin Region
>0,650	7	Republic of Karelia, Arkhangelsk Region, Novgorod Region, Nizhny Novgorod Region, Tomsk Region, Primorsky Territory, Chukotka Autonomous District

Source: authors, based on official data of the Federal state statistics service [19]

For the generalizing statistical characteristic of the developed level in information society, the assessment of information economy development is given. This problem was solved by constructing and calculating the Index of the information economy development. The method of its calculation: the initial and derivative elements of the composite characteristics are determined, the system of indicators is constructed, and weights are assigned. The composition of the developed index includes the following sub-indexes: computerization of workplaces, availability of network access, software applications, energy security.

Sub-indexes are formed from indicators groups of "infrastructure" and "applied" system blocks of statistical indicators of developed results in the sphere of information and communication technologies. In addition, the structure of the created Index introduced a sub-index, the developed level of regional energy. This is appropriate, as the availability of affordable and inexpensive electricity resources is a necessary "infrastructural" condition for the information economy. Sub-index of energy security" - performed a corrective function, through the use of potential indicators and volume of energy capacity, determining the technical capabilities within the information economy. According to the calculated values of the Index, a multidimensional typological grouping of the subjects in Russia was carried out. This allowed us, based on the processes that were taken into account in the model, in turn, to identify the following regional types of the information economy (Table 2).

Table 2. The multidimensional typological grouping of the regions according to the development types of the information economy

Type of development	Number of regions	Interval of Index values	Characteristics of development
Progressive	21	0,701–1,000	All aspects of the information economy have reached (or almost reached) the target level
Promising	24	0,601–0,700	The formation of the information economy is targeted on development
Neutral	30	0,501–0,600	The development of the economy has no pronounced tendency to informatization
Alternative	5	0,000–0,400	Non-information economy that is being formed

Source: authors, based on official data of the Federal state statistics service [19]

With regard to the method calculating the information economy development Index, it should be noted that all the initial data to achieve comparability were given by the Pattern method in a standard form. This action was also carried out in order to obtain an overall result, the values of which vary in the range [0; 1]. This distinguishes the developed indicator from international statistical indicators that do not have such calculated limits, which makes it difficult to compare their values. The choice of weight coefficients for proposed indicators was determined on the basis of expert assessments in direct proportion to the processes relevance of information economy development.

We propose a system of statistical indicators comparable at the international level and designed to assess the socio-economic results of the information and communication technologies development in Russia as part of the global information society and the information economy.

On the methodological basis the calculation of the ICT development Index designed by the International telecommunication Union, the method calculating the generalizing indicator - the Summary index of infrastructure development in information society is presented. Its construction made it possible to take into account the specifics of the Russian statistical information base. The main methodological problems dealt with constructing a Composite index of development IIS are identified and solved.

A method of constructing an integral indicator in information and communication technological processes with economic orientation in the form of a composite statistical indicator-the Index of information economy development is proposed and tested. In general, the study should be considered as an information and analytical tool to develop solutions for the effective management of information and communication technology in the context of the digital economy.

4 Discussion

The interaction of complex social engineering systems in Russia in the global information society is crucial to improve the competitiveness of the national economy in the global economic system as a result of increased productivity and structural economic modernization. In the context of forming policy management of multitude phenomena and processes occurring in complex socio-technical systems, a special role is given to the scientific statistical methodology. As an indispensable analytical tool, a distinctive feature of which is the objectivity of probabilistic mathematically based assessment, statistics is able to reveal the development patterns of these heterogeneous phenomena and processes in a single information space. The integration process of the Russian Federation into the international statistical community is objectively accompanied by the harmonization of the national statistical methodology with international standards.

Interest to the current research in this field of science and the desire to compare the expected results with the world level has necessitated the appropriate analysis. Currently, in Russia and in the rest of the world a lot of research is carried out on various aspects of the interaction of economic and information technology processes and received the general name of the "digital economy". In particular, the influence of information technology factors on the overall performance [5, 20, 21], general quality management [17], the development of specialized economic activities, in particular those engaged in the provision of information technology services [18, 20]. It is covered both macro-and microeconomic levels [3, 8, 16]. In this regard, the experience of studying the digital economy in the near abroad - the Republic of Kazakhstan [2] and far – South Korea [13] is interesting. Special attention is paid to the phenomenon of shadow digital economy [7], analysis of trends and needs in the global digital economy [15], as well as licensing issues [21], urban [22], environmental [14], and demographic [4] the digitalization aspects of the economy and society.

Today the impact of digital technologies on the development of Russia is the subject for active scientific interest [6, 12, 23]. A separate area of scientific research is the statistical approach to the study of socio-economic results of the information and communication technologies development [9–11]. A wide range of issues in this area and methodological approaches to their solution were raised at the International scientific and practical conference "Statistics in the digital economy: training and use" [1], held under the auspices of the Russian Association of statisticians and St. Petersburg state University of Economics.

The hypothèsis that the socio-economic situation of the territory directly depends on the results of the complex socio-technical systems development in a common information area has been confirmed. Thus, managing the results of this development, expressed in the form of the digitalization process of the economy, it is possible to determine to a large extent the economic growth and the population life quality.

5 Conclusions

There is no doubt that the creation of a scientifically based concept of the study interacting between complex socio-technical systems, which contributes to improving the quality of information and analytical support and the validity of management decisions aimed at the formation of common information area, it is a significant scientific and practical problem. In this regard, the development of indicators to assess the interaction patterns of complex social engineering systems in a common information area is associated with the identification of the main interaction types of complex social engineering systems in order to form a common information area and develop management strategies. Theoretically and experimentally justified choice of a set of statistical and sociological methods successfully solved the problem dealt with developing a theoretical and methodological basis for monitoring the interaction process of complex socio-technical systems in a common information area on the basis of a unified criteria approach.

References

1. Babich, S.G.: The comparative analysis of use of information and communication technologies in regions of the Russian Federation. Statistics in the digital economy: Learning and use, pp. 31–34. St. Petersburg state University of Economics, St. Petersburg (2018). (in Russian)
2. Berdykulova, G.M.K., Sailov, A.I.U., Kaliazhdarova, S.Y.K., Berdykulov, E.B.U.: The emerging digital economy: case of Kazakhstan. Procedia Soc. Behav. Sci. **109**, 1287–1291 (2014). https://doi.org/10.1016/j.sbspro.2013.12.626
3. Chen, Y.H., Lin, W.T.: Analyzing there relationships between information technology, inputs substation and national characteristics based on CES stochastic frontier production models. Int. J. Prod. Econ. **120**(2), 552–569 (2009). https://doi.org/10.1016/j.ijpe.2008.07.034
4. Chistik, O.F., Blinova, S.V.: Component statistical analysis of factors of life expectancy in the conditions of information technologies introduction in Russia. Bull. Samara State Univ. Econ. **6**(164), 62–69 (2018). (in Russian)
5. Chou, Y.-C., Chuang, H.H.-C., Shao, B.B.M.: The impacts of information technology on total factor productivity: a look at externalities and innovations. Int. J. Prod. Econ. **158**, 290–299 (2014). https://doi.org/10.1016/j.ijpe.2014.08.003
6. Garifova, L.: The economy of the digital epoch in Russia: development tendencies and place in business. Procedia Econ. Financ. **15**, 1159–1164 (2014). https://doi.org/10.1016/S2212-5671(14)00572-3
7. Gaspareniene, L., Remeikiene, R., Navickas, V.: The concept of digital shadow economy: consumer's attitude. Procedia Econ. Financ. **39**, 502–509 (2016). https://doi.org/10.1016/S2212-5671(16)30292-1
8. Gunasekaran, A., Nath, B.: The role of information technology in business process reengineering. Int. J. Prod. Econ. **50**(2–3), 91–104 (1997). https://doi.org/10.1016/S0925-5273(97)00035-2
9. Karasev, M.Y.: The Theory of Neural Network Modeling the Integration of Information and Communication Technology and Business Processes. Monograph. Publishing House of Samara State University of Railway Engineering, Samara (2011). (in Russian)
10. Karyshev, M.: The Monograph of the Sphere of Information and Communication Technology. Publishing House of Samara state University of Railway Engineering, Samara (2010). (in Russian)
11. Karyshev, M.: Socio-Economic Efficiency of Information and Communication Technologies: Methodology of International Statistical Comparisons: Monograph. Finance and Statistics, Moscow (2011). (in Russian)
12. Kharchenko, A.A., Konyukhov, V.Y.: Digital economy as the economy of the future. Youth Bull. ISTU **3**(27), 17 (2017). http://mvestnik.istu.irk.ru/journals/2017/03. Accessed 3 Apr 2019. (in Russian)

13. Kim, J.: Infrastructure of the digital economy: some empirical findings with the case of Korea. Technol. Forecast. Soc. Change **73**(4), 377–389 (2006). https://doi.org/10.1016/j.techfore.2004.09.003
14. Kostakis, V., Roos, A., Bauwens, M.: Towards a political ecology of the digital economy: socio-environmental implications of two competing value models. Environ. Innov. Soc. Transit. **18**, 82–100 (2016). https://doi.org/10.1016/j.eist.2015.08.002
15. Lacey, S.: Trade Rules for the Digital Economy. UNSW Sydney (2017). https://doi.org/10.13140/rg.2.2.14067.02088
16. Lin, W.T., Chiang, Ch.-Y.: The impacts of country characteristics upon the value of information technology as measured by productive efficiency. Int. J. Prod. Econ. **132**(1), 13–33 (2011). https://doi.org/10.1016/j.ijpe.2011.02.013
17. Martinez-Lorente, A.R., Sanchez-Rodriguez, C., Dewhurst, F.W.: The effect of information technologies on TQM: an initial analysis. Int. J. Prod. Econ. **89**(1), 77–93 (2004). https://doi.org/10.1016/j.ijpe.2003.06.001
18. Perunovic, Z., Mefford, R., Christoffersen, M.: Impact of information technology on vendor objectives, capabilities, and competences in contract electronic manufacturing. Int. J. Prod. Econ. **139**(1), 207–219 (2012). https://doi.org/10.1016/j.ijpe.2012.04.009
19. Regions of Russia, Socio-economic indexes 2018: Official site of Federal State Statistics Service (2018). http://www.gks.ru/. Accessed 4 Apr 2019. (in Russian)
20. Shao, B.B.M., Lin, W.T.: Assessing output performance of information technology service industries: productivity, innovation and catch-up. Int. J. Prod. Econ. **172**, 43–53 (2016). https://doi.org/10.1016/j.ijpe.2015.10.026
21. Teece, D.J.: Profiting from innovation in the digital economy: enabling technologies, standards, and licensing models in the wireless world. Res. Policy **47**(8), 1367–1387 (2018). https://doi.org/10.1016/j.respol.2017.01.015
22. Tranos, E., Reggiani, A., Nijkamp, P.: Accessibility of cities in the digital economy. Cities **30**, 59–67 (2013). https://doi.org/10.1016/j.cities.2012.03.001
23. Zubarev, A.E.: The digital economy as a form of manifestation of regularities of development of the new economy. J. Pac. Natl. Univ. **4**(47), 177–184 (2017). (in Russian)

Development of International Production Cooperative Relations in the Digital Economy

E. M. Pimenova(✉), A. V. Streltsov, and G. I. Yakovlev

Samara State University of Economics, Samara, Russia
pimenova-elena@rambler.ru

Abstract. The participation of enterprises in global production chains (especially in terms of equal interaction with partners) requires a certain, respectively high level of its technical and economic development. Therefore, the main purpose of this study is to analyze problems associated with the development of international production cooperation in the digital economy. The authors consider development directions and prospects for international production cooperation relations of Russian enterprises in the digital economy, taking into account the fact that international cooperation projects are growing both in scale and in total number. Special attention is paid to the fact that the positive development dynamics of international corporations and advanced digital technologies is influenced by the level of risks, as well as ratings (investment, favorable conditions for business), support of the state and regional authorities.

Keywords: International cooperation · Business climate ratings · Foreign investment · Digital economy

1 Introduction

An important constituent part of the world economy is the Russian economic system which tries to find its rightful place in the international division of labor. In this context, it is necessary to improve the state of domestic institutions of innovative development and foreign economic activities, geo-economic status, as well as change methods and techniques of organizing successful entrepreneurial activities at the micro level. Nowadays, it is largely influenced by the introduction of the digital economy achievements in the daily business practice in order to improve the favorable investment and business climate.

The purposeful efforts of the government and the business community of the Russian Federation have begun bringing significant success: at the end of 2018, Russia achieved the 31st position in the authoritative world ranking of Doing Business (DB) (this rating is regularly published by the World Bank to assess the level of favorable business climate in various countries) [12]. It should be noted that Russia (on this position, by creating a successful business environment) was between such developed countries as France and Spain with a slight margin from the major world economies – the rating leaders – and all this is despite the policy of full containment by the West.

S. I. Ashmarina et al. (Eds.): ISCDTE 2019, LNNS 84, pp. 466–472, 2020.
https://doi.org/10.1007/978-3-030-27015-5_56

As a result, the plans to overcome the hundred steps in this rating by 2018 (from the initial 124th position) set by the President of the Russian Federation V. V. Putin in 2012 (in his "May Decree") were practically fulfilled. One of the most important reasons that led to such a remarkable result is the fact that international cooperation (industrial cooperation of international organizations in the production sphere) is widespread in our country. The main advantage of international cooperation is that it not only provides additional jobs, but also is one of the primary reasons for attracting foreign direct investment to our country, which in turn allows the Russian Federation to receive and further disseminate advanced production and management technologies. It is extremely important for sustainable economic development in the context of digitalization.

According to Forbes data, in 2018, the net outflow of foreign direct investment in the capital of Russian companies was more than $6 billion. It is a record result in the history of observations (since 1997). However, the total investment in the non-financial sector of the economy was only $5.938 billion (it is the minimum since 2002). It should also be noted that the analysis of the payments balance of foreign companies shows a similar negative trend: companies with headquarters in the OECD countries showed a drop in profits from foreign investment by 17% for the last five years. USA corporations were affected to a lesser extent (a decrease of 12%). For non-USA companies, this figure was 20% [5].

According to statistics from the Central Bank of Russia, in 2018, $15.994 billion was invested in the capital of Russian companies from abroad, while foreign investors withdrew $ 22.407 billion. At the same time in 2017, (according to the Russian direct investment Fund), the volume of foreign direct investment in the non-financial sector of the country exceeded $30 billion [11] which was the most significant result of the last four years, even despite the relatively low investment rating (BB+ according to S&P). This success was achieved due to the state support of innovative activities of enterprises, especially in the implementation of digital technologies in real business practice. The empirical strategy is aimed at identification of the role of innovation intensity in four different areas (technology, organization, staff training and information and communication technologies) and the role of production cooperative relations at the enterprise level in the working environment [1]. On the one hand, innovations generally have a positive impact on working conditions but this effect is weak and sometimes negative for specific organizational aspects. On the other hand, production cooperative relations are always positively and firmly linked to the welfare of workers. Therefore, it is necessary to study development directions for international production cooperation in the context of the economy digitalization and find out problems associated with this aspect. This research is devoted to the solution of this problem.

Although, technological changes have an undoubted impact on international relations, the changing nature of world politics affects the tempo and speed of technological changes too. The nature of the technology itself and the extent by which the public sector stimulates innovations have different impacts on international cooperative relations [2]. These relations are aimed at the involvement of practitioners, the introduction of diverse knowledge and experience in a way that, at least theoretically, generates new knowledge that is socially sustainable and socially controlled [3].

Supply chains, production and other complex inter-organizational relationships are the defining features of modern business organizations [7]. The unstable economic situation in Russia in terms of falling domestic demand affects foreign business oriented to the domestic market. But large companies understand that there will be growth period after the period of recession, so they maintain their presence reducing the volume of investment. In order to activate the process of inflow of investment capital into the country, it is necessary to consider this issue in the context of the economy informatization when the key factor of production is data in digital form. That allows processing more information than by traditional forms of management which significantly increases the efficiency of various types of production, technologies, equipment, etc.

In the framework of this study, the authors consider directions and prospects of development of international production cooperation relations of Russian enterprises in the digital economy as the primary basis for attracting foreign direct investment in order to increase the tempo of innovative development of the Russian Federation and strengthen the economy of the country.

2 Research Methods

While carrying out this study, the authors took into account that systematic statistical information on the participation of industrial enterprises in global reproduction chains is often rather limited because of the classification of data on this type of activity to the closed commercial data of enterprises. At the same time, according to the indirect indicators available in the open statistical information sources, it is possible to get a fairly complete assessment of how the foreign economic activity of domestic enterprises is carried out, what is their place in the international division of labor, how effective is the participation of Russian enterprises in the development of international production cooperation. Therefore, it became possible to use the methods of system and factor analysis, empirical, diagnostic, retrospective, predictive, stochastic and others in this study to achieve the research goals.

3 Results

Currently, the situation with the development of international cooperative relations of enterprises of the Russian Federation is quite complicated. According to the Ministry of Economic Development, since 2018, direct investment in Russia has decreased significantly. Despite this fact, foreign investors are showing great interest in investments into the Russian economy and are ready to increase them as soon as a new period of stability comes.

Kormnov described the mechanism of launching the economic growth along the technological chain by expanding the participation in the international cooperation [10]. In his opinion, even in the conditions of fierce political and sanctions confrontation and mutual restrictions, the key point in understanding the problem of forming a development strategy of foreign economic relations of the Russian Federation is the fact that only participation in international economic processes, the liberalization of foreign economic activity of enterprises and attraction of direct foreign investments give the possibility to form the international competitiveness of the country.

It has been repeatedly proved that the strategy of openness to cooperation, the formation of international production alliances allows to ensure the economic growth above the world average indicators using innovations and effective technical solutions created everywhere in the world (through the import of machines containing advanced technology, hiring competent foreign managers and borrowing ideas on the organization and management of production [17]). Today, international production chains and cooperation alliances cover thousands of links (individual firms-co-operators). These links are formed into chains, which are called global value chains. At the world level, the elements of such chains are industrial enterprises located in different countries and on different continents, connected with each other by the end-to-end process of the technology of production of a product.

The modern world economy has begun to experience a slowdown in the processes of globalization, which can be especially clearly seen in the dynamics of international trade, which grew from 1960 to 2008 by 35%, and over the past five years – by only 0.2% [4]. At the same time, while a number of experts put forward a hypothesis about the end of globalization under the influence of unprecedented growth of protectionism, increase in labor costs in developing countries, reshoring and new industrialization of developing countries, Kondratyev says about its new model in accordance with the formation of new proportions of forces of its main players, the disintegration of the material component of the value chains stimulated by the technologies of the industry 4.0, the differentiation of growth ways of developed economies and the increase in the share of services in the international trade [9]. Modern companies need to find new drivers for growth, as their participation in the previous global value chains is canceled because of the phenomena of decentralization and exclusivity of production processes of goods and services. It is important to find new segments of the potential growth in the production of goods and services, which can be potentially found primarily in developing countries, where the growth rate of production has begun to exceed similar indicators of developed countries. Digital and additive technologies, robotics and a high level of communication enable to build enterprises of any scale as close as possible to consumers, minimizing problems of cost savings due to their production scale [6].

The level of involvement of industrial enterprises in foreign economic relations determines the extent of their contribution to the global value chain. At the same time, the contribution of a country's industrial enterprises to global value chains depends on some factors. One of the most significant for Russia factors is the technical level of production and its provision with raw materials and energy resources. The assessment of alternative costs of high-tech production by options (production entirely on their own or with the use of broad international cooperation) shows that the import substitution strategy contains an internal dichotomy for industrial enterprises. The implementation of international scientific and industrial cooperation projects allows its participants from a certain country to understand the limits of their competencies in time and not to spend their resources for achieving an ambiguous result. The business digitalization helps them to decide where they should not waste limited resources but buy finished products or components from leading companies in the industry.

International cooperation relations are impossible without the state support: only the state can force foreign companies (that use the wealth of Russian mineral resources in accordance with production sharing agreements) to buy domestic equipment, raw materials and materials within the framework of localization programs and conclusion of special investment contracts [8]. Therefore, it is obvious that the mechanism for ensuring and maintaining the international competitiveness of industrial enterprises should organically combine both corporate and state regulation instruments and purely market methods. Despite the high risks, earning in Russia is scary but possible, according to foreign investors, because "here there is very profitable jungle" [14].

In the conditions of rapid development of information technologies, one of the key directions of ensuring competitiveness is the high quality of products and their proximity to consumers. Therefore, not all global companies consider cheap labor and unstoppable market growth, as in China and India, undeniable advantages. For example, the Japanese "Canon" is sure that the development of "hi-end" class production at home is a more effective way therefore the company continues to work generally in Japan. The company decided not to save money on cheap labor but to spend it on innovations and to maintain its leadership in the segment of high-quality electronics.

Despite the complex geopolitical background, German business maintains its business activity in Russia at a high level (as well as Chinese, Korean, Japanese, etc.) and still leads in the production localization [13]. German, Japanese, French, even American business has serious and long-term investments in the Russian economy. And it is not only about investing money, but also innovation, transfer of know-how and technology. Despite the sanctions, almost no major German company has left the Russian market. So are the Japanese companies: in November 2018, "Mitsubishi Motors" restarted production of SUVs Pajero Sport in Kaluga and returned the crossover ASX for the sale. According to experts, the Russian version of Pajero can be cheaper by 20%. Other foreign automakers are also increasing production in Russia [16]. The Far East is a special region: over 26% of the total volume of foreign direct investment in the country (especially from China and Japan) are invested here in shipbuilding, petrochemicals, agriculture and fertilizer production. Taking into account that about 80% of RDIF (Russian Direct Investment Fund) investments are made in projects located outside Moscow and St. Petersburg, the thesis is confirmed that there are many attractive places in regions for projects using innovative and digital technologies. Formats of special economic zones and territories of advanced development, actively used today in Russia, are considered as the most important drivers of FDI inflows, as these zones can offer the most favorable tax conditions and benefits for their residents [15]. Now companies that grow together with developing industries (chemical industry companies, manufacturers of medicines, clothing, textiles and furniture) feel in Russia the most comfortable.

4 Conclusion

In order to strengthen the development of international cooperation between enterprises in the Russian Federation, special attention should be paid to the specific organizational form of the strategic alliance of enterprises. The example of PJSC "AVTOVAZ" shows that such a structure allows taking into account both the interests of foreign partners and the interests of the Russian enterprise. In contrast to the creation of a branch or production under the full control of a foreign company, this form allows to combine positive characteristics of activities of both Russian and foreign enterprises. So, the Russian enterprise contributes to this cooperation through its knowledge of the market, the own brand awareness on the market, the qualified personnel, partners, support of authorities. Foreign companies can offer advanced technologies, availability of competitive developments, modern management organization. The use of digital technologies will significantly reduce the time needed for formation of joint cooperation projects and costs on development of an integrated technological chain.

References

1. Antonioli, D., Mazzanti, M., Pini, P.: Innovation, working conditions and industrial relations: evidence for a local production system. Econ. Ind. Democracy **30**(2), 157–181 (2009). https://doi.org/10.1177/0143831X09102418
2. Drezner, D.W.: Technological change and international relations. Int. Relat. (2019). https://doi.org/10.1177/0047117819834629
3. Hinchliffe, S., Levidow, L., Oreszczyn, S.: Engaging cooperative research. Environ. Plan. A: Econ. Space **46**(9), 2080–2094 (2014). https://doi.org/10.1068/a140061p
4. Bhattacharya, A., Khanna, D., Schweizer, C., Bijapurkar, A.: The new globalization: going beyond the rhetoric. BCG Henderson Institute (2017). https://www.bcg.com/publications/2017/new-globalization-going-beyond-rhetoric.aspx. Accessed 25 Apr 2019
5. The Retreat of the Global Economy: The Economist (2017). https://www.economist.com/briefing/2017/01/28/the-retreat-of-the-global-company. Accessed 25 Apr 2019
6. Keller, R.: The rise of manufacturing marks the fall of globalization. Geopolitical Weekly (2016). https://worldview.stratfor.com/article/rise-manufacturing-marks-fall-globalization. Accessed 25 Apr 2019
7. Wright, C.F., Kaine, S.: Supply chains, production networks and the employment relationship. J. Ind. Relat. **57**(4), 483–501 (2015). https://doi.org/10.1177/0022185615589447
8. Volsky, A.I.: From general words to specific actions (about the industrial policy of Russia). Modern National Industrial Policy of Russia: Collection of Materials. Akademkniga, Moscow (2004). (in Russian)
9. Kondratyev, V.: A new stage of globalization: features and prospects. World Econ. Int. Relat. **62**(6), 5–17 (2018). https://doi.org/10.20542/0131-2227-2018-62-6-5-17. (in Russian)
10. Kormnov, Yu.: Cooperation as the factor of overcoming the crisis of the economy. Economist **7**, 28–36 (1999). (in Russian)
11. Koroleva, A.: Where do foreign investors go in Russia? [Expert online] (2017). https://expert.ru/2017/12/28/vo-chto-zahodyat-inostrannyie-investoryi-v-rossii/. Accessed 25 Apr 2019. (in Russian)
12. Krasnushkina, N., Butrin, D.: Their capitalism is not in the system: Russia is gradually rests in your ceiling of the doing business ranking. Newspaper Kommersant, no. 201, p. 1 (2018). https://www.kommersant.ru/doc/3787035. Accessed 25 Apr 2019. (in Russian)
13. Lindt, M.: Sanctions are sanctions, and business is business. J. Expert **43**(1094), 29–37 (2018). (in Russian)

14. Onegina, A.: Very profitable jungle. Vedomosti Newsp. **185**(1225), 1 (2004). https://www.vedomosti.ru/newspaper/articles/2004/10/11/ochen-pribylnye-dzhungli. Accessed 25 Apr 2019. (in Russian)
15. Streltsov, A.V., Yakovlev, G.I.: Features of doing business in special economic zones of the Russian Federation. Russ. Entrepreneurship **4**, 895–906 (2018). https://doi.org/10.18334/rp.19.4.38973. (in Russian)
16. Tsinoeva, Ya.: The foreign car industry is repatriating to Russia. Mitsubishi will return to the Kaluga edition of Pajero Sport. Newspaper Kommersant **173**, 11 (2017). https://www.kommersant.ru/doc/3414891. Accessed 25 Apr 2019. (in Russian)
17. Yakovlev, G.I.: Problems of development of international production cooperation relations. Bull. Samara State Univ. Econ. **7**(141), 21–28 (2016). (in Russian)

Development of Corporate Digital Training

M. V. Lovcheva[1], V. G. Konovalova[1], and M. V. Simonova[2(✉)]

[1] State University of Management, Moscow, Russia
[2] Samara State University of Economics, Samara, Russia
est-samara@mail.ru

Abstract. The authors consider digitalization of corporate training, the prospects for the development of corporate training using immersive technologies - virtual reality, augmented reality and mixed reality, summarize the main advantages of using immersive learning technologies, highlight the factors hindering their use, present examples of successful application of immersive technologies in corporate training of modern organizations. The authors use both general scientific (systemic and functional approaches) and special research methods – sociological (method of expert assessments, observation and analysis) and comparative. The conclusions substantiate the need for further research and solving the problems of intra-corporate digital staff training as a condition for improving the effectiveness and efficiency of business organizations.

Keywords: Augmented reality · Corporate training · Immersive technology · Mixed reality · Training management systems · Virtual reality

1 Introduction

Over the past decade, corporate training has undergone dramatic changes—the formats, models and technologies of education are changing: from traditional corporate universities to e-learning, blended learning, talent-based learning and continuous learning [14].

Most companies are starting to look at digital learning [4] primarily as a more effective way of spreading learning within an organization, making it available "at any time" and "anywhere", reducing direct and indirect costs.

New perspectives for the development of corporate digital education opens up with the use of immersive technologies - virtual reality (VR), augmented reality (AR) and mixed reality (MR) [6].

In the context of this study, the authors use the following terminology:

Virtual reality (VR) is a term used to describe the world created by technical means, transmitted to a person through his sensations: sight, hearing, touch, and others. Virtual reality mimics both impact and response to impact.

Augmented reality (AR) is a term used to describe the result of introducing any sensory data into the field of perception in order to supplement information about the environment and improve information perception. Sensory modalities most often include vision, but may include sound, touch and smell.

S. I. Ashmarina et al. (Eds.): ISCDTE 2019, LNNS 84, pp. 473–479, 2020.
https://doi.org/10.1007/978-3-030-27015-5_57

Mixed reality (MR) is a term used to describe the user environment in which physical reality and digital content are combined in such a way as to ensure interaction with and between real objects and virtual objects.

2 Methodology

General scientific research methods are: systemic and functional approaches; special research methods: sociological (method of expert assessments, observation and analysis) and comparative methods.

3 Results

The demand for AR and VR in corporate training has caused a sharp increase in the global market and, according to forecasts, its volume will reach $ 2.8 billion by 2023 [17]. The advantages of using VR/AR for professional development can be the following:

- make training more innovative and interesting for students (especially for generation Z);
- provide experience in solving complex tasks that cannot be obtained in any other form of training (innovative computer systems and 3D simulators are actively used throughout the world in training professions that involves risk or difficult conditions, VR technologies allow simulating standard and abnormal situations with complete immersing a person in the course of events);
- allow you to participate in training anywhere in the world (with a dispersed labor force, sending trainees to one centralized training place can be costly, and distance learning can lead to a breakdown of communication with remote trainees);
- students are taught through practical modeling, and not on the basis of theoretical concepts (due to the effect of presence, a person deeply assimilates information and gains real practical experience, and the level of skills retention is significantly increased compared with traditional learning technologies);
- offer a training ground that allows students to make mistakes and receive feedback in real time during the training process, which reduces the number of errors in the workplace;
- encourage students to work at their own pace and in their own way [10, 13].

As advantages of VR-based learning technologies, the following can also be noted:

- increased student involvement in the process, the ability to change scenarios, influence the course of events;
- full concentration on learning due to the absence of external stimuli;
- high visibility (with the help of VR-technology, you can demonstrate any process, both for a holistic perception and to simulate any detail);
- flexible application (development for specific educational tasks);

- ability to scale possible situations, create emotional experience and broadcast it to any number of people [8].

Industrial companies use VR to teach highly specialized professional skills:

- interact with sophisticated equipment or production lines;
- work with hazardous production processes as part of training on various aspects of industrial safety;
- work with critical processes in which the error of the unqualified employee can lead to significant financial or property losses (the developer creates virtual copies of the equipment, digitizes the rules for working with it, simulates working situations in the virtual space, and the learner goes through multiple training sessions that develops a steady skill) [5, 16].

So, in the automotive industry, VR is used to train engineers and minimize the cost of "learning from your own mistakes". At the expense of modeling, employees evaluate the functionality, limitations and suitability of certain solutions even before their industrial implementations [2]. In the power industry, VR solutions allow employees of a nuclear power plant to imagine what could happen if a reactor melts at a nuclear power plant, so you can respond to the emergency with less panic and more control [19].

In transportation, the VR solution helps ground staff dock the plane to reduce the cost of fuel, logistics and equipment by offering a safe learning environment for employees. In modern practice, there are already quite a few examples of successful application of VR technologies in corporate training [7, 9, 11, 12]:

- Linde, a transnational chemical company, uses VR technology to teach dangerous work: for example, drivers who transport industrial gases (in virtual reality, students can go through as many repetitions as necessary to prepare them for the first shipments in the real world). The company also created the immersive virtual environment reality to train operators for a plant under construction (students can practice emergency procedures or dangerous tasks, explore the environment, look into existing equipment to see better the installation for which they will soon be responsible);
- The International Air Transport Association launched the virtual reality training module RampVR for ground service workers, which improves the training process while avoiding the risks and logistical costs associated with training employees during normal airport operations (training module uses visual and sound accompaniment to provide a feeling of real working conditions in day and night shifts and in different weather);
- Employees of the largest US oilfield service company Schlumberger are using VR to work on oil rigs (poor visibility, equipment breakdowns and other abnormal situations, operators work in safe conditions);
- The Russian company Gazprom Neft uses VR technologies to train its employees in emergency response plans, technical skills, and also in field exercises, where several people are simulated in real time and in a single virtual space (this allows reproducing complex and dangerous processes, it does not freeze production sites for training specialists and gaining practicing skills);

- The Russian Railways uses a virtual simulator - electricians are learning to repair a machine electric drive.

In the banking sector and retail business, VR technology is used to train staff in communication skills (soft skills), aspects of customer interaction (virtual reality effectively helps immerse a person in a certain emotional environment), for example:

- In 2018 the retail giant Walmart acquired 17,000 Oculus Go headsets to train more than a million employees in almost 5,000 stores around the world (with the help of trainings in virtual reality, sellers, including new Walmart employees, expect to quickly serve customers during the "Black Friday", when the number of visitors increases dramatically, and to "work out" the daily scenarios that employees have to face) and expects to increase the efficiency of training by 10–15%;
- The KFC fast-food chain uses virtual reality simulation to complement its multi-level employee training program (virtual reality training is based on a game that you are not able to leave until all five stages of training are executed flawlessly); The VR version reduced the training time from 25 min per person to 10 min, which should provide significant savings in personnel training costs;
- The corporate university of Sberbank uses several empathy simulations offering situations in which the client of the company may be (a person with disabilities, a citizen of old age). This training allows the specialist to understand what difficulties this category of clients faces in the branches of the Bank and how to interact with them so that the communication process is effective for all parties.

Augmented reality (AR) technologies can be used for informational support when servicing and repairing equipment, making detours, regardless of whether they occur on-site within the company or by technical specialists on-site making service calls at customer sites (it is possible to display a process map or video - instructions for augmented reality glasses, a tablet, any other device, etc.), as well as when learning the relevant skills.

Benefits of AR-technologies [15]:

- Informative content, enrichment of visual and contextual learning;
- Preparation for interaction with real objects without consequences (simulators, virtual laboratory work);
- "Just-in-time" training, when users receive the information they need in real-time as it is needed at work;
- Interaction with objects in real life inaccessible.

We can give some examples of AR-technologies in corporate training, namely:

- The technology giant CISCO Systems (the production and sale of network and telecommunications equipment and other high-tech products) uses AR-technologies in the workplace; applying AR to real physical devices allows technicians to see what they need to do when installing various parts. It eliminated the need to read manuals, increased installation efficiency by 20% and increased installation accuracy by 90%;
- Boeing uses Microsoft HoloLens to show engineers how to assemble pieces of equipment together using 3D models (additional information appears in text form

with voice overlays to guide the engineer on smart glasses). Due to AR-technology, Boeing expects to reduce the time for training engineers by 75%;
- Honeywell, a multinational engineering, industrial and aerospace conglomerate, taking into account the tendency of aging personnel and the need for high-quality transfer of knowledge to young professionals, uses the new generation of VR/AR training tools for personnel training in the most important activities in production. In the Connected Plant Skills Insight Immersive Competency Vocational Training System, Microsoft uses the HoloLens holographic computer and the Windows Mixed Reality headset to simulate various Honeywell C300 controller scenarios (main controller failure and failover, cable failure, power failure, etc.). Simulating these scenarios allows you to train personnel and test acquired skills. This approach allows a 100% increase in the effectiveness of maintaining skills compared to traditional teaching methods and reduces the duration of technical training by 66%, using the experience of leaving workers and sharing it with the next generation in a form that they prefer;
- Lockheed Martin uses the AR app for ship maintenance tasks. Naval engineers who must perform maintenance are very young and they have been trained in many procedures only once or twice. If they need help with a specific technical task, then using an application and an AR viewer, companies can easily call out a hologram of the equipment in question, as well as use video or audio instructions to step through tasks;
- Bosch provides technical training through mobile equipment using AR- technology (AR-Image provides a detailed understanding of the structure and functions of high-voltage motors, students are shown the difference between hybrid and all-electric vehicles, the functions and characteristics of high-voltage components and various troubleshooting methods);
- BIOCAD, an international biotechnology company, uses virtual and augmented reality technologies to practice skills of operators working in biopharmaceutical plants. For example, it trains staff to fill up a bioreactor (this procedure takes more than four hours in real time and access to the bioreactor for educational purposes is difficult; the training of one employee may take several months until he can cope with refueling without the help of the instructor). VR/AR-simulator gives around-the-clock access to the practical development of all necessary skills and regulations.

Mixed reality (MR) is a deeper level of augmented reality, but it does not isolate the user from the external environment, as is the case with VR. Virtual holograms that the user sees on the screen with the help of special equipment (3D glasses or a helmet) are three-dimensional images that are virtually indistinguishable from real objects.

According to experts, although VR has the effect of full immersion, it is likely that AR and MR will find more use in training (including because they are more affordable) [3].

4 Discussion

The results of the Accenture Technology Vision 2018 show that 36% of managers highlight the elimination of distance barriers between people (as a result, information barriers) as a driver for using immersive technologies in training. These technologies reduce the risks and costs associated with personnel training, solve the problem of training narrow specialists, provide quick adaptation of an employee to work tasks and opportunities to preserve and digitize practical skills and knowledge [18].

At the same time, the penetration of VR/AR/MR technologies into corporate training and workflows is hampered by objective reasons: the lack of high-tech infrastructure, the need to develop industry standards, the lack of awareness of potential users about the benefits of new tools [1]. One of the limiting factors is also the constant dependence of many companies on the learning management system (LMS) (too many LMS options are extremely outdated compared to what is needed to use VR/AR/MR in corporate training programs, including advanced web analytics and interfaces application programming (xAPI)). In addition, companies must create a corporate culture that perceives new technologies, such as VR/AR, as a means to achieve the goal, which is to improve the company's bottom line and promote the success of individual employees.

5 Conclusion

Currently, digital staff training is becoming the fastest growing segment of spending on digital technology. At the same time, we should note a radical change in training from internal corporate programs aimed at developing personnel to virtual platforms, which can help employees develop independently. This conclusion necessitates further research and solving the problems of intra-corporate digital staff training as a condition for improving the effectiveness and efficiency of business organizations.

References

1. Alexander, T.: Virtual and augmented reality: innovation or old wine in new bottles? Advances in Intelligent Systems and Computing, vol. 822, pp. 233–239 (2019)
2. Bennett, S.: Advances in engineering and design technologies. Pet. Technol. Q. **18**(3), 47–49 +51 (2013)
3. Bessa, B.R., Santos, S., Duarte, B.J.: Toward effectiveness and authenticity in PBL: a proposal based on a virtual learning environment in computing education. Comput. Appl. Eng. Educ. **27**(2), 452–471 (2019). https://doi.org/10.1002/cae.22088
4. Chen, S.Y., Wang, J.: Human factors and personalized digital learning: an editorial. Int. J. Hum.-Comput. Interact. **35**(4–5), 297–298 (2019). https://doi.org/10.1080/10447318.2018.1542891
5. Ciuffini, A.F., Di Cecca, C., Ferrise, F., Mapelli, C., Barella, S.: Application of virtual/augmented reality in steelmaking plants layout planning and logistics. Metallurgia Italiana **7**, 5–10 (2016)
6. Cooper, G., Park, H., Nasr, Z., Thong, L.P., Johnson, R.: Using virtual reality in the classroom: preservice teachers' perceptions of its use as a teaching and learning tool. Educ. Media Int. **56**(1–2), 1–13 (2019). https://doi.org/10.1080/09523987.2019.1583461
7. De Amicis, R., Riggio, M., Shahbaz Badr, A., Fick, J., Sanchez, C.A., Prather, E.A.: Cross-reality environments in smart buildings to advance STEM cyberlearning. Int. J. Interact. Des. Manuf. **13**(1), 331–348 (2019). https://doi.org/10.1007/s12008-019-00546-x

8. Fertleman, C., Aubugeau-Williams, P., Sher, C., Lim, A.-N., Lumley, S., Delacroix, S., Pan, X.: A discussion of virtual reality as a new tool for training healthcare professionals. Front. Public Health, **6**(FEB) (2018). https://doi.org/10.3389/fpubh.2018.00044
9. Grogna, D., Stassart, C., Servotte, J.-C., Bragard, I., Etienne, A.-M., Verly, J.G.: Some novel applications of VR in the domain of health. In: 20th Congress of the International Ergonomics Association, IEA 2018. Advances in Intelligent Systems and Computing, Florence, vol. 827, pp. 426–427 (2019). https://doi.org/10.1007/978-3-319-96059-3_49
10. Huang, Y.-C., Backman, S.J., Backman, K.F., McGuire, F.A., Moore, D.W.: An investigation of motivation and experience in virtual learning environments: a self-determination theory. Educ. Inf. Technol. **24**(1), 591–611 (2019). https://doi.org/10.1007/s10639-018-9784-5
11. Khandelwal, K., Upadhyay, A.K.: Virtual reality interventions in developing and managing human resources. Hum. Resour. Dev. Int. (2019). https://doi.org/10.1080/13678868.2019.1569920
12. Kornilov, Y.: VR-technologies in education: experience, tools overview and application prospects. Innov. Educ. **8**, 117–129 (2018). (in Russian)
13. Martín-Gutiérrez, J., Mora, C.E., Añorbe-Díaz, B., González-Marrero, A.: Virtual technologies trends in education. Eurasia J. Math. Sci. Technol. Educ. **13**(2), 469–486 (2017). https://doi.org/10.12973/eurasia.2017.00626a
14. Mitrofanova, E.A., Simonova, M.V., Tarasenko, V.V.: Potential of the education system in Russia in training staff for the digital economy. Advances in Intelligent Systems and Computing, vol. 908, pp. 463–472 (2020). Conference on Digital Transformation of the Economy: Challenges, Trends and New Opportunities, Samara (2018). https://doi.org/10.1007/978-3-030-11367-4_46
15. Porter, M.E., Heppelmann, J.E.: Why every organization needs an augmented reality strategy. Harvard Bus. Rev., 1–13 (2017)
16. Ranger, B.J., Mantzavinou, A.: Design thinking in development engineering education: a case study on creating prosthetic and assistive technologies for the developing world. Dev. Eng. **3**, 166–174 (2018). https://doi.org/10.1016/j.deveng.2018.06.001
17. Schiffeler, N., Plumanns, L., Stehling, V., Haberstroh, M., Isenhardt, I.: Technology acceptance in corporate learning processes of production companies: a qualitative pre-study. Paper presented at the Proceedings of the International Conference on Intellectual Capital, Knowledge Management and Organisational Learning, ICICKM, November 2018, pp. 293–300 (2018)
18. Technology Vision 2018 Intelligent Enterprise Unleashed (2018). https://www.accenture.com/_acnmedia/Accenture/next-gen-7/tech-vision-2018/pdf/Accenture-TechVision-2018-Tech-Trends-Report.pdf Accessed 8 Mai 2019
19. Whisker, V.E., Baratta, A.J.: Use of virtual environments to reduce the construction costs of the next generation nuclear power reactors. Paper presented at the Societe Francaise d'Energie Nucleaire - International Congress on Advances in Nuclear Power Plants - ICAPP 2007, "the Nuclear Renaissance at Work", vol. 4, pp. 2407–2416 (2008)

Staff Responsibility as Efficiency-Driven Factor of ERP-Systems

V. M. Svistunov[1], V. V. Lobachev[1], and M. V. Simonova[2(✉)]

[1] State University of Management, Moscow, Russia
[2] Samara State University of Economics, Samara, Russia
m.simonova@mail.ru

Abstract. It is impossible to improve the competitiveness of Russian companies without advanced business and management technologies. The study contains the implementation and operation results of ERP-systems in a number of Russian companies. The study made it possible to determine the level of performance efficiency of these systems and formulate the main problems that prevent companies from carrying out this process as efficiently as possible. One of the important problems that significantly reduce the performance efficiency of ERP-systems is a high level of resistance of workers to implement and use these systems. The earliest elimination of contradictions between the company's management, who are for ERP-systems, and the staff involved in the system is the way to improve the performance efficiency of the system as a whole. The study presents the level of resistance among various categories of staff in a number of Russian companies. The obtained data allowed us to identify the reasons that increase the level of resistance of company staff to ERP-systems, and to develop areas that increase the level of commitment, or the number of employees supporting relevant procedures.

Keywords: Commitment · ERP-system · Implementation · Organization · Resistance to implantation · Staff

1 Introduction

Current trends in globalization of the world economy and the expansion of markets for goods and services inevitably lead to toughed competition and complicate the processes and management tools implemented and used by company management. The active digitalized production [9, 13] and the increasing problems of management make the companies' management solve the problem of automating their business processes by introducing modern corporate information systems (CIS).

The modern CIS is based on the so-called Requirements/resource planning, which consists of a number of logically related rules and procedures that transform production planning into a chain of requirements synchronized in time and material coverings of these requirements for each unit of resource required to complete the production schedule. The first software implementation of this concept was focused on material management (production and supply), becoming the basis for the development of automated enterprise management systems since the mid 80s of the last century, and

S. I. Ashmarina et al. (Eds.): ISCDTE 2019, LNNS 84, pp. 480–486, 2020.
https://doi.org/10.1007/978-3-030-27015-5_58

was called the Materials Resource Planning System (MRP) [15]. Further improvement in planning led to the transformation of the MRP system into an extended modification, later called the MRP-II system (Manufacturing Resource Planning System). The development of these systems has led to the emergence of the ERP (Enterprise Resource Planning System) system, which not only supports functional business planning and integrated resource planning, but also ensures the company's "on-line" interaction with its suppliers and customers. This problem in ERP-systems is solved using SCM - Supply Chain Management [3, 4] and CRM - Customer Requirements Management.

Modern ERP-systems have the following features:

- The system is designed and capable of collecting (receiving), analyzing and storing information about the company's markets of interest and the "players" on them in volumes significantly larger than even 5–10 years ago [5, 11];
- The system develops and supports "big data" storage and also has efficient and numerous algorithms for their automated processing [2, 6];
- The system efficiency requires the involvement of staff with professional competencies for maintenance and for its further development. At the same time, the company's full-time employees must have special knowledge, tools and skills for searching, selecting, analyzing, correct and targeted transmission of information within the ERP-system [6–8, 14];
- The current organizational structure of the management must comply with the information exchange within the company with a clear definition of responsibility centers to optimize costs, prevent possible loss of information and improve the management efficiency [8, 9].

The introduction of ERP-systems, most often, leads to fundamental organizational and methodological changes in the company. At the same time, workers perceive changes ambiguously, and it divides them into appropriate groups - from open opponents to overt supporters. Causes of concern: retraining to obtain new professional knowledge and skills, a possible change of profession, changes in pay and motivation principles, fear of losing your job, etc. [1, 5, 10, 11]. Under emerging conditions, the task of the company's management is to achieve staff commitment to changes and reduce staff resistance to changes caused by introduced automated technologies [4, 7, 12, 13].

2 Methodology

The effective methods of resistance management allows reducing the negative reaction of all company staff, certain groups of it or individual employees, which significantly complicates the implementation and practical use of the ERP-system. At the same time, resistance is the expected response of staff to organizational and methodological changes in business processes that accompany the implementation of new information and communication technologies into the practice of the company. Negative reaction of staff is a signal of unwillingness to change the established regulations, an attempt to "protect" from the need to learn new methods and tools to fulfill official duties. Most often, the resistance to the ERP-system is reflected in the passive participation of employees in this process, unwillingness to participate in ongoing training programs or professional retraining, the use of disagreement tactics and hidden sabotage in relation to changes.

This transformation, taking place under existing stereotypes and prejudices of workers, allows us to determine the level of resistance among various categories of staff and the need to develop and implement practical actions aimed at increasing the level of staff commitment in the context of implementing and operating the ERP-system. At the same time, the level of commitment characterizes the number of company employees who are ready to support and actively participate in the implementation and operation of the corporate ERP-system.

3 Results

The authors assess the current implementation of ERP-systems in Russian companies. Is the expert community satisfied with the implementation of ERP-systems in domestic companies? How do experts assess the progress achieved and the existing problems? Each expert was asked to rate these questions on a 5-point scale. The survey results are presented in Table 1.

Table 1. Satisfaction with the implementation of ERP-systems in Russian companies (% of the number of experts in this category)

Expert category	Fully satisfied	Rather satisfied, than none	Something satisfied, something none	Rather not satisfied, than satisfied	Dissatisfied
Rating scale	5	4	3	2	1
Top managers	42,4	29,4	28,2	0,0	0,0
Middle managers	17,6	31,0	38,8	12,6	0,0
Low-level managers	8,7	20,6	36,5	27,0	7,2
Specialists	6,9	21,3	40,8	25,2	5,8
Workers	2,7	18,5	55,1	23,7	0,0
TOTAL	12,6	23,6	40,2	19,9	3,7

Source: compiled by the authors.

The average aggregate assessment of satisfaction with the implementation of ERP-systems was 3.22%; assessments generally correspond to the normal Gaussian distribution. At the same time, assessments for individual expert groups show that the highest among them are among top and middle managers (4.14 and 3.54, respectively), due to a small proportion of “routine” operations with information files.

More critical, in their assessments, are groups of low-level managers and specialists (2.97 and 2.98, respectively). For these groups of experts, the satisfaction curve is shifted to the negative part of the scale, which indicates a critical attitude to the implementation of ERP-systems. For workers, the index value is 3.0.

Experts of the older age group (over 55 years old) have the least degree of adherence to implementation processes of modern CIS, and young experts (up to 35 years old) are the most positive. The satisfaction index of the first group is 2.24, and the index of the second group - 3.77. Analysis of the gender parameter shows that men are more preferable to processes introducing ERP-systems than women (index values are 4.18 and 2.34, respectively).

The most critical view on the problem under study is among low-level managers - men over 55; and the most optimistic group of experts are men under 30 who perform the duties of functional specialists.

4 Discussion

The study used expert assessments representing seven domestic companies from four regions of the Russian Federation that are currently introducing or having previously introduced ERP-systems into the practice of their management activities: Moscow city, Perm city, the Moscow and Tula regions. 246 people were interviewed, representing five categories of experts: top managers (6% of the total number of experts), middle managers (27%), low-level managers (17%), specialists (39%) and workers (11%).

The sample included deputy directors of companies; heads of departments and services; line managers; members of project teams and groups (specialists) and workers who are currently taking part or having previously participated in project activities implementing and operating the ERP-system in their company. In total, experts directly involved amounted to 62.7% in the sample. The number of experts who use the KIS in their daily work practice is 37.3%. Interviews and surveys conducted to collect baseline information were anonymous.

The ratio of men and women in the sample was 78.0% and 22.0%, respectively. The age structure of the sample: the youngest expert is 23 years old and the oldest is 71 years old. The average age of experts is 56 years old.

To assess the implementation level of ERP-systems in Russian companies, the satisfaction index was used, which was calculated as a weighted average of a 5-point satisfaction scale for each category of experts.

Having studied the problems in Russian companies when implementing and operating ERP-systems, an interview method was used with an open-ended question, answering which the experts called the problems without the external influence and the influence of the proposed response options and their sequence. The method used allowed us to collect material characterized by experts' open statements who are directly involved the implementation and practical operation of ERP-systems in their daily activities. The data obtained in the interview were subjected to quantitative and qualitative content analysis.

5 Conclusion

The content analysis of experts' open statements in answering the question of what implementation and operational problems seem to be the most acute for them allowed us to form twenty-three groups of answers. Table 2 presents generalized groups of answers expressing expert opinion on the most significant problems causing active resistance among various categories of staff in domestic companies implementing or operating ERP-systems.

Table 2. The main reasons for the resistance of various categories of staff in domestic companies implementing ERP-systems

Reason of occurrence	Reason of resistance	Possible consequences	Staff category
Misunderstanding the main goals of ERP-systems, lack of criteria for achieving goals and assessing project results	Low level of staff awareness of basic principles, tools and capabilities of ERP-systems in the automation of business processes	Artificial "slowdown", passive participation of employees in the implementation or hidden sabotage	Low-level managers, specialists
Function mismatch of ERP-systems to the declared opportunities	Increase in labor productivity	Decrease in the level of motivation due to the decrease in labor productivity	Middle managers, specialists
Weak consideration of the company specifics in the course of implementation	Implementation of unified solutions that complicate management processes	Reduced efficiency of management processes	Low-level managers, specialists
Lack of directed staff motivation in the course of implementation	Lack of incentives to participate in the implementation of ERP-systems	Artificial "slowdown" or passive participation of employees in the implementation	Middle and lower-level managers, specialists, workers
Misunderstanding the project goals and personal goals of employees	Fears of change in working conditions and payment	Artificial "slowdown" of personal participation	Lower-level managers, specialists, workers

Source: compiled by the authors.

The position of top managers improves the implementation of ERP-systems (42.9% of them are completely satisfied, and 29.4% are almost satisfied with the implementation in their companies). Using their authority and official powers, they "make" the implementation process quite manageable and inevitable. The results obtained in the course of the study indicate that there are serious difficulties faced by the management of Russian companies in the course of implementing modern ERP-systems. First, it is a low level of staff commitment to modern trends in digitalization of management procedures. We can solve this problem by:

- Increasing the level of organizational and methodological trust between staff and management of the company;
- Developing effective incentives aimed at creating necessary conditions for the conscious involvement of staff in projects introducing EIS;
- Convincing the company's employees in the necessity and inevitability of these systems;
- Increasing staff awareness of the goals, objectives and progress of the corporate ERP-system.

References

1. Budiningsih, I., Dinarjo, T., Ashari, Z.: Improvement of employees' performance through training intervention in digital era. Eur. Res. Stud. J. **20**(4), 637–654 (2017)
2. Arce Cuesta, D., Borges, M., Gomes, J.O.: Planning the combination of "Big data insights" and "Thick descriptions" to support the decision-making process (2019). https://doi.org/10.1007/978-3-030-11890-7_8
3. Evtodieva, T.E., Chernova, D.V., Ivanova, N.V., Protsenko, O.D.: Business analytics of supply chains in the digital economy. In: Ashmarina, S., Mesquita, A., Vochozka, M. (eds.) Digital Transformation of the Economy: Challenges, Trends and New Opportunities. Advances in Intelligent Systems and Computing, vol. 908, pp. 329–336. Springer, Cham (2020)
4. Guryanova, A.V., Smotrova, I.V., Makhovikov, A.E., Koychubaev, A.S.: Socio-ethical problems of the digital economy: challenges and risks. In: Ashmarina, S., Mesquita, A., Vochozka, M. (eds.) Digital Transformation of the Economy: Challenges, Trends and New Opportunities. Advances in Intelligent Systems and Computing, vol. 908, pp. 96–102. Springer, Cham (2020)
5. Klopova, O., Komyshova, L., Simonova, M.: Professional development in the field of human resource management of heads and specialists of the innovative organizations. Probl. Perspect. Manag. **16**(1), 214–223 (2018). https://doi.org/10.21511/ppm.16(1).2018.21
6. Korneeva, T.A., Potasheva, O.N., Tatarovskaya, T.E., Shatunova, G.A.: Human capital evaluation in the digital economy. In: Ashmarina, S., Mesquita, A., Vochozka, M. (eds.) Digital Transformation of the Economy: Challenges, Trends and New Opportunities. Advances in Intelligent Systems and Computing, vol. 908, pp. 66–78. Springer, Cham (2020)
7. Kot, M.K., Spanagel, F.F., Belozerova, O.A.: Problems of digital technologies using in employment and employment relations. In: Ashmarina, S., Mesquita, A., Vochozka, M. (eds.) Digital Transformation of the Economy: Challenges, Trends and New Opportunities. Advances in Intelligent Systems and Computing, vol. 908, pp. 227–234. Springer, Cham (2020)
8. López, M.A., Ros-Garrido, A., Fluixá, F.M.: Professionals supporting employment: training and accompaniment in work integration enterprises. [Profesionales de apoyo a la inserción: Formación y acompañamiento en empresas de inserción]. CIRIEC-Espana Revista De Economia Publica, Social y Cooperativa, vol. 94, pp. 155–183 (2018). https://doi.org/10.7203/ciriec-e.94.12698

9. Mitrofanova, E.A., Konovalova, V.G., Mitrofanova, A.E.: Opportunities, problems and limitations of digital transformation of HR management. In: Mantulenko, V. (ed.) GCPMED 2018 – International Scientific Conference Global Challenges and Prospects of the Modern Economic Development, Samara, vol. LVII, no. 174, pp. 1717–1727. The European Proceedings of Social & Behavioural Sciences EpSBS (2019). https://doi.org/10.15405/epsbs.2019.03.174
10. Nahipbekova, S., Kuralbayev, A.: Methodical aspects of job satisfaction measure of employees in hotel business quality improvement in Kazakhstan. Afr. J. Hosp. Tour. Leis. **7**(3), 1–12 (2018)
11. Shafigullina, A.V., Akhmetshin, R.M., Martynova, O.V., Vorontsova, L.V., Sergienko, E.S.: Analysis of entrepreneurial activity and digital technologies in business. In: Ashmarina, S., Mesquita, A., Vochozka, M. (eds.) Digital Transformation of the Economy: Challenges, Trends and New Opportunities. Advances in Intelligent Systems and Computing, vol. 908, pp. 183–188. Springer, Cham (2020)
12. Sho, R.B.: Keys to Trust in the Organization: Effectiveness, Decency, Care Manifestation. Business, Moscow (2000). (in Russian)
13. Svistunov, V.M.: Four questions about digital economy. Pers. Intellect. Resour. Manag. Russia **1**(40), 5–14 (2019). (in Russian)
14. Van der Togt, J., Rasmussen, T.H.: Toward evidence-based HR. J. Organ. Effectiveness **4**(2), 127–132 (2017). https://doi.org/10.1108/JOEPP-02-2017-0013
15. Zimin, V.V., Mit'Kov, V.V., Zimin, A.V.: Calendar planning of it-services of the enterprise's ERP-project. Izv. Ferrous Metall. **61**(4), 319–325 (2018). https://doi.org/10.17073/0368-0797-2018-4-319-325

Evaluation of Event Marketing in IT Companies

D. V. Chernova[1], N. S. Sharafutdinova[2], E. N. Novikova[2], I. T. Nasretdinov[2], N. G. Xametova[3], and Y. S. Valeeva[4](✉)

[1] Samara State University of Economics, Samara, Russia
[2] Institute of Management, Economics and Finance, Kazan Federal University, Kazan, Russia
[3] Institute of International Relations, Kazan Federal University, Kazan, Russia
[4] Kazan Energy University, Kazan, Russia
valis2000@mail.ru

Abstract. The study considers evaluation of event marketing in IT companies. The competitiveness of IT companies differs from the competitiveness of manufacturing companies: the quality of services provided, innovation, security, flexibility. Tele2 Telecommunications Company has gained popularity due to a non-standard approach to business, a well-thought-out marketing strategy, accounting and anticipation of customer needs. The authors analyzed the existing evaluation methods of event marketing in IT companies and substantiated quantitative methods based on the results of surveys before and after the event, which makes it possible to most accurately show the degree of influence of event marketing on customer loyalty. This mechanism is aimed at attracting the target audience, the intensive development of IT companies and infrastructure development. The existing methodologies evaluate the communicative or only economic effect. Therefore, it is proposed to use a generalized method that allows determining each of the stages of marketing communications in event marketing and, if necessary, make appropriate adjustments. The method proposed quantitative indicators for evaluation of event marketing, and the results are evaluated in certain periods.

Keywords: Competitiveness of the telecommunications company · Digitalization · Evaluation of event marketing in IT companies · Event marketing · Infrastructure

1 Introduction

Due to the increasing supply and requirements for IT players, the transformation of ways to search for, obtain and analyze information about goods and services by end users, changes in attitudes towards classical advertising and communication policies of companies and increasing costs of marketing activities of companies, marketing tools become the most effective marketing tools which include activities aimed at strategic and tactical promotion of companies, their services, products and brands. Tele2 is a well-known cellular operator and provider in the Russian Federation, offering cellular services, mobile Internet and television. At the same time, competitive strategies that

S. I. Ashmarina et al. (Eds.): ISCDTE 2019, LNNS 84, pp. 487–493, 2020.
https://doi.org/10.1007/978-3-030-27015-5_59

should be aimed at obtaining and strengthening the company's essential advantages for successful business and improving the quality of life of the population allow the company to determine the most priority market trends.

The purposes of the study are to develop and systematize theoretical positions, generalize practical experience and formulate methodological foundations for the systematic use and management of event marketing of IT companies:

- Identify and systemize methods for evaluation of event marketing activities;
- Develop and test a methodology for evaluation of event marketing in IT companies;
- Develop recommendations for evaluation of event marketing in IT companies.

Some researchers consider the indicator of brand awareness as an indicator of the success of an advertising campaign in marketing communications, and accordingly, in event marketing [8]. Perepelkin applies indicators for evaluating the cost-effectiveness of events [4].

2 Methodology

Taking into account the diverse nature of various marketing communications tools, a synergistic effect and effects of various tools, it is advisable to use a generalized method. In the first case, effectively implemented communications and a certain corporate reputation increase the length of time for obtaining maximum income. In terms of return on investment and economic result, evaluation of event marketing is not comprehensive and sufficient. The second effect implies that as a result of the communication policy, the time it takes for a company to achieve financial indicators comparable to the average level in the industry can be significantly reduced. The foregoing leads to a number of conclusions. First, evaluation should be carried out using an integrated approach, i.e. it is necessary to analyze and evaluate not individual elements and tools of communicative policy, but their integrated system. Secondly, evaluation of event marketing should also include consideration of communicative or economic effects and their combination.

It is important to know that the analysis of the degree of satisfaction and trust in the company, readiness for dialogue or cooperation also largely determines the effective interaction with the target audience, which at the present stage should be an active participant in the communicative process.

Offline event marketing is becoming increasingly popular. As large amounts of data from social networks based on location, such as Foursquare, Gowalla and Facebook, using this data to analyze user behavior in social networks becomes an important issue for offline event marketing [3].

Schematically, the technique is presented in Fig. 1. Let us consider in more detail each of the stages. The first and fundamental step is to determine key indicators in assessing the impact of event marketing on a brand. The authors Romanov and Panko identified in the methodology one single, but at the same time, key indicator of Brand Awareness with and without a hint [6]. In the literature, its conventional designation is Brand Awareness. In general, the effect of event marketing tools is determined by the degree of impact of events on changes in sales or turnover. When calculating the cost-effectiveness, the costs of organizing special events and increasing the profit that the company received through the implementation of the event are compared. It is often difficult to unequivocally calculate the increase in profits, since special events have a prolonged effect of impact and do not always directly affect the increase in profits.

This is also justified by the fact that the profit gained by the company during an event rarely covers the cost of its holding. In addition, supporting events aimed at an external marketing environment, the company uses the full range of capabilities of integrated marketing communications tools.

As part of the methodology, the following indicators of the direct impact of events are applied: the number of new customers or the increase in the client base; the number of units sold during the event and at the event (or the number of new contracts entered into).

To assess Tele2's integration into the Red Bull Air Race international event, the authors took three main indicators from the method proposed by Perepelkin [4].

As one of the main indicators, we calculate the "pay-off point of the event". Having information about sales of goods for a certain period of time, it is easy to transform this indicator into a temporary one by entering the ratio of time (period) to price and variable costs. Another indicator of the cost-effectiveness of communication event marketing is the growth rate of turnover. When launching a product or focusing an event on a particular product, you can determine the profitability of the product promoted within this event.

Among the relatively universal and illustrative metrics, there are indicators of the communicative cost-effectiveness, the leading role of which is played by the ratio of the invited (or calculated) number of participants in the event and the actual number of participants, the coverage ratio of the event.

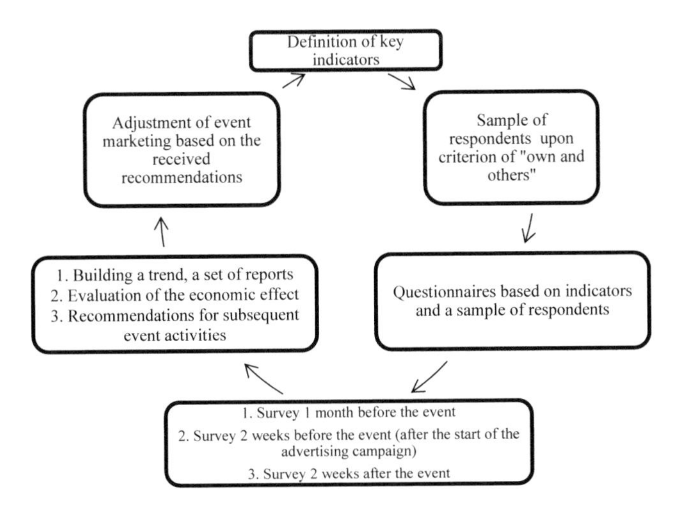

Fig. 1. The impact of event marketing on a brand (Source: compiled by the authors)

This ratio shows how successful the advertising and informational support of the event was, and can also speak about the interest of the contact audience in the topic (product) of the event. Potentially, this ratio can also talk about brand strength and customer loyalty to the company or its product.

3 Results

Based on the data obtained, the following conclusions can be made: the event pays off in the very first month, but the product used at this event has a rather low profitability. Also, the authors noted, based on the coverage of the event, that the actual and projected number of participants directly in the Tele2 zone has strong differences. It suggests that there was a low degree of interest of subscribers due to a weak announced advertising campaign. Input data for the calculation of indicators and their results are presented in Table 1. Thus, a comparative analysis of indicators of events clearly demonstrates the need for a comprehensive evaluation of these indicators in order to form a real picture of the economic and qualitative benefits of using event marketing tools in companies.

Table 1. Indicators of Tele2's integration into the Red Bull Air Race international event*

Payback point of the event (PPE, units)			Growth rate of turnover from the event (GRTE, %)			Profitability of the promoted product (PPP, %)			Coverage rate of the event (CRE, %)		
P (RUB)	272	2400	ATBEP	272	5,14%	P	57 285	8,7%	NPE (people)	350	50%
VC (RUB)	270		ATAEP	286		CI	650 000		NGE (people)	700	
CI (RUB)	650 000										

Source: compiled by the authors.

* P – price of the realized goods (services), average (RUB); CI – costs of implementation (RUB); VC – variable costs, average due to the diversity of the range (RUB); ATAEP –average turnover after the event for the period (RUB); ATBEP – average turnover before the event for the period (RUB); P – profit from the sale of the promoted product (RUB); NPE – number of participants of the event (people); NGE – number of guests invited to the event - estimated number of participants (people).

Also, we can state that the possible cost-effectiveness of events and ways of its evaluation directly depend on the goals that are put before event marketing in the process of planning an event marketing program.

Based on the data obtained, the following conclusions were made:

1. The weak announcement of Tele2's integration in the Red Bull Air Race international event;
2. Low profitability of the product being promoted directly at the event.

We have analyzed such indicators as the coverage and cost of the used announced channels: radio (live on Bim-radio channel); social networks (popular public and bloggers); targeted advertising (VK, Yandex); internal SMS sending. Taking into account the fact that the population in Kazan exceeds 1 million people, the authors concluded that the use of only these channels is essentially small and does not cover even one third of the city residents. It is also necessary to take into account the fact that there can be duplication of announcement channels per person, thus the total coverage can actually be much lower.

Therefore, to announce the event, the authors proposed the following information channels on indicators of coverage and costs: radio record (drawing tickets, on the eve of the event); 4 radio stations (broadcasting audio clips in the period of 2 weeks before the event); social networks (popular public and bloggers); outdoor advertising on the media carrier (in the period of 3 weeks before the event); targeted advertising (vk, yandex); internal SMS sending.

According to the authors, the use of these channels could increase the number of visits to the Tele2 zone, and since the minimum ticket to a closed zone cost 2000 rubles, this advertising campaign would pay off on the day of the event. The second important criterion for the subsequent adjustment of work is the use of the product for sales directly at the event. The profitability ratio of the product used was 8.7%.

The authors suggested using the service instead of the tariff plan with a bonus (the first month of the monthly subscription fee as a gift) - the usual package of services (My online tariff).

4 Discussion

Various systems of evaluation, developed by foreign and domestic authors can be divided into two groups: a system based on the use of economic indicators, and a system based on indicators of perception. Information for evaluation is usually collected by two methods. Most companies do simple research during or immediately after the event. More sophisticated research on brand preferences, brand loyalty and awareness is much less frequent, although more meaningful.

When planning and subsequent control of event marketing, it is necessary to take into account the presence of the so-called spill-over effect, when it is impossible to determine individual communication tools in their integrated use. The authors compared the differential impact of event marketing and advertising costs on brand value and company revenue. Using the longitudinal dataset for 74 real estate companies in China (2006–2013), the authors found that both event marketing and advertising costs had a positive impact on the company's revenue and brand value [2].

Brand awareness is considered in the works of various authors as an indicator of the success of an advertising campaign in marketing communications. Awareness Index, proposed in the early 80 s of the XX century by Millward Brown, records how much brand awareness is increasing with an increase in advertising exposure of 100 GRP. According to the authors Romanov and Pankov, an indicator of brand awareness can be evaluation of event marketing [6].

The study of consumer change is not evaluation of the event, it only allows you to notice and guess the consumer's preferences in order to react to the increased demand for a product or service earlier. Ultimately, a change in consumer attitudes is reflected in an increase in sales. It seems paradoxical that a change in consumer attitudes is not always associated with the company's marketing activities. Thus, according to the authors, the use of quality management system tools will increase customer loyalty to the company through a systematic approach to assessing customer satisfaction [7].

We agree with the opinion of Perepelkin [4], who offers a specific set of economic indicators for evaluation of event marketing. The authors have developed a useful practical set of success factors in event management to provide the optimized evaluation for live communication. A scientifically based study explained how the introduction of event management via the Internet influences these success factors with impressive results [1].

Syhre and Luppold describe the necessary basic technical elements that determine the success of lighting, sound, stage and media technology [9]. The authors explain the different event areas and production technologies that can be used at events.

It is also worth noting that taking into account the specifics of the industry, the development of a module for assessing the level of customer satisfaction with the quality of services will allow companies to increase loyalty to the company [5]. In our opinion, this set of criteria fully describes the economic component of activities in the framework of event marketing, but there is no psychological component in evaluation.

Our methodology is based on conducting surveys before and after the event, which allows us to most accurately show the degree of influence of event marketing on the brand. The following indicators are taken as a basis: brand awareness, brand loyalty, attitude to brand quality, attitude of consumers to brand value.

5 Conclusion

The methodology was tested on Tele2's integration in the Red Bull Air Race international event, where telephone interviews of respondents were conducted, the degree of influence of the event as a whole on the brand was assessed, and the economic effect of integration into a large-scale event was evaluated.

Based on the analysis of the event under study, the authors identified the problems in the preparation and implementation of the event were and gave recommendations to deal with these problems in order to more efficiently use such marketing tools as event marketing.

The study developed and proposed a universal methodology that allows companies to adapt to current market conditions and increase their market sustainability and the overall cost-effectiveness.

Thus, a comparative analysis of indicators of events clearly demonstrates the need for a comprehensive assessment of these indicators in order to form a real picture of the economic and qualitative benefits of using event marketing tools in companies. Also, we can state that possible events and ways to evaluate them directly depend on the goals that are put before event marketing in the process of planning an event marketing program.

References

1. Coppeneur-Gülz, Ch., Rehm, S.-V.: Event-Resource-Management mit digitalen Tools. Springer Fachmedien Wiesbaden GmbH. Springer Gabler, Wiesbaden (2018). https://doi.org/10.1007/978-3-658-22331-1
2. Liu, L., Zhang, J., Keh, H.T.: Event-marketing and advertising expenditures: the differential effects on brand value and company revenue. J. Advert. Res. **58**(4), 464–475 (2018). https://doi.org/10.2501/JAR-2017-043
3. Liu, Y., Liu, A., Liu, X., Huang, X.: A statistical approach to participant selection in location-based social networks for offline event marketing. Inf. Sci. **480**, 90–108 (2019). https://doi.org/10.1016/j.ins.2018.12.028
4. Perepelkin, N.A.: Event marketing. Basics of planning corporate events. Econ. Entrep. **10**(1), 681–684 (2015). (in Russian)
5. Rolbina, E.S., Novikova, E.N., Sharafutdinova, N.S., Martynova, O.V.: The study of consumer loyalty services. AD ALTA-J. Interdiscip. Res. **7**(2), 248–253 (2017)
6. Romanov, A.A., Pankov, A.B.: Marketing Communications. Eksmo, Moscow (2006). (in Russian)
7. Sharafutdinova, N., Valeeva, J.: Quality management system as a tool for intensive development of trade organizations. Mediterr. Ean J. Soc. Sci. **6**(N1S3): 498–502 (2015). https://doi.org/10.5901/mjss.2015.v6n1s3p498
8. Skorobogatyh, II., Selenite, I.R.: Effectiveness of event marketing (on the example of organization of applications, the legendary marketing guru Professor Philip Kotler in Plekhanov Russian University of Economics. Sci. J. "Initiatives XXI Century" **4**, 45–49 (2014). (in Russian)
9. Syhre, H., Luppold, S.: Event-Technik. Technisches Basiswissen für erfolgreiche Veranstaltungen. Gabler Verlag, IX, 75 (2018). https://doi.org/10.1007/978-3-658-19798-8

Trends in the Effectiveness of Russian Logistics in the Digital Economy

L. A. Sosunova[1(✉)], S. V. Noskov[1], K. P. Syrova[1], and I. G. Bakanova[2]

[1] Samara State University of Economics, Samara, Russia
kafedra-kl@yandex.ru
[2] Samara State Transport University, Samara, Russia

Abstract. The relevance of the research topic is due to the positive influence of logistics on the growth of the efficiency of the use of resources of national economies and, accordingly, the high requirements for methods of assessing the effectiveness of logistics. There are many methods for assessing the effectiveness of logistics at the national level. The choice of the method for determining and increasing the scientific basis for assessing the effectiveness of logistics is a problem. Its solution is possible on the basis of the application of theoretical, expert and economic-mathematical methods. The research objective is to more rigorous scientific justification of logistics costs and to determine the logistics efficiency of national economies, in particular, the economy of the Russian Federation. Research methods include methods of economic theory, expert and economic-mathematical methods. The authors formed a new structure of expenses for business logistics of the Russian Federation (in the structure of costs are costs of fixed capital for transportation and storage), developed a method to bring capital costs to current logistics costs, carried out expert evaluation of logistics performance indicators of the Russian Federation and developed a method for calculating a generalized logistics performance index at the national level.

Keywords: Costs · Efficiency · Generalized index · Logistics · National economy · Valuation methods

1 Introduction

1.1 Establishing a Context

The key role of functional logistics areas, its business processes and operations in the companies' management is to reduce their production, commercial costs and the profit growth. The analysis of cause-and-effect relations of the main economic indicators of the companies' activity on the markets of goods and services allows to draw a conclusion about the logistics impact, its strategic and operational decisions on the stability of economic entities in a changing competitive environment, improvement of the use of resource efficiency and customer satisfaction.

In terms of macroeconomics, efficient logistics enables the production of competitive goods and services with high added value and global customer service. The

S. I. Ashmarina et al. (Eds.): ISCDTE 2019, LNNS 84, pp. 494–505, 2020.
https://doi.org/10.1007/978-3-030-27015-5_60

experience of developed national economies shows a certain and positive logistics impact on the growth of the efficient use of capital, labor and investment resources, the reduction of energy and transport costs, the increase of gross domestic product (GDP) and living standards. High micro-and macro-efficiency of logistics management in a highly competitive environment is associated with significant current and capital costs for freight transportation, warehousing and storage of goods and material resources, logistics services, management of logistics business processes. In 2016, logistics costs in the US decreased by 1.5% compared to 2015 [8]. The dynamics and proportion of the individual components of logistics costs in total costs changed little with the exception of batch deliveries (10% increase), carload shipments by rail (13.8% decrease), the supply water transport (10% decrease) (Table 1).

Table 1. The structure and dynamics of business logistics costs in the USA

The structure of business logistics costs	2016	2016/2015, %	Average annual growth rate, %
Transport costs			
Truck transportation	595,5	−0,4	4,3
Batch deliveries	86,3	10,0	6,4
Railroad transportation, including	71,9	−11,0	−1,1
- Carload shipments	52,6	−13,8	−1,4
- Intermodal transportation	19,3	−2,5	−0,5
Air transport, including domestic, import, export, express delivery	66,9	1,5	2,4
Water transport, including domestic, import and export	40,6	−10,0	−0,1
Pipeline transport	33,6	1,1	4,2
Total transportation costs	894,8	−0,7	3,6
Inventory costs			
Storage cost	143,5	1,8	3,6
Reduced working capital	143,4	−7,7	−2,2
Other (obsolescence, loss, insurance)	122,9	−3,2	0,5
Total inventory costs	409,8	−3,2	0,5
Other costs			
Costs of transport service providers	44,7	0,7	4,2
Administrative costs	43,3	−4,6	2,8
Other total costs	88,0	−2,0	3,5
The total cost of the business logistics	1392,6	−1,5	3,6

Source: compiled by the authors based on data of CSCMP's 28th Annual State of Logistics Report [1, 8]

The total logistics costs included three components: inventory costs, consisting of warehousing costs (29,4% of total logistics costs in 2016), transport costs (64,3%) and administrative costs connected mainly with the dispatching costs and logistics information technology resources (6.3%).

According to the annual report of the logistics state, conducted by the supply chain management professionals (CSCMP), the total US logistics costs increased by 6.2% in 2017 compared to the previous year up to 1.5 trillion dollars [9]. It was 7.7% of the US GDP. Logistics costs have increased for a number of reasons, including higher infrastructure capacity, higher fuel prices and a growing shortage of truck drivers. Logistics costs actually declined in 2016 for the first time since 2009, but were higher in all directions in 2017. Cargo transportation increased by 6.6%, road transport costs by 9.5%, and parcel and Express delivery, a segment which is closely linked to e-Commerce, grew by 7%. In the future, stable macroeconomic growth is expected on the one hand, but on the other hand, higher logistics costs due to higher interest rates, tougher labor market and higher fuel prices.

High logistics costs in developed and developing national economies require the implementation of programs to reduce their relative value, that is, the share of GDP. The decline in the absolute value of the logistics costs is possible only due to constant total effective indicators and more private, for example, the indicators of service level, inventory turnover and customer satisfaction. The absolute and relative value of logistics costs for national economies in 2016 is shown in Table 2.

Table 2. Logistics costs of national economies in 2016 (billion dollars)

Country	GDP	The share of logistics costs in GDP, %	Logistics costs
Russian Federation	1 281	16,1	206,2
Austria	387	9,3	36,0
Germany	3 467	8,8	305,5
Spain	1 233	9,7	119,3
Britain	2 629	8,8	230,4
USA	18 569	8,2	1 523,0

Source: compiled by the authors based on data of International Monetary Fund, Australian Logistics Council, NESDB, Vietnam Business Forum, Logistics Viewpoints and Indonesia Investment, and Armstrong & Associates, Inc. Databases.

The logistics performance index of individual national economies is calculated at the macroeconomic level by the World Bank. The logistics performance index (LPI) is an interactive benchmarking tool for identifying the challenges faced by national economies in their logistics activities and identifying opportunities to improve their efficiency. LPI is based on a worldwide survey of operators (global freight forwarders and Express carriers) about specific parameters of logistics activities. LPI consists of both qualitative and quantitative indicators and helps to create logistics profiles of individual national economies [17]. The index includes the measurement of some indicators, such as customs characteristics, infrastructure quality and timeliness of deliveries.

This approach uses the main determinants of overall logistics performance and includes customs procedures, infrastructure, international transport, logistics competence, tracking and timeliness. The comparative logistics performance of national economies according to the World Bank in 2018 is presented in Table 3.

Table 3. Logistics performance index of national economies

Country	Rating	LPI index	Customs	Infrastructure	International transport	Logistics competence,	Tracking	Timeliness
Germany	1	4,20	4,09	4,37	3,86	4,31	4,24	4,39
Sweden	2	4,05	4,05	4,24	3,92	3,98	3,88	4,28
Belgium	3	4,04	3,66	3,98	3,99	4,13	4,05	4,41
Austria	4	4,03	3,71	4,18	3,88	4,08	4,09	4,25
Japan	5	4,03	3,99	4,25	3,59	4,09	4,05	4,25
Netherlands	6	4,02	3,92	4,21	3,68	4,09	4,02	4,25
Singapore	7	4,00	3,89	4,06	3,58	4,10	4,08	4,32
Denmark	8	3,99	3,92	3,96	3,53	4,01	4,18	4,41
Britain	9	3,99	3,77	4,03	3,67	4,05	4,11	4,33
Finland	10	3,97	3,82	4,00	3,56	3,89	4,32	4,28
Russian Federation	75	2,76	2,42	2,78	2,64	2,75	2,65	3,31

Source: compiled by the authors based on [17].

The Russian Federation takes the 75th place according to the logistics performance index. If we consider the individual index components, the Russian Federation lags behind the developed countries because of the customs efficiency and border control, the quality of trade and transport infrastructure, the competence and quality of international transport.

Thus, taking into account the high value and share of logistics costs in the Russian Federation, its low rating of the logistics performance index, as well as its positive impact on the rational use of resources, it is necessary to have evidence-based methods for logistics performance assessment.

1.2 Literature Review

The work is based on the study of European experience in logistics and export competitiveness [15], the logistics impact on the efficiency in the context of supply chains and services [12]. The empirical analysis of the logistics performance and global competitiveness was held by Kurganov [11], Yildiz [21], Yergaliyev and Raimbekov [20]. Liu researched the competitiveness of logistics service providers and international management practices in China and the UK, and Spillan et al. compared the effect of logistics strategies in the US and China [13, 19].

The importance of logistics performance especially such items as delivery times, costs and quality of services for international trade and economic growth in developing countries is discussed in the studies of Hummels and Schaur [6], Faria, Souza and Vieira [3].

The development of trade logistics, the study of the logistics impact on trade in the context of business was studied by Rutner, Aviles and Cox [16], Korinek and Sourdin [10], Hausman, Lee and Subramaniam [5], Portugal-Perez, Wilson [14], Arvis et al. [2], Gani [4].

Effective organization and the management of international forwarding operations, timely tracking of deliveries, high quality of transport infrastructure and information technologies were investigated by Korinek and Sourdin [10] and Jacyna [7]. However, there is a lack of research in the cost estimation methodology and the determination of the national logistics performance.

1.3 Establishing a Research Gap

The estimation of logistics costs in the Russian Federation is difficult due to the lack of detailed data on individual cost items in the methodology adopted by many developed countries. As a rule, such cost items for domestic business logistics as transportation costs, storage inventory costs, working capital costs, order processing costs and administrative costs are allocated. However, such an important element of logistics costs as the fixed capital cost for transport and warehouse purposes is not presented in the list of cost items for business logistics. In addition, there is a problem of bringing the cost of logistics fixed capital to its current costs.

The macroeconomic assessment of logistics performance in the Russian Federation is not accurate enough to be used in practical recommendations to improve the logistics performance. A more rigorous and evidence-based approach of the macroeconomic efficiency assessment of national logistics is to introduce weighting factors of importance and develop a generalized logistics performance index for any national economy.

1.4 Stating the Purpose

The purpose of the study is to provide a more rigorous scientific justification for national-level logistics expenditure and a generalized logistics performance index. The implementation of this goal has required the following tasks:

- the introduction of fixed capital for transport and warehouse purposes into the structure of costs for national logistics as an important factor of production;
- the development of a method of bringing the cost of fixed capital for transport and warehouse purposes to the current logistics costs;
- the employment of the expert evaluation method of private logistics performance indicators and economic-mathematical methods to determine their importance.

2 Methodology

At the first stage, the total value analysis, the analysis of structure and he share of logistics costs of national economies in GDP, including the economy of the Russian Federation, was carried out. Statistical methods were used. At the second stage, the necessity to include fixed capital costs for transport and warehouse purposes in the structure of logistics was proved and a method of bringing them to the current logistics costs was developed. Economic theory and statistical methods were used. At the third stage, a more rigorous approach was applied to determine the logistics performance index of the Russian Federation on the basis of expert assessments and mathematical methods.

3 Results

1. Besides traditional expenditure items (transport costs, inventory costs, administrative costs and others), fixed capital costs for transport and warehouse purposes should be included in the structure of national logistics costs. Its value is established according to the Federal state statistics service (the specific structure of fixed assets of the Russian Federation by the type of economic activity "transportation and storage"). At the beginning of 2017, the fixed capital cost for transport and warehouse purposes amounted to 17,421 trillion rubles.
2. The fixed capital of the transport and warehouse destination is a moment value and is measured in monetary terms at the evaluation date. The fixed capital is a one-time cost in comparison with the current costs of the national logistics. The refinancing rate of the Central Bank of the Russian Federation equal to 7.5% (0.075/year) has been used to bring the fixed capital costs to the current ones. Then the fixed capital costs for transport and warehouse purposes will be 1.31 trillion rub/year: 17,421 trillion rub 0.075/year = 1.31 trillion rub/year (Table 4).

Table 4. The total amount and the structure of logistics costs of Russia

Cost items	Amount, trillion rub	Specific weight, %
Transport costs	6,12	45,6
Inventory storage costs	2,87	21,4
Working capital costs	2,66	19,7
Fixed capital costs	1,31	9,8
Administrative costs	0,47	3,5
Total costs	13,43	100,0

Source: compiled by the authors.

3. The assessment of the national logistics performance is a comparison of its results and costs. The logistics performance index can be taken into consideration as a result of the logistics development. The assessment of the macroeconomic logistics performance on the basis of LPI is insufficient and simplified, which complicates its practical use. A more rigorous approach to the assessment of the national logistics performance is to modify the LPI. The LPI modification has included the expert assessment procedure of the index indicators on a ten-point scale, their later rating on the interval from 0 to 1 and calculating their average value (Table 5).

The next procedure for the LPI modification and forming a generalized national logistics performance index was to determine the weighting factors that characterize the contribution of each private logistics performance indicator to reduce the share of logistics costs in GDP. The calculation of the weighting factors of private logistics performance indicators allows us to give a more accurate and reasonable assessment of the logistics performance. This assessment is associated with the use of various expert and economic-mathematical methods.

Table 5. Expert evaluation of private logistics performance indicators in the Russian Federation in 2018

Indicators	Scoring	Rated score	Average value
1. Level of payment and fees			
Port costs	6	0,6	0,56
Airports	5	0,5	
Shipping costs	7	0,7	
Railway transportation costs	7	0,7	
Warehousing/reloading	4	0,4	
Agency fees	5	0,5	
2. Infrastructure quality			
Ports	7	0,7	0,61
Airports	7	0,7	
Automobile transport	5	0,5	
Railway transport	7	0,7	
Warehouse/reloading facilities	6	0,6	
Telecommunications and IT	5	0,5	
3. Competence and quality of services			
Automobile transport	5	0,5	0,59
Railway transport	6	0,6	
Air Transport	6	0,6	
Sea transport	5	0,5	
Warehouse/reloading and distribution	6	0,6	
Forwarding	7	0,7	
Customs agencies	6	0,6	
Quality Assurance/Standards Agencies	5	0,5	
Customs brokers	6	0,6	
Trade and transport associations	6	0,6	
Cargo receiver/cargo shipper	7	0,7	
4. Process efficiency			
Import shipping	5	0,5	0,56
Delivery for export	6	0,6	
Transparency of customs clearance	5	0,5	
Providing adequate and timely information on regulatory changes	6	0,6	
Accelerated Customs Clearance for High Compliance Traders	6	0,6	
5. Changes in the logistics environment since 2015			
Customs clearance procedures	5	0,5	0,56
Other official clearance procedures	5	0,5	
Trade and transport infrastructure	6	0,6	
Telecommunications and IT infrastructure	6	0,6	
Logistics services	6	0,6	
Logistics regulatory	5	0,5	
Extortion of unofficial payments	6	0,6	
6. Availability of qualified personnel			
Operational logistics staff	7	0,7	0,63
Administrative Logistics Employees	6	0,6	
Logistics controllers	6	0,6	
Logistic managers	6	0,6	

Source: compiled by the authors.

The hierarchy analysis method (HAM) is a pairwise method for comparing alternatives that characterize the degree of a goal achievement [18]. The first stage of this method was to use expert evaluation of the comparative significance of private performance indicators according to their degree of a goal achievement. At the next stage, economic and mathematical calculations of the contribution coefficients of each indicator were carried out and the coherence of expert assessments of their significance was determined. The matrix of the pairwise comparison of six private logistics performance indicators in the earlier numbering is presented in Table 6.

Table 6. The matrix of pairwise comparison of performance indicators

№	1	2	3	4	5	6
1	1	1/2	1/3	1/2	3	2
2	2	1	1/4	1/3	5	6
3	3	2	1	1/2	7	8
4	4	3	2	1	9	9
5	1/3	1/5	1/7	1/9	1	1/2
6	1/2	1/6	1/8	1/9	2	1

Source: compiled by the authors.

The comparison of private logistics performance indicators according to their degree of a goal achievement has been carried out on a priority scale from 1 (equal importance of compared indicators) to 9 (absolute superiority of this indicator over others). The priority matrix has indicated both forward and reverse priorities in the form of fractional values.

The developed matrix of pair comparisons of private indicators has allowed us to use a computer program to determine the importance coefficients of private logistics performance indicators.

All calculations have been carried out in the matrix form and are presented below:

– matrix A has been formed according to Table 6.

$$A := \begin{pmatrix} 1 & \frac{1}{2} & \frac{1}{3} & \frac{1}{2} & 3 & 2 \\ 2 & 1 & \frac{1}{4} & \frac{1}{3} & 5 & 6 \\ 3 & 2 & 1 & \frac{1}{2} & 7 & 8 \\ 4 & 3 & 2 & 1 & 9 & 9 \\ \frac{1}{3} & \frac{1}{5} & \frac{1}{7} & \frac{1}{9} & 1 & \frac{1}{2} \\ \frac{1}{2} & \frac{1}{6} & \frac{1}{8} & \frac{1}{9} & 2 & 1 \end{pmatrix}$$

– calculations of intermediate results have been carried out in the matrix form

$$n := cols(A) \quad i := 0 \ldots n-1 \quad v_i := \sum A^{\langle i \rangle} \quad N^{\langle i \rangle} := (v_i)^{-1} \cdot A^{\langle i \rangle}$$
$$MN := N^T \quad i := 0 \ldots n-1 \quad NA_i := \frac{1}{n} \cdot \sum MN^{\langle i \rangle}$$

– the matrix of importance coefficients of private logistics performance indicators (NA) has been determined

$$NA := \begin{pmatrix} 0.106 \\ 0.156 \\ 0.264 \\ 0.398 \\ 0.033 \\ 0.043 \end{pmatrix}$$

– the concordance coefficient of the importance of private indicators (CR) has been calculated

-

$$w1 := A \cdot NA \quad \sum w1 = 6.276$$
$$nm = \sum w1 \quad CI := \frac{nm - n}{n - 1} \quad RI := \frac{6.276 \cdot (n-2)}{n}$$

-

$$CR := \frac{CI}{RI} \qquad CR = 0.013.$$

Thus, the concordance coefficient (CR) is equal to 0.013 (1.3%). This means that the expert assessment of the importance of private logistics performance indicators is good. According to the calculations the highest importance in the contribution to the goal achievement belongs to such private business logistics performance indicator as the process efficiency (0.398), and the lowest significance has been in the logistics environment changes since 2015 (0.033).

4. At the final stage, the calculation of the generalized logistics performance index of the Russian Federation in 2018 was carried out (Table 7).

Thus, the generalized logistics performance index of the Russian Federation in 2018 is 0.59, which indicates the importance to improve the efficiency of logistics processes, as well as the competence and the service quality as its main private performance indicators.

Table 7. The calculation of the of the generalized logistics performance index

Private indicators	Average	Importance coefficient	The product of columns 2 и 3
1	2	3	4
1. Level of payment and fees	0,56	0,106	0,06
2. Quality of infrastructure	0,61	0,156	0,10
3. Competence and quality of services	0,59	0,264	0,16
4. Process efficiency	0,56	0,398	0,22
5. Changes in the logistics environment since 2015	0,56	0,033	0,02
6. Availability of qualified personnel	0,63	0,043	0,03
Total	–	1,000	0,59

Source: compiled by the authors.

4 Discussion

The results of the study justify the structure of national logistics costs. The introduction of the fixed capital cost for transport and warehouse purposes corresponds to the economic theory, where the fixed and working capital is one of the factors of production. In addition, the introduction of the fixed capital value in the national logistics cost allows to strengthen the coordination of calculations of its effectiveness with the calculations of the logistics companies efficiency, business processes and activities of logistics management. The reduction of the fixed capital cost to the current logistics costs can also be carried out by other methods (based on economic efficiency factors, cost recovery periods).

The calculation of the generalized national logistics performance index determines the importance (significance) of the individual components of its indicators. This makes it possible to establish their contribution in the share reduction of the logistics costs in GDP and justify its value. The identification of the importance of individual indicators of logistics performance can be carried out by other methods: the definition of relationships between private indicators, correlation and regression analysis, etc.

The proposed justification and calculating methods of the national logistics costs and its generalized logistics performance index can be used in the activities of state statistics bodies and researchers, as well as in the development of measures to reduce the share of logistics costs in GDP and increase its performance index.

The direction of the future research may be the development of goals and strategies to improve the logistics performance of the Russian Federation in the strategic planning of the national economy.

5 Conclusion

The national logistics performance is determined by its results in LPI and the share of costs in GDP. The Russian Federation ranks the 75th place in the logistics performance index and had a high share of logistics costs in GDP (16.1%) in 2016.

A more rigorous scientific justification of the absolute value and the cost structure of the national logistics determines the expenses costs of the basic capital transportation and storage purposes. Their reduction to the current logistics costs can be carried out on the basis of the refinancing rate of the Central Bank of the Russian Federation. The LPI modification is necessary for a more accurate assessment of private logistics performance indicators on the basis of expert and economic and mathematical methods. The low value of a generalized logistics performance index of the Russian Federation (0.59 or 59%) requires to improve logistics performance processes, competence and service quality. Further research can be aimed at setting goals and developing strategies to improve the national logistics performance in the strategic planning of the national economy.

References

1. Armstrong & Associates, Inc.: Global 3PL Market Size Estimates (2017). https://www.3plogistics.com/3pl-market-info-resources/3pl-market-information/global-3pl-market-size-estimates/. Accessed 24 Apr 2019
2. Arvis, J.-F., Mustra, M., Ojala, L., Shepherd, B., Saslavsky, D.: Connecting to compete 2012: trade logistics in the global economy. The logistics performance index and its indicators (English). World Bank Group, Washington (2012)
3. Faria, R.N., Souza, C.S., Vieira, J.G.V.: Evaluation of logistic performance indicators of Brazil in the international trade. RAM. Revista de Administração Mackenzie **16**(1), 213–235 (2015)
4. Gani, A.: The logistics performance effect in international trade. Asian J. Shipping Logistics **33**(4), 279–288 (2017)
5. Hausman, W.H., Lee, H.L., Subramaniam, U.: The impact of logistics performance on trade. Prod. Oper. Manag. **22**(2), 236–252 (2012)
6. Hummels, D., Schaur, G.: Time as a trade barrier [Working Paper N° 17758]. National Bureau of Economic Research, January 2012. https://www.nber.org/papers/w17758.pdf. Accessed 24 Apr 2019
7. Jacyna, M.: Cargo flow distribution on the transportation network of the national logistic system. Int. J. Logistics Syst. Manag. **15**(2–3), 197–218 (2013)
8. Kearney, A.T.: CSCMP's 28th Annual State of Logistics Report (2017). https://www.penskelogistics.com/pdfs/2017-CSCMP-SOLReport.pdf. Accessed 24 Apr 2019
9. Kearney, A.T.: CSCMP's 29th Annual State of Logistics Report (2018). https://www.penskelogistics.com/insights/industry-reports/state-of-logistics-report/. Accessed 24 Apr 2019
10. Korinek, J., Sourdin, P.: To what extent are high-quality logistics services trade facilitating? OECDTrade Policy Papers, No. 108. OECD Publishing (2011). https://www.oecd-ilibrary.org/docserver/5kggdthrj1zn-en.pdf?expires=1556104238&id=id&accname=guest&checksum=4A8168B58BCDD1BA9D2B994B7B33BCD5. Accessed 24 Apr 2019

11. Kurganov, V.M.: Efficiency of logistics and competitiveness of Russia. Transp. Russ. Fed. **1** (44), 19–23 (2013). (in Russian)
12. Li, D., Hanafi, Z.: A study of eco-performane of logistics services in food supply chains. In: Zhang, Z., Zhang, R., Zhang, J. (eds.) Proceedings of 2nd Conference on Logistics, Informatics and Service Science, LISS 2012, pp. 223–228. Springer, Heidelberg (2013)
13. Liu, X.: Competitiveness of logistics service providers: a cross-national examination of management practices in China and the UK. Int. J. Logistics-Res. Appl. **14**(4), 251–269 (2011). https://doi.org/10.1080/13675567.2011.636736
14. Portugal-Perez, A., Wilson, J.S.: Export performance and trade facilitation reform: hard and soft infrastructure. Policy Research working paper, no. WPS 5261. World Bank, Washington, DC (2012)
15. Puertas, R., Marti, L., Garcia, L.: Logistics performance and export competitiveness: European experience. Empirica **41**(3), 467–480 (2014)
16. Rutner, M.S., Aviles, M., Cox, S.: Logistics evolution: a comparison of military and commercial logistics thought. Int. J. Logistics Manag. **23**(1), 96–118 (2012)
17. The World Bank Global Rankings 2018 (2018). https://lpi.worldbank.org/international/global. Accessed 24 Apr 2019
18. Saaty, T.L., Vargas, L.G.: Models, methods, concepts & applications of the analytic hierarchy process. In: International Series in Operations Research and Management Science, vol. 175. Springer, Heidelberg (2012)
19. Spillan, J.E., McGinnis, M.A., Kara, A., Yi, G.L.: A comparison of the effect of logistic strategy and logistics integration on firm competitiveness in the USA and China. Int. J. Logistics Manag. **24**(2), 153–179 (2013)
20. Yergaliyev, R., Raimbekov, Z.: The development of the logistics system of Kazakhstan as a factor in increasing its competitiveness. Proc. Econ. Financ. **39**, 71–75 (2016)
21. Yildiz, T.: An empirical analysis on logistics performance and the global competitiveness. Bus.: Theor. Pract. **18**, 1–13 (2017)

Modeling Enterprise Architecture Using Language ArchiMate

L. A. Opekunova[1], A. N. Opekunov[2], I. N. Kamardin[2], and N. V. Nikitina[3(✉)]

[1] SAP SE, Mannheim, Germany
[2] Penza State University, Penza, Russia
[3] Samara State University of Economics, Samara, Russia
nikitina_nv@mail.ru

Abstract. In today's turbulent and rapidly changing world, companies are seeking a coherent description of the enterprise that provides a deep insight into a company, enables communication among stakeholders and guides complicated change processes.

Enterprise architecture comprises different aspects and viewpoints on the structure of an enterprise, and is used as an instrument to align business, information systems and technology domains.

This paper investigates how the ArchiMate enterprise architecture modeling language can be used to model component-based model systems, such as Common Component Modeling Example (CoCoME). The resulting model is based on ArchiMate, as the most popular approach applicable to most enterprises. The resulting enterprise architecture of CoCoME describes the organization structure, business roles, processes, services, and functions, IT and technology infrastructure, as well as their interrelationships.

Keywords: Business process · Data · Enterprise architecture · Digitalization · Information systems · IT infrastructure

1 Introduction

The beginning of the application of business analysis as a knowledge system is difficult to establish, primarily due to the diversity of approaches to its understanding, disordered terminology, as well as due to the fact that its tools were born from the acute needs of business and locally began to be implemented in various sectors of the Western economy in the 70 s of the last century [7].

During this period, many companies faced the need to shift from production to the consumer, what required switching analysts' attention from the internal environment to the external, to market segmentation, control of the value chain for target consumers, diagnostics of business processes and other new aspects of business activity analysis.

The problem was that the necessary information was either not available because of its novelty, or was generated in different accounting systems, with the harmonization between such data, and the more so as, their integration and performance were absent.

S. I. Ashmarina et al. (Eds.): ISCDTE 2019, LNNS 84, pp. 506–513, 2020.
https://doi.org/10.1007/978-3-030-27015-5_61

Only the elaboration and introduction of new computer and information technologies in business activities allowed to solve this problem.

In our opinion, the origins of business analysis were formed in modern management concepts and in the course of development of corporate governance practices. The basic methodological approaches in business analysis include system, situational, process and behavioral approaches, as well as the concept of business performance management (BPM), the concept of management by goals, management by deviations, value-oriented management, partnerships, stakeholders and social responsibility of the business [2].

The system approach, which has been developing since 60–70-ies of XX century, considers business as a holistic phenomenon and reveals its system integration properties. The situational approach realized on the basis of the system analysis reveals features of dynamics, environment and other characteristic features of business in their concrete spatio-temporal and cause-effect dependence, i.e. in a certain business situation [2].

The streamlining of business processes, refined formalization and monitoring of their effectiveness will ensure the development of the organization even with the complexity of management and increasing the tension of the external environment. A clear orientation of the staff to a certain business process and its quality performance increase the interest in meeting the requirements of customers, suppliers, partners, which makes the organization more flexible and customer-oriented, increases its organizational effectiveness [9].

The integration function of business intelligence is manifested in the combination of initial concepts. Modern management concepts have led to the formation of both a system of analytical indicators, significantly different from the traditional one, and a system of analytical procedures (clustering, neural networks, fuzzy logic, genetic algorithms, data maps, decision trees, etc.), performed with the use of developing computer technologies, and the continuous complication of analytical platforms (for example, Deductor Academic) [9].

2 Methodology

There are several interpretations of the concept of “business analysis”. One of the most common – business analysis is considered as a symbiosis of financial analysis and analysis of economic activity. Another interpretation, which we adhere to, arises in the automation of companies, where the main stage-the introduction of information systems for the formation of enterprise architecture, description and analysis of business processes taking place.

In our opinion, there are several main approaches to the formation of the enterprise architecture:

- The Zachman Framework for Enterprise Architectures (John A. Zachman) [11].
- The Federal Enterprise Architecture Framework (FEAF) (Allen Sayles) [10].
- The Open Group Architectural Framework (TOGAF) (Andrew Josey) [3].

A widely accepted standard in the enterprise architecture development is the TOGAF ADM, the architecture development method, designed by the Open Group. TOGAF ADM is a process for developing an enterprise architecture which consists of eight development phases, complemented with the preliminary phase where framework and principles are defined. The ADM is set up as an iterative process model, where the final stage indicates a start of a new iteration.

The main activities of ADM are establishing an architecture framework, developing an architecture content, transitioning, and governing a realization of architectures. An ADM iteration cycle includes all these activities, which aim in continuous architecture definition and realization that allows organizations to transform their performance in response to business goals and objectives. All phases of the ADM are presented in the work of Andrew Josey [3].

A modeling language is a way to express how information, data, people, and systems are structured, based on a defined set of rules. Specifically for enterprise architecture modeling, a language is mean to specify how organizational actors, processes, and systems should operate and interact together. The modeling language should also describe and visualize business, information systems and IT infrastructure by providing alignment among them at the same time. In order to succeed in this, it is necessary to define semantics of modeling concepts as well as a set of architectural views in order to relate these concepts to the enterprise architecture context [8].

UML is one of the widely accepted modeling languages which focuses on how to model software designs. However, the scope of UML is more suitable for designing actual systems and applications. UML is a very popular modeling language, which is aimed to design software projects.

ArchiMate is very useful when trying to model enterprise architecture, especially with business services mapping to software services and IT infrastructure. ArchiMate has a standard abstract syntax, a concrete syntax and semantic definitions for all its constructs, which make it very convenient to use. ArchiMate is the only modeling language which complies with enterprise architecture. ArchiMate is now an Open Group Standard and with its latest version being 3.0, it is fully aligned with TOGAF and follows same definitions. Moreover, UML and BPMN complement the ArchiMate language in several areas.

The Zachman, TOGAF, and the FEAF enterprise architecture frameworks were chosen because they are the most accepted and used frameworks for enterprise architecture analysis and modeling. However, none of these frameworks support modeling and analysis of component-based systems.

A modeling language is necessary to define and represent concepts and relationships between different architectural domains. Moreover, the enterprise architecture modeling approach has to consider different viewpoints for the development and/or communication of various architecture layers as well as for addressing stakeholders requirements. In this paper, ArchiMate modeling language will be used as an enterprise architecture language, since it satisfies described criteria, and thus, can be used to model the CoCoME example.

The ArchiMate language was designed between 2002–2004 in the Netherlands by a project team from the Telematica Institut in cooperation with several partners from government, industry, and academia, including Ordina, Radboud University Nijmegen, the Leiden Institute for Advanced Computer Science (LIACS), and the Centrum Wiskunde & Informatica (CWI). In 2008, ownership rights were transferred from the ArchiMate Foundation to The Open Group. Since then, The Open Group has developed different versions of ArchiMate and officially published them [4].

In this paper, the ArchiMate 3.0 specification is used, which was released in June 2016, and can be considered as a major update to the ArchiMate 2.1 specification.

New features included in the Version 3.0 provide elements for modeling a strategic level of an enterprise, such as stakeholders, goals, and outcome. It also supports modeling of the physical world of materials and equipment. Moreover, the structure and relationships of ArchiMate have been improved, many definitions have been aligned with other standards, and its usability has been enhanced in various ways [5].

3 Results

The ArchiMate core framework describes a structure of most generic elements and their relationships, which can be further represented in different layers. Figure 1 illustrates three layers, which are defined within the ArchiMate core language as follows [5]:

- Business Layer. This layer represents business services offered to customers, which are realized in an organization by business processes performed by business actors.
- Application Layer. This layer describes application services that support business, and applications that realize them.
- Technology Layer. This layer shows technology services, for instance, processing, data storage, and technology communication services, required to run applications, and system software that realize those services. In last version of ArchiMate, physical elements have been added for modeling physical equipment, materials, and distribution networks to this layer. The general structure of models within different layers is similar. Provided that, the same types of elements and relationships are used, although their exact nature and granularity differ.

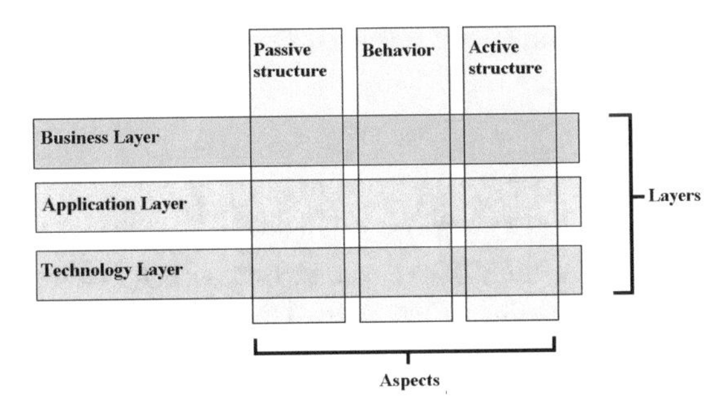

Fig. 1. ArchiMate core framework (Source: compiled by the authors)

Aspects of the core, as defined by three types of elements at the bottom, combined with layers create a framework of nine cells, as illustrated in Fig. 1. This is known as the

ArchiMate Core Framework. The structure of these elements allows to model an enterprise from different viewpoints, where the position within cells highlights concerns of stakeholders. A stakeholder typically can have concerns that cover multiple cells.

The Common Component Modeling Example (CoCoME) was chosen as an example of a component-based system that allows to use it for modeling an enterprise architecture. The CoCoME describes a Trading System that operates in a supermarket handling sales. All descriptions of use cases and component architecture are delivered by a business company as it could possibly be in reality and were taken from [6].

Figure 2 (Appendix A) illustrates interdependencies between critical business functions and business processes presented in CoCoME. Business functions work best when they overlap, which means that employees and business object in an enterprise work towards common goals. Each business function area depends on the support of other business functions and business processes. These high-level business functions can be associated with corresponding business processes that can be at the same way decomposed into lower level sub-processes.

This section describes the Component view on CoCoME, as CoCoME is an example of a component-based system and components have significant roles in the system. As systems play more and more complex, component-based design techniques play a more and more important role in research and practical solutions.

Components in many frameworks are considered not as physical elements, but rather as logical components that represent logical building blocks of software systems [1].

Components in the CoCoME example include logical elements, but some of components also include hardware devices, which means that components in CoCoME are distributed over many layers and represented with various viewpoints.

The layered viewpoint is aimed to picture several layers and aspects of the CoCoME example as a one view. The underlying idea is that each layer exposes a layer of services (realization relationship) which are used by the next layer.

Cash Desk and Inventory applications are collaborating together via communication service, or, in other words, the Cash Desk Connector If interface. In that way, application services (payment service, store statistics, sales and etc.) can communicate with each other. Figure 3 (Appendix B) illustrates the cooperation of Business, Information Systems and Technology layers of CoCoME on a high level of detail.

4 Discussion

The modeled architecture of CoCoME follows viewpoints of the ArchiMate language and does not follow all phases of TOGAF framework, as TOGAF includes steps which do not cover component modeling at all. Also, TOGAF is time consuming, provided that you follow all phases and steps. ArchiMate is very helpful in developing enterprise architectures, but in some architecture domains it is a bit incomplete, especially when modeling complex systems and components. Therefore, it makes sense to tailor the existing enterprise architecture approaches to become component-enabled. To avoid this shortcoming, it would be interesting to combine the ArchiMate modeling language with other frameworks or methodologies, e.g., KobrA, and configure it in a way to be more specific in a design of complex systems.

5 Conclusion

Main objectives of the architecture vision for CoCoME example are to design a high-level vision of business goals, strategic drivers, stakeholder concerns and objectives that have to be delivered as a result of the modeled enterprise architecture. Viewpoints demonstrated in this section are a part of the last ArchiMate 3.0 version.

As part of the Architecture Vision, main stakeholders in the architecture engagement and their concerns modeled as internal drivers in the ArchiMate standard have to be identified. In the ArchiMate standard, this can be expressed using the Stakeholder viewpoint. The ArchiMate standard defines the Stakeholder viewpoint to model stakeholders, internal and external drivers for change, and assessments (regarding strengths, weaknesses, opportunities, and threats) of these drivers. Also, links to initial high-level goals that address these concerns and assessments can be described. These goals form the basis for requirements engineering process, including goal refinement, contribution and connect analysis, and the derivation of requirements that realize goals [5].

Appendix A

See Fig. 2.

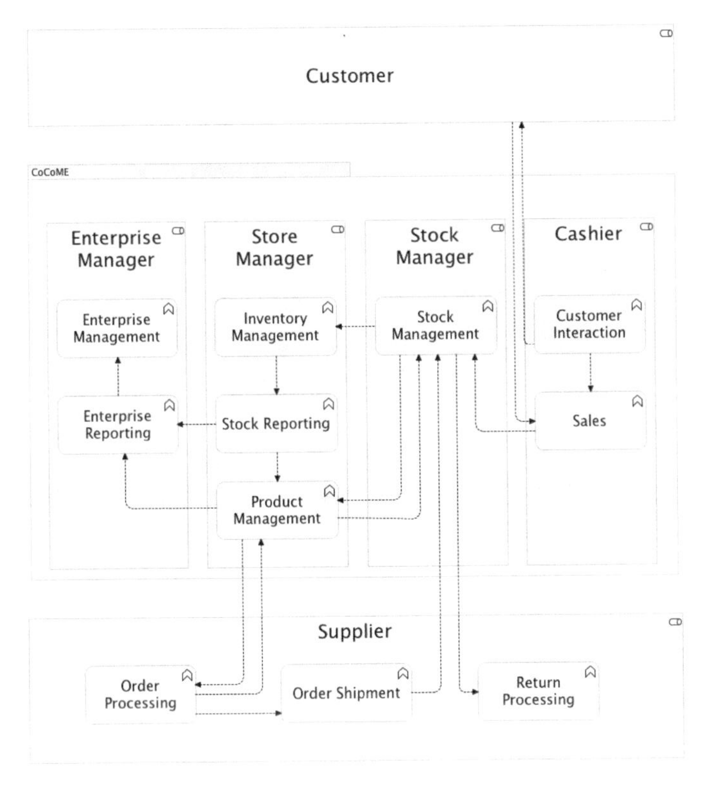

Fig. 2. CoCoME's core business functions (Source: compiled by the authors)

Appendix B

See Fig. 3.

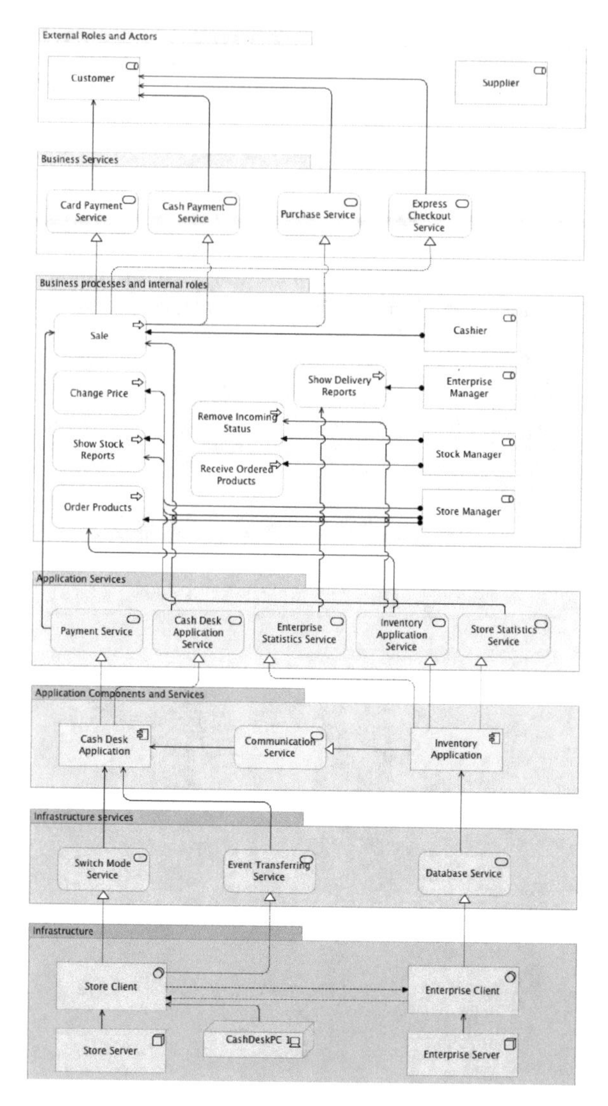

Fig. 3. Layered viewpoint of CoCoME (Source: compiled by the authors)

References

1. Atkinson, C., Bayer, J., Bunse, C., Kamsties, E., Laitenberger, O., Laqua, R., Muthig, D., Paech, B., Wüst, J., Zettel, J.: Component-Based Product Line Engineering with UML. Component Based Development Series. Addison-Wesley Professional, Berlin (2002)
2. Barilenko, V.I., Chugumbaev, R.R.: The development of business analysis and the role of training business analysts in providing digital transformation of the Russian economy. Manag. Bus. Adm. **2**, 146–155 (2018). (in Russian)
3. Josey, A., et al.: TOGAF Version 9.1. A Pocket Guide. Van Haren Publishing, Berkshire (2011)
4. Josey, A., et al.: ArchiMate 2.1 A Pocket Guide. Van Haren Publishing, Berkshire (2013)
5. Josey, A., Lankhorst, M., Band, I., Jonkers, H., Quartel, D.: An introduction to the ArchiMate. 3.0 specification (2016). https://www.vanharen.net/blog/enterprise-architecture/introduction-archimate-3-0-specification/. Accessed 24 Apr 2019
6. Herold, S., Klus, H., Welsch, Y., Deiters, C., Rausch, A., Reussner, R., Krogmann, K., Koziolek, H., Mirandola, R., Hummel, B., Meisinger, M., Pfaller, Ch.: CoCoME – the common component modeling example. In: Lecture Notes in Computer Science, vol. 5153, pp. 16–53. Springer, Heidelberg (2008)
7. Carlberg, K.: Business Analysis Using Microsoft Excel. Publishing House "Williams", Moscow (2012). (in Russian)
8. Lankhorst, M.: Enterprise architecture at work: modelling, communication and analysis. The Enterprise Engineering Series. Springer, Heidelberg (2009)
9. Paklin, N.B., Oreshkov, V.I.: Business-Analytics: From Data to Knowledge. Publishing House: Peter, St. Petersburg (2013). (in Russian)
10. Sayles, A.: Development of Federal Enterprise Architecture Framework using the IBM Rational Unified Process and the Unified Modeling Language. U.S.A. IBM Corporation Software Group (2003)
11. Zachman, J.A.: The Zachman Framework for Enterprise Architecture and Rational Best Practices and Products. USA. Zachman International, Cupertino (2002)

Students' and Their Parents' Choice of Higher Education Institution in the Era of Digitalization

S. I. Ashmarina(✉), L. G. Lebedeva, Yu. A. Tokarev, and A. M. Izmailov

Samara State University of Economics, Samara, Russia
asisamara@mail.ru

Abstract. The relevance of the study is caused by great challenges facing the educational system. First of all, we mean the degree of compliance between the level and quality of training provided by the education system and real requirements of the labor market where the skillful knowledge of digital technologies comes to the fore. Important in this aspect is how applicants and their parents choose the university for training. In this regard, this article aims to analyze the issue of applicants' and their parents' choice in relation to higher education. The leading method for this study is the sociological method with the subsequent interpretation of results, that allows to make a comprehensive review of the structure of the main factors affecting the choice of an educational institution for training. Taking into account that parents of applicants play a significant role by the decision making process in this sphere, their position was also studied. The article presents results of the aggregated digital data obtained in the course of sociological research (questionnaire in the electronic form). The authors describe the most significant factors influencing the choice of an educational institution, as well as the main differences in relation to the factors affecting the choice of an educational institution by parents and applicants. The materials of the article are of practical value for employees of higher educational institutions, scientists, applicants, students and parents of future students.

Keywords: Applicants · Higher education · Sociological research · Values · Digitization · Digital technology

1 Introduction

At the present stage of society development, digitalization is characterized by deep penetration into all social spheres including the education system. The changing under the influence of modern technologies world has an impact on a huge number of processes, including the choice of applicants and their parents in relation to higher education. The views of these social groups on the choice of education direction are largely determined by their environment which is becoming more digitalized every day. If the choice of parents can still be based on the way of thinking formed in the pre-digital era, the applicants themselves have usually preferences based on the surrounding realities. The system of higher education in Russia faces a number of large-

S. I. Ashmarina et al. (Eds.): ISCDTE 2019, LNNS 84, pp. 514–525, 2020.
https://doi.org/10.1007/978-3-030-27015-5_62

scale challenges, in particular, with regard to competitiveness in the international arena and innovative development within the country. Accordingly, the answers to these challenges imply an objective consideration of certain technological changes in the society, especially digital technologies [1]. For example, OECD countries (OECD) have committed to providing inclusive and equitable quality education, as well as lifelong learning opportunities for all by 2030 [11]. Generally speaking, "equitable quality education" is a global goal in the educational field put forward by developed economic countries.

Similar tasks are set in Russia. So, in the document "Bases of the State Youth Policy of the Russian Federation for the Period till 2025" in Article 4, it is emphasized that we need to use the innovative potential of development effectively and productively taking into account that the main carrier of this potential is our youth [4]. Researchers (following education practitioners and employers) pay attention to "employability" as a problem reflecting the quality of education [3, 5, 9]. The paradox is that higher education does not guarantee employment. At the same time, the sphere of professional self-realization is one of the most important and the social well-being of modern Russian youth depends on the effective inclusion in it [6].

Obviously, "professional self-realization", "employability" and "ability to find a job" concern not only graduates of educational institutions, but also those who already have work experience. At the same time, "employability" is considered as a certain competence of a person assuming the orientation of the individual (specialist) for continuous training and improvement of the own professional qualification level [10]. An important prerequisite for the implementation of this competence and for successful employment in real practice is the skillful knowledge of modern information technologies (which, in turn, is one of the specific modern competencies of specialists and employees).

More flexible forms of education with a focus on lifelong learning meet the requirements of the modern economy with its needs for more qualified personnel and the de-standardization of labor [7]. In many cases, researchers discuss certain inconsistencies in the development of higher education to existing and expected needs in the society, the crisis in the sphere of higher education. There are, for example, statements that the real crisis in the higher education is that we teach students the past, not the future. Based on the discussion of higher education problems in the developed countries, it is increasingly concluded that "more traditional education, which was the answer to many problems in the 20th century, is not necessarily the right answer in the 21st century" [2].

Thus, it is necessary to search for such ways of higher education development that would meet the objective present and future needs of our society. It requires an objective analysis and appropriate assessment of real problems and needs of the main stakeholders, contact audiences regarding to the development, provision and use of higher education. The main aim of this work is to study factors and reasons of applicants' and their parents' choice in relation to higher education institutions for training.

2 Methodology

The main method of the presented study is a sociological survey of applicants and their parents on issues relating to the main aspects of their choice of professional training areas. The sample consisted of 138 respondents. Applicants entering the universities of the Samara region in 2018 were interviewed. At the first stage of the study, a questionnaire was prepared and agreed, which included 48 questions, both general and special, aimed at identifying gaps. In the second phase, the questionnaire was converted into electronic form and supplied free to students giving application forms to the Samara State University of Economics. In parallel with applicants their parents were interviewed too. At the final stage, we have received a set of data in the electronic form subjected to subsequent statistical processing and interpretation of the obtained results.

The main groups and categories of respondents for the sociological survey are presented in Tables 1 and 2. The choice of the main groups of respondents (representatives of the main stakeholders) and categories of respondents (representatives of contact audiences) for the sociological survey was determined taking into account the main research goal.

Table 1. Groups and categories of respondents for sociological survey

№	Groups of respondents (representatives of the main stakeholders)	Categories of respondents (representatives of contact audiences)
1.	Applicants and applicants' parents	Applicants
		Applicants' parents

Source: compiled by the authors.

Thus, the survey covered representatives of all major stakeholders, contact audiences regarding the development, provision and use of higher education. A more specific distribution of respondents by categories (and subsamples) is presented in Table 2.

Table 2. Categories of respondents in the sample for the sociological survey

№	Category of respondents	Survey tool	Sample (people)	Survey period
1.	Applicants	The questionnaire (45 questions)	115	June–July 2018
2.	Applicants' parents	The questionnaire (45 questions)	23	June–July 2018
Total:		2 questionnaires	138	

*Note: if the respondent could give several answers to a question, the sum of specific weights (relative frequencies) in the table can exceed 100%.
Source: compiled by the authors.

The sample (138 people) by categories of respondents for the survey is quite representative. The composition of respondents (applicants and their parents) by age shows that the modal value is the age up to 18 years (70,18%) in the group of applicants and the age of 40–49 years (77,27%) in the group of the applicants' parents (Table 3).

Table 3. Data on the age (full years) of the Groups of Respondents

Applicants		Parents	
Age	%	Age	%
Up to 18 years (inclusive)	70,18	Up to 39 years (inclusive)	13,64
19–20	23,68	40–49	77,27
21–22	2,63	50–59	9,09
23 and older	3,51	60 and older	0
Total	100,0	Total:	100,0

Source: compiled by the authors.

Thus, the difference between the modal values in the age of two groups of respondents was, on average, about 26–28 years. We can say that parents of applicants are basically those who entered adulthood at the turn of the Soviet and Post-Soviet eras, and applicants themselves are those who were born mainly at the turn of XX–XXI centuries. It means that parents went to school mostly in the Soviet period, and their children (applicants) are graduates of modern Russian schools of recent years.

To identify gaps in education and professional choice within the generational line, a survey was conducted among applicants and their parents. The main questions for applicants and their parents were intended, first of all, to determine the degree of coincidence of the desired and actually chosen direction of education (specialty) and the university, as well as to identify main reasons for the coincidence or discrepancy between them. The survey provided an opportunity to assess the degree of independence of applicants and the degree of influence of older generations, their parents, on the choice of the applicants.

In general, the survey revealed interests, expectations and the degree of satisfaction of applicants and their parents in relation to the selected university and the direction of training (specialty). In particular, it was found out to what extent applicants and their parents take into account the situation and trends in the labor market, employment prospects after graduation.

3 Results

First of all, there is the question of how important is the fact of higher education (university diploma) to solve the task of employment, getting a decent, well-paid, desirable work in the eyes of representatives of different generations.

What are the values of material and non-material nature that the respondents associate with the chosen profession? More than half of the respondents in both groups (parents and applicants) consider the presence of a university diploma a great chance for a high-paid job. But the distribution of specific answers of applicants and r parents are markedly different.

The undisputed leader in the responses of respondents was the option that the chosen profession should ensure the "material well-being" of the applicant. This is the unanimous choice of more than half of the respondents in each target group (55.65% of applicants and 56.52% of parents).

Let us turn to the problem of the quality of secondary education received by the applicant (Table 4).

Table 4. Distribution of answers to the question about the degree of satisfaction with the quality of secondary education received by applicants? (%)

№	Essence of the responses	Applicants	Parents
1	Quite satisfied	47,83	60,87
2	More satisfied(a) than not	23,48	17,39
3	Can't say definitely	17,39	4,35
4	Rather not satisfied	10,43	13,04
5	Not satisfied at all	0,87	4,35
Total:		100	100

Source: compiled by the authors.

Only slightly less than 50% of applicants and slightly more than 60% of parents are "Quite satisfied" with the quality of the received secondary education. If we add to these groups those who are "More satisfied(a) than not", then, in general, those who are more or less satisfied with the quality of secondary education received by applicants, are slightly more than 70% of applicants and slightly less than 80% of parents. Thus, not everyone is satisfied with the quality of secondary education.

A very significant part of the respondents is not satisfied with the quality of secondary education. This is sufficiently consistent with the fact that a significant proportion of respondents (more than one third of applicants and more than one quarter of parents) show concern and anxiety about the applicants' ability to cope with learning difficulties.

Regarding the factors that affect the level of concern among respondents, in the first place of all that is "most disturbing at the moment", it is the question "will the applicant cope with the difficulties of learning?" (37.39% of applicants and 26.09% of parents). Even the problem of admission to the university was not so significant: for parents it is in the second-fourth places (17.39%), and for applicants even in the fifth place (6.09%). This sociological fact probably reflects a complex of problems related to the quality of previous (school) education, as well as the shortcomings of career guidance work.

Modern higher education is also not considered by respondents to be a "continuation of the dynasty" or "family traditions". It also does not stand out as a significant factor in the majority of surveyed parents.

For representatives of different generations, an absolutely priority requirement of employers seems to be "knowledge of the newest technologies in the specialty" (for 63.48% of applicants and for 56.52% of parents).

"High level of theoretical training" was in the absolute first place for applicants (63.48%) and for their parents (65.22%). This choice is consistent with the fact that, in their opinion, potential employers expect from graduates, first of all, the knowledge of the newest information technologies in their branch.

The survey was conducted on the basis of Samara State University of Economics, so it is impossible to absolutize the results of this survey, automatically transferring the preferences of students of this university to the entire population of students of the Samara region. But some trends in these preferences (discussed above) can still be talked about.

Further, we consider the preferences to a particular character (profile) of the higher education (Table 5).

Table 5. Distribution of answers to the question of what nature higher education is desirable (profile), the opinion of applicants and their parents (several answers), in the rating order

Place	Applicants		Place	Parents	
	Character (profile) of higher education	%		Character (profile) of higher education	%
1	Financial and economic	53,04	1	Financial and economic	39,13
2	Management	37,39	2	Law	21,74
3	Humanities (philosophy, psychology, etc.)	28,7	3	Natural sciences (mathematics, biology, chemistry, etc.)	17,39
4	Social science (history, sociology, political science, etc.)	19,13	4	Social science (history, sociology, political science, etc.)	17,39
5	Pedagogy	9,57	5	Management	17,39
6	Natural sciences (mathematics, biology, chemistry, etc.)	8,7	6	Pedagogy	13,04
7	Law	7,83	7	Humanities (philosophy, psychology, etc.)	8,7
8	Engineering	6,09	8	Engineering	8,7
9	Medical science	1,74	9	Medical science	4,35
10	State and municipal administration	0,87	10	Agricultural	4,35
11	Agricultural	0	11	State and municipal administration	0
Total:		173,06	Total:		152,18

Source: compiled by the authors.

As for the choice of vocational training, the opinions of the older and younger generations coincide – they choose the "Financial and economic direction" of education (applicants – 53.04%, parents – 39.13%). In other areas, such (complete) unanimity is not observed. Direction "Management" is in the second place among applicants (37.39%), and by parents this direction is on the third – fifth place with a number of other areas (17.39%). The second place is the direction of "Law" by the parents (21.74%). There are differences in opinions of both generations about such areas as "Humanitarian (philosophy, psychology, etc.)", which took the third place (28.7%) by applicants and only the seventh place by parents (8.7%).

In modern Russia, there are various assessments of the problems and consequences of digitalization of society, the spread of information and computer technologies in all social spheres. Relevant is the issue of knowledge of information and computer technologies in relation to employment, especially for older generations. Naturally, the so-called "digital divide" between generations [8, 13, 14] is actively discussed nowadays.

The degree of knowledge of information and communication technologies can, to some extent, be judged by the example of the survey "The Choice of the Applicants and their Parents 2018" (Table 6).

Table 6. Distribution of answers to the question: "How well do you know a computer?" (several answers), in %

Answers	Applicants	Parents
I know the computer well and often use it	48,70	56,52
All school tasks with the computer I did successfully	44,35	–
My work is connected with the computer, I cope with all tasks successfully	–	8,70
I know the computer well enough, often surf via the Internet	43,48	21,74
I know the computer well enough, but rarely use it	7,83	21,74
I have programming skills	5,22	4,35
I don't know the computer well and don't use it	6,96	13,04
Difficult to answer	0,87	4,35
Total:	157,41	130,44

Source: compiled by the authors.

On the basis of the data of Table 6, it is possible and necessary to speak not so much (and not only) about the degree of knowledge of information and communication (computer) technologies among young people or older generations, but about the insufficiently wide actual application, use of these technologies, in particular, in education and work among all groups of the population, including young people. So, the answer "I know the computer well and often use it" was chosen by 48.70% of applicants and 56.52% of parents (by the opportunity to select several answers from the options). At the same time, only 6.96% of applicants and only 13.04% of their parents chose the answer "I don't know the computer well and don't use it". These data show, first of all, the insufficient actual use of computer technology, and not the lack of computer literacy at all. In fact, we had the similar data in our previous sociological survey in 2016 [8].

Data from the sociological survey also show that for educational purposes, the computer is also not very actively used. Many researchers point out that the vast majority have experience with a minimum set of "basic" programs, especially for study and entertainment. A significant part of people, including the older generation, use Skype. Many, but primarily young people, periodically play computer games. But advanced users are only 2% [12].

At the same time, trends in the labor market indicate that the knowledge and skillful usage of modern information technologies in employment is becoming increasingly popular. It is useful to take into account the warning (forecast) of E. Toffler in the book "Shock of the future" about the rapid arrival of new technological systems due to which huge information flows fall on people and require even greater ability to penetrate into the essence of things.

Forecasts and warnings about the accelerating pace of technological development of our society are important for both older and younger generations. This problem should be increasingly taken into account and reflected in the improvement/reform of the educational system. This problem is important not primarily for young people, but for almost the entire population. It may be associated with re-training, different directions of professional training, especially using information and computer technology.

4 Conclusion

The solution to the problem of employment depends on many factors, but, of course, it also depends on the competence, knowledge and experience of the applicant, especially in the most modern specialties and types of work. Accordingly, it is important to ensure the fight direction (profile) of education and the high quality of vocational training at the university: first of all, a set of competencies (knowledge, skills) which are universal (primarily humanitarian and socio-psychological), general and special (according to the profile of education).

What competencies are important for the employment of current and future graduates, on the one hand, for employers, on the other hand, for applicants and their parents? Do the expectations and interests of employers and applicants/their parents coincide? In this article we analyzed the answers of applicants and their parents in this context (Table 7).

Table 7. Distribution of answers of applicants and their parents to the question: "How do you imagine what potential employers expect from graduates?" (multiple answers), in rating order

Place	Applicants		Place	Parents	
	Answer options	%		Answer options	%
1	Knowledge of the latest technologies in the specialty	63,48	1	Knowledge of the latest technologies in the specialty	56,52
2	Foreign language skills	56,52	2	Ability to take initiative at work	47,83
3	Ability to take initiative at work	53,04	3	Good computer training	34,78
4	Strong basic education	42,61	4	Social skills to work in a team	34,78
5	Desire for self-development and self-education	40,87	5	Strong basic education	34,78
6	The ability to apply innovations in their work	38,26	6	Desire for self-development and self-education	34,78
7	Social skills for business communication	38,26	7	Foreign language skills	34,78
8	Good computer training	32,17	8	The ability to apply innovations in their work	30,43
9	Social skills to work in a team	29,57	9	Social skills for business communication	30,43
10	Knowledge of legislation	21,74	10	Knowledge of legislation	17,39
Total:		416,52	Total:		356,5

Source: compiled by the authors.

For representatives of both generations, an absolutely priority requirement of employers seems to be "knowledge of the latest technologies in the specialty" (for 63.48% of applicants and for 56.52% of their parents).

The first "three" in the rating order includes "the ability to take initiative at work": on the third place among applicants (53.04%) and on the second place among parents (56.52%).

On the second place by applicants is the "knowledge of a foreign language" (56,52%). But by the parents, this item is in the group of answers from the third to the seventh place with a relatively average percentage of choice 34.78%.

In the middle zone of the applicants' rating, there is "strong basic education" (42.61% and the fourth place). For parents, this item is in the zone "from the third to the seventh place" too (34.78%).

In general, the focus on the fact, that potential employers will expect from graduates of actual knowledge and the ability to take the initiative at work, is quite consistent with the requirements of the time. But the fact that "knowledge of the legislation" is in the last place for applicants (only 21.74%) and for their parents (only 17.39%) deserves additional attention and research.

And what opportunities, conditions for learning and getting necessary competencies do applicants and their parents expect from the chosen university? (Table 8).

Table 8. Distribution of answers of applicants and their parents to the question: "What opportunities are students provided with in the chosen educational institution?" (multiple choice answers), in rating order

Place	Applicants		Place	Parents	
	Answer options	%		Answer options	%
1	High level of theoretical training	63,48	1	High level of theoretical training	65,22
2	Opportunities for obtaining two diplomas	49,57	2	High level of practical training	43,48
3	The opportunity to study at the Translation Department	33,91	3	Opportunities for obtaining two diplomas	30,43
4	Instilling students the desire for self-development and self-education	32,17	4	Well-organized educational process	30,43
5	Well-organized educational process	30,43	5	Formation of students' readiness to work in their chosen specialty	30,43
6	High level of practical training	28,70	6	Instilling students the desire for self-development and self-education	26,09
7	Formation of students' readiness to work in their chosen specialty	21,74	7	Good organization of educational and industrial internship	17,39
8	Good organization of educational and industrial internship	15,65	8	Good material and technical base of the university	17,39
9	Good material and technical base of the university	13,91	9	High level of teaching staff	17,39
10	High level of teaching staff	12,17	10	The opportunity to study at the Translation Department	13,04
11	Opportunities for leisure time	10,43	11	Good computer training	13,04

(continued)

Table 8. (*continued*)

Place	Applicants		Place	Parents	
	Answer options	%		Answer options	%
12	The possibility of a real choice of some academic disciplines	6,09	12	Opportunities for leisure time	8,7
13	Good computer training	5,22	13	Difficult to answer	8,7
14	Difficult to answer	2,61	14	The possibility of a real choice of some academic disciplines	0
Total:		326,08	Total:		321,73

Source: compiled by the authors.

"High level of theoretical training" was in the absolute first place for applicants (63.48%) and for their parents (65.22%). This choice is quite consistent with their belief that potential employers expect from university graduates, first of all, "knowledge of the latest technologies in the specialty" (Table 7). "The possibility of obtaining two diplomas" is the first "three" in the rating order: the second place among applicants (49.57%) and the third place among parents (30.43%).

"High level of practical training" is in the second position among parents (43.48%) and only on the sixth place with a relatively average percentage of choice (28.70%) by applicants.

"Good computer training" in the parents' evaluation took the eleventh place from thirteen (13.04%), and in the evaluation of applicants – the thirteenth place (5.22%). But perhaps it's not that computer training is not so important, but the fact that other things are really no less important? Moreover, as it can be seen from the Table 6, the situation with the overall computer literacy is generally relatively well.

It may be surprising, at first glance, that the option of "High level of teaching staff" took only the ninth or the tenth place. But this point is "overlapped" by the fact that in the first position "High level of theoretical training" is, of course, impossible without a sufficiently high level of teaching staff competences.

At the same time, options "Well-organized educational process", "Instilling students the desire for self-development and self-education", "Education of students readiness to work in their chosen specialty" are on relatively high (though not leading) places. These are signs of a very positive attitude towards education and self-education, as well as future work. At the same time, the solution of the above mentioned problems objectively implies a "High level of the teaching staff."

Acknowledgements. The study was carried out within the framework of the State Task of the Ministry of Education and Science of the Russian Federation № 26.940.2017/PM, the project "Change Management in the Higher Education System on the Basis of the Concept of Sustainable Development and Harmonization of Interests".

References

1. Analytical Centre of the Government of the Russian Federation: Bulletin on Education. Reform of higher education: domestic and foreign experience. Issue № 12, June 2017, Moscow (2017). (in Russian)
2. Armstrong, L.: The real crisis in higher education? (2011). https://www.changinghighereducation.com/2011/02/three-articles-from-among-many-that-contain-important-questions-for-higher-education-california-visual-effects-firms-faci.html. Accessed 23 May 2019
3. Baidenko, V.I., Vorozheykina, O.L., Karacharova, E.N., Selezneva, N.A., Tarasyuk, L.N.: Bologna process: Glossary (based on the experience of the monitoring study). Moscow: Research center of the quality of training (2009). http://fgosvo.ru/uploadfiles/npo/20120408235619.pdf. Accessed 23 May 2019. (in Russian)
4. Bases of the state youth policy of the Russian Federation for the period till 2025 (Approved by the order of the Government of the Russian Federation of November 29, 2014 No. 2403-p). http://static.government.ru/media/files/ceFXleNUqOU.pdf. Accessed 23 May 2019. (in Russian)
5. Bolotin, I.S., Basalay, S.I., Mikhailov, A.A.: Employment and employability of graduates of Russian universities: past and present. Sociol. Educ. **2**, 16–29 (2016). (in Russian)
6. Eflowa, M.Y., Ishkinyeeva, F.F., Fursova, V.V.: Social well-being and value orientations of student youth in the context of social changes. Bull. Inst. Sociol. Online J. **3**(10), 34–44 (2014). (in Russian)
7. Erofeeva, P.: Individual and structural in social mobility in the context of individualization: a review of empirical studies. Sociol. Rev. **14**(2), 107–150 (2015). (in Russian)
8. Lebedeva, L.G.: Intergenerational gaps and risks in the virtual space. Bull. Saratov State Socio-Econ. Univ. **1**(70), 147–151 (2018). (in Russian)
9. Minin, M.G., Nomoklinova, A.V.: Employability as a psycho-pedagogical category. Vestnik of FGOU VPO MGAU **4**(1), 33–35 (2012). (in Russian)
10. Mozgovaya, A.V., Yaichnikov, A.Y.: Employability as a resource of professional adaptation of the person. Bull. Inst. Sociol. **9**(3), 143–157 (2018). https://doi.org/10.19181/vis.2018.26.3.530. (in Russian)
11. OECD: Education at a Glance 2018: OECD Indicators, OECD Publishing, Paris (2018). http://dx.doi.org/10.1787/eag-2018-en
12. Toffler, A.: Shock of the Future. Moscow: Publishing House AST (2002). (in Russian)
13. Volchenko, O.V.: Dynamics of digital inequality in Russia. Public Opin. Monit.: Econ. Soc. Changes **5**, 163–182 (2016). https://doi.org/10.14515/monitoring.2016.5.10. (in Russian)
14. Zherebin, V.M., Makhrova, O.N.: Digital split between generations. FES: finance. Econ. Strat. **4**, 5–9 (2015). (in Russian)

Digitization, Digital Technology, and Importance of Digital Technology in Teaching

I. Kmecová(✉)

The Institute of Technology and Business in České Budějovice,
České Budějovice, Czech Republic
kmecova@mail.vstecb.cz

Abstract. This study focuses on digitization in the education process and on the importance of using digital technologies in the teaching process. The introduction sketches out the need of leading young people towards developing their digital competences and digital as well as information literacy. The teaching process should lead towards gaining key and practically oriented knowledge and skills according to Industry 4.0. The importance of the role of the teacher in the education process is stressed who is a managing adviser influencing the effectiveness of the education process. The first chapter is dedicated to digital literacy and other literacies and connected terminology. The next chapter focuses on case studies and evaluating the opinions of students on the importance of using digital technologies during class. The results were gathered via a questionnaire survey. The objective is to point out specifics of the education process. Gathered data can be valuable and inspirational for the decision-making process when planning and preparing teachers for the education process. The conclusion of the entry stresses the importance of innovative study programs and introducing new study programs, which focus on supporting the development of digital literacy and working with digital technologies.

Keywords: Digitalization in teaching · Digital technology · Digital literacy · Digital media · Digital competences · Education process

1 Introduction

The issue of computerization is currently a discussed topic in all branches of human activity – in education as well. The role of the teacher changes based on the dissemination of computers and information networks etc. While in traditional teaching, the teacher plays the role of the "source of information", in new teaching trends, the teachers is "a guide through the information environment". That is why nowadays all developed countries, including the Czech Republic, try to construct such educational systems that would significantly speed up the development of education and contribute to the further development of the national population. Currently, it is called [2] the "learning society" in which not only compulsory education but also all other forms of formal and informal education gain more importance for the lives of everyone and the entire society. Koníček [13] writes about the importance of using computers in

S. I. Ashmarina et al. (Eds.): ISCDTE 2019, LNNS 84, pp. 526–537, 2020.
https://doi.org/10.1007/978-3-030-27015-5_63

education. In the current society, which is full of information, the ability to gain, sort, and process information by using computers is one of the most crucial skills. In the future, the importance of this skill with only increase since all jobs will require good skills when working with computers and digital technologies. Developed countries realize that it is crucial for students on all levels to become accustomed to new technologies because skills and abilities gained this way will increase their future competitiveness as well as the competitiveness of the entire economy. The education concept corresponds with the stated trends. The dynamic development of information society is transferred to teaching as well and places new requirements on current and future teachers. Knowledge grows in a large scale and can be disseminated into the entire world in a moment thanks to information communication technologies (ICT). Teachers as disseminators of knowledge cannot keep up with this trend. Their role shifts rather to a position of a guide, adviser, partner. This new look on tradition roles of student and teacher is necessary if the ones who teach are supposed to take on more responsibility for their own education and have the option of showing more independence in teaching. Students have to be, however, lead towards using ICT in all subjects, which requires increased efforts from teachers on all levels.

2 Theoretical Foundation

Thanks to the dynamic development in all possible areas of human activity, trends and methods in teaching must also develop and adapt to new needs. At the current time when technology and particularly computing develop rapidly on a day-to-day basis, it is crucial to try to understand new things quickly and use them in other fields. The same situation applies to education [12].

The current teaching process at universities includes technology and must include new technologies. By using such technology, some communication and decision-making activities are transferred from humans to technology, thereby creating a larger space for creative human activities at the expense of mechanical and reproductive activities [3].

An individual (student, pupil) participates in society via online activities, looking for opportunities to personally develop and increase qualifications via digital technologies, and at the same time develops his/her ability to using digital technologies in the current digital environment. He/she perceives and evaluates potential and risks of including digital technologies in various processes and situations and acts accordingly. An individual thereby identifies issues and possibilities for their solving via digital means. He/she thinks about and critically evaluates several resolutions and if needed adapts digital tools for concrete approaches [14]. For example, the development of the Internet introduced new possibilities to the process of education.

Digital Technology
Consists of primarily equipment: computing technology – computer and peripherals, digital camera, video camera, kinetic audio-visual projection (film, video, DVD player), audio reproduction (CD player), simulators, trainers, computer rooms, mobile phones [3, 6].

Didactic Technologies and Digital Literacy
These have become one of the fundamental pillars of current society – these technologies can be understood not only as products of our culture but also as important factors that co-create our culture, society, and everyday lives. Contemporary youth grows up since birth in an environment in which digital technologies are commonly available and omnipresent. That is one of the important differences in comparison to the life experiences of persons born previously that only encountered digital technologies during the course of their lives [4].

Digital technologies, including the rise of social networks and online gaming, make easier the access to friends, colleagues that can be disseminated across the world. Digital literacy makes easier the processes of interacting and participating, it enables students to become active in interpersonal contexts. Apart from that, some people use these technologies to design and create their own media. They can e.g. create pages, create and edit music, videos, or share information. Digital literacy supports this process of people becoming active creators of digital media. It is important to focus on participating in digital literacy in education via expanding and increasing their use for creativity and self-expression, and develop better understanding of the intricacies that the technologies and media enable. Education systems need to help young people to understand and benefit from their participation in digital technologies and cultures [6, 7].

Digital Technologies in Education. The fact that digital technologies increasingly influence the environment in which we live and imminently also our lives, is without a doubt. The same goes for other areas of human activity – observing technological trends that influence teaching, education, and the school system [20].

According to Burďák [6], digital technologies also include: wireless technologies (USB, Bluetooth, WiFi, GMS, etc.), navigational technologies, operating systems, mutual communication devices (synchronization of PDA with PC), etc.

Digital Literacy
What does it mean to be digitally literate? To be digitally literate means to have knowledge and skills to effectively work, critically evaluate, and create information using a large scale of digital technologies. A digitally literate person strategically uses technologies to find and evaluate information, to analyze and synthesize them, cooperates with others, all for the purpose of reaching professional and personal goals [16].

Ferrari [9] states that digital literacy is a set of digital competences (knowledge, skills, stances, values) that an individual needs for a sage, self-confident, critical, and creative use of digital technologies when working, learning, during his/her free time, and when incorporating them into society.

Digital literacy can be put into context with ICT literacy, information literacy, media literacy, or Internet literacy.

The importance of ICT literacy as a set of competences for life and employment is steadily increasing.

ICT literacy is a set of competences that an individual needs in order to be able to decide how and why to use available ICT and then how to purposefully use them when solving various situations when learning or in life and in a changing world [1]. ICT literacy can be perceived as one of the key competences. The development of ICT competences must be a part of teaching from early childhood. The development of ICT competences cannot happen in an isolated subject but must be anchored in teaching all subjects, it must be integrated into teaching programs and strategies of life-long learning, as described by Russow [21] and Bučková [5].

Digital Competences
What does the term digital competences mean? As stated by Burďák [6, 17], digital competences in the sense of key competences can be understood as a set of knowledge, skills, abilities, stances, and values that an individual needs to use digital technologies and digital media for activities such as building: task solving, communication, handling information, problem solving, cooperation, creating and sharing content, and building knowledge. These competences are applied when working, during free time or social and citizen activities, when learning, and for personal growth, or when determining our own life needs – effectively, adequately to our intended or given purpose, critically, independently, flexibly, and ethically. We understand digital competences as certain and critical use of technologies of the information society when working, during free time, and when communicating. They are based on basic ICT skills: using computers to gather, evaluate, save, create, and exchange information, and to communicate and cooperate within the networks of the Internet environment. In his work, Burďák [6] describes that the development of digital literacy is conditioned by the option of a stable and physical access to digital technologies of a certain minimum quality. To develop digital literacy, there are according to the "Strategy of digital literacy Czech Republic for the period 2015–2020" three ways: informal learning on an individual level (e.g. via the trial and error method), learning via informal communities (group of friends, family, hobby club, library, online communities), formal education (using manuals, school education, or official courses).

Digital competences include the following characteristics [10]:

- Information: being able to identify, localize, gather, save, organize, and analyze digital information, and evaluate their relevancy.
- Communication: being able to communicate in a digital environment, sharing sources via online tools, connect with others, and cooperate via digital tools, communicate in communities.
- Creating content: ability to create new and edit old forms of content, using media, creatively create, and respect copyright.
- Security: securing data and own identity, safe use of technologies, and own presence in the digital world.
- Problem solving: using digital sources for decision-making, using technologies for solving problems, to identify digital needs.

2.1 Characteristics of Selected Teaching Tools and Didactic Technologies

In school practice, it is possible to separate the following basic groups of *teaching tools* [3]:

- *Audio* (hearing), such as audio recordings od CDs, video recordings, movies, computer programs for acoustic outputs, etc.

- *Visual* (seeing), such as models (three-dimensional artificially created objects that replace examined real objects, depictions (graphs, schemes, diapositives, maps, diagrams, photographs, etc.)
- *Audio-visual teaching tools*, such as dynamic sound projects via film, TV, video, computer program, etc.
- *Multimedia teaching tools*, such as teaching programs for computers (e.g. simulation programs), programs for teaching tools, hypertext (electronic teaching books), the Internet, etc.

Internet as an Environment for Education. The development of the Internet provided new possibilities for the teaching process.

Benefits of the Internet in relation to education [7, 18]: Connecting the school environment to the real practice, relatively easy and quick sharing of educational content with students, simpler preparation of teachers for classes (teachers can use video, audio, pictures, photographs, create materials not only as text), easy incorporation of interactive applications, developing creative and presenting skills of students, support of self-study and self-education of students, using multimedia content in education, etc.

Digital Camera

It enables to present digital photographs using a data projector. It is possible to edit a digital photograph on a computer and use it at any time, e.g. in a presentation program, or to place it on a website.

Visualizer

It is used for the teacher to be able to project presentations of a smaller three-dimensional objects (samples of products, technical parts, etc.) or even planar pictures (photographs, schemes, technical sketches, etc.).

Interactive Board

The board combines a special projection area with an electronic pen, a slide projector, computer and special software, securing the functions of the board. It projects pictures from the computer onto the projection screen, on which it is possible to write and paint notes written by hand using the electron pen. The board functions as a touch screen as well via which we can control the opened and projected program [19].

Video Technology

It is a set of technical tools including video cameras (digital cameras and web cameras), recording devices, broadcast devices, slide projectors, TVs, and computer monitors [3].

Digital Media

Recently, the so-called e-education has been experiencing a fast development in the area of modern education systems. Electronic education includes various forms in the context of used technology and methodology of education. The terms are not exactly defined and develop based on new technologies. According to L.C. Russow [21], electronic education includes several forms, e.g. e-learning, blended e-learning, video conferences, virtual class rooms, etc.

E-learning
It is one of the most commonly used ways for adult education. Teachers and students usually do not meet in person. This approach to education is primarily appropriate for short-term courses. High motivation of the student is crucial.

Blended E-learning
It is an approach to education during which longer periods of self-study alternate with short consultations with a teacher. High motivation of the student is also crucial.

Video Conferences
During video conferences, work places can be connected via video signals in real time, mostly using the Internet. Students and teachers can be in the same building or in different cities. When video conferencing, both sides can ask questions. A simple version of this is the program Skype that is used primarily for Internet calls, and via connecting a simple web camera, it is possible to broadcast a video signal between the participants of the video conference.

Virtual classroom enables to connect two or more participants connected via the Internet. Using a whiteboard tool, the participants can from a distance share a space that runs as an application on a computer (presentation, text editor, etc.).

Computer
It has an irreplaceable role in multimedia teaching. It enables communication and transfer of messages via different media over distance. Computer assisted learning can be implemented not only for groups but also for individual teaching, frontal instruction, or e-learning practiced by individual students [8]. Electronic forms of education experience rapid development in the context of growing availability of the Internet and the development of Internet technologies.

Andres and Vališová [2] describe it as "*a learning society in which not only compulsory education but also all other forms of formal and informal education gain more importance for the lives of everyone and the entire society*".

Modern technologies primarily include according to the same authors:

- networks (local computer networks, Internet and available online libraries, databases, and other sources of information, video conferences, etc.),
- multimedia that connect different forms of information presentations (hypertext, picture, and animated picture, sound, etc.) on different types of carriers (online, CD/DVD-ROM, etc.),
- mobile tools and access, supporting flexi-schooling and other forms of distance education, including wireless networks, notebooks, tablets that students borrow to work from home, etc.

Besides that, all types of schools are starting to apply the trend called BYOD (bring your own device), in which students and employees have the possibility to access wireless school networks and use a number of education services, such as the access to school LMS (Learning Management System) [2].

The current world of technology develops very rapidly which requires teachers to constantly educate themselves and understand how to use new technologies in teaching.

2.2 Case Studies on the Topic of Teaching Digitalization

An interesting study by the authors Fransson, Holmberg, Lidberg and Olofsson [11] *"Digitalise and capitalise? Teachers selfunderstanding in 21st-century teaching contexts"* explores how digitization can influence teachers by focusing on (a) the way how teachers manage to use digitization, and (b) how it can influence the digitization of their own self-knowledge. They observed two colleagues – English teachers – in the same teacher team. They focused on the same characteristics: self-defense, self-respect, work motivation, and perception of tasks. While the first teacher was able to use digitization purposefully and professionally, the other colleague felt pressured. They concluded that limited or expanded use of digital technologies should not be considered an indicator of teaching quality.

Another study called "The Impact of Digitqalization of Network Space on Journalism Education! by the authors Melnik a Teplyahina [16] focuses on the nature of the terms "multimedia" and "multimedia technologies" and describes educational requirements for introducing multimedia technologies to the educational process and discusses the possibilities of using them in class. Active introduction of multimedia technologies in education programs of journalism in a crucial factor in the European Union regarding the modernization of education. Currently, these technologies are one of the most dynamic and promising areas of digital technologies.

Another example is the study "*ICT literacy*" (Available at: http://clanky.rvp.cz/clanek/c/Z/13397/ICT-GRAMOTNOST-V-RVP-ZV.html) in which it is stated: the field of education that forms its own requirements and at the same time supports the development of ICT literacy is for some perhaps surprisingly physical education. A first-grade student understands information sources about physical activities and sporting events at the school and his/her living area, independently gathers needed information. At the end of the second grade, he/she processes gathered data and information about physical activities and participates in presenting them. It is important to note that in this case, even if the students do not use any ICT (and find needed information on boards at school and local print, or process information on paper, e.g. in the form of a table with the results of the school ping-pong tournament), they will develop their skills that are needed for the development of ICT literacy.

2.3 Opinions of Students of VSTE on Digitalization in Teaching

The case study presents the opinions of VŠTE students on the benefits of digital technologies in education. Table 1 shows partial results (from the first 10 respondents). The data was gathered via a survey that 74 students (combined study form) participated in (from the overall number of 115 students) who were approached by the author as part of the subject Corporate Information Systems. A questionnaire survey was used. The students voiced their opinions on a 1–5 scale (1 – definitely yes/I agree with the statement, 2 – yes, I agree, 3 – sometimes possible, 4 – rather disagree with the statement, 5 – not at all/I do not agree with the statement). The survey was implemented by the author on May 19, 2019.

Table 1 shows that students evaluated given benefits on the scale mostly with 1–3. That means that they agreed or partially agreed with them.

Table 1. Classification of the benefits of using digitization in education as seen by the VŠTE students

Respondent	Benefits of digitization in education/evaluating scale														
	V1	V2	V3	V4	V5	V6	V7	V8	V9	V10	V11	V12	V13	V14	V15
R 1	3	3	2	1	2	3	2	1	2	2	2	1	3	2	1
R 2	2	2	3	1	1	3	1	1	2	2	1	1	1	2	1
R 3	4	3	3	1	2	3	3	4	2	2	2	1	3	2	2
R 4	2	2	4	1	2	4	2	3	2	2	1	1	3	2	2
R 5	2	3	3	1	1	2	3	3	3	1	1	1	2	2	2
R 6	2	3	3	2	3	2	2	3	3	2	1	1	3	2	3
R 7	4	3	3	3	3	2	2	1	2	2	2	2	3	2	2
R 8	2	3	3	2	3	3	2	2	2	1	3	2	2	2	2
R 9	2	2	3	3	3	2	3	2	3	2	2	2	2	1	2
R 10	1	2	3	2	3	2	3	2	2	3	2	1	3	3	2

Note 1: R- respondent, V- value DT – Digital technology
V1 more effective education using DT
V2 option of finding information quicker
V3 option of carrying out exercises on a PC
V4 option of watching videos
V5 fun form of learning
V6 easier presentation of information
V7 option of a better sharing of information and thereby easier to learn from them
V8 DT enable easier access to data gathering
V9 education using interactive board
V10 option of online exam registration
V11 option of projecting prepared materials
V12 option of sharing information available worldwide on the Internet
V13 digitization of documents/sharing document
V14 online education courses
V15 modeling components

Note. 2: scaling
1 definitely yes/I agree with the statement
2 yes, I agree
3 sometimes possible
4 rather disagree with the statement
5 not at all/I do not agree with the statement

Source: compiled by the author.

The illustration shows some of the selected opinions of the students:

(a) Digitization brings new sources of information on a large scale of
(b) Digitization enables the implementation of modern methods into the education process, the sharing of information online, not only lectures but also feedback (voting systems, etc.).
(c) Digitization enables easy finding of current information, saving paper, students do not have work with paper books, everything is faster, simpler.
(d) Digitization enables sharing of information between subjects.
(e) Digitization in education/digital technologies significantly influences the possibility of teaching technical subjects. These are primarily new options of screening (for laboratories), e.g. digital microscopes, including records, etc. For technical subjects, these are 3D visualizations, the option of modelling parts, etc. Using 3D modelling of software. SketchUp to teach engineering subjects, 3D scanning and 3D printing enable practical experience that is simulated during class but is almost identical to real practice, etc.

Especially the last onion of one student is interesting. The student points out concrete benefits of using digital technologies when teaching technical subjects. He based this opinion on his own experience of using modelling of technical parts.

The author of the study also presents (see Table 2) summary of the results of the scale and the entire research sample.

Table 2. Expressing the views of respondents (74 students) on the importance of using digital technologies in education - the overview

Rota DT (digital technology)		Average on scale
Value 1	More effective education using DT	2,40
Value 2	Option of finding information quicker	2,67
Value 3	Option of carrying out exercises on a PC	3,06
Value 4	Option of watching videos	1,74
Value 5	Fun form of learning	2,43
Value 6	Easier presentation of information	2,60
Value 7	Option of a better sharing of information and thereby easier to learn from them	2,63
Value 8	DT enable easier access to data gathering	2,54
Value 9	Education using interactive board	2,36
Value 10	Option of online exam registration	1,79
Value 11	Option of projecting prepared materials	1,76
Value 12	Option of sharing information available worldwide on the Internet	1,38
Value 13	Digitization of documents/sharing document	2,56
Value 14	Online education courses	2,00
Value 15	Modeling components	2,13

Source: compiled by the author.

The students of the subject Corporate Information Systems worked on a group project in group of 4–5 students in which they solved a team project (Using modules of corporate information systems in a selected enterprise). The students showed during the presentations that they mostly have very good information and digital skills.

Currently, digital technologies are often used in many schools. They are an option how to share information with students that are available on the Internet faster than when these technologies were not used. Digital technologies definitely are beneficial for students (not only for teachers), a fact supported by the results of the research.

3 Conclusion

In connection with the above-mentioned information, I add that digital literacy influences all other areas of literacy. If we only have a command of one area, we would be lost in the current information society. That is why I think that digital literacy is the fundament of life-long learning. Nowadays, without using digital technologies, no one can effectively master any other literacy. We have to, therefore, educate ourselves during the course of our life. It is not enough to be able to use a computer or a mobile phone, we have to understand what they mean in practice: ICT literacy, Internet literacy, information or media literacy. To be able to use our functional literacy, it is important to develop creative skills and the ability to critically and innovatively think. Schools are expected to support students' effective use of information literacy that develops the ability to look for information and to sort them, evaluate them, and verify them. Digital literacy should be consolidated, starting in primary schools. However, it is needed that each student (pupil) develops his/her information skills and working with digital technologies as part of self-study.

As stated by Lozano [15] and Burďák [6], currently the role of the teacher increasingly shifts from the provider of information to an advisor, creator of appropriate learning environment for students' own activities. The intent of the strategy of digital education is to naturally merge formal education with informal learning and information learning. The goal then should be an open school environment whose common part will be the non-school environment which will become a crucial and important part of the learning experience of students. That is why it is crucial to strive to interconnect or synergize the world of school education and the outside world, connecting the learning experiences of students in and outside of school. A school as a closed or isolated institution will no longer be able to survive in the digital and online world of the 21st century.

4 Recommendation

It is necessary to pay attention to the area of education (introducing new teaching programs to support the development of digital literacy) with implementing digital technologies and non-traditional teaching methods into the process of education so that students can gather technical and expert skills in connected with Industry 4.0 (fourth industrial revolution, term for the current digitization trend) that brings with it many changes on the labor market and demands for the working force.

Schools must constantly take the necessary steps so that the graduates are well prepared for the changes of the information society and are able to effectively adapt, thereby being a contribution to society.

References

1. Altmanová, J., et al.: Gramotnosti ve vzdělávání: příručka pro učitele. Výzkumný ústav pedagogický v Praze, Praha (2010)
2. Andres, P., Vališová, A.: Elektronizace ve vzdělávání, fenomén současné doby (2019). http://jaroslavbalvin.eu/wp-content/uploads/2014/10/Andres_Valisova.pdf. Accessed 05 May 2019
3. Bajtoš, J.: Didaktika vysoké školy. Iura Edition, Bratislava (2013)
4. Bawden, D.: Origins and concepts of digital literacy. In: Lankshear, C., Knobel, M. (eds.) Digital Literacies: Concepts, Policies & Practices, pp. 17–32 (2008). https://litmedmod.ca/sites/default/files/pdf/bawden-lankshear-knobel_et_al-digitalliteracies_lr.pdf. Accessed 06 Oct 2019
5. Bučková, H.: Readiness of elementary school teachers to curriculum changes in the development of teaching informatics in Czech republic. J. Technol. Inf. Educ. **10**(2), 5–15 (2018)
6. Burďák, A.: Digitální gramotnost žáků ZŠ na 2. Stupni. Vydavatel. Univerzita Palackého Olomouc (2016)
7. Carniero, R., Kastis, N.: Digitální gramotnost aneb vývoj gramotností ve 21.století. eLearning Papers. P.A.U. Education, S.L. (2009)
8. Chromý, J.: Materiální didaktické prostředky v informační společnosti. Verbun, Praha (2011)
9. Ferrari, A.: Digital Competence in Practice: An Analysis of Frameworks. Publication Office of the European Union, Luxembourg (2012)
10. Ferrari, A.: DIGCOMP: A framework for developing and understanding digital competence in Europe (2013). http://bit.ly/1pm1qya. Accessed 06 Oct 2019
11. Fransson, G., Holmberg, J., Lindberg, O.J., Olofsson, A.D.: Digitalise and capitalise? Teachers' self-understanding in 21st-century teaching contexts. Oxf. Rev. Educ. **45**(1), 102–118 (2019). https://doi.org/10.1080/03054985.2018.1500357
12. Kolář, M.: Optimální využití didaktických technologií ve výuce odborného předmětu. Masarykova univerzita v Brně, Brno (2008)
13. Koníček, L.: Počítačem podporovaná výuka a experiment. Ostravská univerzita v Ostravě, Vydavatel a tisk (2003)
14. Lessner, D.: Analýza významu pojmu. Computational thinking. J. Technol. Inf. Educ. **6**(1), 71–88 (2014)
15. Lozano, M.T.U., Haro, F.B., Diaz, C.M., Manzoor, S., Ugidos, G.F., Mendez, J.A.J.: 3D digitization and prototyping of the skull for practical use in the teaching of human anatomy. J. Med. Syst. **41**(5), 1–5 (2017)
16. Melnik, G.S., Teplyashina, A.N.: The impact of digitalization of network space on journalism education. Media Educ. (Mediaobrazovanie) **59**(1), 86–92 (2019). https://doi.org/10.13187/me.2019.86
17. MŠMT, Ministerstvo školství mládeže a tělovýchovy: Strategie digitálního vzdělávání do roku 2020 (2014). http://www.msmt.cz/file/34429. Accessed 20 Mar 2015

18. Payton, S., Hague, C.: Digital literacy across the curriculum. National Foundation for Educational Research (2010). http://www.nfer.ac.uk/publications/FUTL06. Accessed 05 May 2019
19. Petlák, E., Fenyvesiová, L.: Interakcia vo vyučovaní. Bratislava, Iris (2009)
20. Pleskač, L.: Využití digitálních technologií ve výuce matematiky. Pedagogická fakulta Univerzity Hradec Králové, Hradec Králové (2015)
21. Russow, L.C.: Digitization of education: a panacea? J. Teach. Int. Bus. **14**(2–3), 1–11 (2003)

Institutional Transformation of the Legal Environment in the Context of Digitalization of the Economy

Subject Structure of the Offense in Artificial Intelligence (AI) and Robotics

A. V. Sidorova(✉)

Samara State University of Economics, Samara, Russia
an.sido@bk.ru

Abstract. Digitalization and digital technologies are a source of revolutionary changes affecting all aspects of human activity without any exception. Adaptation of society to digitalization is not only a technological, economic, socio-psychological, but particularly a legal problem. No one doubts that in the period of the fourth technical revolution, the regulatory framework should be aimed at providing a fertile and healthy ground for the progressive development of various technological innovations, preventing offenses. The development of robotics and artificial intelligence (AI) and their widespread use in all areas of the economy suggests the need to revise the traditional views on legal personality and offense. Without determining the legal status of work technicians and AI, the resolution of the question of their responsibility for the offenses committed and the damages caused, the further qualitative development of this area is impossible. Modern research in the field of law pays great attention to particular issues of the legal responsibility of robots and AI. In this regard, a number of questions arise directly related to the general theoretical concept of "offense", namely, can robots and AI be recognized as subjects of an offense and, as a result, bear legal responsibility, if so, what are the scope and consequences of this responsibility? The results of this study can be used in the practice of legislative bodies of the Russian Federation and international organizations. The theoretical conclusions of the study can be used in research and development institutes, in higher educational institutions for teaching not only legal but also economic disciplines.

Keywords: Accountability · Artificial intelligence · Pseudo-legal entity · Offense · Robotics

1 Introduction

The rapid development of digital technologies and the Internet in the last ten years have affected almost all spheres of human life and activity. Communication is becoming faster, the electronic view is increasingly preferable, which, on one side, simplifies our life, and on the other, complicates it from the point of view of the culprit, since an increasing number of operations are performed using robotics, AI and cyber-physical systems. The commercial value of AI in the economy is not subject to doubt: it is made up of additional revenue, volume of reduced costs, as well as revenues received as a

S. I. Ashmarina et al. (Eds.): ISCDTE 2019, LNNS 84, pp. 541–547, 2020.
https://doi.org/10.1007/978-3-030-27015-5_64

result of improving the quality of customer service through the introduction of such technologies [8]. One cannot underestimate the impact of robotics and AI on taxation and state income, since their introduction reduces jobs and the corresponding consequences in personnel wages.

The volume of investments in AI grows annually, and the companies' expenditures for development in this area are constantly increasing. However, the head of Sberbank of Russia has repeatedly stated that AI errors cost his financial organization millions of rubles [12]. But an employee for an error can be punished, but what to do with the robot is not clear. Complicated legal relations when using AI raise new questions in the study of the general theoretical category of "offense" and oblige to revise the traditional views on its manifestations. In the newest Russian law science, the number of signs of an offense have decreased in comparison with the Soviet period. Scientists most often mark illicitness, danger or harmfulness, culpability, punishment or legal liability as signs.

Crisis situations in the economy entail the emergence of new types of offenses and changes in the signs of an offense. In particular, the sign of guilt, which until recently was mandatory for the notion of an offense, is currently losing its position. There are more and more offenses, the guilt of the subject of which is not mandatory for bringing to legal responsibility. Thus, a legal entity re-emerged during the merger of several legal entities is brought to administrative responsibility for committing an administrative offense by a legal entity participating in the merger (Article 2.10) [2]. A legal entity (organization) operating a hazardous production facility is responsible for causing harm to such an entity without any fault.

In the modern literature it is noted that a promising direction for the development of AI in the economy is the use of the blockchain, which is defined as "a multifunctional and multi-level information technology designed to reliably account for various assets" [10]. Blockchain as a kind of AI creates new opportunities for finding, organizing, evaluating and transferring any discrete units, and, therefore, can harm its activities, that is, in fact, commit an offense. In the legal literature it is noted that in the era of digitalization and mixing of the market of goods and services in the "online" functioning on the Internet, it is necessary to study "offense" taking into account the influence of digital factors and the international nature of the legal relations of participants [14]. For this reason, it will be necessary to determine the legal status of AI, with the goal of recognizing it as an offender and the possibility of compensating for the damage caused to them, since without clear criteria for compensating property damage, the development of a modern economy is hardly possible.

2 Methodology

The author has used the historical-legal, formal-legal and structural-legal method. A special role is given to the comparative legal and systematic method. The basis was the theoretical position as a general theory of law, and in individual branches of law.

3 Results

Considering the concept of "offense" in relation to ordinary human life, it is necessary to pay attention to such a sign of it as the wrongfulness of the act. The essence of such a multidimensional legal phenomenon as wrongfulness can be more accurately and fully disclosed through the definition of a wrongful act as one of the elements of public life. Social life can be defined as follows: Social life is a process occurring in society, including the genesis, functioning and development of both the entire social body and its individual elements. A person, being born in a certain society and in a certain time period, finds the established system of social relations, which he cannot ignore. But he can and must define his place and role in this life, find his destiny in it as a living being and acting person. The power of the objective laws of society is not fatal in this sense either. Consequently, a person is not thinking outside public relations. Between all individuals from the moment of their birth until the moment of death, there are "threads" by which they are interconnected among themselves and with society as a whole. This interrelation concerns various issues: economic, political, religious, social legal and many others. Here the question arises: "What is a legal relationship?" First of all, this is of course one of the types of public relations. Secondly, it is a kind of social relations that are regulated by the rules of law. Thirdly, the participants of legal relations as one of the types of social relations have subjective legal rights and obligations. And since the offense does not exist outside of legal relations (legal field), which in turn are a type of social relations, in our opinion, it will be correct to define it as a social relation.

In short, social relations can be defined as a combination of people's activities. By activity we shall mean active actions of a person aimed at achieving consciously set goals, aimed at satisfying one's needs and interests, as well as fulfilling the requirements set by the society and the state. Human activity as an active action carried out exclusively in the presence of motive is focused and, among other things, its formation occurs when social experience is assimilated, and nothing else. Based on the foregoing, an offense is an activity of a subject of public relations that violates objective and (or) subjective rights protected by the state and entails the emergence of law enforcement relations.

Such a conclusion is absolutely applicable in classical social relations between people. However, in the modern world, robots and technologies called AI are increasingly being introduced into daily life. In this regard, the question arises whether robots and AI are subjects of social relations, can they commit unlawful acts, that is, offenses?

4 Discussion

First of all, it is necessary to separate the concepts of robot and AI. In the legal literature it is noted that by robots it is necessary to understand a certain device or mechanism [9]. In 1983, it was proposed to define AI as the ability of a device to perform functions that are commonly associated with human intelligence, such as reasoning, learning, and self-improvement [18]. The modern concept of AI is broader and it is believed that it differs from the robot in that it has special characteristics: ability to communicate, knowledge of oneself, knowledge of the external world, ability to achieve goals and a certain level of creativity [5]. Of course, these characteristics and capabilities are the result of human-written code that programs or determines the action of AI, but its specific actions are carried out without human influence.

The world community drew attention to the need for legal regulation in the sphere of "smart robots" ten years ago. The first normative legal act regulating the sphere of robotics was the Law of the Republic of Korea "The Law on Promoting the Development and Distribution of Smart Robots" [6], which determined the definition of "smart robot" as "a mechanical device that is able to perceive the environment, recognize the circumstances in which it functions, and purposefully move independently" (Article 2). The specified regulatory legal act does not define any elements of the legal capacity of robots, but establishes the liability of legal entities only in the field of robotic investments.

The year 2017 in Europe was marked by the adoption of a conceptual regulatory act affecting the sphere of robotics - European Parliament Resolution No. 2015/2103 (INL) of February 16, 2017 "Civil Law Standards on Robotics", which reinforced the need to determine the legal responsibility of robots [3]. The normative act declares that "today the responsibility for the harm caused by the robot should lie on the person, not the robot". Under the existing legal framework, robots themselves cannot be held responsible for actions or inactions that caused harm to third parties. The existing rules of liability provide for cases where the actions or inactions of robots are in causal connection with the actions or omissions of specific individuals - manufacturers, operators, owners or users, and they could foresee and avoid the behavior of robots, as a result of which damage was caused (p. AD Article Z). In this case, the responsibility of these persons is divided into the responsibility of the manufacturer of the robot and its user.

Meanwhile, p. AF Article Z explicitly states that "the standard rules of liability are not sufficient in the case when the damage was caused due to the decisions made by the robot itself, since in this case it is impossible to identify a third party who will be obliged to pay compensation or compensate for the damage caused". It is proposed to abandon the principle of individual responsibility of a person for the harm caused, and place it on a person who, under certain conditions, could minimize the risks and take into account negative consequences (paragraph 56).

In the Russian Federation, the system of legal regulation of robotics is at the very beginning of its development. Currently, no legal acts have yet been adopted in any way concerning issues or defining the concept of the legal responsibility of robots. Nevertheless, active work is underway on a draft law regulating the sphere of robotics. In December 2016, the law firm Dentons commissioned Grishin Robotics to develop the concept of Russia's first draft law on robotics, tentatively called the "Grishin Law". The draft law proposes amendments to part one of the Civil Code of the Russian Federation [1], in particular, introduces the concept of a "robot agent" - "a robot that, by decision of the owner and by virtue of its design features, is intended to participate in civilian circulation. The robot agent has separate property and is liable for them by its obligations; it can, on its own behalf, acquire and exercise civil rights and bear civil obligations. In cases established by law, a robot agent may act as a participant in a civil process" (Article 1).

Based on this approach, the robot is actually equal to legal entities and has a specific legal personality. In this case, the measures of legal responsibility of the robot agent are equal to the measures established for legal entities. But, as mentioned earlier, despite the fact that the harm is directly caused by the robot, there is currently a tendency to recognize individuals and legal entities responsible for it - the creators, owners and other persons influencing the robot. That is, the robot itself is not destructive and cannot be held liable for the harm caused by it, but can only be considered as a human tool [7, 13, 17].

AI is a technology that has the ability to autonomous, that is, without additional human participation, self-learning based on the information available in this process [11]. Unlike robots, which in any case depend on the owner and the pledged code, AI can carry out cognitive processes (self-study and make quasi-dependent decisions) independently. It results in multiple questions about the legal personality of AI: whether to treat it as in the case of a robot, dependent on owners, creators and other persons, or to give it a pseudo-legal entity. With the prefix "pseudo" for the reason that legal personality in the full sense carries not only duty and responsibility, but also rights that it is doubtful to link to AI.

The question of the AI accountability, in our opinion, is much more complicated compared to a robot. It is noted that the appropriation of the legal personality of AI in its willing sense is important because it can restore the chain of cause-and-effect relations and the limitation of the responsibility of the owner [5]. However, with regard to AI, it is impossible to say for sure that he has free will, which can lead to the performance of prohibited actions in order to achieve his own goals. Thus, one cannot attribute to him a degree of guilt, such as negligence or recklessness [4].

Meanwhile, in modern Russian literature, "accountability" is often distinguished as a sign of an offense. Sabirova argues that "the offense is the act of the destructive person, i.e. subject capable of bearing legal responsibility. In order to be responsible for the appropriate act, a person must reach a certain age and be sane. For example, a person who has reached the age of 16 by the time of committing an administrative offense may be brought to administrative responsibility" [15]. In certain cases defined by law, the subject of the offense may be exempt from legal liability because of his lack of ability - the ability of the subject to be solely responsible for the harm caused by his unlawful act (act or omission). As a rule, it is the youngest in comparison with the age of the offender defined in the law, the incapacity of the person, etc. That is, the offense is actually committed - the act was illegal, caused harm to the rights and duties, property of other legal entities, violated public order, etc., but due to the identity of the subject of the offense, the measures of legal responsibility for such an offense will not be applied to the offender, or responsibility for the offender will be borne by other persons (parents, legal representatives, etc.).

On the example of AI, capable of making decisions and causing harm, it is clear that neither age nor sanity and accountability as a whole can be a sign of an offense, a perfect robot. At the same time, his activity may be illegal, but only from an objective point of view. This completely changes the approach to the sign of wrongfulness that has emerged in the domestic literature, according to which wrongfulness can be viewed from two aspects - objective and subjective. On the objective side, wrongfulness is a violation of objective law, its rules containing legal obligations and prohibitions.

The subjective side of wrongfulness is characterized by a conscious will of the person, that is, from the point of view of law, there is only a subjective-objective wrongful act, but not an objectively wrongful act.

In the case of AI, in our opinion, there is no subjective component, since it is not a person and does not perform conscious actions of its own will. This or that act can really be done by the will of AI, however, it is not conscious, because consciousness exists only in a person who is able to socialize his experience, create joint knowledge that is fixed in speech, samples of material and spiritual culture. For this reason, we should agree with the point of view that the legal personality of AI should not be conceptually different from a legal entity. If we perceive AI as a rational being, then it would be our moral duty to give them rights [16].

5 Conclusion

Based on the above mentioned, it is possible to draw unambiguous conclusions that the formation and development of modern legal science, the development of views on the nature of unlawful acts as a legal phenomenon, the level and technique of fixing the definition of "legal capacity" in the legislation is influenced by quite a number of factors, among which the latest factor is the achievements of science and technology in the form of robotics and AI and their active use in the economy. It is necessary to determine the legal status of robots and AI, having considered the possibility of integrating them with a pseudo-legal entity, within which there will be no rights.

References

1. Civil code of the Russian Federation (part one) No. 51-FZ dated 30 November 1994. http://www.consultant.ru/cons/cgi/online.cgi?rnd=B5456AE0380E62C5D41CC8549669FB2F&base=LAW&n=300822&dst=4294967295&cacheid=35143974863E68B0CF2F83ABE17FC387&mode=rubr&req=doc#1hlnasd374t. Accessed 24 Apr 2019. (in Russian)
2. Code of administrative offences of the Russian Federation of December 30, 2001 No. 195-FZ. http://www.consultant.ru/document/cons_doc_LAW_34661/. Accessed 24 Apr 2019. (in Russian)
3. European Parliament Resolution No. 2015/2103 (INL) of February 16, 2017 "Civil Law Standards on Robotics". http://www.europarl.europa.eu/doceo/document/TA-8-2017-0051_EN.html?redirect
4. Krainska, A.: Legal personality and artificial intelligence. newtech.law (2018). https://newtech.law/en/legal-personality-and-artificial-intelligence/. Accessed 24 Apr 2019
5. Kritikos, M.: Artificial intelligence ante portas: legal and ethical reflections. Scientific Foresight Unit (STOA), March 2019. http://www.europarl.europa.eu/RegData/etudes/BRIE/2019/634427/EPRS_BRI(2019)634427_EN.pdf. Accessed 24 Apr 2019
6. Law of the Republic of Korea "The Law on Promoting the Development and Distribution of Smart Robots". http://www.elaw.klri.re.kr/eng_mobile/viewer.do?hseq=39153&type=lawname&key=robot
7. Leenes, R., Lucivero, F.: Laws on robots, laws by robots, laws in robots: regulating robot behavior by design. Law Innov. Technol. **6**(2), 193–220 (2014)

8. Lipinsky, D.A., Evdokimov, K.N., Musatkina, A.A.: Actual problems of international cooperation of Russia in the sphere of cyber security. Lect. Notes Netw. Syst. **57**, 495–504 (2019). https://doi.org/10.1007/978-3-030-00102-5_52
9. Neznamov, A.V., Naumov, V.B.: Regulation strategy for robotics and cyber physical systems. Law **2**, 69–75 (2018)
10. Openkov, M.Y., Varakin, V.S.: Artificial intelligence as an economic category. Bull. North. (Arct.) Fed. Univ. Ser.: Humanit. Soc. Sci. **1**, 73–83 (2018)
11. Rawlinson, P., Arievich, E.A., Ermolina, D.E.: Intellectual property objects created with the help of artificial intelligence: features of the legal regime in Russia and abroad. Law **5**, 63–71 (2018)
12. RBC: G. Gref admitted the loss of billions of rubles due to artificial intelligence (2019). https://www.rbc.ru/finances/26/02/2019/5c74f4839a7947501397823f. Accessed 24 Apr 2019
13. Revina, S.N., Sidorova, A.V.: Transformation of general-theoretical category "offense" in the internet era. In: Mantulenko, V. (ed.) GCPMED 2018 - International Scientific Conference "Global Challenges and Prospects of the Modern Economic Development", The European Proceedings of Social and Behavioural Sciences EPSBS, vol. LVII – GCPMED 2018, pp. 1672–1679. Future Academy, London (2018). https://doi.org/10.15405/epsbs.2019.03.169
14. Revina, S.N., Paulov, P.A., Sidorova, A.V.: Regulation of tax havens in the age of globalization and digitalization. In: Ashmarina, S., Mesquita, A., Vochozka, M. (eds.) Digital Transformation of the Economy: Challenges, Trends and New Opportunities. Advances in Intelligent Systems and Computing, vol. 908, pp. 88–95. Springer, Cham (2020)
15. Sabirova, L.L.: Lawful conduct, offence, legal responsibility. In: Bakulina, L.T. (ed.) Problems of the Theory of Law and Law Realization. Statut, Moscow (2017). (in Russian)
16. Singh, Sh.: Attribution of legal personhood to artificially intelligent beings. In: Bharati Law Review, pp. 194–201, July–September 2017
17. Solaiman, S.M.: Legal personality of robots, corporations, idols and chimpanzees: a quest for legitimacy. Artif. Intell. Law **25**(2), 155–179 (2017)
18. Willick, M.S.: Artificial intelligence: some legal approaches and implications. AI Mag. **4**(2), 5–16 (1983)

Analysis of Legal and Economic Risks for Entrepreneurs in Digital Economy

F. F. Spanagel, O. A. Belozerova, and M. K. Kot(✉)

Samara State University of Economics, Samara, Russia
mkroz@mail.ru

Abstract. The urgency of the issue under study stems from the incompleteness of the regulatory legal definition of concepts and the legal regime for new objects of digital economy or even complete absence of this definition, as well as the need to determine the effects of digital technologies impact on legal and economic relations with the participation of entrepreneurs. The purpose of research is to identify and reveal general trends in legal and economic risks for entrepreneurs during the period of digital economy formation. Research methods of this issue are formal legal analysis and synthesis, which provide a comprehensive review of legal and economic risks' interaction in the activities of entrepreneurs. Results: the article describes general trends of legal and economic risks' impact on business entities. It reveals the multidirectional influence of digital technologies on legal, economic and other risks and justifies the necessity of developing a new toolkit for identifying and analyzing risks both at the state level and by individual entrepreneurs. The data of the article could be useful for experts in jurisprudence, economics, risk management, and for those involved in law-making activity.

Keywords: Digital economy · Digital technologies · Economic risks · Entrepreneur risks · Paternalism

1 Introduction

In the conditions of 'digital revolution', the Russian Federation and domestic businesses face the task of actively engaging in elaboration, introduction and development of digital technologies in such important areas as big data, blockchain, including smart contracts, telemedicine, unmanned vehicles, Internet of Things (IoT), cryptocurrency, new platform companies and many others. This will drastically improve the lives of people, multiply increase labor productivity, diversify the national economy, freeing it from 'carbon dependence', and enable entrepreneurs to enter foreign markets with competitive products (primarily intellectual property).

At the same time, the introduction and widespread use of digital technologies in addition to many benefits and amenities entails numerous threats that not only significantly change the numerous previously existing business risks, but also create new risks for participants.

Thus, in the near future, millions of entrepreneurs and the bulk of the population of our country will actively use digital services, the range of which will expand much

S. I. Ashmarina et al. (Eds.): ISCDTE 2019, LNNS 84, pp. 548–556, 2020.
https://doi.org/10.1007/978-3-030-27015-5_65

faster than other services, goods and jobs, which will reduce the control of relevant services and may cause a sharp increase in the number of fraudulent activities and other cybercrimes. These are, for example, extortion through blocking access to databases (using the 'Petya' virus, for example), illegal acquisition of commercial information from entrepreneurs, theft of non-cash funds from bank clients, hacked IoT devices, and others.

Studying legal, economic and other risks, their appearance in the digital reality will allow us to identify them, determine patterns of their impact on legal and economic relations in business sphere, help entrepreneurs and government bodies to predict their changes, including scale and intensity, to organize their management (risk management).

2 Methodology

2.1 Research Methods

In the process of research, the following methods were used: analysis, synthesis, comparison, generalisation, they allow to thoroughly review the process of forming an audit institution as an element of market relations and an object of legal regulation in Russia and a number of foreign countries.

2.2 Research Base

The research is based on scientific studies, publications by Russian and foreign lawyers and economists who study various aspects of risks in business sphere, and on the latest legislation in digital economy.

2.3 Stages of Research

The study of the issue was carried out in two stages:

The first stage: analysis of literature on the subject and digital economy legislation; highlighting the problem, purpose, and research methods.
The second stage: formulation of conclusions obtained after analysis of literature and legislation, preparation of the publication.

3 Results

The long-term policy of sanctions unleashed by the developed and affiliated countries against the Russian Federation (and above all against the national economy), restrains its development, intensifies its lag from the developed economies of the world in technological sphere. This circumstance refers to the objective factors of national economy development.

The subjective nationwide factor is the great capability of the Russian state to ensure legal regulation (if not advancing, at least adequate) of business processes, with the participation of national businesses and foreign entrepreneurs on the Russian Federation territory. Timely and full realization of such capabilities will create an additional attractiveness of the business environment, in particular, its legal component.

At the same time, it should be pointed out that an important feature of digital technologies is that they weaken legal constraints for businesses when crossing state borders with goods and services (primarily digital technologies), and other intellectual property, as well as mandatory national jurisdictions. They are increasingly allowing not only consumers, but also entrepreneurs to 'vote with their feet', that is, to choose the most acceptable jurisdiction for them, transferring their production activities to neighbouring and other countries, which significantly intensifies competition between states and entails multidirectional consequences for entrepreneurs: for some, significantly increasing and exacerbating, and for others, reducing or even eliminating legal and other risks.

In recent years, entrepreneurs have increasingly actively and massively selected countries with the most favourable business climate, primarily including the legal regime for relevant operations, for their entrepreneurial and investment activities. Thus, a significant increase in the tax burden on the 'rich' in France led to an active transfer of production and capital by entrepreneurs to neighbouring countries, which led to a negative effect from the initiated tax reform.

With the multidirectional and constantly boosting development of digital technologies, patriotism in business is becoming less and less obvious factor in entrepreneurship. It is becoming increasingly difficult for states even with a highly developed economy and well-organized patriotic education to use the opportunities and strength of their own sovereignty to retain entrepreneurs and investments within the borders of their jurisdiction.

Actively discussed aggressive attempts by the President of the United States to preserve American business under its exclusive jurisdiction can confirm our conclusion. A typical and very illustrative example of such actions is his conflict with a large domestic motorcycle company Harley-Davidson. This company, a symbol of the United States of America, has chosen as its mission the realization of riders' dreams of personal freedom and it positions itself as an international company that has 'excellent prospects in both developed and growing foreign markets, where its position is as strong, as in the USA' [11].

Due to the European Union's introduction of tariff anti-sanctions in response to US 'tariffs barrage', Harley-Davidson decided to withdraw part of motorcycles production from the United States to the territory of the European Union in order to exclude threatening losses. In response to the company's legitimate actions, Donald Trump publicly promised to use his 'political capabilities' to destroy it, threatening to strangle it with taxes [4] (obviously 'personal').

Despite the measures taken (unprecedented for a liberal state with an advanced market economy), it can be seen that even illegal threats against a lawfully operating company and accusations of national betrayal did not help the President. At the same time, the new trend in the government of the United States and a number of other countries in the context of a new, digital reality should not go unnoticed: instead of applying the numerous legal means available to the President, he as an individual actively uses his twitter account for political purposes [4].

Donald Trump and other politicians more often use unofficial methods and forms of nonlegal influence (and in our case even illegal open pressure) on politicians, businessmen and even sovereign states through unilateral announcement to the world of their position on significant issues, including legal ones, and statements of intent with non-veiled threats. Russian politicians so far rarely use public pressure on entrepreneurs, but in the process of performing control and supervisory functions, such methods of influence on entrepreneurs are not rare.

In addition, the US-launched trade wars with Russia, the EU, China, and other countries form a dangerous and highly risky practice of adopting restrictive, and basically 'barrage' measures with disregard and gross violation of generally accepted norms of international trade law. Such voluntarism creates instability, entails justified retaliatory actions, multiplying the legal and economic risks for the world and national economies in the context of their globalization.

Any state seeks to ensure the stability and sustainability of the national economy [7], taking into account the condition and trends of the global economy. Thus, South Korea, when forming the economic program, proceeds from the thesis of the 'increasing instability of the world economic system' [4], which seems to us accurate and objective. Judging by the decisions taken and the documents approved, the government of the Russian Federation and the Central Bank of Russia in their activities lean on the same pattern.

With all the importance of foreign investment and foreign business experience, it should be recognized as unacceptable providing foreign participants with more favourable conditions for doing business than domestic entrepreneurs, while creating incentives for doing business in our country.

The following pattern can be seen here. On the one hand, the globalization of most segments of world, regional and national markets is accelerating and deepening, also due to activities of the World Trade Organization, EU member states, APEC, EAEU and other associations that have created about 300 free trade zones. In some cases, the economies of neighbouring countries have become interconnected to concentration levels that are dangerous for national sovereignty, for example, in the share of South Korean exports, China's share is about 40% [9]. On the other hand, as a reaction to the strengthening of the first tendency and noticeably increased risks of the participants in this regard, the policy of paternalism is actively and even aggressively pursued by a number of countries, this policy can be seen in the activities of most states of the world and it is supported by their citizens and first of all by their entrepreneurs.

On the one hand, the world economy is moving towards deeper integration, and on the other hand, fearing the loss of economic sovereignty, countries within and outside the norms of international law strengthen protectionism, setting barriers and restrictions, even unleashing 'trade wars' using digital technologies, and setting 'electronic boundaries' (restrictions in the virtual sphere). In the long-term plan, the 'victory' of the globalization processes is inevitable, however, in the near future, the strengthening of protectionism can be predicted in state policies in favour of national business.

In the current circumstances, the optimal legal regulation of economic relations is the selection of legal tools adequate to the digital economy, their fine tuning. It is necessary to form sets of legal means for achieving specific goals of legal impact on various participants of business activities, to work out their interaction, which can reflect the influence of an increasing number of various risk factors on business participants.

When the government conducts economic and legal policies, it is necessary to recognize as positive the practice of increasingly widespread use of legal experiments within a single subject or constituent entities of the Russian Federation [6], which makes it possible to work out technologies for introducing innovations with less social, political, legal, economic, managerial and information risks.

In the conditions of rapid changes in the digital economy and the resulting uncertainty, the readiness of state bodies and local governments to reconfigure their activities, to quickly train officials and other workers in new professional skills and mastering legal tools for changing economic relations, to adequate response to new risks and challenges is also very important.

The current system of economic state management and, above all, business management, does not fully correspond to the capabilities of updated technologies and the speed of changes in business processes.

Legal regulation is clearly lagging behind the needs of the digital economy, which contributes to the increasing negative impact of digital reality on business processes in our country and may further lead to an increasingly noticeable technological lag behind the world leading countries and even create risks of the so-called 'digital slavery', which has reasonably become a subject of discussion in the scientific literature [5]. It is necessary to mention the absence of a regulatory framework or its inconsistency, the late establishing of the state's position on legal regulation of relations in the virtual sphere: in information and telecommunications network, including electronic commerce, on use of blockchain technology, including cryptocurrency. Many legal risks may appear in the sphere of creation and operation of robotics, 3D printers, bioengineering, and others.

More and more types and varieties of informational and legal risks are revealed in cybersecurity at all its levels: the state, the society, individual citizens, and the latter (especially entrepreneurs) are most vulnerable to them. These risks increase significantly due to the lack of unity between countries in their fight against cyber threats, and even more so due to the states' numerous mutual accusations of hostile activity against each other and the threats that follow. For example, if the Danish government only calls for an attack on our country in cyberspace, the British military command declares its readiness to commit a cyber-attack on the power system of Moscow [8].

Both Russian entrepreneurs (especially micro and small businesses) and consumers have weak security in cyberspace and bear enormous risks that their rights and legally protected interests can be violated. Entrepreneurs' attempts to protect their violated rights often end at the initial stage of identifying offenders hiding in the vast expanses of information and communication networks.

Without introduction of adequate methods for identifying Internet users and, in particular social networks users, the number of frauds in sphere of digital technologies will only increase. The numerous cases of unlawful refusal to deliver goods paid by the buyer, when it is extremely difficult to fix such facts and initiate criminal proceedings, especially to investigate such and other crimes committed in a virtual environment can be an example here. This necessity of reducing legal and economic risks is reasonably mentioned in the literature [3].

Digital technologies in the field of state and municipal control demonstrate the original combination of new opportunities with new threats. Thus, recently relatively time-consuming and expensive, procedures of registering legal entities and individual entrepreneurs, transferring rights to real estate and transactions with it, reporting to the tax authorities and others have become fast and convenient for citizens thanks to digital technologies.

In addition to ensuring transparency and speeding up registration procedures as forms of preliminary control over the actions of economic entities, digital technologies also make it possible to significantly increase the effectiveness of follow-up control, for example, over the compliance with antitrust laws.

Thus, in past years, the implementation of norms for countering cartel agreement in oil and other country's markets was not effectively put into action, and business entities that didn't belong to such illegal agreements, as well as millions of consumers, suffered from it.

The Federal Antimonopoly Service, in conjunction with other government agencies, is actively using digital technologies, which helps in fight with cartels and has led to an annual increase in the number of identified and investigated schemes aimed at violating competition law. The positive results of the subsequent antitrust control were obtained due to use of new software that helps to identify malicious computer programs (robots) that allowed entrepreneurs to enter and implement cartel agreements. FAS revealed them, in fact, organizing the fight of robots and bringing it to successful for the state and consumers court decisions.

At present, the country's antimonopoly authority has organized electronic monitoring of monopolized markets and halts offenders' activities in the process of their interaction and use of appropriate software (a robot with robots), which could not be done several years when government control bodies didn't have these programs.

In this regard, the head of the FAS said: '… instructions from the head of state to actively fight against cartels have been issued… if someone wants to go to prison, let them create cartels' [1]. This FAS statement, on the one hand, is a clear signal to market participants about strengthening control over them and the inevitability of administrative and criminal liability of offenders, and on the other hand, it shows the state of law and order in the business environment, confirming that even the federal anti-monopoly body, besides having proper software, needs the instructions of the head of state to implement the rules on countering cartels, without which the requirements of the legislation would apparently remain unfulfilled.

In the Russian Federation, the role of subjective factors, the selectivity in bringing business entities to administrative and even criminal responsibility remains crucial, which increases risks' impact on them and does not allow to eradicate legal nihilism, which appears at various levels of both state and corporate governance. Different types and methods of digital control can provide the necessary transparency of management and significantly reduce the existing legal risks for entrepreneurs and consumers, facilitate bringing of offenders to criminal, administrative and civil liability.

In addition to reducing or even completely eliminating the aforementioned legal risks, digital technologies will entail many other positive consequences for business and people. As an example, it is appropriate to refer to the expected use of the blockchain technology, which can simplify and cheapen the realisation of a large number of different legally significant actions.

The most important thing is to deprive intruders of the possibility to falsify or destroy various entitling and endorsing documents, which will minimize legal risks and reduce relevant offences, also it's important to abandon current procedures and minimize paperwork to significantly save financial and other resources.

Much is being done to create advanced domestic digital technologies to improve management efficiency in Russia, for example, the range of electronic services provided to entrepreneurs and others is expanding. To accelerate and expand the use of digital and other new technologies in all spheres of society, the Russian government approved a comprehensive longterm program [6], which includes measures to create legal, technical, organisational and financial conditions for development of digital economy in Russia. Its successful implementation can radically change entrepreneurs' activities and citizens' lives, significantly reducing the number and intensity of existing legal and other risks, as well as create conditions for new risks and significantly complicate all types of social ties, giving rise to global and comprehensive dependence of state, society and people on the reliability of electronic communications, including protection from possible offences and abuse of rights by participants of digital environment.

4 Discussion

The risk is present in almost all spheres of human activity, has many types and subtypes, and, while studying it, it is necessary to see various aspects of its adverse effects. The most common direction of risk research is the economic one, it includes primarily banking, insurance and other financial activities, which is reflected in their legal regulation.

An important and socially significant type of threats and challenges is the entrepreneurial risk, the most important types of which are legal and economic risks. The strong connection and interaction of legal and economic adverse effects arising from the realisation of risks make their comprehensive study necessary. At the same time, there are no grounds for distinguishing the term 'economic and legal risks' as an independent scientific category.

Various aspects of economic and legal audit activity are studied and analyzed by economists and lawyers both in Russia and in foreign countries. Lawyers and economists both in the Russian Federation and in foreign countries study legal and economic risks in their researches.

The category of risk was studied by the philosopher Beck [4], by lawyers: Oygenzicht [10] - in the aspect of general correlation between risk and liability in civil law, Tikhomirov [13] - to risks and prediction in law, Arkhipov [2] and others - to issues of distribution of contractual risks.

The economic risks and their management were analyzed more fully, in particular, in the national studies by: Barbaumov [3], Avdiysky and Bezdenezhnykh [1], and others, as well as in studies of foreign scientists: Roubini and Mime [12] and others.

In connection with formation and functioning of digital economy, remarkable changes can be observed both in the number of risks and in the nature and intensity of their appearance, taking into consideration the characteristics of particular risks.

In the conditions of increasing uncertainty inherent to modern market relations, the majority of their participants are interested in receiving objective information about legal and economic risks they take making decisions.

In this article, the impact of legal and economic risks on entrepreneurial activity is considered in the aspect of their appearance in conditions of rapidly changing digital technologies.

5 Conclusion

A comprehensive study of the impact of digital economy development in the Russian Federation on the legal and economic risks of entrepreneurs needs to be carried out more widely and actively in order to simultaneously change the legislation both on use of the latest technologies and on the legal regime of business. Such work should be carried faster, taking into account accelerating changes in business. So, the introduction and use of digital technologies, planned by the Russian government, depends on extensive and well-coordinated work of all government branches: legislative, executive and judicial – at all levels of state and municipal government; and on ability and desire of millions of people, both entrepreneurs and consumers, to master constantly updated digital and other technologies on user level.

References

1. Avdiysky, V.I., Bezdenezhnykh, V.M.: Risks of Economic Entities: Theoretical Foundations, Methodology of Analysis, Forecasting and Management: Educational book. Alpha-M; INFRA-M, Moscow (2013). (in Russian)
2. Arkhipov, D.A.: Distribution of Contractual Risks in Civil Law: An Economic and Legal Study. Statut, Moscow (2012). (in Russian)
3. Barbaumov, V.E. (ed.): Encyclopaedia of Financial Risk Management. Alpina Publisher, Moscow (2013). (in Russian)
4. Beck, U.: Risk Society. Toward a New Modernity. Translated from German. Progress-Tradition, Moscow (2000). (in Russian)
5. Kommersant Newspaper No. 174: Everyone was happy and robbed the consumer. FAS Head Igor Artemyev About State-Monopoly Capitalism (2018). https://www.kommersant.ru/doc/3751719. Accessed 3 June 2019. (in Russian)
6. Point of support. Challenges, threats and prospects of digital economy (2017). https://www.to-inform.ru/. Accessed 2 Oct 2018. (in Russian)
7. Glazyev, S.: The great digital economy. Challenges and prospects for the economy of the XXI century (2017). http://ruskline.ru/opp/2017/sentyabr/14/velikaya_cifrovaya_ekonomika_vyzovy_i_perspektivy_dlya_ekonomiki_xxi_veka. Accessed 10 Oct 2018. (in Russian)
8. Danilov, I.: Donald Trump fell under motorcycle. What threatens Harley-Davidson (2018). https://ria.ru/analytics/20180627/1523465099.html. Accessed 1 Oct 2018. (in Russian)

9. Kasperskaya, N.: Digital economics and risks of digital colonisation (Detailed theses of a speech at the Parliamentary hearings in the State Duma) (2018). https://ivan4.ru/news/traditsionnye_semeynye_tsennosti/the_digital_economy_and_the_risks_of_digital_colonization_n_kasperskaya_developed_theses_of_the_spee/. Accessed 1 Oct 2018. (in Russian)
10. Oygenzicht, V.A.: The Problem of Risk in Civil Law. Irfon, Dushanbe (1972). (in Russian)
11. The official Internet portal of legal information: meeting of the legislation of the Russian federation, No. 32, art. 5138. http://www.pravo.gov.ru. Accessed 10 May 2017. (in Russian)
12. Roubini, N., Mime, S.: Nouriel Roubini: How I Predicted the Crisis. EKSMO, Moscow (2011). (in Russian)
13. Tikhomirov, Y.A.: Law: Forecasts and Risks: Monograph. INFRA-M, Moscow (2015)

Personal Data and Digital Technologies: Problems of Legal Regulation

N. V. Deltsova(✉)

Samara State University of Economics, Samara, Russia
natdel@mail.ru

Abstract. The urgency of the research issue is due to the trends of social and economic development in the context of digitalization. Big Data has become an answer to the modern social challenges. It is the technology for processing large amounts of data, which allows you to quickly process large and diverse data, systematizing them. This raises the question of ensuring the legal protection of personal data when they are processed by these technologies. The purpose of this article is to identify the problems of legal regulation, which arise in the processing of personal data in Big Data systems. The research objectives are to determine approaches to the legal regulation of the processing of personal data and Big Data systems, identify gaps in the current legislation governing these relations, analyze project legislation, as well as to identify the main current trends of legal regulation in this field. The following methods were used in the research: general scientific methods – analysis, synthesis, comparison and generalization; private and scientific formal and legal method. As a result of the research, it is concluded that the digital transformation of the development of society has put forward new requirements for the legal protection of legally protected information and personal data. The current Russian legislation does not have special legal provisions related to the processing of information using Big Data technologies. The materials of the article are of practical value for experts and scientists in the field of information processing and legal protection of personal data.

Keywords: Big Data technologies · Digitalization · Legal protection of personal data · Legal regulation · Personal data

1 Introduction

Constitution of the Russian Federation protects the right to privacy, to personal and family secrets [6]. In pursuance of these constitutional provisions, the Russian sectoral legislation establishes rules on the protection of intangible goods and liability for their violation. A special place among the regulatory acts aimed to protect private information is taken by Federal Law "On Personal Data" (hereinafter referred to as the "Personal Data Law") No. 152-FZ dated July 27, 2006, which regulates relations related to various forms of processing of personal data of individuals [7].

Under the influence of the digital environment, a significant amount of information about citizens is processed (collected, systematized, stored, etc.) daily on the Internet. Today, the everyday life of citizens is not without the use of search systems, "smart

S. I. Ashmarina et al. (Eds.): ISCDTE 2019, LNNS 84, pp. 557–563, 2020.
https://doi.org/10.1007/978-3-030-27015-5_66

things", which are usually linked to personal databases, social networks, the use of which certainly leaves a digital footprint [17]. Modern information technologies make it possible to identify, compare and analyze user behavior on the network, thereby processing a variety of personal information and personal data.

According to the data given in the list of references, there were six trackers on Avito.ru, seven – on HH.ru and nine – on Gismeteo in 2018. In addition, the trackers were in private offices of several large banks, in files of arbitration courts, as well as on the pages of ministries and law enforcement agencies. The most popular service that collects information about the user behavior on Runet is the Yandex.Metrica web analytics service, and the trackers of Mail.Ru Group take the second place [20]. The same is occurred in other countries with a steady tendency to increase the scale of this phenomenon [8]. This is largely due to the commercial value of data collected and transmitted in the digital environment. The researchers note that the most successful business projects are based on a personal-oriented approach to the customer needs, taking into account his/her preferences, attachments, wishes, field of activity, sex, age, etc. [20].

Big Data has become a kind of answer to modern challenges. It is the technology for processing large amounts of data, which allows you to quickly process large and diverse data, systematizing them. Big Data technologies significantly exceed the capacity of conventional computers and allow you to provide information in a user-friendly format. These technologies are universal by nature and can be used in various fields: "to create maps, digitize large amounts of information in funds and libraries, for weather forecasting, public transport systems and processing information about customer transactions in banks", including the automated production; they are also related to the Internet of Things (IoT) and Internet of Everything (IoE) [2].

The use of Big Data technologies produces a good economic effect, but so far it does not fit into the modern system of legal protection of personal data, being in the "gray zone" in the absence of proper legal regulation.

The purpose of this article is to identify the problems of legal regulation, which arise in the processing of personal data in Big Data systems. The research objectives are to determine approaches to the legal regulation of the processing of personal data and Big Data systems in Russia and foreign countries, identify gaps in the current legislation governing these relations, analyze project legislation, as well as to identify the main current trends of legal regulation in this field.

2 Methodology

The following methods were used in the research: general scientific methods – analysis, synthesis, comparison and generalization; private and scientific formal and legal method. The research basis was scientific researches, publications of Russian and foreign scientists and lawyers studying the problems of legal regulation of personal data protection during processing in Big Data systems in the context of digitalization. The research also used analytical materials on issues related to digitalization, data processing based on them, as well as ensuring the safety of the use of personal data. The research of the problem was carried out in two stages: The first stage: an analysis of the existing scientific and analytical literature on the research subject, as well as legislation of legal regulation of personal data protection during processing in Big Data systems was carried out; the research problem, purpose and methods were highlighted. The second stage: the conclusion obtained in the analysis of scientific and analytical literature and legislation was made, and the publication was prepared.

3 Results

The current Russian legislation defines the concept of "personal data" as follows: personal data shall mean any information relating to an individual who can be identified or designated directly or indirectly (personal data subject). The current Russian law assumes that the personal data subject should maintain control over his/her personal data and their processing by third parties to the most extent within the frameworks of the concept of informed consent to process personal data.

In connection with the above legislative provisions, the question arises about classifying specific information as personal data. Some studies have noted the excessive broadside approach to the definition of personal data, and it is proposed to interpret this concept restrictively, including only that information sufficient to identify a person [18]. This approach is explained by the practical need to use the personal data law. Indeed, from the point of view of the personal data processor, the availability of a list of personal data (which was also previously enshrined in the law) greatly facilitates the solution of law enforcement problems, for example, in terms of defining the list of personal data in a bylaw defining the policy for processing relevant data, consent for the personal data processing, definition of a threat model, etc. However, the situation changes when there is a more problematic issue in the global sense – personal data processing using Big Data technologies.

At the present day, there is no legislative definition of the technology in the Russian Federation. In the literature, Big Data is defined as "a set of tools and methods for processing structured and unstructured data of large volumes from various sources, subject to constant updates, in order to improve the quality of management decision-making, create new products and improve competitiveness" [17]. The researchers note that information collection and processing in Big Data technologies is carried out "in relation to all personal information in all possible ways, sometimes without an exact purpose, in the hope that someday the accumulated arrays of information will be used; they are compared with data from other sources, expecting to receive new information, etc." [9].

This approach to data processing, including personal data, is significantly different from what the current personal data law offers. It is impossible to specify the list of processed information and the list of processing objectives, to obtain specific, informed and explicit consent to their processing, etc. within its frameworks.

In this connection, it seems justified to include a broad concept of the "personal data" legal category into the personal data law. Legal scholars emphasize that "only a broad definition can achieve the purposes of regulating the personal data processing, the most important of which is to protect privacy, personal and family secrets" [11]. It should be additionally noted that since May 2018, the new General Data Protection

Regulation (hereinafter referred to as the "GDPR") [16] has been applied in the European countries, which determines the rules for personal data processing taking into account the challenges of the modern digital environment, which also uses a broad approach to the definition of personal data. However, just one broad definition of the personal data concept is inadequate to determine the rules of legal regulation for personal data processing in Big Data technologies.

The California Consumer Privacy Act of 2018 (hereinafter referred to as the "CCPA") adopted in 2018 in the state of California, the USA, is also of interest [5]. It defines the procedure for personal data protection when carrying out business processes and using the Internet of Things (IoT) [3]. In accordance with the CCPA that will come into effect on January 1, 2020, the consumer shall be provided with the right to require the company to disclose the categories and certain parts of personal information it collects about the consumer, the source categories from which the information is collected, the commercial purposes of collecting or selling information, as well as the categories of third parties to whom the information is transferred [5]. Thus, European and American legal acts establish the principle of transparency of personal data processing for their subjects and increased responsibility on the part of processors for non-compliance with the corresponding processing rules.

For Russia, Big Data information processing technologies are noted in Government Decree of the Russian Federation "On Approval of the Strategy for the Development of the Information Technology Industry in the Russian Federation for 2014–2020 and for the Long Term until 2025" No. 2036-p dated November 1, 2013, designed to ensure global technological competitiveness of Russia [10].

In October 2018, bill draft "On Amendments to the Federal Law "On Information, Information Technologies and the Protection of Information" (hereinafter referred to as the "bill draft") No. 571124-7 [4] was submitted to the State Duma of the Russian Federation in order to develop the above document. According to the Explanatory Note to this bill draft, it is focused to improve the efficiency of protection of information collected from various sources, including the Internet, the number of which exceeds one thousand network addresses, about individuals and (or) their behavior, which does not allow determining a specific individual using additional information and (or) additional processing.

The bill draft proposes to enshrine the concept of large user data, defining them as a set of non-personal data information about individuals and (or) their behavior, which makes it impossible to determine a specific individual without the use of additional information collected from various sources, including the Internet, and (or) additional processing, the number of which exceeds one thousand network addresses.

In the bill draft, processing of large user data shall mean "any action (operation) or a set of actions performed with or without the use of automation tools with large user data, including the collection, recording, systematization, accumulation, storage, clarification, retrieval, use, transfer, deletion and destruction of large user data" [4].

According to the Explanatory Note "the bill draft proposes to prohibit the processing of large user data aimed at determining (identifying) a particular individual, except the processing at the request of federal executive authorities engaged in law enforcement intelligence operations in order to prevent violations of the constitutional human and civil rights, including privacy, personal and family secrets".

The bill draft defines two forms of personal data processing: for one's own purposes and for the purposes of third parties. Processing for one's own purposes can be carried out only after notifying users about the implementation of such processing under the form established by the Federal Service for Supervision of Communications, Information Technology and Mass Media, either by posting relevant information on the website or by sending information directly to the device user. With regard to processing for the purposes of third parties, the bill draft determines that it can be carried out either for a fee or without it, and must be carried out on the basis of the informed consent of users, the form of which will also be established by the Federal Service for Supervision of Communications, Information Technology and Mass Media. As a regulator of relations, the Federal Service for Supervision of Communications, Information Technology and Mass Media is responsible for the creation and operation of the Register of Processors of Large User Data, the federal state information system.

Evaluating the norms of the submitted bill draft, experts note their groundlessness and similarity with the norms of the personal data law in terms of user data processing, which raises questions about the relationship between the norms of the bill draft and the personal data law [1]. At the present day, the bill draft has been delivered to the subject of legislative initiative to fulfill a number of mandatory requirements of the Constitution of the Russian Federation, and no substantive discussion of the norms of the bill draft was carried out by the legislative body of the Russian Federation. We believe that the criticism of the bill draft should be recognized as fair. It is submitted that the rules of personal data processing in Big Data systems should not contradict the legislation regulating the protection of personal data and ensuring the protection of privacy against encroachments. Data to be processed must be received in an impersonal form, using adequate data encryption tools that ensure the security of data use. In this connection, it is necessary to further improve the current legislation that creates the basis for the personal data processing in Big Data systems.

4 Discussion

The problems studied within the frameworks of this article are multifarious and relevant. The nature and features of the Big Data technology are presented in the works of Laney [13]. In his works he proposed the theory of "three V" (Volume, Variety, Velocity), which describes the features of the Big Data technology [13]. The study of the features of this system was carried out by Ramanathan [15].

The problems of regulating the Big Data technology are analyzed in the article by Shaidullina [17]; her work reveals the concept, describes the genesis and advantages of the technology; it studies the analyst on the scale of distribution of big data in different sectors of the economy.

Some aspects of the use of Big Data technologies in modern political practice are considered in the scientific work of Volodenkov, which is devoted to the analysis of potential threats and challenges associated with the use of digital arrays of information in the processes of social and political development [19], which undoubtedly have an impact on the formation of the regulation system. Such aspect as the use of data in the Internet of Things is studied by Peppet [14], Janeček [12].

5 Conclusion

The digital transformation of the society development has put forward new requirements on the legal protection of legally protected information and personal data. Under the influence of the digital environment, a significant amount of information about citizens is processed (collected, systematized, stored, etc.) daily on the Internet. At the same time, the modern information technologies such as Big Data allow you to identify, compare and analyze user behavior in the network, thereby processing variety of personal information and personal data.

The current legislation of the Russian Federation does not have special legal provisions related to the processing of information with Big Data technologies. However, the relevant rules must be determined by virtue of the state policy in the field of information technology and ensuring the rights and legitimate interests of personal data subjects.

The research has showed that for the purposes of legal regulation of personal data during their processing by Big Data technologies, it is necessary to preserve a broad definition of personal data, because this allows achieving the purposes of regulating the personal data processing, the most important of which is to protect privacy, personal and family secrets. It is submitted that the rules of personal data processing in Big Data systems should not contradict the legislation regulating the protection of personal data and ensuring the protection of privacy against encroachments. In this connection, it is necessary to further improve the current legislation that creates the basis for the personal data processing in Big Data systems.

References

1. Alekseichuk, A.: On the new bill draft on Big Data processing (2018). https://zakon.ru/blog/2018/10/25/o_novom_zakonoproekte_ob_obrabotke_big_data. Accessed 03 Mar 2019. (in Russian)
2. Analytical statement: Big Data and Information Security (2017). https://securenews.ru/big_data. Accessed 03 Mar 2019. (in Russian)
3. Bahar, M., Nolan, F., Satnick, T., Sutherland, E.: Right Out of the Box: California Enacts First-of-its-Kind Statute Regulating Internet-of-Things. Legaltech News (2018). https://www.law.com/legaltechnews/2018/11/14/right-out-of-the-box-california-enacts-first-of-its-kind-statute-regulating-internet-of-things/. Accessed 13 Mar 2019
4. Bill draft № 571124-7 On Amendments to the Federal Law "On Information, Information Technologies and the Protection of Information". http://asozd2.duma.gov.ru/main.nsf/(Spravka)?OpenAgent&RN=571124-7. Accessed 03 Mar 2019. (in Russian)
5. California Consumer Privacy Act of 2018. https://privacylaw.proskauer.com/2018/07/articles/data-privacy-laws/the-california-consumer-privacy-act-of-2018/. Accessed 10 June 2019
6. Constitution of the Russian Federation. http://www.constitution.ru. Accessed 03 Mar 2019. (in Russian)
7. Federal Law "On Personal Data" (hereinafter referred to as the "Personal Data Law") No. 152-FZ from 27 July 2006. http://base.garant.ru/12148567/. Accessed 10 June 2019

8. Gantz, J., Reinsel, D.: The Digital Universe in 2020: Big Data, Bigger Digital Shadows, and Biggest Growth in the Far East. In: IDC 2012 (2012). http://www.emc.com/collateral/analyst-reports/idc-the-digital-universe-in-2020.pdf. Accessed 09 Mar 2019
9. Gapotchtnko, D.: BIG DATA 2017: how to process big, but personal data (2017). https://www.computerworld.ru/articles/BIG-DATA-2017-kak-obrabatyvat-bolshie-no-personalnye-dannye. Accessed 01 Feb 2019. (in Russian)
10. Government Decree of the Russian Federation "On Approval of the Strategy for the Development of the Information Technology Industry in the Russian Federation for 2014–2020 and for the Long Term until 2025" No. 2036-p dated November 1, 2013. http://www.consultant.ru/document/cons_doc_LAW_154161/. Accessed 10 June 2019
11. Gribanov, A.A.: The general data protection regulation: ideas for improving the Russian legislation. Law **3**, 149–162 (2018). (in Russian)
12. Janeček, V.: Ownership of personal data in the Internet of Things. Comput. Law Secur. Rev. **34**(5), 1039–1052 (2018). https://ssrn.com/abstract=3111047. Accessed 23 Mar 2019
13. Laney, D.: 3-D Data Management: Controlling Data Volume, Velocity and Variety. Application Delivery Strategies. META Group (2001). https://www.bibsonomy.org/bibtex/263868097d6e1998de3d88fcbb7670ca6/sb3000. Accessed 18 Feb 2019
14. Peppet, S.R.: Regulating the Internet of Things: first steps toward managing discrimination, privacy, security and consent. Tex. Law Rev. (2014, forthcoming). https://ssrn.com/abstract=2409074. Accessed 03 Feb 2019
15. Ramanathan, S.: Data to Big Data - a paradigm shift and a professional challenge. CSI Commun. 36–37 (2014). http://citeseerx.ist.psu.edu/viewdoc/download;jses-sionid=3305A447A27E06092C5E76E7CFAD2975?doi=10.1.1.588.5588&rep=rep1&type=pdf. Accessed 10 June 2019
16. Savelev, A.I.: Problems of application of personal data legislation in the era of "Big Data" law. J. High. Sch. Econ. **1**, 43–66 (2015). (in Russian)
17. Shaidullina, V.K.: Big Data and personal data protection: the main theoretical and practical issues of legal regulation. Soc.: Polit. Econ. Law **1**, 51–55 (2019). (in Russian)
18. Uden, L., Aho, A.-M.: Developing data analytics to improve services in mechanical engineering company. In: Uden, L., Fuenzaliza Oshee, D., Ting, I.H., Liberona, D. (eds.) Knowledge Management in Organizations. KMO 2014 Lecture Notes in Business Information Processing, vol. 185, pp. 99–107 (2014)
19. Volodenkov, S.V.: Big Data technologies in the modern political processes: digital challenges and threats. Tomsk State Univ. Bull. Philos. Sociol. Polit. Sci. **44**, 205–212 (2018). (in Russian)
20. Zakharov, A.: Total surveillance: how the world of user data trading works. Technologies and Media (2018). https://www.rbc.ru/magazine/2018/04/5aafdfc99a7947654297214d. Accessed 28 Mar 2019. (in Russian)

The State Sovereignty in Questions of Issue of Cryptocurrency

S. P. Bortnikov(✉)

Samara State University of Economics, Samara, Russia
serg-bortnikov@yandex.ru

Abstract. The analysis of powers and order of monetary issue by the Central Bank of Russia in the system of the relations of the state sovereignty is carried out. Is considered contradictions of independent position of the Central Bank and the Government of the Russian Federation in questions of the state economic policy. Is defined that the fiat monetary system is based only on the authority of the Central bank and government whereas cryptocurrency – on the system of own safety and a way of its generation: blockchain. Is defined that in the independence and activity of the Central Bank of the Russian Federation can not consider "wishes" and plans of the government and Senate of Russia. Positive and negative properties of cryptocurrency in a modern financial system are considered. The cryptocurrency is considered out of communication with a monetary system, but only as subject to financial manipulations. It is necessary to change ideology of monetary issue. Non-cash money, special drawing right and so forth are considered as cryptocurrency prerequisites, but controlled by the currency center. Whereas the sense of cryptocurrency consists in its decentralization. The release of money has to be provided with a national product. The offer on establishment of legal bases of regulation of process of monetary issue is made.

Keywords: Monetary issue · Cryptocurrencies · Fiat money · Money supply · Credit policy

1 Introduction

Penetration of digital technologies into all spheres of public life is created by dependence of social and economic communications on the power of the computer, safety of transaction, a decision-making vector. The principles of the organization of society, the public relations and business contacts will be transformed. Crisis of 2008 undermined belief of people in the developed financial system, the traditional centers of decision-making. Two counter trends: globalization and multipolarity create new models of administration of modern processes in economy, finance, the taxi, the household relations through self-organizing electronic trading platforms and forums at which locals offer for rent the various tool, selecting thereby bread at household shops and so forth. For example, in power "… we are witnesses of the beginning of massive transition to a new distribution model of the electric power – from the "central station – network of consumers" model which arose in the 1980th years to model of "wide

S. I. Ashmarina et al. (Eds.): ISCDTE 2019, LNNS 84, pp. 564–573, 2020.
https://doi.org/10.1007/978-3-030-27015-5_67

distribution" with solar panels on roofs of houses, local cellular batteries and micronetworks" [11].

The distributed economy or economy of cooperation (when people through computer networks distribute information on own free resources, offer temporarily free tools, exchange the ideas without intermediaries, create alternative media space) led to emergence of new ways of interaction both in society, and in economy. However, whether "everything what gives in to decentralization, will be decentralized" [13]? Any economy needs financial service which value and the principles tried to prove in various monetary theories. How will processes of decentralization affect the state sovereignty?

Questions of the state sovereignty reveal by means of many aspects. Sovereignty … - a natural and necessary condition of existence of statehood of Russia having centuries-old history, culture and the developed traditions - can be characterized, including and as exclusive maintaining on certain questions to which number it is necessary to carry also questions of monetary issue.

According to Article 71 of the Constitution of the Russian Federation questions of financial, currency, credit, customs regulation, bases of the price policy, federal banks, monetary issue, the federal budget … and other questions concerning the organization of monetary circulation (Article 71 of the Constitution of the Russian Federation) are under authority of the Russian Federation [3].

2 Methods

The research methods are the system analysis, expert evaluation methods, polling and interviewing, modeling. The main research methods

- the first stage was implemented by means of analysis of the existing approaches to entrepreneurial values and entrepreneurship as values from the finnational, economic and legal points of view;
- the second stage included monitoring and diagnostic work, development of the model of readiness of entrepreneurs to changes, and the complex of the most significant values in the enterprise environment is investigated;
- at the third stage the results of diagnostics are processed and analyzed, objectives for a further investigation phase are formulated.

3 Results

3.1 Monetary Issue as Sovereign Function of the Central Bank

Monetary issue in Russia is referred to exclusive competence of the Central Bank of the Russian Federation (Paragraph 2 of Article 4, Article 29 of the Federal law of 10.07.2002 N 86-Federal Act "About the Central bank the Russian Federation (Bank of Russia)"). And, according to Article 29 of the specified law issue of cash (the banknote and coins) [6], the organization of their address and retirement in the territory of the

Russian Federation is within the exclusive competence of the Bank of Russia. The lack of a regulation of release non-cash and cybercash and also withdrawal and destruction of money in the country is paradoxical.

Except a traditional form of money and the calculations which are carried out with their help in a financial turn also other mechanisms are used: cryptocurrencies, electronic money and so forth. According to the Federal law of June 27, 2011 No. 161 - Federal Act "On the National Payment System" "Electronic money is money which are previously provided by one person (the person which provided money) to other person considering information on the amount of the provided money without opening of the bank account (the obliged person) for execution of liabilities of the person which provided money before the third parties and concerning which the person which provided money has the right to transfer orders only with use of electronic payment instruments" (Paragraph 18 of Article 3 of law No. 161 - Federal Act) [7].

The smart the gadget which became the priority instrument of mobile bitcoin calculations at the same time turns into subject to close attention of a number of the payroll technical companies aiming to revolutionize ways of carrying out calculations. The way of data transmission used by cybermoney a blockchain is the independent phenomenon and can be used in any communications, and not just in calculations. The Sberbank of Russia and VTB already held the sessions with use of this technology. Its attractions deserve the closest attention.

Financial activity of the state, one of the main by function of which it is possible to determine monetary emission, is a special type of the state activity [1] which should be carried out in a legal form.

De jure the Bank of Russia is independent, it carries out budget and financial and own economic activity independently, and, vyplayachivy taxes from the profit. On the other hand, the Bank of Russia works in the system of norms of the international financial law with participation of such organizations as IMF, the World Bank, the Financial Stability Forum (SFS), the Bank for International Settlements (BIS), the Basel committee on bank supervision and so forth.

Many instructions and acts of the international bodies and a number of the international economic organizations (for example, FATF), are not formally obligatory, but failure effects to follow the models developed by them for the state and its central bank can be very notable both in political, and in material sense.

By approximate calculations, for the last 5 years the USA poured in world economy in about 2.75 trillion dollars which FRS generated that is called "from air". A fate of buyers of US Treasuries, unfortunately, did not pass also the Russian monetary authorities: according to the latest data, the Russian Federation owns the American debt obligations for the sum of 81.7 billion dollars and is the 16th largest creditor of the USA. The official ratio between ruble and gold or other precious metals is not established. Monetary issue is carried out on security reserves of the Central Bank of the Russian Federation or without those. But all this belongs to monetary issue. As the above-stated principles belong to cryptocurrency, than it is provided. It is possible to tell that only the mechanism of the safety, and now – a speculative agiotage. Direct fraud at the cryptoexchanges, price manipulation in pools (Pump & Dump) are widespread in the world of cryptocurrencies. Thus, the current cost of cryptocurrencies is not result of fair market process of pricing and is very conditional [9].

Fundamental issue of traditional currency is that its functioning requires trust. The Central Bank needs trust that it will not weaken currency, however the history of fiat currencies is full of cases of abuse of this trust. The main advantage of bitcoin over dollar call limitation of its issue while issue of dollar is not limited. The value of an inflation for cryptocurrency – to the main evil of a fiat system is not investigated yet. Turns out opposite: the last years the Central Banks of the developed countries are taken with fight against deflation – the more important evil for economy, than moderate inflation, and adherents of a crypt – start of new cryptocurrencies (for 2017 the bitcoin share in their general capitalization fell from 90 to 35%).

The legislation of Russia does not regulate an order of "placement" of the emitted money by the Bank of Russia. Moreover, at the existing ban to finance the Government of the Russian Federation of the Central Bank operates independently and on other, not Russian markets. For financing of own government by the Central Bank the State Duma of Russia has to make the special decision.

There is a gap in adoption of the motivated decision and a source of mobilization of funds for its realization. The Government of the Russian Federation realizes financial and economic policy, develops strategy of its realization, chooses tactical means. The Central Bank of the Russian Federation participates in this process along with all authorities, but makes the decision on monetary issue how will consider necessary. In the course of definition of a source of mobilization of means for implementation of economic policy the Government and the Central Bank of the Russian Federation can have competing interests or views.

3.2 Issue Preferences of the Central Bank Concerning Cryptocurrency as Instrument of Monetary Policy

Issuing paper money (and coins not from precious metals) under own reserves, the Central Bank of the Russian Federation as the organization working with profit starts money in a turn on a commercial basis (sells them), but can transfer them and to the government on special conditions. Under usual conditions the Bank of Russia has no right to grant the loans to the Government of the Russian Federation for financing of federal budget deficit, to buy the state securities at their primary placement unless it is provided by the federal law on the federal budget. Financing (investment of funds) of the governments of the foreign states is not forbidden to the Bank of Russia.

It is considered to be that the bitcoin is a step forward in the device of monetary circulation, however in own way it is, on the contrary, a step backwards. Traditional money is not counters, not coins and not records on accounts, and the credits and debts. Money in economy is delivered not by the central bank which only regulates monetary circulation through norms of reservation and key interest rates, and commercial banks. Nakamoto's protest against "our money" were given to someone else, is irrational and destructive for global economy. "The monetary system" of bitcoin is constructed on mistrust – that at the price of expenditure of big resources to exclude need for trust as an element of a modern economic paradigm, having replaced it with a technology solution. The world of the future with the won bitcoin is the world where nobody trusts anybody, and the safety of money is guaranteed by cryptography. The medal back – from such monetary system the animator is thrown absolutely out. Central bank of bitcoin is distributed on all knots of network and emits money on time, despite of demand, limiting all transactions by monetary base. It is similar to the monetary system of the Middle Ages based on calculations exclusively gold. However world economic blossoming of the second half of the 20th century is in many respects connected with refusal of the gold standard and creation of a system of fiat currencies which kind of are provided with nothing, except monetary policy.

At the beginning of 2014 the CEO of MasterCard Adzhai Banga gave an interview to journalists and publishers of Wall Street Journal in which expressed the point of view of those who considered that just it is better to ignore untimely innovations. In particular, he told the following about bitcoin: "In the world and there are so enough currencies. What problems the bitcoin is capable to solve?" [12].

Emergence of the central bank comes true desire to limit the national governments in questions of the state credit, to create the system of a counterbalance to free injection of money in economy, improvements of the market and correction of its shortcomings. "Good" undertaking comes true control of inflation, prevention of a covering of budget deficit monetary issue and so forth. It is supposed, as public opinion has to protect freedom of the Central Bank from political and tactical pressure.

At the same time, "games" with credit policy which conducted the central banks of almost all countries in the 20th century became the constant source of economic misinformation causing the global discoordination of economy which is shown in particular in artificial fluctuations of business activity [13].

Creation of cryptocurrency is not something unique in a financial system. It is a logical stage in formation of both units of account, and the mechanism of functioning of the financial sphere. SPZ, the credits, non-cash money and so forth created prerequisites of emergence of cryptocurrency.

Emergence of non-cash money in a turn as special records has no standard regulation, the banking system, especially against the background of its reforming, can generate them. Now the prevailing point of view in the sphere of understanding of the nature and emergence of this money in civil circulation is the thesis that this money is the rights of the requirement, and in civil circulation they function by means of non-cash payments.

Thus, non-cash money is still legally uncertain risk asset. Still it is not clear what mint issues such money in modern Russia.

Thus, disappearance of non-cash money is not regulated in any way, not clearly where this money what goods they provided how there can long be a monetary debt leave. From proofs there is a contract of purchase and sale (or other contractual obligation), an extract from the register, the certificate on the death of the natural person, the certificate on the right for inheritance, the arbitration court ruling on elimination of the debtor from the state registry and some other documents. In such situation application of provisions of the Federal law "About Counteraction of Legalization (Washing) of Income Gained in the Criminal Way and to Terrorism Financing", Article 173, 173.1, 419, 421 Civil Code of the Russian Federation [2], Article 174, 174.1 Criminal Code of the Russian Federation [4] have estimated character.

3.3 Alternative System of Sources of Mobilization of Srelstvo of Payment and Accumulation

"The independent central bank" to the name of the scheme of the organization of a credit and monetary system most widespread today. Leadership handover by monetary policy of the country is ideally provided in hands of a small group of professionals to the heads of the central bank protected from influences of group interests and observing the interests of the country and its economy. That is in the expectations society has to rely on personal qualities of staff of the Central Bank, on hope of their independence and so forth. Such approach contradicts the elementary beginnings of management, need of removal of a human factor and so forth.

The independent central bank is successful when holds a certain position in society, possesses trust and the authority, cooperates with the power in achievement of strategic objectives. On the contrary, the modern central banks are conductors of the ideas of the international organizations and what lead such actions to is visible on examples of Ukraine, Greece, and the USA.

The prereform system of the address of the last period of existence of the USSR was in fact multi-currency, that is barter between regions and between the enterprises practiced, cash and non-cash rubles with the variable course relation between them, cash and non-cash foreign currency, and, at last, last not least - exchange of administrative services - bureaucratic trade were used as separate currencies [13]. It is possible to claim that the negative role of concentration of monetary issue function at the State Bank was leveled, and competences of the Government and the Supreme Council could correct to Rabat of the State Bank.

However use of digital currencies is not deprived of shortcomings. First of all, digital currencies assume high degree of anonymity; use of digital currencies in illegal activity grows "in a geometrical progression", and the complexity of crimes increased too. In this regard, according to experts, at legalization of cryptocurrencies it is necessary to get rid of anonymity by input of identification of participants of operations. Besides, today such currencies cannot be exchanged on cash, and the rate of digital currencies is characterized by high variability.

Experts carry to other shortcomings that issue and use of digital currencies are not regulated by the either the international, nor national legislation, at least, in the majority of the countries yet.

Now in Russia the bill of the Federal law "About Digital Financial Assets" which defines a concept of digital transaction – action or the sequence of the actions directed to creation, release, the address of digital financial assets is offered. However, this bill assumes regulation of the address, but not issue of cryptocurrency and calculations with its help. Still, the ruble remains the only means of payment in the territory of the country.

Destruction of monetary systems of the countries of the world by means of a virus of "cryptocurrencies" can lead to the fact that the U.S. Federal Reserve will be the only emission center in the world.

The number of supporters grew in the economic theory in the last quarter of the 20th century to refer matters of monetary issue to the market, the private organizations, privatization of the central banks was offered. "Our own conclusion… consists that transfer of monetary and bank establishments to the discretion of the market would result in satisfactorier result, than that which is nowadays reached thanks to the state intervention" [8].

The idea of denationalization of money was stated also by young erudite economists in the seventies of the 20th century [12]. Cryptocurrencies are first of all transition from centralized to private monetary systems. Both the feature and the main problem of regulation consists in it – in fact cryptocurrencies assume the monetary circulation alternative state, that is their value is defined not centrally. If to speak simply, then cryptocurrencies are an alternative to traditional money. They represent the currency existing only in a digital form, and all calculations with it are made by means of cryptography. Cryptocurrencies are decentralized – that is there is neither central bank, nor administrators, nor any bodies of supervision which could trace for what money is used or whom and when they were sent; it does operations with cryptocurrencies very attractive.

It is necessary to change ideology of monetary issue. The gold and exchange equivalent (historically to refusal of a bimetallic monetary system) sharply reduces quantity of money in economy. The release of money has to be provided with a national product. Money has to be issued by the state from such goods which had not to be borrowed [14]. Nechvolodov wrote at the turn of the century that only 18% of paper money in Europe are provided with gold [17].

To restriction of a monetary arbitrariness can find the regulation establishment of criteria of growth of money supply gain of GDP, increase in productivity of work, the needs for money. There is the main unresolved problem a legal regulation of market determination of required volume of money in circulation.

4 Discussion

4.1 Possibilities of Existence of Digital Currencies of the Central Banks

The last one or two years against the background of attempts of legalization of cryptocurrencies the subject of digital currencies of the central banks – CBDC is quite widely discussed. It is in our opinion wrong to consider CBDC as a new financial instrument. The key economic innovation in case of CBDC will consist in expansion of access to obligations of the Central Bank, but not in use of some concrete technology or creation of an alternative to money. CBDC have to and can be considered as an innovative form of monetary issue.

We see that CBDC is generally already realized in the form of reserves of commercial banks in the Central Bank. Now limited access to liabilities of the Central Bank is realized – for commercial banks it is reserves and deposits in the Central Bank. Money on the basis of the distributed register for similar calculations among participants of the money market would become a new form of money.

Now questions of monetary issue remain the mystery of the central banks: what quantity of money is emitted what quantity is withdrawn and liquidated, all these questions are secret. CBDC will reveal this secret, any participant of these relations will be able to obtain full information. Moreover, the central banks will be forced to transfer a part of the functions in the system of widespread data to commercial banks or other financial settlements centers. From the point of view of users, CBDC can have a set of characteristics and enter a turn both as an alternative to cash, and as an alternative deposits.

Introduction of retail digital currencies can have wide consequences for all participants of a monetary system, that is for economic agents (the population and the companies), a banking system and also for monetary policy of the Central Bank and its policy in the sphere of financial stability.

CBDC can be attractive for a number of reasons: the obligation of the Central Bank can become the least risk and most liquid asset available to a wide range of persons; the new currency will allow to lower transaction expenses in economy. Whether the last can be reached by means of new technologies, for example DLT, – still open question. In case CBDC is rather liquid and simple in use, it can quite become a full equivalent of cash. Nevertheless It should be noted that CBDC de facto will not be able to provide the same level of anonymity which is provided by cash. For CBDC commercial banks – it is the type of money competing with accounts in these banks. In recent years banks began to pay much attention to a possibility of use of the accounts for implementation of money transfers including between natural persons. Growth rates of money transfers are comparable to growth rates of volumes of the cash given in ATMs and payment transactions of cards. Competitive possibilities of CBDC in comparison with banking systems can depend on design of the most retail digital currency of the central bank.

Growth of volume of assets on balance of the Central Bank can lead to changes in the debt market and the capital market as it occurred in a number of the countries during QE. Now there is no analysis of the competition of new money of the Central Bank to money of commercial banks. Permanent decrease in size of deposits (in favor of CBDC) can reduce the size and change structure of liabilities of commercial banks, so, and funding cost [13]. It can lead to some compression of volume of assets of commercial banks, in particular crediting of natural persons and the companies.

Whether CBDC will influence crediting of the real sector by banks? In work Kumhof & Noone it is shown that at certain prerequisites and a way of release the CBDC size of the credit of banks to the private sector not necessarily has to decrease [15]. Realization of CBDC may contain many options: providing with government bonds (relevant how to provide design of the competition sovereign powers of the Central Bank on monetary issue and inquiries of the Ministry of Finance of the country), gold currency reserves, liquid goods or repo transactions, the option and so forth.

4.2 Investment and Investment Preferences of CBDC

Undoubtedly, CBDC can affect competitiveness of various credit institutions differently. CBDC can become an alternative to many investments with the fixed profitability, so, will make the Central Bank by the competitor of commercial banks in respect of attraction of obligations.

On the one hand, the possibility of digital access to liabilities of the central bank can increase trust of all economic entities to a financial system in general. Perhaps, CBDC issue, has to be provided with highly liquid assets, but is not clear whether users have to have an opportunity to acquire CBDC directly through the account in commercial bank [16] or only in exchange for the assets serving as providing to again issued money [15]. In the first case free convertibility, in the second – automatic security of new money with reliable assets will be provided.

Dyson and Hodgson, indicate what emergence of new currency will allow to redistribute сеньораж from commercial banks in favor of citizens [5]. Emergence of CBDC in one country can lead to reduction of transaction costs for currency conversion that can create prerequisites for fast inflow of the capital. However, experience of Venezuela forces to think of the system of factors which can level positive CBDC value.

In general, regulators, banks and the financial organizations invest already now in alternative payment service providers, technologies and services which can become replacement to cash and to hypothetical CBDC. In April, 2018 Visa and MasterCard payment service providers announced start of money transfer from the card to the card by the phone number. In 2019 the Bank of Russia started own system of fast payments to which commercial banks can be connected, and next year plans start of the Marketplace project for access to financial services of many financial organizations at once.

5 Conclusion

Questions of monetary issue have to is under authority of public authorities of Russia. It is necessary to withdraw the ban on financing of own government by the Bank of Russia and investment of funds in assets of the country. The release of paper money has to be made taking into account GDP, and not just gold and foreign exchange reserves. The release of paper money has to have a legal regulation (in the light of the main beginnings and restrictions), but not to depend on will and vision of individuals. There is a wish to complete saying article by Dr. S. Glazyev. "Our monetary authorities should not read tea leaves, and to take real measures for restoration of control of the financial market. Today, as showed experience and a collapse of ruble exchange rate last year and extreme volatility this year and also the actual failure with all invention of targeting of inflation proves that the financial market remains uncontrollable" [10].

References

1. Ashmarina, E.M.: Some aspects of expansion of a subject of the financial right in the Russian Federation: problems and prospects: monograph. Press, Moscow (2004). (in Russian)
2. Civil Code of the Russian Federation. http://www.consultant.ru/document/cons_doc_LAW_5142/. Accessed 25 Apr 2019
3. Constitution of the Russian Federation. http://www.consultant.ru/document/cons_doc_LAW_28399/. Accessed 25 Apr 2019
4. Criminal Code of the Russian Federation. http://www.consultant.ru/document/cons_doc_LAW_10699/. Accessed 25 Apr 2019
5. Dyson, B., Hodgson, G.: Digital Cash: Why Central Banks Should Start Issuing Electronic Money. Positive Money, London (2016). https://positivemoney.org/wp-content/uploads/2016/01/Digital_Cash_WebPrintReady_20160113.pdf. Accessed 15 Apr 2019

6. Federal law of 10 July 2002 N 86-Federal Act "About the Central bank the Russian Federation (Bank of Russia)". http://www.consultant.ru/document/cons_doc_LAW_37570/. Accessed 25 Apr 2019
7. Federal law of 27 June 2011 No. 161 - Federal Act "On the National Payment System". http://www.consultant.ru/document/cons_doc_LAW_115625/. Accessed 25 Apr 2019
8. Friedman, M., Schwartz, A.J.: Has government any role in money? J. Monet. Econ. **17**(1), 37–62 (1986). https://doi.org/10.1016/0304-3932(86)90005-X
9. Gandal, N., Hamrick, J.T., Moore, T., Tali, O.: Price manipulation in the bitcoin ecosystem. J. Monet. Econ. **95**, 86–89 (2018). https://doi.org/10.1016/j.jmoneco.2017.12.004
10. Glazjev, S.: Central Bank wonders on the coffee grounds instead of work. Politicus (2015). https://politikus.ru/v-rossii/51811-sergey-glazev-cb-gadaet-na-kofeynoy-gusche-vmesto-raboty.html. Accessed 17 Mar 2019. (in Russian)
11. Gore, A.I.: The turning point: new hope for the climate. Rolling Stone (2014). https://www.rollingstone.com/politics/politics-news/the-turning-point-new-hope-for-the-climate-81524/. Accessed 25 Apr 2019
12. Guerrera, F.: Bitcoin's Crisis is Turning Point for Currency, Wall Street Journal, MoneyBeat Blog (2014). https://blogs.wsj.com/moneybeat/2014/02/17/bitcoins-crisis-is-turning-point-for-currency/. Accessed 25 Apr 2019
13. Hayek, F.A.: Choise in Currency: A Way to Stop Inflation. Institute of Economic Affairs, London (1976)
14. Juks, R.: When a Central Bank digital currency meets private money: the effects of an e-Krona on banks. Sveriges Riksbank Econ. Rev. **3**, 79–99 (2018). https://www.riksbank.se/globalassets/media/rapporter/pov/engelska/2018/economic-review-3-2018.pdf. Accessed 25 Apr 2019
15. Kumhof, M., Noone, C.: Central Bank Digital Currencies - Design Principles and Balance Sheet Implications (2018). https://www.bankofengland.co.uk/-/media/boe/files/working-paper/2018/central-bank-digital-currencies-design-principles-and-balance-sheet-implications. Accessed 30 May 2019
16. Meaning, J., Dyson, B., Barker, J., Clayton, E.: Broadening Narrow Money: Monetary Policy with a Central Bank Digital Currency (2018). https://www.bankofengland.co.uk/-/media/boe/files/working-paper/2018/broadening%20narrow%20money%20monetary%20policy%20with%20a%20central%20bank%20digital%20currency. Accessed 30 May 2019
17. Nechvolodov, A.D.: From Ruin to Prosperity. Printing house of headquarters of troops of Guard and Petersburg military district, St. Petersburg (1906). (in Russian)

Digitization: The Bar's Aspect

J. A. Dorofeeva(✉)

Samara State University of Economics, Samara, Russia
logl612@yandex.ru

Abstract. The article is directed at researching the application of new laws on digital rights, digital money, electronic transactions, digitalization of legal proceedings, advocacy, including the part of the rules on managing of law files in lawyer practice. The use of digital technologies in the implementation of advocacy should be correlated with the protection of attorney-client confidentiality, personal data and privacy. The ratio of private law and public interest is particularly pronounced in the sphere of realization of non-property rights, where there is no intermediary in the form of rights to property, which has the equivalent of the right enshrined in its value terms. the conclusion is made the general direction when considering issues of digitalization of legal assistance - non-entrepreneurial activities is the establishment of a model of legal assistance of an attorney who observes the rights of the principal - both in terms of providing qualified assistance and in terms of the opportunities contained in electronic resources. As a result of the analysis of advocacy concluded the topic of applying digital technologies in advocacy is divided into two clusters - on the one hand, advocates should know what legal norms should be applied tomorrow to relations related to digital products as objects of civil rights and participate in creating adequate practice on this subject, on the other hand, to have skills in handling digital technology objects, to be able to use legal systems, and official Internet resources of the legislative, executive authorities, courts, specialized computer programs and databases.

Keywords: Bar · Digitization · Access to justice · Digital rights · Digital technologies · Electronic resources

1 Introduction

The relevance of the study is based on two aspects of advocacy: the substantive aspect and the procedural component of the legal profession. Providing legal assistance to citizens and legal entities is the goal of a lawyer, its subsistence and nature, which requires knowledge of the norms of current legislation and the practice of its application. Therefore, changes in the norms of positive law, which include the incorporation of digital rights and digital money, as well as the rules of their circulation, electronic transactions, into the norms of civil legislation of new legal institutions, require close study. From the point of view of the procedure and the process of implementation of advocacy, a number of issues related to the provision of legal assistance through electronic resources are relevant from the point of view of the introduction of digital technologies. The advantages and disadvantages of providing legal assistance with the

S. I. Ashmarina et al. (Eds.): ISCDTE 2019, LNNS 84, pp. 574–580, 2020.
https://doi.org/10.1007/978-3-030-27015-5_68

use of digital technologies are also the subject of consideration in this article and are essential to the practice of advocacy at the present stage.

2 Methods

The descriptive method is fundamental to this study, it includes the method of observation, interpretation, comparison, and generalization. Also, analysis was used in the study, as it allows us to divide the elements of legal assistance into individual components related to the use of digital technologies and to give a qualitative assessment of their use in the practice of advocacy.

3 Results

The conclusions that were drawn as a result of this research summarize into provide qualified legal assistance one must know and apply the latest legislation on digital rights, digital money, electronic transactions, digitalization of legal proceedings, advocacy, including in the area of lawyers filing. The use of digital technologies in the field of legal assistance should be correlated with the protection of legal professional privilege, personal data, privacy secrets, although there are discussions in the scientific literature regarding the boundaries and limits of these categories.

We should agree with the comment that "Law, whether understood as a profession, a method of solving disputes, a tool to achieve justice, a superstructure in the hands of the powerful to protect their interests, or, more simply, an instrument to guide human behaviour, is at its core an intellectual endeavour that depends on handling, storing, interpreting, and sharing knowledge as well as information. Recent advancements in digital technology are precisely transforming the ways in which information is created, stored, and conveyed. Moreover, these developments are making inroads into artificial knowledge production, thereby potentially entering the intellectual and human aspect of law" [1]. Digitalization of respect as subjects of law and serves as the object of the law and the right to have access to individual documents, typical of the pre-digitization era, obsolete under the onslaught of new information technologies.

The ratio of private law and public interest is particularly pronounced in the sphere of realization of non-property rights, where there is no intermediary in the form of rights to property, which has the equivalent of the right enshrined in its value terms. Therefore, the general direction when considering issues of digitalization of legal assistance - non-entrepreneurial activities is the establishment of a model of legal assistance of an attorney who observes the rights of the principal - both in terms of providing qualified assistance and in terms of the opportunities contained in electronic resources.

The main piece of legislation defining the procedure, terms and content of the legal assistance provided by a lawyer, is the Federal law forced on 31.05.2002 N 63-FZ "On Advocacy and the Legal Profession in the Russian Federation" [11]. Paragraph 1 of Article 1 of the regulation advocacy is defined as a qualified legal aid rendered on a professional basis by persons who obtained the status of the lawyer in the order established by the present Federal law, individuals and legal entities (hereinafter - the client) in order to protect their rights, freedoms and interests as well as ensuring access to justice. Thus, the law establishes the obligation to provide lawyer principals only qualified legal assistance. The latter is sometimes identified with the legal assistance provided by a lawyer, or is regarded as a category that generalizes the legal counsel and legal assistance provided by paralegals or non-professionals in the field of advocacy. But only for the legal assistance provided by a barred lawyer, the legislator has established its characteristics, in indication of the need to comply with the proper qualifications. Legal entities and individuals who apply for legal counsel and legal aid are entitled to have a lawyer of adequate knowledge of both previous legislation and newly adopted law, which requires lawyers to study substantive law before they are enacted by the legislator.

No less important is the procedure of conducting legal practice. The inability to use modern information resources of the courts (the “Data bank of Arbitration Cases”, “State automated system Justice”) will strip a lawyer of the ability to inform the client about the initiation of legal disputes concerning such client, as on the changes in cases, it will deprive the client of the opportunity to receive fast and effective protection, and in cases of summary procedure it could completely deprive the client of the right to legal protection.

The lack of necessary skills in dealing with digital legal systems shall not lead to the consequences listed above, but it will create additional difficulties in preparing a case for court, in creating procedural documents, and will not allow a lawyer to resolve arising issues quickly in the process of rendering legal assistance.

Thus, the topic of applying digital technologies in advocacy is divided into two clusters - on the one hand, advocates should know what legal norms should be applied tomorrow to relations related to digital products as objects of civil rights and participate in creating adequate practice on this subject, on the other hand, to have skills in handling digital technology objects, to be able to use specialized computer programs and databases, legal systems, and official Internet resources of the legislative, executive authorities, courts.

Analyzing the norms of substantive law, which was forced in October of 2019 and will regulate the sphere of digital technology products as objects of civil rights, every lawyer will note the unusual specificity, dissimilarity to other objects and legal norms that establish the rules of their circulation in civil circulation based on the provisions of The Federal Law №34-FZ of March 18, 2019 “On Amendments to Parts One, Two, and Article 1124 of Part Three of the Civil Code of the Russian Federation” [12].

Thus, the norms of positive law will contain such category as “digital rights”, the definition of which will be determined by art. 141.1 of the Civil Code of the Russian Federation [9] as the law of obligations and other rights named in such capacity in the law, the content and conditions for the implementation of which are determined in accordance with the rules of the information system that meets the criteria established by law. Implementation, disposal, including transfer, pledge, encumbrance of a digital right by other means or limitation of disposal of digital law are possible only in the information system without recourse to a third party; in the new edition of art. 128 of the Civil Code of the Russian Federation digital rights will be assigned to property rights, which, in conjunction with things and other property, are objects of civil rights.

The law defines the holder of a digital right as a person who, in accordance with the rules of the information system has the opportunity to dispose of this right, establishes the rules for transferring a digital right on the basis of a transaction and does not require the consent of the person obligated under such digital law. On the circulation of digital rights in the context of their legal regulation by the rules of law, which will take effect from 10/01/2019, there is controversy in the scientific literature. So Fedorov points to a number of inaccuracies made by the authors of the bill signed by the President of the Russian Federation on March 18, 2019: identification of a digital code and a digital right, the creation by the legislator of the right to right construction, and also in the legislation defining the circulation of digital rights, there is still no construction of the fair acquisition of digital rights as a result of the appearance of the right [2]. Guznov, Mikheeva, Novoselova et al. consider the norms of legislation that are planned to be forced, aiming at regulating digital rights, are positive and accepted with regard to the experience of regulating such objects of civil law as non-documentary securities with their transformation from the institute of real right to the institute of law of obligation and indicates the purpose of the token: digital technology: this is a digital designation of the right to an object of law, utilitarian tokens make it possible to use some services without creating obligations; in addition, there are calculated tokens, for example, crypto-currency [3]. The authors note that introduced, in particular, in Art. 128 of the Civil Code of the Russian Federation provisions on digital rights have technological neutrality determining the place of digital assets in the system of civil rights objects as she indicates that criticism of the a not too successful term "digital rights" due to the cumbersome design of the "right to rights" is a matter of terminology and not the essence. The authors point out that the introduction of new categories of civil law, regulating various digitized objects, makes it possible to introduce transactions with such objects that entail tax into the legal field, that causes financial and other consequences [3].

It is necessary to agree with the opinion that changes in legislation concerning the introduction of the concept of digital rights into circulation transform habitual and well-established categories and institutions of both civil law and other branches of legislation [8] Thus, the concept of digital money introduced into circulation is a collection of electronic data (a digital code or designation) that does not certify the right to any object of civil rights, created in an information system that meets the statutory features of a decentralized information system and is used by users of this system to make payments (Article 141.2 of the Civil Code of the Russian Federation) is very far from the traditional concept of money as a generic thing. The notion of a written form of a transaction has also been changed, which, starting October 2019 if a person makes a transaction using electronic or other technical means to reproduce the contents of the transaction on a tangible medium unchanged, the requirement to have a signature will be considered fulfilled if any method to reliably determine the person who expressed the will to transfer was used. These changes to the law will nullify all disputes relating to the invalidity of the contract due to non-compliance with the form, if the transaction was made using technical means. These, as well as other changes in legislation related to the introduction into civil circulation of such objects of civil rights as digital rights, electronic transactions, electronic money, etc., are vital and so significant that they can be called a revolution of established and ordinary norms and the rules, the coup system of civil rights, their implementation and protection.

Therefore, knowledge of the substantive law on digital rights is necessary for lawyers to provide qualified legal assistance already from October 2019 - from the date of commencement of changes in legislation that govern the concept, place, procedure for the implementation and circulation of digital rights. Nonetheless in lawyer's practice the importance of knowledge of the content of legislation relating to the protection of information, secrecy, as well as the boundaries and limits of these legal institutions, which are also subject to change by the processes of digitalization of advocacy is inevitable.

Protection of personal data in the Russian Federation is regulated by the Federal Law of 27.07.2006 N 152-FZ "On Personal Data" [14], yet the practice of its application is only shaping. At the same time, even in the countries of the European Community, where the protection of personal data received legal protection earlier than in our state, legislation is not perfect and there is a discussion in the literature about its change. Thus, the issue of the correlation of human rights, including the protection of personal data as a manifestation of the human right recognized by the state to protect privacy secrets, is relevant and important for resolving both within the jurisdiction of a region or a state, as it is a need of modern society. Besides, when studying the norms of substantive law, which regulate issues of digitalization of the economy, one should take into account the experience of foreign law, as well as explanations of foreign lawyers on the practice of its application [1]. Among knowledge of the norms of substantive law, the skills of a lawyer in drafting documents, drafting legal texts, and working with electronic resources are also important.

Article 15 of the Federal Law of December 22, 2008 No. 262-FZ "On Ensuring Access to Information on the Activities of the Courts in the Russian Federation" [13]; Decree of the Presidium of the Supreme Court of the Russian Federation of September 27, 2017 "On approval of the Regulations on the procedure for placing texts of judicial acts on the official websites of the Supreme Court of the Russian Federation, courts of general jurisdiction and arbitration courts in the Internet information and telecommunications network" [10] state: all judicial acts are subject to digitization; participants in the process have the opportunity to familiarize themselves with the case materials by accessing the electronic resource by entering the code provided by the court; Audio recording is carried out automatically in the courtrooms, documents can now be submitted to the court in electronic form. Digitization - translation to the platform-oriented code of legal texts, as well as re-digitization - printing of digital documents [7], are necessary actions for a lawyer who drafts legal documents in electronic form with a subsequent printout. Thus, the lack of skills in dealing with electronic resources, digital technologies makes the work of a lawyer practically impracticable; which allows us to conclude that the requirement to provide qualified legal assistance, arising from the concept of advocacy established by Part 1 of Art. volume, content of digital rights and their turnover, as well as skills in possession of digital technologies used when working with legal resources. It is not by chance that the Federal Chamber of Lawyers has taken a focus on the digitization of advocacy, which is what the IX All-Russian Bar Congress held on April 18 of this year. As the adviser of the Federal Antimonopoly Service of the Russian Federation E. Avakian, stressed the digitalization of state power sets important tasks for the legal profession [4]. These tasks include the implementation of an automated system for the distribution of orders between lawyers on the appointment of inquiry bodies, preliminary investigation or court authorities, data protection in accordance with the legislation on personal data protection, the need to keep a lawyer's file, taking into account the need to observe lawyer secrets. The digitization of these processes and procedures is the task of the legal profession, as they relate to issues related to the provision of legal assistance to the public by a lawyer.

Lawyers for the company, as well as serving the business lawyers, must possess the skills and other software products that provide distribution management company overseeing the Organization and execution of orders. As rightly pointed out by Nantem, the existing business model is digitally mapped and/or new digital products are developed [6]. Information, communications, processes and services are networked via digital platforms [5]. Therefore, the possession of skills of modern digital technologies and knowledge of legislation for the protection of digital rights has for a lawyer working in the Organization, essential.

4 Conclusion

The result of the study is the reached confirmation of the fact that modern lawyers need to expand the digitalization of lawyers' actions related to the provision of legal assistance to citizens and legal entities to resolve issues of balanced and sufficient use of digital technologies in legal practice, including the need to comply confidentiality in relation to any information. In addition, a modern lawyer needs knowledge of changes in legislation related to the inclusion of digital rights and related legal institutions, the circulation of digital rights and the conditions for their implementation in law.

References

1. Caserta, S., Madsen, M.R.: The legal profession in the era of digital capitalism: disruption or new dawn? Laws **8**(1), 1 (2019). https://doi.org/10.3390/laws8010001
2. Fedorov, D.: Tokens, cryptocurrency and smart contracts in domestic law projects from the perspective of foreign experience. Zakon.ru (2018). https://zakon.ru/blog/2018/05/21/tokeny_kriptovalyuta_i_smart-kontrakty_v_otechestvennyh_zakonoproektah_s_pozicii_inostrannogo_opyta_73792#comment_464488. Accessed 9 May 2019. (in Russian)
3. Guznov, A.G., Mikheeva, L.Yu., Novoselova, L.A., et al.: Digital assets in the system of objects of civil law rights. Zakon **5**, 16–30 (2018). (in Russian)
4. Katanyan, K.: Advocacy should keep pace with the times. Advocacy of Dagestan Republic (2019). http://advokatrd.ru/news/news_625.html. Accessed 9 May 2019. (in Russian)
5. Melnichenko, R.: Electronic advocacy. Eurasian Advocacy **3**(28), 38–44 (2017). (in Russian)
6. Nanteme, P.: Definition: what is digitalization? Successful digitalization of business models. AOE (2018). https://www.aoe.com/en/digitalization.html. Accessed 9 May 2019
7. Plöger, I., et al.: Legal Challenges of Digitalization. An Input for the Public Debate. Federation of German Industries-Noerr, Berlin (2015)

8. Talapina, E.V.: Law and digitalization: new challenges and prospects. J. Russ. Law **2**(254), 5–17 (2018). (in Russian)
9. The Civil Code of the Russian Federation (part 1) approved by the state Duma of the Russian Federation dated 30 November 1994 N 51-FZ (1994). https://codex-online.ru/codex/gk_rf/chast1.php. Accessed 9 May 2019. (in Russian)
10. The Decree of the Presidium of the Supreme Court of the Russian Federation approved by the Supreme Court of the Russian Federation dated 27 September 2017 "On approval of the Regulations on the procedure for placing texts of judicial acts on the official websites of the Supreme Court of the Russian Federation, courts of general jurisdiction and arbitration courts in the Internet information and telecommunications network" (2017). https://www.garant.ru/products/ipo/prime/doc/71684972. Accessed 9 May 2019. (in Russian)
11. The Federal Law of the Russian Federation approved by the state Duma of the Russian Federation dated 31 May 2002 N 63-FZ. "On Advocacy and the Legal Profession in the Russian Federation" (2002). http://www.consultant.ru/document/cons_doc_LAW_36945/. Accessed 9 May 2019. (in Russian)
12. The Federal Law of the Russian Federation approved by the state Duma of the Russian Federation dated 18 March 2019 N 34-FZ. "On Amendments to Parts One, Two, and Article 1124 of Part Three of the Civil Code of the Russian Federation" (2019). http://www.consultant.ru/document/cons_doc_LAW_320398/. Accessed 9 May 2019. (in Russian)
13. The Federal Law of the Russian Federation approved by the state Duma of the Russian Federation dated 22 December 2008 N 262-FZ "On Ensuring Access to Information on the Activities of the Courts in the Russian Federation" (2008). https://base.garant.ru/194582/. Accessed 9 May 2019. (in Russian)
14. The Federal Law of the Russian Federation approved by the state Duma of the Russian Federation dated 27 July 2006 N 152-FZ "On Personal Data" (2006). https://base.garant.ru/12148567/. Accessed 9 May 2019. (in Russian)

Using Source Code Escrow as a Way to Develop Information Technology Industry

M. A. Tokmakov(✉)

Samara State University of Economics, Samara, Russia
maxim.tokmakov@gmail.com

Abstract. Source code escrow has been successfully used in many countries for many years. However, it is still not implemented in Russia. This research is aimed at identifying the benefits and drawbacks of institution of source code escrow in order to consider the possibilities and prospects for its implementation into the legal frame of Russia. This paper compiles the principal arguments of the opponents of the said scheme presented in foreign materials that are duly assessed. The review of the legislation with regard to the possible use of source code escrow in Russia is presented.

Keywords: Escrow · Source code · Computer program · Software · Escrow agent · Information technologies

1 Introduction

Since 1 June 2018, due to successive reforming of civil legislation in Russia, a full escrow institution appeared which until then existed in Russian law just in the form of particular fragmentary provisions. Escrow structure has the potential to be used in different fields through dealing with problems associated with confidence among the parties to business transactions. Such potential has been already examined [7]. To date, some of the proposed applications of escrow have been used in practice[1], others are still in the state of improving the legal model of escrow use [15].

However, some capabilities of escrow have still not been used in Russian realities, one of which as it was mentioned [7] is the use of escrow for the source code of computer program.

Such mechanism is well known and has been successfully used in many foreign countries. Best practices in this field belongs certainly to the USA where escrow has been legislatively established in the mid 20th century, and the services with regard to the name source code escrow have been rendered since 1982[2]. For instance, since

[1] Amendments to the Federal Law 214-FZ “About participation in shared construction” have actually changed the financing scheme of housing construction in Russia from the use of funds of the participants in shared construction in favor of project financing of construction by banking sector. After 1 July 2019, fund raising under the contracts for participation in shared construction will be possible only upon the use of escrow accounts.

[2] Iron Mountain Inc. - the company who commercialized the concept of source code escrow in 1982. https://www.ironmountain.com/information-management/software-escrow.

S. I. Ashmarina et al. (Eds.): ISCDTE 2019, LNNS 84, pp. 581–588, 2020.
https://doi.org/10.1007/978-3-030-27015-5_69

1995, the law in the State of California has included the detailed regulation for the use of source code escrow applied in the electoral process on a mandatory basis[3]. Thus, the USA has half a century of experience using source code escrow not only in business but also in government sector. Use of source code escrow, however, is often criticized on the part of scientific and professional community [6, 8].

Development of information technology is now one of the priorities for Russia. At the same time, the Strategy for the Development of Information Technology Sector in the Russian Federation [13] (hereinafter referred to as "the Strategy") has presented the unpromising data with respect to the state of this sector as of 2012. The prepared project [10] on the realization of the Strategy includes the data concerning the state of this sector as of 2017 that can hardly be considered as satisfying. However, the Strategy highlights the need for increase of attractiveness of Russian jurisdiction for IT companies as against CIS and East Europe countries. It is also suggested that special attention should be given to the intellectual property protection services in the field of information technology that should be affordable, convenient and ensuring integration into international reporting systems of intellectual property. It appears that the implementation of source code escrow is aligned with the objectives of improving the competitiveness of Russian legal order and it can potentially solve some sector's problems, thereby facilitating its development.

2 Methods

The main objective of this research is to identify the benefits and drawbacks of institution of source code escrow in order to consider the possibilities and prospects for its implementation into the legal frame of Russia.

In this regard, the framework of studies comprises the comparative and legal method in terms of studying international practices in the use of source code escrow, as well as modelling method in terms of application of these practices to Russian realities. This paper compiles the principal arguments of the opponents of the said scheme presented in foreign materials that are duly assessed. In conclusion, the review of Russian law with regard to the possible use of source code escrow is presented.

3 Results

3.1 Source Code Escrow

Generally, a computer program is written in programming language (such as C++ or Java) that can be read, understood and changed when having relevant knowledge. Final version of computer program is called "source code" that is translated into computer language known as the "object code" or "machine code" which can be installed or

[3] A software whose source code was not placed into escrow should not be used in the electoral process in California [14].

launched on the computer. The machine code is a command system recognized directly by computer hardware but unreadable for a man.

2 CCR § 20621 of the California Code of Regulations contains the definition of source code as a "version of computer program where the source program statements of a programmer are expressed in source language to be compiled or collected and combined in an equivalent object code to be executed by the computer which results in an executable program" [14].

According to Freedman, the programmers can hardly read or understand the object code and even harder change it. Therefore, in most cases, the programmers shall have access to the source code of computer program, as well as other materials related to it (such as design documentation, manuals) in order to change (for example, error correcting) or improve (addition of new features) the program [3].

Software providers usually offer users the copies of licensed software in the version of object code. As noted by Overly and Karlyn, a "source code, generally, is a very valuable commercial confidentiality of the provider and it should be carefully protected. In most cases, none of customers is provided with source code" [9]. This approach has a lot of grounds including the following:

- the source code of the designed software product is an intellectual property of software company. Thus, the transfer of source codes along with the product itself to the customer by software company shall be considered the transfer of its intellectual property, know-how. Instead, the customer receives a compiled code (object code) but not a source code, since illegal use or disclosure of the source code can lead to losses or irreparable damage to the writer;
- those customers who have no access to the source code have usually to pay for the software maintenance and other services rendered by the source code author. Most of the profit on software sale is based on licensing, while this restricts the possible actions of software user who has a compiled code. This scheme is not only beneficial to the author of program code but it can also be useful for customers who use commercial software since the author is able to provide service to it effectively.

Nevertheless, in some cases (for example, when the licensed software is expensive to be bought from and serviced by the author or its fast update is essential for routine transactions of the customer), the customer can refuse to use services of the software provider in full. The customer can demand the right to access and use of source code for the purposes of service provision to licensed software in case the provider refuse to or cannot do it.

This trend is attributed to risk factors associated with the long-term operation of software in life cycle, i.e. there are issues of long-term relationships with providers.

The fact that customer depends on software company is not a new problem. A software company can fail, disappear from the market or just lose interest in the product sold. Thus, the customer who has added it in its technological chains cannot simply exclude it through the replacement.

Most often, the customer is concerned about the possibility of further work with the product in case of bankruptcy or, for example, breach of obligations by software company, in particular when buying expensive and complex products. In case when the software company withdraws from the market or fails to fulfill its obligations for the maintenance of the product, the customer loses the possibility to effectively use the product and has no access to the source code to take on the maintenance functions or transfer them to another company.

The most common situation is that the customer specifies the requirements to provide guarantees of software operability. In this case, the software company is not willing to transfer the source code to the customer. Another case is a violation of technical support contract.

In case when the customer company pays the software company for the software development, the risks of provider sustainability (in particular, possible bankruptcy) can be minimized through the source code escrow.

Source code escrow is the deposit of the source code of software. In case of bankruptcy, death or failure to fulfill obligations by the licensor, the source code should be transferred to the licensee, thus allowing it to avoid difficulties with program modification.

The possibility to obtaining access to the product source code eliminates this problem, and the transaction of source code escrow addresses a sensitive issue of confidentiality and business interests of the licensor (software provider). The escrow agent provides the licensee with restricted access to the source code and prevents the disclosure of information, as well as its sale to other parties.

In case of condition for the transfer of source code to the licensee, the latter shall make a corresponding request to the escrow agent who, in turn, shall inform the licensor and, if there are no objections of the latter, provides the licensee with access to the source code. Generally, a software escrow agreement allows the licensor to object to the disclosure of source code within a certain time. In case of licensor's objections, the escrow agent shall keep the deposited code until the resolution of the dispute. In some cases, the licensor and the licensee may agree on a mandatory procedure for the immediate transfer of the source code.

Thus, as Freeman rightly points out, "if the escrow agreement is negotiated and drafted properly, it can be effective in granting rights and protection to the software customer" [4].

The use of escrow also provides for a number of implicit benefits. First, the source code escrow enhances the status of the licensor as a business partner; second, the software protected by an escrow agreement is of great value for the market, since the licensor provides the licensee with the continuous use of software.

In addition, the source code escrow is also an advantage for the software end users whose investments in software are protected by the stability of its work and maintenance.

3.2 Prospects for the Use of Source Code Escrow in Russia

To date, Russian legislation is generally prepared to the establishment of an institution of source code escrow. The source code is recognized as a software element in the legislation and literature [for example, 11, 12], i.e. it is a form of expression (presentation) of a program. However, some sources also present another approach that is more consistent with the global one, in which the source code is recognized as an "original computer program written in a readable form (using programming language) that should be translated into the machine-readable form before its execution by the computer" [2].

Anyway, in the context of Russian law, the transfer of the source code should be rather considered as the transfer of the program itself expressed in the form of source code (text), i.e. escrow can be applied to the computer program presented on a tangible medium in the form of source code.

As already mentioned, the Civil Code of the Russian Federation has been recently supplemented by a separate chapter concerned with escrow. However, in order to implement the source code escrow, some adjustments will be required. The point is not only in the need to increase the number of escrow objects, which is now limited to the movable things, monetary funds and uncertificated securities in Russian law. This problem can be potentially bypassed through the tangible medium (disk, USB drive, etc.) escrow and specifying in the escrow agreement the obligations of the escrow agent to verify the source code recorded on it.

The nature of structure of source code escrow is somewhat does not fit into the existing model of escrow. In Russia, escrow is presented only as a way to fulfill the obligation, i.e. the licensor under the escrow agreement shall place the property into escrow in order to fulfill the obligation for its transfer to the beneficiary, and this obligation shall be considered to be fulfilled from the moment of the property transfer to the escrow agent.

However, in the situations under review, the software provider has no obligation for the source code transfer (otherwise, the escrow would be meaningless) and, consequently, no software supply agreement is executed through the escrow. Such obligation of software company may arise in the event of breach of obligations resulting not from the license agreement, but from the maintenance (technical support) agreement.

With this approach, the conditions of the source code transfer can be considered as the accord and satisfaction agreement, i.e. in case of obligation default, the software company (licensor) undertakes to grant the licensee the right to use the source code (in addition, the transfer of the source code to the customer in case of release from escrow should be followed by the amendment to the license agreement with regard to the change of limits to the rights and allowed uses of the program that can be executed through the suspensive condition in the license agreement).

Thus, when the source code is transferred to the escrow agent, an inconsistent situation occurs, i.e. the compensation is provided which means that the obligation of the software provider is terminated.

Other issues are also possible in the introduction of source code escrow institution into Russian legislation (for example, the issue of maintaining the secrecy of know-how when transferring the source code to the escrow agent), which surely require the more careful assessment.

4 Discussion

Notwithstanding all the above benefits of source code escrow, there are also opposing views in the literature with regard to the inefficiency of such mechanism. Thus, in particular, Mezrich notes that "the protection received by the customer at the time of source code escrow is amazingly poor, and the standards required of the escrow agent are much lower and have less legal remedies in case of violation as compared to the standards required of other escrow agents (for example, banks or law firms)" [8].

With regard to the possibility of introducing the source code escrow in Russia, it seems reasonable to assess the principal arguments of opponents of using such structure in the law and order, in which it is used.

The main disadvantages of source code escrow are generally as follows:

1. *A small number of cases of the deposited source code release.*
   Experts note that most of the deposited source codes are never released from escrow due to non-occurrence of events for release. Thus, Shawn Helms and Alfred Cheng adduce the statistics of Iron Mountain as an argument, according to which the company released the source code only in 96 cases during the period of 1990–1999 having several thousand escrow accounts [6].
   More recent data presented by Iron Mountain show that in 2017, the number of releases per year increased 3.6 times as compared to the 90s [1]. However, the article does not provide data on the number of open escrow accounts during this period, and assuming that their number also increased, the figures remain insignificant.
2. *Defect of deposited source code.*
   In fact, the source code after release can often be defective, out-of-date, incomplete or otherwise unable to ensure the software operation (for example, a tangible medium is damaged).
   The recent data from Iron Mountain showed that over 70% of all deposited source codes were considered incomplete, 92% required additional information from the developer, 38% did not contain any instructions for adjustment or construction[4]. Thus, most of the deposited source codes are simply unusable after release.
   In this regard, the source code to be transferred shall be subject to complete check, which ensures that the transferred source code is complete and, if necessary, can be effectively used. Mark Grossman notes that this issue is often neglected; therefore, the verification of the source code by an escrow agent in this area is of particular importance [5].
3. *Inability of the beneficiary to use the source code due to the lack of its own specialists.*
   In many cases, even after the release of serviceable source code, the beneficiary is unable to figure it out on its own. According to Mezrich, even with access to the source code, one needs to find programmers who understand it and can quickly adapt to it that can be consistent with the costs for a new program [8].
   Helms and Cheng note, in turn, that in most cases, the source code is deposited only because the software provider provides technology and experience that the customer does not possess, and thus the customer itself is not able to introduce the software properly and train its staff to support it [6].

[4] The data are presented on the official website Iron Mountain. https://www.ironmountain.com/resources/data-sheets-and-brochures/e/escrow-verification-services-faq.

Thus, the time and financial expenditures of the beneficiary for finding an appropriate specialist, retraining him or outsourcing may exceed the costs for the development or purchase of new software, which makes the source code escrow meaningless.

4. *Efficient defense of the beneficiary's rights through other means.*
   One of the key issues addressed by escrow, as previously mentioned, is the bankruptcy of the software provider (licensor). At the same time, some US researchers believe that such risks of a provider's bankruptcy are now eliminated by the protection available under the US Bankruptcy Code (USBC)[5].

Mezrich points out that, even though many escrow options are not available within the framework of USBC (for example, technical inspection services, source code release for a reason other than the provider's bankruptcy, etc.), such limited protection provided for in USBC may be quite adequate for many software users [8].

High cost of escrow services, lack of common standards among various escrow agents, risks related to the activities of the escrow agent (in particular, its bankruptcy), as well as increasing cases of using software with open source code are also mentioned at times among the shortcomings. However, it seems that all these arguments are implicit and do not directly relate to the issues of escrow institution, but center around the competition between escrow agents or software providers.

The analysis of the arguments presented by opponents of using source code escrow shows that most of the arguments are outside the escrow institution and they should to a large extent be overcome by free will of the parties who wish to use escrow in a particular situation. "To be or not to be", that is the choice of the parties to business transactions. Certainly, the source code escrow is not a cure-all, so that clever participants in legal relations should evaluate the potential risks and benefits of using this scheme in each specific case.

The arguments presented above would worth noticing in case of mandatory need to use escrow, as is the case with software escrow in the electoral process. However, it is particularly referred to the free market, which sets its own causes and conditions for the use of source code escrow. Moreover, as international practice shows, the interest of the parties to business transactions in the source code escrow continues and the demand for relevant services keeps growing.

In such conditions, the lack of control of source code escrow does not have the best impact on both the competitiveness of Russian law and the information technology industry itself (as a result, the industry players shall move to another jurisdiction).

5 Conclusion

Source code escrow has been successfully used in many countries for many years. However, it is still not implemented in Russia, thus having no the best impact on both the competitiveness of Russian law and the information technology industry itself. It

[5] The Law provides for the opportunity for the software customer (licensor) to obtain access to the source code after provider's bankruptcy notice.

appears that the implementation of institution of source code escrow is in line with modern demands and it can potentially solve some problems of the sector, thereby facilitating its development.

References

1. Boruvka, J.: Technology escrow releases: trends from three decades of data. Info go to (2017). https://www.infogoto.com/technology-escrow-releases-trends-from-three-decades-of-data-2/. Accessed 15 April 2019
2. Eurasian Economic Commission Council Resolution dated 03.11.2016 No. 81 "Concerning adoption of the Good laboratory practice of EAEU in the field of drug circulation". Official website of the Eurasian Economic Union (2016). http://www.eaeunion.org/. Accessed 15 Apr 2019. (in Russian)
3. Freedman, B.J.: Software license agreements: a practical guide. Borden Ladner Gervais LLP (2014). http://docplayer.net/8442431-Software-license-agreements-a-practical-guide-by-bradley-j-freedman.html. Accessed 15 Apr 2019
4. Freeman, E.H.: Source code escrow. Inf. Syst. Secur. **13**(1), 8–11 (2004). https://doi.org/10.1201/1086/44119.13.1.20040301/80429.2
5. Grossman, M.: Technology Law: What Every Business (and Business-Minded Person) Needs to Know. Scarecrow Press Inc., Maryland (2009)
6. Helms, S., Cheng, A.: Source code escrow: are you just following the herd? CIO.com (2008). https://www.jonesday.com/files/Publication/09e73d7d-7240-450f-ae18-07970d03977f/Presentation/PublicationAttachment/7cbcbfeb-8cf4-4056-9c89-11c07b903a73/JD_Source%20Code%20Escrow.pdf. Accessed 15 Apr 2019
7. Medentseva, E.V., Tokmakov, M.A.: Escrow: international experience and perspectives of application in Russia. J. Adv. Res. Law Econ. **8**(3), 906–909 (2017)
8. Mezrich, J.L.: Source code escrow: an exercise in futility? Marquette Intellect. Property Law Rev. **5**(1), 118–131 (2001)
9. Overly, M.R., Karlyn, M.A.: A Guide to It Contracting: Checklists, Tools, and Techniques. CRC Press Taylor & Francis Group, Boca Raton (2013)
10. Project on the Strategy for the Development of Information Technology Sector in the Russian Federation for 2019–2025 forward looking up to 2030 (2018). http://www.tadviser.ru/images/f/f8/2019_strategy_9January.pdf. Accessed 15 Apr 2019. (in Russian)
11. Rozhkova, M.A., Glonina, V.N., Bogustov, A.A. (eds.): Civil Concept of Intellectual Property Within the Framework of Russian Law. Statut, Moscow (2018). (in Russian)
12. Saveliev, A.I.: Software Licensing in Russia: Legislation and Practice. Infotropic Media, Moscow (2012). (in Russian)
13. Strategy for the Development of Information Technology Sector in the Russian Federation for 2014 – 2020 forward looking up to 2025 (approved by the Resolution of the Government of the Russian Federation on 1 November 2013 No. 2036-r) (2013). https://digital.gov.ru/ru/documents/4084/. Accessed 15 Apr 2019. (in Russian)
14. The California Code of Regulations. https://govt.westlaw.com/calregs/Index?bhcp=1&transitionType=Default&contextData=%28sc.Default%29. Accessed 15 Apr 2019
15. Tokmakov, M.A.: Model of the internet escrow's legal regulation as a factor of efficiency of its use in e-commerce. In: Ashmarina, S., Mesquita, A., Vochozka, M. (eds.) Digital Transformation of the Economy: Challenges, Trends and New Opportunities. Advances in Intelligent Systems and Computing, vol. 908, pp. 368–375. Springer, Cham (2019). https://doi.org/10.1007/978-3-030-11367-4_36

Responsibility in the Telemedicine Area

S. N. Revina(✉)

Samara State University of Economics, Samara, Russia
29.revina@mail.ru

Abstract. The development of technological innovations has led to development in medicine of a new field - telemedicine. Worldwide, telemedicine has changed the quality of medical services. Every year the use of telemedicine technologies is growing. However, a number of scientists speak about its insufficient reliability and accuracy. But it cannot be denied that this technology helps simplify the administrative process, personal and material savings, provides a flexible workplace design, speeds up the entire treatment process and at the same time accelerates critical sub-processes (provision of information), provides a strategic competitive advantage for those involved. One of the positive effects of the use of telemedicine that is rarely spoken is the saving of funds not only of health organizations, but also of patients. Savings for patients consist of reducing transportation costs or wages for missing days at work. Health authorities can save money by reducing the costs of equipment for patient rooms, reducing the number of medical staff, which will reduce the cost of medical services. Expanding the scope of telemedicine and providing it at the expense of OMS will reduce costs for commercial enterprises. Telemedicine is a wide discussion topic. And today there is no single and universal definition of telemedicine. This also leads to other controversial issues, such as liability issues in the use of communication technologies in providing medical care.

Keywords: Electronic health · Health care · Legal regulation · Medical care · Telemedicine · Information and telecommunication technology

1 Introduction

The use of telecommunication technologies in healthcare has profoundly influenced the quality of medical services. Telemedicine provides better access to treatment, reduces costs and improves interactions with the patient. As a reactive form of treatment, telemedicine can actually serve as a center for "informed virtual assistance". In this regard, the development of telemedicine has a direct impact on the economies of all countries [15]. However, despite all of these benefits currently available to medical professionals, the courts are confronted with the issue of legal liability when using telemedicine. The purpose of this study is to analyze the legal framework governing the responsibility of health authorities and doctors in the provision of medical services using innovative technologies. The authors set the following tasks: development of recommendations for improving the legal regulation of the responsibility of medical organizations when using telemedicine.

S. I. Ashmarina et al. (Eds.): ISCDTE 2019, LNNS 84, pp. 589–596, 2020.
https://doi.org/10.1007/978-3-030-27015-5_70

2 Methodology

The authors in the implementation of this study applied the following methods: logical, method of analysis and synthesis, system and modeling. And since each separately taken research method does not allow recreating the "complete" picture of the actual state of affairs in the area of this study, the above methods were applied in a comprehensive manner.

3 Results

In the foreign literature, the term "telemedicine" is proposed to mean: medical care through information and communication technologies, which includes "tele-health", "e-health", "mobile health" and "associated health" [2], the use of telecommunication technology to provide clinical medical care at a distance [20]; the use of advanced communication technologies in the context of clinical health in the provision of care at a considerable physical distance [11]; transferring information about patients, their health through text, sound, images and other forms of data for the prevention, diagnosis, treatment and monitoring of the patient [10], etc. Analysis of these definitions indicates the absence of a unified approach to the content of the concept of telemedicine [14]. This is typical for domestic science. The World Health Organization has indicated that the risks of medical liability of specialists providing telemedicine services are one of the main obstacles to its implementation around the world, along with the lack of an international legal framework allowing the provision of services by specialists from different countries [22]. Improper delivery of medical care refers to actions (or inaction) of medical workers that do not meet the established provisions of medical science and practice, accepted legal norms, medical rules, ethics and deontology. Russian legislation provides for administrative, criminal, and civil liability for inadequate provision of medical services using telemedicine technologies: for carrying out such activities without a license; for violation of the terms and volume of medical services using telemedicine technologies; for providing incorrect information about the service; for violation of the rights of patients and harm to their health and life. Today, a number of issues of responsibility in the field of telemedicine are not regulated. These include the allocation of responsibility between the consultant and the consulting physician or consultation.

4 Discussion

Inadequate telemedicine services should be considered on basis of their characteristics. First, the telemedicine service can be provided only by a medical organization licensed to the appropriate type of medical care and connected to EGISZ or another information system. The provision of such services without a license for medical activities constitutes an administrative offense under Art. 14.1. Code of the Russian Federation on Administrative Offenses [5]. Administrative liability is subject to the provision of telemedicine services in violation of the requirements and conditions provided for by a special permit (license). Moreover, for gross violation of such requirements and conditions, the legislator establishes higher measures of responsibility up to administrative suspension of activity for up to ninety days. The note to article 14.1 of the Code on Administrative Offenses of the Russian Federation states that the Government of the Russian Federation in relation to a specific type of activity to be licensed establishes the concept of gross violation. In medical activity, a gross violation is understood as the licensee's failure to comply with the requirements of clause 4 and sub-clauses "a", "b" and "c (1)" of clause 5 of the Resolution of the Government of the Russian Federation No. 291 of April 16, 2012 [7] that entailed the consequences established part 11 of Art. 19 of the Federal Law "On licensing certain types of activities". Note that the licensing requirements for the licensee in carrying out medical activities are the requirements for the license applicant, including compliance with the procedures for providing medical care, compliance with the established procedure for providing paid medical services, etc. Therefore, violation of the procedure for organizing and providing medical care using telemedicine technologies, as well as other procedures for providing medical care, is grounds for bringing the medical organization to administrative responsibility under Art. 14.1. Administrative Code. The current legislation of the Russian Federation establishes criminal responsibility for the illegal occupation of medical and pharmaceutical activities and involves punishment in the form of a fine, forced labor, or restriction or imprisonment. In this article 235 of the Criminal Code of the Russian Federation [6] introduces criminal liability for the implementation of medical activities by a person who does not have a license for this type of activity, only in those cases where, through negligence, caused harm to human health (part 1) or his death (part 2).

Secondly, the telemedicine service can be provided only by a medical professional who is entered in the Federal Register of Medical Professionals and has an appropriate level of qualification and a specific specialty (specialization). Otherwise, this is the basis for bringing the medical organization to administrative responsibility. Telemedicine service should be provided in a timely manner and volume. When rendering medical assistance using telemedicine technologies within the framework of the state guarantees program, consultations (doctors' consultations) using telemedicine technologies in a planned form are carried out taking into account the observance of the established requirements for the consultation periods. The terms are determined in accordance with the terms of contracts, including contracts of voluntary medical insurance, unless other requirements are provided for by federal laws or other regulatory legal acts of the Russian Federation. Thus, the late provision of telemedicine services should be regarded as inappropriate services. Paragraph 19 of the Procedure for the organization and provision of medical care with the use of telemedicine technologies establishes the time limits for the consultation (consultation of doctors) in emergency and emergency situations. In telemedicine, the terms of providing services and consultations may be violated for reasons that are not found in case of full-time medicine: technical failures on the side of the communication operator of the medical consulting organization or the medical organization requesting the consultation; technical failures in the information system. In our opinion, in such cases the medical advisory organization is relieved of the responsibility for non-compliance with the consultation period, unless it independently maintains the medical information system. This rule also applies to the inappropriate volume of telemedicine services provided.

The volume of telemedicine services, as well as the term, can be established in accordance with the procedures for providing medical care and taking into account the standards of medical care in force in the Russian Federation, as well as by agreement of the parties to the contract for the provision of telemedicine services. Do not forget that the telemedicine service provided to patients should meet the goals established by the legislation of the Russian Federation and cannot include diagnosis and prescription of treatment. The provision of telemedicine services in the amount less than that stipulated by the procedure, standards and contract is a significant violation. At the same time, the implementation of an overestimated amount of medical measures carried out in the absence of appropriate medical indications can also cause harm to the patient's health [21]. If the time or amount of medical care provided using telemedicine technology is established by the procedure or standard of medical care, then in addition to compensation for harm caused by improper provision of a service, a medical organization may be held administratively liable. The legal bases of civil liability in connection with the infliction of harm in improper provision of medical care are the norms of Ch. 59 of the Civil Code. Article 1064 of the Civil Code of the Russian Federation expresses the principle of a general delict, according to which the harm caused to a subject of civil law is subject to compensation in full by the person who caused the harm.

Harm caused to the health and life of a citizen is reimbursed in the manner specified in paragraph 2 of Ch. 59 of the Civil Code of the Russian Federation [4]. In this case, the responsibility of the medical organization for the harm caused to the life or health of the patient as a result of deficiencies in the service provided, as well as inadequate or insufficient information about it, in accordance with Art. 1095 of the Civil Code does not depend on the availability of evidence of guilt. And it does not matter whether the contract was concluded between the contractor and the patient [3]. This means that in the provision of telemedicine services in the form of consultations between medical organizations, each of them is responsible for the harm caused. Clause 9 of the procedure for organizing and providing medical care using telemedicine technologies explicitly states that the consultant (doctors - participants of the consultation) is responsible for the recommendations provided as a result of the consultation (consultation of doctors) using the telemedicine technologies within the medical certificate given to them. First of all, let us draw attention to the inconsistency of this norm from the point of view of the legal (legislative) equipment of others to the norms of the legislation of the Russian Federation. In accordance with Paragraph 2 of Art. 98 Principles of health protection, medical organizations, medical workers and pharmaceutical workers are responsible in accordance with the legislation of the Russian Federation for violation of rights in the field of health protection, causing harm to health and (or) life in the provision of medical care to citizens. Article 1064 of the Civil Code of the Russian Federation says about the compensation of harm caused to a person or property of a citizen, as well as damage caused to the property of a legal entity. The norms of administrative and criminal legislation in the field of medicine also establish responsibility for harm to the health or life of the patient. It follows from this that liability may arise for the violation of rights and the infliction of harm. The wording of clause 9 of the procedure for organizing and providing medical care using telemedicine technologies establishes the responsibility of a physician (doctors' consultation) for recommendations that do not comply with the norms of civil, administrative or criminal law, as well as the Basics of Health Protection. The idea of the legislator to limit the responsibility of consultants in the provision of telemedicine services is understandable and correct, but the rule of the clause should be stated differently. The introduction of the special responsibility of the medical consultant has been discussed for a long time. Ryzhov noted that "the health worker who prepared the information for the telemedicine service is responsible for the completeness and quality of the telemedicine information provided. A medical consultant is responsible for the issued opinion, its quality and timeliness" [16]. A number of scientists say that the final diagnosis is made by the attending physician is questioning the legality of the responsibility of the consultant (a member of the council) [19]. According to the norms of the Basics of Health Protection, the attending physician is a physician charged with organizing and directly providing the patient with medical care during the period of observation and treatment. Point 2 of Art. 70 of The Basics of Health Protection explicitly provides that recommendations of consultants are implemented only in consultation with the attending physician, except in cases of emergency medical care. Judicial practice also pays attention to this [1]. A consultation of doctors is a meeting of several doctors of one or several specialties necessary to establish the patient's state of health, diagnosis, determine the prognosis and tactics of medical examination and treatment, expediency of referral to specialized departments of a medical organization or other medical organization and to resolve other issues in cases provided for in The Basics of Health. The legislator did not directly indicate the obligatory decision of the consultation for the attending physician, the choice of patient treatment tactics. In accordance with the Methodological Guidelines on the procedure for organizing consultations and consultations in medical institutions, recommendations and a consultant and consultation are mandatory for attending physicians [12].

In case of disagreement with the appointment of a consultant, the head of the department is obliged to make an entry with the appropriate justification in the medical history, which apparently does not exclude his obligation. Responsibility for the implementation of the decisions of the consultation is the head of the department (in a clinical hospital - and the head of the relevant department). It is believed that the decisions of the consultation of doctors are advisory in nature and are not mandatory for the attending physician. At the same time, the convocation of a consultation of doctors in itself cannot be the basis for exempting a medical organization and (or) a medical worker from liability, but, nevertheless, can be an important criterion indicating proper measures taken by the attending physician in order to understand in a difficult situation, solve the patient's problem [8]. If the legislator has approved a norm on the responsibility of consultants and consultations of doctors based on the results of consultation with the use of telemedicine technologies in order to limit the responsibility of the attending physician, shifting its part to those who gave recommendations, this is a violation of the rights of medical workers, a kind of discrimination. It turns out that the decision of the council and the recommendations of the consultants data in person or using telemedicine technologies have different legal consequences. If the purpose of the introduction of the norm was the division of responsibility for harm caused to the health and life of the patient, between the attending physician and the consultant (consultation of doctors), then it is necessary first of all, to improve the basics of health protection by establishing the mandatory decision of the consultation or recommendation of the consultant for the attending physician. According to paragraph 9 of the Procedure for Organizing and Providing Medical Care with the use of telemedicine technologies, the responsibility of consultants and doctors council is established only in relation to the medical organization that received the consultation. And the consulting organization, without directly providing medical assistance to the patient, participates in the provision of such assistance remotely as a person engaged by another medical organization [13]. Here the question arises - how exactly should counselors be responsible for the harm done. The harm caused to health and (or) life the of citizens in the provision of medical care to them is reimbursed to medical organizations (Clause 3 of Article 98 of the Principles of Health Protection). Consequently, a claim for compensation for harm caused to health or life due to inadequate medical care must be submitted to the medical institution in which medical services were provided [9]. The actual causes of harm (health workers) are responsible in the order of recourse in accordance with Art. 1081 of the Civil Code of the Russian Federation [4]. In legal relations in the field of telemedicine on the side of the provider of medical services also serves a medical organization, whose employees provide counseling using telemedicine technologies. Consequently, the medical organization will also act as a debtor on the part of the consultants, but only in relation to the organization providing medical services to the patient. Shevchuk reasonably singled out three mandatory conditions for the onset of joint liability of medical organizations: the harm was caused to health in the provision of health services by two or more entities (medical organizations or individual entrepreneurs) carrying out medical activities on the basis of a license; the behavior of each subject in the provision of medical services is in causal connection with the onset of general harm to health; if the harm to health in the provision of medical services is the indivisible result of the actions of these persons, regardless of whether they were simultaneous or performed sequentially [17]. As noted in the legal literature, "in case of joint harm to life or health in the provision of medical services, responsibility will always be shared and the rules of Art. 1080 of the Civil Code of the Russian Federation on the joint character of responsibility cannot be applied" [18] due to the lack of a common will to cause harm, a direct causal link between the harm and the behavior of each medical organization, the indivisible nature of the behavior when causing harm to the patient's health.

5 Conclusions

In our opinion, telemedicine is the provision of medical services using telemedicine technologies and other electronic means of exchanging information. For convenience, the ratio of the legislation of the Russian Federation and foreign countries along with the concept of "telemedicine" may use the concept of "telemedicine service." In order to settle the responsibility during consultations, it is necessary to mend the following provision at the legislative level: the consultant (doctors - participants of the council) is responsible for the harm caused to the health and (or) life of the patient, within the medical advice (recommendations) given to them based on the results of the consultation (Consilium of Doctors) using telemedicine technologies. Additionally, it is necessary to amend the binding decision of the consultation or the recommendation of the consultant for the attending physician. Due to the direct limitation of consultants' responsibility (doctors' consultation) in the volume of recommendations given by them using telemedicine technologies, responsibility for harm caused to the patient's health and life is shared with the medical organization where the patient is undergoing treatment, and the medical organization whose employees gave recommendations. If medical assistance was rendered poorly through the fault of the consulting medical organization, the medical organization that applied the recommendations in treating the patient, after reimbursing the required amount to the patient, has the right to turn to the consulting medical organization with a claim for recovery of the sums paid to the patient.

References

1. Appeal definition of the St. Petersburg City Court of 01.12.2017 No. 33a-19489/2017 in case No. 2a-1425/2017. http://www.consultant.ru/cons/cgi/online.cgi?req=doc&ts=49512852902098596386291003&cacheid=1ED2B385213C1E17F36482A919CF8300&mode=splus&base=SARB&n=117568&rnd=0.3140442874916849#21bc3752d4s. Accessed 20 Apr 2019. (in Russian)
2. Bashshur, R., et al.: The empirical foundations of telemedicine interventions for chronic disease management. Telemed. J. E Health **20**(9), 769–800 (2014). https://doi.org/10.1089/tmj.2014.9981770
3. Boytsov, G.: Responsibility of medical institutions for errors in the treatment of patients under an agreement with the company. Disputes and advice to companies. Labor Law **6**, 75–94 (2017). (in Russian)
4. Civil Code of the Russian Federation (part two) No. 14-ФЗ dated January 26, 1996 (as amended on July 29, 2017). http://www.consultant.ru/cons/cgi/online.cgi?req=doc&base=LAW&n=300853&fld=134&dst=1000003545.0&rnd=0.7862144070598782#06015148787286839. Accessed 20 Apr 2019. (in Russian)
5. Code of administrative offences of the Russian Federation of December 30, 2001 No. 195-FZ (as amended on 12/27/2018). http://www.consultant.ru/document/cons_doc_LAW_34661/. Accessed 20 Apr 2019. (in Russian)
6. Criminal Code of the Russian Federation dated 13.06.1996 N 63-FZ (as amended on 04.23.2019). http://www.consultant.ru/cons/cgi/online.cgi?req=doc&base=LAW&n=323343&fld=134&dst=100439.0&rnd=0.6927525888582#05921051078324175. Accessed 20 Apr 2019. (in Russian)
7. Decree of the Government of the Russian Federation of April 16, 2012 No. 291 (as amended on December 8, 2016) "On licensing medical activities (with the exception of this activity carried out by medical organizations and other organizations within the private health care system in the territory of the Skolkovo Innovation Center)". http://www.consultant.ru/document/cons_doc_LAW_128742/. Accessed 20 Apr 2019. (in Russian)
8. Functions of the doctors' consultation, working order. https://www.kormed.ru/baza-znaniy/pravila-okazaniya-meduslug/konsilium-vrachei/funktsii-konsiliuma-vrachey-poryadok-raboty/. Accessed 20 Apr 2019
9. Grishaev, S.P.: Responsibility for failure to perform or improper performance of the contract for the provision of medical services (2019). http://www.consultant.ru/cons/cgi/online.cgi?req=doc&base=CJI&n=109309#042900529408739074. Accessed 20 Apr 2019

10. Gusarova, A.: Data protection in telemedicine. In: SHS Web of Conferences, vol. 2, p. 00013 (2012). https://doi.org/10.1051/shsconf/20120200013
11. Latifi, R. (ed.): Current Principles and Practices of Telemedicine and E-Health. Studies in Health Technology and Informatics, vol. 131 (2008)
12. Methodical instructions on the procedure for organizing consultations and consultations in medical institutions (approved by the Ministry of Health of the USSR 14.11.1982 No. 06-14/14). Collection of official documents on the organization of hospital work. GUZM, Moscow (1984). (in Russian)
13. Morozov, S.P., Vladzimirsky, A.V., Varyushin, M.S., Aronov, A.V.: Distribution of responsibility for poor-quality medical care when using telemedicine technologies. J. Telemed. E-Health **1–2**, 10 (2018). (in Russian)
14. Revina, S.N., Blinov, S.V., Kuzmina, N.M., Sidorova, A.V.: Features of the legal regulation of telehealth in the United States of America. Russ. Justice **4**, 70–72 (2019). (in Russian)
15. Revina, S.N., Sidorova, A.V., Zakharov, A.L., Tselniker, G.F.: Economic relations and economic systems (2018). https://doi.org/10.1007/978-3-319-75383-6_69. (in Russian)
16. Ryzhov, R.S.: Actual problems of legal support for the accumulation of confidential information about citizens in telemedicine. Theory Pract. Soc. Dev. **7**, 249 (2011). (in Russian)
17. Shevchuk, E.P.: Subjects of obligation as a result of causing harm to the patient's health when rendering medical services. Bull. Zabaykalsky State Univ. **8**, 170 (2014). (in Russian)
18. Shimanskaya, S.V.: Features of civil liability in the implementation of medical activities. Ph. D. thesis abstract, Moscow (2013). (in Russian)
19. Sokolenko, N.N., Bagnyuk, M.E., Bagnyuk, D.V.: Medical care with the use of telemedicine technologies: some problems of legal regulation. Med. Law **4**, 14–17 (2018). (in Russian)
20. Sood, S.P., Negash, S., Mbarika, V.W., Kifle, M., Prakash, N.: Differences in public and private sector adoption of telemedicine: Indian case study for sectoral adoption. Stud. Health Technol. Inf. **130**, 257–268 (2007)
21. Tabunshchikov, A.T.: Compensation of harm caused to life and health in the provision of medical services (2014). http://dspace.bsu.edu.ru/handle/123456789/8244. Accessed 20 Apr 2019
22. Telemedicine: opportunities and development in Member States. Report on the results of the second global e-health survey. Global E-Health Observatory Series. https://www.who.int/goe/publications/goe_telemedicine_2010.pdf. Accessed 22 May 2019

Digital Technologies in Counteracting to Extremism Among Young People

I. E. Milova, M. A. Yavorskiy(✉), and N. I. Razzhivina

Samara State University of Economics, Samara, Russia
yavorm@mail.ru

Abstract. The goal of the research is to analyze and systematize the main directions of counteracting to extremism among young people with the use of digital technologies. The authors analyze the threats of modern digital technologies and the opportunities they can provide for effective counteracting to the manifestations of extremism among young people. The work shows the directions of youth extremism prevention with the use of the latest digital technologies. The authors disclose the contents of digital information support for fighting against extremism; note that digitalization can contribute to the formation of citizenship, patriotic education of students. They assess the significance and role of such opportunities in the field of counteracting to extremism among young people. In the opinion of the authors IT-technologies can contribute to effective information support for counteracting to extremism among young people.

Keywords: Counteraction · Prevention · Propaedeutic measures · Cyber-brigades · Extremism

1 Introduction

In the modern period the problems of the development of innovative technologies in the information sphere acquire exceptional importance, not only in the theoretical aspect, but also in practical implementation. This article is an attempt to look at the relevant problems from the legal, ideological, psychological, social and pedagogical points of view, as the knowledge in these areas, when properly processed, can be used for optimal counteracting to extremism, especially among young people. On the one hand, the word combinations "modern young people" and "information-digital technologies" sound in unison and cannot be considered separately from each other. On the other hand, many researchers of the question are convinced that most of the manifestations of youth extremism are directly related to digitalization and the wide spread of related technologies that provide possibilities for the fastest possible transferring of information to a wide range of users.

The steady growth in the use of social networks has become evident nowadays. Of course, this is an integral feature of progress, which cannot be stopped. At the same time, we are witnessing attempts of some users to make the Internet a place for the distribution of extremist views, propaganda of hate, hostility, aggression, sexual promiscuity, criminal, hooligan and fascist ideologies. Digital technologies, as an

S. I. Ashmarina et al. (Eds.): ISCDTE 2019, LNNS 84, pp. 597–603, 2020.
https://doi.org/10.1007/978-3-030-27015-5_71

element of mass media structure, along with positive aspects led to the increase of aggression in the society. In addition, their use led to the increase of the overall level of anxiety of certain social groups, the emergence of depressive moods, and the formation of suicidal aspirations in some communities.

The Internet and its "derivatives" are often viewed as a "tool for the propaganda of extremism" [8]. Criminal elements use this platform as a means of "extremizing the youth audience" [2], as a way of "online recruiting" of new radicals [1, 3, 12, 13]. In this regard, we would like to highlight the studies devoted to:

- counteracting to extremism and terrorism in the Internet and educational environment [6, 11];
- psychological moments of engaging young people in extremist groups through different Internet sites [7, 9];
- problems of distributing extremist views through the Internet and the influence of this process on young peoples' minds [15];
- revealing of the reasons of communicative verbal aggression and "the language of hostility" in the information sphere [14].

We agree to the position of a number of authors who reasonably note in this connection that "the real-world simulacrum created in front of our eyes thanks to the Internet carries all the vices and contradictions characteristic to its matrix-physical society and works as their amplifier. The negative factor, coming from society into cyberspace, is not simply projected there, but, modifying and amplifying, returns to its prototype on the principle of echo, having a destructive impact on it… Extremism, traditionally achieving its goals through primitive pressure, today increasingly uses the latest intellectual achievements" [4].

The literature reveals the contents of preventive measures that can combat the distribution of the ideology of terrorism among young people, with serious attention being paid to digital technologies and the replication of the ideology of terrorism through the Internet:

- it is used as a communication tool for coordinating extremist activities; in this connection digital content acquires harmful organizational properties;
- the Internet is used for searching new sources of extremist organizations financial support;
- the Internet is used as the means of recruitment of new adherents of radical structures;
- instructions for making means of terror are placed in the Internet [5, 10].

The Internet makes it possible to reach a wide range of users simultaneously, passing information to them in high-speed mode; this method is also cost-effective, both for translators and for recipients of information. The researchers of the question indicate that the main obstacle in the fight against extremism in this segment is the anonymity of the participants of the Internet communities. This factor prevents their rapid identification by specialized services, and also gives rise to their confidence in impunity. Unfortunately, gaps in the legal regulation of the so-called "computer law" still remain. So far, discrepancies in terminology and understanding of a number of basic definitions have not been eliminated. The problem of extremism has long become global. However, the United States and some EU countries demonstrate double standards in evaluating such activities. They justify some terrorists saying that they act with good purposes. Such asymmetry of approaches is unacceptable. There should be a general standard of categorical non-acceptance of such actions, their unconditional condemnation and struggle against them of the entire international community, using all available methods, including information interaction. Representatives of gang formations and their supporters must feel a united opposition of all healthy forces, both at the level of state contacts and in the structures of civil society.

At present extremist organizations are not limited to using traditional channels of social interaction. They actively use the latest digital and information technologies for recruiting young people into the ranks of radicals. The statistics confirms that the third of all the terrorist and extremist crimes are committed with the use of the Internet. Digital technologies, being introduced into social networks, will lead to even faster transferring of information to the increasing number of users. As a result, we can make a pessimistic forecast that criminal elements will try to use the latest achievements in this field for criminal purposes. Obviously, the state and the society must be ready to respond adequately to such challenges as the emergence of new forms of information crime and the "digital" distribution of radical ideas of various kinds.

The researchers of the question talk about the formation in the modern period of such a variety of extremist activities as information extremism. Representatives of this trend are trying to distribute extremist methods and violate mass communications. Recently we have heard a lot about hacker attacks directed against government structures in order to disorganize the life of large societies.

Communicative technologies and digital processes are actively used by information extremists for their illegal purposes, including provoking of ethnic and religious conflicts, destabilizing of the political and ideological situation in a number of countries. For quite a long time many authors tried to link the development of extremism with Islam - one of the world religions. There was a situation when many representatives of this confession preached views with an evident terrorist orientation. The supporters of traditional Islam, on the contrary, took the position of the "repentant", trying to dissociate themselves from those who tried to make the provisions of the Koran justify extremism. The recent events have shown that terrorism has no religious preferences. It can choose Christian temples, mosques or synagogues as its target. So, it is necessary to unite the efforts of all confessions to counteract manifestations of extremism.

On April 24, 2019, Samara State University of Economics held a very significant event - the All-Russian Scientific and Practical Round Table "Counteracting to Extremism among Young People: Theoretical, Legal and Organizational Aspects". The purpose of the meeting was to discuss the issues related to the prevention of extremism among the young. The discussion turned out to be vivid and informal. The awareness of the state and society of the importance of this issue was obvious for the participants of the dialogue. Law enforcement officers, the scientific community representatives, the representatives of various religious confessions and the leaders of public organizations spoke about the need to join efforts to counteract this evil. The students of the university also participated in the conversation. The talk did not have a declarative character and was oriented towards practice. It was concluded by the formulation of specific practical proposals and guidelines.

During the discussion, it was said that the most optimal struggle with extremism among the young is that with the involvement of youth organizations. For this purpose, the Internet should be used where relevant information will be placed. It is necessary to talk with young users in a language they understand suggesting other values instead of extremist ideology, making the desire for active, spiritually filled life more appealing than suicidal moods. Representatives of the Cossack formations, voluntary law enforcement groups spoke about the creation of cyber-brigades, which, on a voluntary basis, will identify sites with an extremist orientation, timely reporting this material to law enforcement agencies and blocking such sites.

The participants of the round table told us about the "Cool Lawyer" program, during the implementation of which the best human rights defenders of the region came to schoolchildren to tell them about the masks of extremism, about how it tries to manipulate teenagers whose personalities are not forms yet.

Such events are held constantly in Samara State University of Economics, the university platform makes it possible to integrate theoretical approaches with the definition of the vectors of their practical implementation.

It is important that the round table was held at the university of economic profile because it is obvious that terrorism as a deeply negative and harmful background phenomenon impedes economic growth, the development of the information type of society and its digitalization. All the participants of the discussion stated that extremism and terrorism have become the main challenges to the man and society putting the existence of civilization at risk. Extremism carries a threat to the economic security of states. As a result of extremist activity, citizens suffer huge moral and material damage. The economic damage from both violent and non-violent forms of extremism cannot be calculated. The annual damage to the global economy from terrorism alone is estimated as more than $ 53 billion. And it is impossible to cover human losses from manifestations of extremism.

So, at the present stage, there is a variety of extremist activities, and in recent years there has been a high level of development of economic and information extremism. Actions of the adherents of this destructive activity are focused on eliminating the diversity of ownership forms; they propagandize the rejection of the principles of state regulation of the economic sphere, the transition to anarchic and uncontrolled economic system and the rejection of competition in the economic sphere. Extremism threatens the very existence of humanity; it is no coincidence that it is orientated towards young people trying to leave civilization without future. All the above-mentioned makes it possible to affirm that economy, digital information technologies and extremism, can be viewed in the context of a single topic.

2 Methods

The research was carried out with the use of general scientific methods of analysis, synthesis, deduction, induction, generalization, which were supplemented by the use of a number of private scientific methods, including comparative legal method, system-structural method and formal-logical method. Comparative legal approach was used to identify the specific character of counteracting to extremism as no uniform universal approaches were worked out by the researchers of this problem. System-structural method has become the basis for generalizing data on counteracting to extremism among the young with the use of modern digital technologies, which is a principle difference of the present period of knowledge development in this field. Formal-logical approach was applied in order to find a causal link between the unfair use of the Internet space for the dissemination of extremist ideology in the youth segment and the identification of possibilities to counteract to it with the use of the newest information-digital technologies.

3 Results

The use of IT-technologies contributes to the effective information support for counteracting to extremism among young people. When preparing measures to prevent extremism among the young the realities should be taken into consideration, namely, the progress in the digital information sphere, the growing importance of the Internet space to young people. Information counteracting to extremism in the Internet is carried out in such areas as restricting access to extremist materials, educational and information activities, propaganda support for anti-terrorist activities, using volunteers for timely identifying extremist websites, bringing information to law enforcement agencies, attracting representatives of various religious confessions and public formations to performing the above mentioned activities. At the same time, this work should be comprehensive and systemic in nature, based on the idea of active interaction of public authorities and civil society institutions.

4 Discussion

A very important area of application of IT-technologies is information and digital support for extremism counteracting. The strategy of counteracting to extremism in the Russian Federation until 2025 among the main objectives of state policy in the field of counteracting to extremism mentions the organization of information support for the activities of authorities at all levels, institutions of civil society and organizations counteracting to extremism, in the media, information and telecommunication networks, including the Internet, as well as the implementation of effective measures of informational counteracting to the distribution of the ideology of extremism.

It is obvious that preventive activities are of special importance. Within their framework the events should be held that show young people and society in general how the Internet is used for distributing of extremist ideology. A special role is played by media education intended to prepare young people for life in new information conditions, and for right perception of various information; to teach a young person to interpret the information adequately, to be aware of the consequences of its impact on mind, to master ways of communicating on the basis of non-verbal forms of communication with the use of technical means. The main thing is to form the basic skills of filtering the information they receive in the Internet.

A significant factor in combating extremism among the young is the increase of the level of digital literacy, information culture, including that of the students of educational institutions. We mean the ability of young people to express their information needs, not only to seek quantitative perception of material, but to find and evaluate the quality of information, to use it safely and ethically for socially useful purposes.

We think that in order to counteract to extremism among young people effectively, modern society should learn to use the latest methods of counter-propaganda in cyberspace with the use of digital technologies and not only respond to extremist actions, but also act in advance (The Federal List of Extremist Materials 2018). There is no doubt that it is impossible for one, even a very developed state, to fight extremism and its extreme manifestation – terrorism alone. Obviously, the efforts of the entire world community should be integrated.

One of the main tasks of the state youth policy is the creation of conditions for the development and realization of the intellectual potential of modern young people, including their information resource. An important role should be played by not formal but systematic propaganda of legal knowledge on youth sites and in social networks [10].

The use of new information and digital opportunities can contribute to the patriotic education of the younger generation, because only tolerant, encyclopedically educated young people with a high level of legal culture are ready to solve complex economic and political problems of the state level, to resist various manifestations of extremism, to create necessary social innovations for life and to build a decent society.

5 Conclusion

1. Activities aimed at counteracting to extremism among the young should be based on the active use of information-digital technologies, taking into account the growing importance of the Internet in the youth segment. 2. Information counteracting to extremism in the Internet should go not only be carries out in the direction of restricting access to such materials but also within the framework of cyber education and propaganda of antiterrorist actions. This kind of activities should be systemic in nature and based on the active interaction of state authorities with civil society institutions. 3. Media education progress should be widely used. Within the framework of media education young people should gain skills of adequate perception and interpretation of the information received with its proper filtering. 4. There is a positive experience of interregional youth public movement “Cyber-brigade”, the goals of which are to detect malicious content in the Internet and carry out events to improve computer literacy among young people. This volunteer movement should be actively developed at universities, working in close contact with law enforcement agencies, public structures and religious scholars.

References

1. Burayeva, L.A.: Radicalism and online recruiting on the internet. Soc. Polit. Sci. **4**, 149–151 (2017). (in Russian)
2. Bykadorova, A.S., Churilov, S.A., Shapovalova, E.V.: The extremization of youngsters in the internet. Kazan Pedag. J. **3**, 181–184 (2016). (in Russian)
3. Cossack Cyber-Brigades are Coming to Regions. The Official Site of Moscow State University of Technologies and Management named after K.G. Razumovskiy (PKU). http://www.mgutm.ru/content/news/13559/. Accessed 8 Oct 2018. (in Russian)
4. Dibirov, N.Z., Safaraliev, G.K.: Modern Political Extremism: Concept, Origin, Reasons, Ideology, Problems, Organizations, Practice, Prevention and Counteraction. Dagestan Institute of Economics and Politics, Makhachkala (2009). (in Russian)
5. Extremism Notifications. The Official Site of the General Prosecutor Office of the Russian Federation. https://genproc.gov.ru/contacts/extremism/send/. Accessed 13 Oct 2018. (in Russian)
6. Frequently Asked Questions. The Official Site of the Federal Service on Supervision in the Sphere of Communications, Information Technologies and Mass Communications. https://rkn.gov.ru/treatments/p459/p750/. Accessed 13 Nov 2018. (in Russian)
7. Garkin, I.N., Medvedeva, L.M.: Methods of prevention of extremism and terrorism ideology penetration in university educational environment. Youth Sci. Rev. **6**, 81–85 (2017). (in Russian)
8. Gladyshev-Lyadov, V.V.: Social networks as the tool of extremism propaganda. Rev. NCPTI **2**, 28–31 (2013). (The National Centre of Information Counteracting to Terrorism and Extremism in Educational Environment and the Internet). (in Russian)
9. Kruzhkova, O.V., Vorobyova, I.V., Nikiforova, D.M.: Psychological aspects of involvement of young people in extremist groups in the Internet environment. Educ. Sci. J. **10**(139), 66–90 (2016). (The National Centre of Information Counteracting to Terrorism and Extremism in Educational Environment and the Internet). (in Russian)
10. Kulyagin, I.V.: Prevention of Terrorism Ideology Distribution in Educational Sphere and Youth Environment. Counteracting to Extremism and Terrorism. Urait, Moscow (2018). (in Russian)
11. Lashin, R.L., Churilov, S.A.: Counteracting to extremism and terrorism in the Internet and educational environment. Rev. NCPTI **7**, 34–39 (2015). (in Russian)
12. Lokota, O.V., Vorontsov, S.A., Blaginin, A.M., et al. (eds.): Counteracting to extremism and terrorism ideology among young people. In: Materials of Interregional Scientific and Practical Conference with International Participation. South-Russian Institute of Management, Russian Academy of Ntional Economy and State Service under the President of the Russian Federation, Rostov on Don (2017). (in Russian)
13. Manoilo, A.V., Shegaev, I.S.: Terrorist recruiting in the post-soviet space: contemporary trends and risks for Russia. Bull. Moscow State Reg. Univ. (Electron. J.) **2**, 129–142 (2018). (in Russian)
14. Nekrasova, E.V.: The information aspect of extremism and terrorism and the destructive trends in the mass media. Bull. RUDN (Peoples' Friendsh. Univ. Russia) Ser.: Sociol. **1**, 57–66 (2013). (in Russian)
15. Panfilova, Y.S.: Extremism in virtual environment as a social problem: reflection in the minds of the young. Anthol. Mod. Sci. Educ. **9**, 101–104 (2014). (in Russian)

Cryptocurrency – Money of the Digital Economy

G. S. Panova(✉)

Moscow State Institute of International Relations (University) of the Ministry of Foreign Affairs, Moscow, Russia
gpanova@mail.ru

Abstract. The study considers modern aspects of the digital economic development based on cryptocurrencies. The above problems are analyzed using qualitative and quantitative methods for assessing the use of cryptocurrencies in the national economy and in international relations. The study covers the period of 2013–2019. The author assesses the influence of positive and negative factors of international economic and financial cooperation. An analysis of the practice of applying digital currencies allowed justifying the need for urgent innovative solutions in the field of macro-prudential regulation of the cryptocurrency turnover. The gradual transition to a new model of economic growth implies a more intensive use of new digital technologies. The study presents debatable issues of the possibility and feasibility of using collective cryptocurrencies and illustrates the author's position on the prospects and risks of using blockchain technologies and global digital currencies, which is a key issue in the international cooperation of countries. In the context of the global information space, the first steps to achieve a balance of currency relations and trade exchange are to develop national monetary systems based on national digital currencies and their gradual integration into the global currency system in the future.

Keywords: Digital technology · Global currency · Global coin · Monetary circulation · National cryptocurrency

1 Introduction

The modern world is in the process of digital transformations in all spheres of public life. Analyzing the genesis of the industrial revolution in the world, we can conclude that the main characteristics of the latter include an increasing number of technical innovations, including neural networks, artificial intelligence (AI) and machine learning (ML), Internet of Things (IoT), 3D printing, robotics, large data volumes (Big Data) and others. Centralized IT platforms began to be used as digital intermediaries. And the use of electronic money, including cryptocurrency has become an important feature of the modern monetary economy. However, the essence of cryptocurrency is discussed throughout the world. Cryptocurrencies have become a new phenomenon, both in the field of computer science and cryptography, and in the field of economics. The emergence of cryptocurrency caused a real stir in society. While adherents of digital currencies declare the goal of renouncing fiat money due to their replacement with

S. I. Ashmarina et al. (Eds.): ISCDTE 2019, LNNS 84, pp. 604–612, 2020.
https://doi.org/10.1007/978-3-030-27015-5_72

cryptocurrency, many supporters of a conservative approach to finance say that the cryptocurrency market is about to collapse. In recent years, cryptocurrencies have gained great popularity among investors, and also attract increased interest from governments and international organizations. Competition in the market leads to the emergence of increasingly convenient and secure cryptocurrencies, while the state is considering the possibility of creating cryptocurrencies emitted by central banks.

2 Methodology

The current situation in the field of the digital economic development based on the use of cryptocurrencies in the world and in Russia is studied. The study covers the period of 2013–2019. The author analyzed the materials of the Expert Council of the Bank of Russia on operations in national currency; results of research projects 2017–2019 of The Eurasian Economic Commission (EEC), the Eurasian Development Bank (EDB), the Interstate Bank, the Chamber of Commerce and Industry of the Russian Federation and the Association of Russian Banks (ARB); statistical materials of higher educational institutions and the results of their own research on the digital economic development. Research methods include: a comparison method, a benchmarking analysis of market practices (benchmarking to market practices), recommendations from international auditing and consulting companies.

3 Results

3.1 Theoretical Approach and Practice of Using Cryptocurrency

The term "cryptocurrency" was introduced into scientific circulation after the publication of Andy Greenberg's article "Crypto Currency" in the Forbes magazine on April 20, 2011. Its appearance is connected with the advent of the Bitcoin project developed by Satoshi Nakamoto. The Bitcoin project is a decentralized system, the operation of which is based on cryptographic methods and blockchain technology, developed to create a means of payment beyond the control of state institutions, which is an alternative to existing fiat currencies and their electronic options [7]. Subsequently, all cryptocurrencies that appeared after Bitcoin became known as "altcoins" (altcoin - short for Eng. alternative coin).

From an economic point of view, cryptocurrency is a special form of exchange of intangible values without intermediaries, not having a physical embodiment, where information about completed transactions exists within the network used. This form of exchange is supported not by the state or currency, but by consensus of network participants implemented within the network. Cryptocurrencies can be quoted on the stock exchange and are used for investment, speculative purposes or for purchasing goods or services.

Cryptocurrencies are a kind of so-called private money, the existence of which was predicted by von Hayek in 1986 [9], and electronic money. The reason for the emergence of private money is the state's monopoly on the issue of money, the very existence of which adversely affects the economic development of society. The stability of such a system should be the product of market competition between existing private currencies. The presence of private currencies could also solve the problem of concealing the real level of inflation by the state, since within the framework of market competition the consumer will be focused on the most open private monetary system.

According to Friedrich A. von Hayek, money can exist (and existed) without the participation and control of the state. But after the appearance of fiat currencies, the shortcomings of state emission became significant. Hayek argued that Gresham's law, according to which "the worst money ousts the best money from circulation," is valid only for the situation when the law establishes fixed exchange rates for currencies. In the case of floating exchange rates, society continues the selection process until there is a preference for the best type of money issued by various economic agents. Economists have long wondered how to make payments safe, confidential and independent. As a result, they began to use electronic money that can circulate outside the banking system, as well as participate in exchanges in state and bank payment systems.

Many international organizations and states participate in discussions regarding the essence of electronic money [5]. In Russia, the Federal Law "On the National Payment System" contains the following definition of electronic money: "This is money that was previously provided by one person (the person who provided the money) to another person who takes into account information about the amount of money provided without opening a bank account, for the fulfillment of monetary obligations of the person who provided the money to third parties and in respect of which the person who provided the money has the right to make orders exclusively using electronic means of payment. At the same time, monetary funds received by organizations engaged in professional activities in the securities market, clearing activities and (or) management of investment funds, mutual investment funds and non-state pension funds and accounting information on the amount of provided funds without opening a bank account in accordance with the legislation controlling the activities of these organizations are not electronic cash" [4]. In the EU, electronic payments are regulated by the European Parliament Directive [2]. The Bank for International Settlements has been researching the phenomenon of electronic money since 1995 and annually publishes materials on this topic.

The key difference between electronic cash (and then cryptocurrency, as its varieties) from the generally accepted forms of electronic money is the absence of the intermediary between the seller and the buyer. Obviously, the emergence and development of cryptocurrency is associated with the creation and expansion of the Internet, as the task of creating a comprehensive and public network, within which such a form of exchange of digital assets – cryptocurrency could exist, required large database management capabilities and large bandwidth so that transactions are carried out as soon as possible.

Another prerequisite that distinguishes cryptocurrencies from well-known forms of electronic money is the use of cryptographic principles in organizing such a form of exchange, since their use allows anonymity to be achieved when conducting transactions. For the first time, the use of cryptographic protocols for this type of system was proposed by the American computer science and cryptography researcher David Chom in 1983 in the annual publication Advances in Cryptology. He introduced "electronic cash" (ecash, digitalcash) into scientific use [8].

Bitcoin project became the first cryptocurrency. However, along with the growth of its popularity and an increase in the market price and the number of network participants, significant shortcomings were identified. Among them: insufficient transaction speed, high power consumption when generating the block, insufficient anonymity of participants in the transaction, the impossibility of creating Bitcoin using the script language Bitcoin with conditional Turing or conditional transactions or smart contracts.

The market price of Bitcoin has increased, but the potential for further growth and utility as cryptocurrencies have been disputed. Many users, businesses and investors preferred altcoins, and the share of Bitcoin in the cryptocurrency market in the first half of 2017 decreased from 88% to 38%. The number of users of cryptocurrency is steadily increasing. The estimated number of unique active users of cryptocurrency wallets has increased more than 5 times since 2013, and currently their number is close to 6 million.

In the process of forming a crypto market, projects and companies have appeared that provide products and services which simplify the use of cryptocurrencies for main users and create an infrastructure for applications running on top of public blockchains. The existence of these services adds significant value to cryptocurrencies, since these services expand the potential range of consumers.

The above data testifies to the high popularity of the cryptocurrency market and the readiness of traders to quickly change it in order to profit from short-term changes in exchange rates. At the same time, the cryptocurrency market is a highly volatile market where many individual traders and few large institutional players operate. However, with the emergence of such institutional entities, the cryptocurrency market may undergo significant changes due to the emergence of new hedging mechanisms, the emergence of high-speed bots that make transactions in microseconds using the developed trading algorithms, and an increase in the impact of financial funds. Goldman Sachs, one of the world's largest financial and investment companies, is already helping its customers buy and sell Bitcoin futures contracts [11].

In the future, the number of derivative financial instruments in the cryptocurrency market will increase, which will attract large funds and will require more precise rules for its regulation.

Investing in cryptocurrencies is highly risky due to their extremely high volatility, with total capitalization growing at a significant pace. Therefore, the cryptocurrency market is interesting from the point of view of its analysis within the framework of the theory of the efficient and inefficient market, where its capitalization growth is associated with the support of millions of people, despite the skepticism of many supporters of traditional finance, who talk about the lack of providing a large number of cryptocurrencies by some real assets, except for electricity and equipment spent for its production.

The cryptocurrency market has features that could help market actors to act rationally and make it efficient: information becomes available to all market actors simultaneously; there are no transaction costs and other factors that prevent transactions; transactions made by participants in settlements may not affect the overall price level.

Thus, theoretically, cryptocurrencies can turn into a widely used global currency or means of accumulation. But at present, digital currency is too volatile to represent a meaningful, practical and reliable exchange rate, with the possible exception of certain high-tech niche markets, where anonymity is more important than price volatility. At the same time, while there is an uptrend in the value of the currency, its owners will not want to spend it. Thus, cryptocurrencies are one example of the inefficient market. The evolution of cryptocurrency also involves the development of the adequate regulatory framework before they can take a worthy place in the money turnover.

3.2 Prospects for the Creation of National Digital Currencies

In recent years, central banks have begun conducting experiments with distributed registry technology. The topic of cryptocurrency has also attracted the attention of governments and international organizations. The Bank for International Settlements identifies three cryptocurrency features that are important when analyzing the prospects for the existence of state cryptocurrencies: they have an electronic form, are not obligations, and support the peer-to-peer exchange function (without intermediaries). Some of these features are common to other types of money [1]. For example, cash is an analogue of the peering exchange of obligations of the central bank not only in electronic form; deposits in commercial banks are liable for them, they are in electronic form and are exchanged in a centralized way, either within one bank (transferred from one balance sheet item to another), or between different banks through the central bank; most commodity forms of money (gold coins, for example) can also be part of a peer-to-peer exchange, but they are not obligations and do not have electronic form.

State cryptocurrencies are inherently electronic liabilities of the central bank, which can be used in the exchange process between equal counterparties. But it ignores an important feature of other forms of central bank money, namely cash. Currently, cash is available to all, while opening accounts in the central bank is usually available only to a limited number of organizations, mainly banks.

State cryptocurrency projects:

- Fedcoin - a concept that was proposed by J. Koning in 2014 and was not approved by the Federal Reserve System. The Central Bank created its own cryptocurrency. It was assumed that the currency could be converted in both directions at par with the US dollar, and the conversion would be managed by Federal Reserve Banks. Fedcoin's offer had to, like cash, increase or decrease depending on the willingness of consumers to keep it. Fedcoin would be the third component of the monetary base, along with cash and reserves. Unlike Bitcoin, Fedcoin would not be competing "private money," it would become an alternative form of sovereign currency.
- CADcoin is an example of a publicly accessible state-owned cryptocurrency, which is used as an experiment for the implementation of the universal payment system based on distributed registry technology. CADcoin was used in simulations performed by the Bank of Canada in collaboration with Payments Canada, R3 and several Canadian banks, but was not applied in practice.
- eKrona. In Sweden, the demand for cash has dropped significantly over the past decade. Currently, many stores do not accept cash, and some branches of banks no

longer issue or collect cash. In response, Riksbank has launched a project for testing eKrona for retail payments. The decision on technology has not been made yet.
- Dinero electronico is a mobile payment service in Ecuador, where the Central Bank provides basic public accounts. Citizens can open an account by downloading the application, registering their national identification number and conducting verification. People enroll or spend money in special places. Thus, this is a rare example of a deposited foreign currency account at the Central Bank. Since Ecuador uses the US dollar as the official currency, accounts are denominated in that currency.
- El Petro. The El Petro cryptocurrency was officially issued in Venezuela in February 2018. The country's oil and gas supply is provided by El Petro. The initiative to use cryptocurrency arose after the release of 100 thousand Bolivars, the largest banknote in the history of the country, which was to stabilize money circulation.
- Crypto-ruble is being developed as a digital analogue of the ruble. It is proposed to ensure secure transactions with it through a closed blockchain network. In this case, the national cryptocurrency will not be mined. National digital money is proposed to be used to attract foreign investment in cryptocurrency. And it will be the only legal scheme for withdrawing funds from cryptocurrency into rubles - crypto exchangers and cryptobirds will be outlawed.

UAE and Saudi Arabian Central Banks have launched a pilot cross-border payment initiative.

The question of whether the central bank should provide a digital cash alternative is debatable. In countries such as Sweden, where the use of cash is rapidly declining, central banks have to decide whether issuing a state-owned cryptocurrency makes sense. When making this decision, central banks can take into account not only consumer preferences regarding confidentiality and possible efficiency gains - in terms of payments, clearing and settlements - but also the risks that this may entail for the financial system and the economy as a whole, as well as any consequences for monetary policy. Some of the risks are currently difficult to assess (for example, the cryptocurrency vulnerability to cyber attacks).

3.3 Regulation of Digital Currency Circulation

The monetary authorities in many countries have introduced cryptocurrency regulation and have expressed their position on this method of attracting investment. China and South Korea officially banned cryptocurrencies in their territories. But China was the first country to test the digital prototype of its national currency.

In contrast, the United States, Canada, Switzerland, Japan, Singapore and the United Arab Emirates have adopted a number of decisions that officially authorize the use of cryptocurrency.

In the United States, bitcoin investors will have to report profits to the United States Internal Revenue Service to pay tax in accordance with Notice 2014–21. In December 2017, UK and EU financial regulators planned to enact a law, according to which cryptocurrency traders and investors in some cases would have to disclose their private information, and cryptocurrency exchangers should provide authorities with access to user information. According to government agencies, these measures are necessary to prevent money laundering and terrorist financing. Reconciliation of various approaches to this issue should be completed soon. The law is expected to come into force by the end of 2019.

Operations with Bitcoin are prohibited in countries such as Bolivia, Ecuador, India, Bangladesh, Iceland, Kyrgyzstan, Morocco, Nepal, Malaysia, Indonesia and Taiwan. At the same time, China, where it is prohibited to trade on local exchange sites and Russia are preparing bills of regulation of cryptocurrencies.

In the EAEU countries, a regulatory framework is being developed for the use of cryptocurrencies [3] at the state level, together with business associations and financial market regulators.

However, the State Duma of the Russian Federation recently postponed consideration of the draft law "On digital financial assets", although President V.V. Putin instructed the Government to accept the necessary regulatory documents governing this area by July 1, 2019. At the same time, the Bank of Russia is considering the possibility of imposing restrictions on cash transactions for Russians. Restrictions may include all operations with blockchain tokens, real estate, securities, shares of companies, etc. totaling no more than 600 thousand rubles a year.

The Ministry of Communications and Mass Media of Russia has developed a draft law to regulate cryptocurrency registration, in which the regulation of cryptocurrency and the turnover of tokens are aimed at providing protection and liquidity for investors. The Bank of Russia is considering using Masterchain, a protocol derived from Ethereum, to send financial messages through SWIFT within the Eurasian Economic Union. Under the control of the Central Bank of the Russian Federation, the Russian Fintech organization has been developing the Masterchain distributed registry since 2016, having tested its first operational version in 2017. On July 28, 2018, the Government of the Russian Federation adopted the "Digital Economy of the Russian Federation" Program and determined the goals for the development of the innovative economy based on the concept "Industry 4.0".

The emergence of new cryptocurrencies may require the development of a new regulatory framework for regulating relations between different cryptocurrencies, as well as between cryptocurrencies and national currencies. At the same time, the development of the regulatory framework should not be redundant in order not to prevent cryptocurrencies from realizing their potential [6].

4 Discussion

This study solved only some problems. At the same time, the status of digital currencies, legislative problems of their use and distribution at the national level remain unresolved, and there are problems affecting the adoption of global currency exchange decisions, which are necessary today to avoid currency risks and improve money circulation in the world.

The first problem is the lack of a single theoretical approach to the terminology and essence of digital currencies. Solving this problem requires further discussion and consolidation of scientific and practical approaches to this issue. The second problem is related to the development and further implementation of the legal framework for the circulation of digital currencies. There are two options to solve this problem. One of them is to increase funding for state-owned national cryptocurrency development projects. And the other is the gradual introduction of new means of payment and investments on the basis of cryptocurrency in international relations (at the initial stage locally on the basis of bilateral and multilateral agreements between countries (for example, in the EU, the CIS and other country associations) and later on the basis of the global currency). Obviously, it is necessary to discuss possibilities and risks of the digital currency turnover. The idea of global currency is in the air. For several years, this issue has been discussed in order to overcome currency risks. For example, N.A. Nazarbayev, the ex-president of the Republic of Kazakhstan, proposed the creation of a global digital currency G-Global in 2017 [10]. This initiative can actually be seen as a challenge to transform the global financial architecture and a way to transform the system of international payments based on digital currency. The first steps in this direction have already been made. In June 2019, the Financial Times reported that Facebook had begun negotiations with the CFTC to create its own digital currency, GlobalCoin, and produce it in 2020.

5 Conclusion

An analysis of theoretical and practical approaches to the problem of digital currencies and prospects for the development of the monetary sphere on this basis, taking into account the results of sociological surveys conducted in 2017–2019, showed that most experts support the role of cryptocurrency in money transfers. However, there are serious technical, legislative and administrative barriers. Therefore, further areas of work on the implementation of cryptocurrencies suggest: (1) Intensify trade and financial flows of global digital currencies between countries; (2) Expand mutual trade, which would save the world from currency wars, speculation, avoid imbalances in trade relations, as well as reduce the volatility of financial markets; (3) Form transparent common markets and transform pricing mechanisms; (4) Integrate currency markets and liberate conditions for access to them. These measures will help to expand the scope of cryptocurrency in international relations and will strengthen the global trend towards the formation of the multipolar world. In this regard, special attention should be paid to the development of intellectual property rights, risk management of banks and other financial institutions.

Acknowledgements. The research was conducted by the authors with the financial support of the Russian Foundation for Basic Research, Project no. 19-011-20091.

References

1. Bech, M., Garratt, R.: Central bank crypto currencies. Bank for International Settlements. BIS Q. Rev. (2017). https://www.bis.org/publ/qtrpdf/r_qt1709f.pdf. Accessed 3 June 2019
2. Directive 2000/46/EC of the European Parliament and of the Council of 18 September 2000 on the taking up, pursuit of and prudential supervision of the business of electronic money institutions. http://eur-lex.europa.eu/LexUriServ/LexUriServ.do?uri=CELEX:32000L0046:EN:HTML. Accessed 3 June 2019
3. Eurasian Economic Union Concept of Cooperation of the States Members of the Eurasian Economic Union in the Monetary Area Law No. 220 (2005). www.evrazes.com/docs/view/66. Accessed 3 June 2019
4. Federal Law N 161 "On National Payment System" Adopted on 27 June 2011. https://base.garant.ru/77663625/. Accessed 3 June 2019
5. Lin, I.-C., Liao, T.C.: A survey of blockchain security issues and challenges. Int. J. Netw. Secur. **19**(5), 653–659 (2017). https://doi.org/10.6633/IJNS.201709.19(5).01
6. Mandeng, O.J.: Cryptocurrencies, monetary stability and regulation: Germany's nineteenth century private banks of issue. Institute of Global Affairs (2018). http://www.lse.ac.uk/iga/assets/documents/research-and-publications/LSE-IGA-WP-5-2018-Ousmene-Mandeng.pdf. Accessed 3 June 2019
7. Nakamoto, S.: Bitcoin: a peer-to-peer electronic cash system (2009). https://www.researchgate.net/publication/228640975_Bitcoin_A_Peer-to-Peer_Electronic_Cash_System/citations. Accessed 3 June 2019
8. Steve, B.: In Brief: InfoSpace Buys eCash Technologies. The American Banker (2002). https://www.americanbanker.com/news/in-brief-infospace-buys-ecash-technologies. Accessed 3 June 2019
9. Von Hayek, F.A.: Private money. Russian Translation: Institute of National Economy (1996). http://www.library.fa.ru/files/Hayek-Money.pdf. Accessed 3 June 2019. (in Russian)
10. Yeremina, N., Faljahov, R.: Digital virus struck the CIS. Gazeta.Ru (2017). https://www.gazeta.ru/business/2017/06/13/10719959.shtml. Accessed 19 June 2019. (in Russian)
11. Zhao, W.: Goldman sachs to begin bitcoin futures trading. Coindesk (2018). https://www.coindesk.com/goldman-sachs-to-begin-bitcoin-futures-trading-within-weeks/. Accessed 3 June 2019

Prospects for the Legal Regulation of Central Bank Digital Currency

E. L. Sidorenko(✉) and A. A. Lykov

Moscow State Institute of International Relations (University) of the Ministry of Foreign Affairs, Moscow, Russia
12011979@list.ru

Abstract. Digital currency has become one of the most talked about products of digital financial technology. Today, the concept of digital currency is not developed even in theory. Only a small number of states expressed a desire to explore this financial asset, which does not reduce the relevance of digital as a new step in the global development of money. For this reason, the study considers the theoretical understanding of the concept of digital currency to determine its possible place in the economic and legal system of the state.

Keywords: Central bank · Digital currency · Cryptocurrency · Digital finance

1 Introduction

The rapid developments of financial technologies and economic digitalization have led to alternative payment methods, and later to digital state assets as alternatives to fiat currency. The first attempt to implement this concept was electronic money and cryptocurrency (the capitalization of which exceeded 255 billion US dollars [23] and only then - central bank digital currency.

Currently, more than 15 financial regulators have raised the issue of introducing central bank digital currency, but most of them have to face a number of serious economic and legal problems that will inevitably accompany the development of the digital payment model at the national and international levels.

In particular, in the near future, we have to determine economic advantages and legal status of central bank digital currency and to develop an adequate regulatory environment for its development and minimize legal risks when using digital economic exchange tools.

Unfortunately, at present these issues are not adequately addressed. States develop their own central bank digital currency models, relying solely on their infrastructure capabilities and geopolitical interests, and deliberately ignore global challenges of creating a safe and uniform digital financial environment. But to implement these tasks, it is important to formulate scientifically based approaches to the definition of central bank digital currency with access to the predictive model.

S. I. Ashmarina et al. (Eds.): ISCDTE 2019, LNNS 84, pp. 613–621, 2020.
https://doi.org/10.1007/978-3-030-27015-5_73

2 Methodology

This study focuses on the prospects for legal regulation of central bank digital currency. It is based on statistical and analytical materials of international organizations and expert centers, reports of analytical agencies and public organizations, as well as scientific papers related to this issue. The research methods were dialectic, systemic, formal legal and comparative legal methods, as well as a whole complex of sociological (statistical, polling and expert assessments) and theoretical methods (abstraction, analysis and synthesis, idealization, analogy, formalization, modeling, hypothesis methods, axiomatic method, etc.).

3 Results

Currently, the issue of the legal status of central bank digital currency is not fully defined. For example, the experts of the International Monetary Fund define central bank digital currency as a new form of money issued in digital form by the central bank and intended to fulfill the role of legal tender [19].

Bech and Garrat describe central bank digital currency as tan electronic form of central bank money that can be transferred decentralized, that is, directly between users without the participation of a centralized intermediary [5].

According to the experts of the Bank for International Settlements, central bank digital currency is central bank's liabilities calculated in the existing measure of value and serving as both a means of exchange and value saving [3].

The experts of the Bank of Russia expressed a similar definition of central bank digital currency as "central bank-denominated liabilities that have a digital representation and are capable of acting as a means of payment, measure and value preservation" [17].

Taking into account the existing positions of other authors [6, 7], central bank digital currency can, with some degree of conditionality, be defined as numerical obligations of the central bank (financial regulator of the country), expressed in the existing measure of value, which, as a rule, is the relevant national currency, and serving as a medium of exchange and value saving based on distributed registry technology (blockchain).

For a better understanding of the nature of central bank digital currency, it is important to determine its main characteristics. Among the latter are the following:

(1) The intangible form of central bank digital currency. Unlike cash and documentary securities, it exists in digital form (that is, it is the result of the work of the relevant program), and its use is determined by constant or periodic access to the Internet [25];
(2) The centralized way of creating and initial placement of central bank digital currency implies that it is created by the central bank (state financial regulator) and is transferred to credit institutions or directly to private individuals. This distinguishes it from private (decentralized) cryptocurrencies (Bitcoin, Monero,

Eferum) and at the same time from the so-called commodity money (raw materials), having any number of manufacturers;

(3) The price stability of central bank digital currency. Its value must be established and provided by the state, in contrast to virtual currencies (WOW-gold, Linden dollars), electronic money (M-Pesa, YandexMoney), cryptocurrencies (including steylcocoins) and securities, the value of which is either determined by individuals who created them or spontaneous market processes [9, 14].

However, the listed characteristics are not enough to fully understand the nature of central bank digital currency. Currently, there are 15 central bank digital currency projects in the world [19]. Let's call the main ones.

CAD-coin is an integral part of the Jasper project, conducted jointly by R3, six Canadian banks, and the Bank of Canada in order to ascertain the possibilities of the distributed registry in the field of clearing and large payments. It is assumed that CAD-coin will be used only by banks for settlements and repayment of obligations between each other [10].

The other currency, El Petro, is the only central bank digital currency created and released into free circulation. It was created by Venezuelan authorities in order to circumvent economic sanctions imposed on the US state and to attract foreign investment. At its core, El Petro is a bond to a certain amount of minerals mined in Venezuela [11].

E-Krona is considered by the Bank of Sweden as a substitute for cash, the amount of which is steadily decreasing in Sweden. According to Swedish experts, the complete disappearance of cash crowns can create citizens' dependence on bank payments and entail other risks. For this reason, E-Krona should become a new "digital" generation of cash.

The Central Bank of the Bahamas Islands published a statement in 2018 that it was planning a test launch of its central bank digital currency. The main objective of the digital Bahamian currency is to ensure equal access to the financial system for all citizens, which is lost due to the reduction in the number of bank branches in the state [1].

E-Shekel was considered by the Bank of Israel as a possible central bank digital currency. It was expected that E-Shekel would increase the efficiency of the state's payment system and help fight the shadow economy. However, shortly after launch, this project was closed [4]. In addition to the universal characteristics typical of all the currencies listed above, central bank digital currency can also have highly individual parameters [3].

There are the following characteristics:

1. Availability of the central bank digital currency payment system may be provided 24/7 or limited by working hours of the Central Bank or other controlling organization.

   In particular, Central Banks of China, Canada and Sweden are considering the possibility of round-the-clock and uninterrupted operation of the settlement system in central bank digital currency, even in the face of a power outage and the Internet.

In their opinion, uninterrupted operation will equalize digital and fiat currency in capabilities [19].

2. The marketing component of contributions may also differ significantly. There may be no interest on central bank digital currency at all or only on the amounts in central bank digital currency that are kept on the accounts of credit institutions. However, as the IMF study showed, all central banks that publicly examined the issue of introducing central bank digital currency rejected the idea of charging interest on savings in central bank digital currency [19].
3. Anonymity of using central bank digital currency. It depends on a number of factors. For example, who has access to the registry of central bank digital currency transactions: only central bank employees serving banks, or any other people. In particular, the People's Bank of China is considering the possibility of creating fully anonymous central bank digital currencies, the use of which, however, will be limited to a small amount available for transfer [19].
4. Transmission of central bank digital currency. It can be completely decentralized and carried out directly between the parties to the transaction (like cryptocurrency transactions). This decentralized method seems to be the most optimal for "cash" central bank digital currencies based on tokens, which will be transferred directly between the parties to the transaction (token-based CBDC). Alternatively, selective central bank digital currency transactions may be conducted under the control of the central bank (account-based CBDC).
5. Finally, the use of central bank digital currency can be carried out by any persons (both in the case of theoretical Fedcoin or E-Krona [22]), and in a limited number of subjects (for example, only credit institutions for the accelerated transfer of funds, as in the case of CAD-coin [10]).

The studies of most states indicate that the creation of such central bank digital currency, which is a constructive alternative to fiat currency, for example, the Swedish E-Krona [22], will be most likely. In fact, it is a digital form of fiat currency and can be used along with cash and non-cash money. This strategy eliminates the constant adjustment of exchange rates and the creation of a separate regulatory regime [20]. The introduction of such an alternative to fiat currency will require significant changes in law. For example, in Russia, the Central Bank is entitled to issue only cash, while it has no authority to issue central bank digital currency. Depending on the technical characteristics, there may be two not exclusive, but rather complementary strategies for the emission of central bank digital currency [22].

Within the framework of the first strategy, central bank digital currencies are similar in their properties to non-cash funds in the bank account. Such "non-cash" central bank digital currency can be stored in accounts at a bank (or other organization authorized by the Central Bank) and then, from a legal point of view, such central bank digital currency will be non-cash funds. Additionally, it should be noted here that if central bank digital currency is allowed to be deposited only by banks, the negative consequences of outflow of funds from "classical" (fiat) bank accounts will be largely neutralized [21].

As for "cash" central bank digital currencies, similar to cash, they are stored in the digital wallet of the owner for the convenience of their use and are an analogue of electronic money, settlements with which, like ordinary cash, can be carried out directly between the parties to transactions with entering the corresponding entries in the central bank digital currency registry [13].

Each model has its own pros and cons. The lack of "cash" central bank digital currency is that when it is transferred, a new entry is created in the registry of the corresponding central bank digital currency, which must be confirmed and fixed for some time and at the expense of a certain amount of energy, after which the increased copy of the registry will be distributed to all central bank digital currency nodes.

On the contrary, when transferring "non-cash" funds, the system simply changes balance sheets of the sender and the recipient, with the result that settlements with "non-cash" central bank digital currencies turn out to be almost instantaneous and gratuitous [8].

The reverse side of "non-cash" central bank digital currency is the need to identify each owner of the relevant central bank digital currency account and control these funds by authorities (for example, for confiscation or seizure in pursuance of a court decision).

4 Discussion

The idea of introducing central bank digital currency is one of the most ambitious and difficult problems of the modern financial system.

The specialists of the International Monetary Fund believe [19] that if central bank digital currencies come into circulation, they will face stiff competition from traditional banking services, electronic money from private companies and cryptocurrencies [21, 24], since they are not significantly different from their competitors.

However, from the point of view of central banks, the introduction of central bank digital currency also has certain social benefits.

First of all, the creation of central bank digital currency will increase the efficiency of the existing payment system due to the fact that payments in central bank digital currency are cheaper, safer and faster than non-cash funds and cash.

In particular, maintaining cash flow creates a significant burden for the economy. Thus, the Visa study showed that India spends 1.7% of its GDP on the production and servicing of cash [2]. Therefore, the experts of the International Monetary Fund believe that replacing cash will have a beneficial effect on the economy [19]. In addition, the disappearance or strong reduction in cash will create a monopoly power for payment operators, thereby damaging consumer rights. Nevertheless, central bank digital currency is far from the easiest way to counter the monopolization of the national payment market. A much simpler alternative is the development of antitrust laws and systems of cheap and fast payments.

The possibility of exercising full control over the circulation of funds in the form of central bank digital currency allows law enforcement agencies to more effectively monitor the compliance of citizens with the law and counteract the shadow economy. In particular, such control will be useful in fighting against corruption and other violations of the law (for example, in the sphere of paying taxes or enforcing court decisions).

Finally, with the complete disappearance of cash, there is a risk of destabilization of the economy and prices. The introduction of central bank digital currency, together with appropriate monetary policy, will help restore lost stability [8]. However, the introduction of central bank digital currency may be associated with a number of risks.

Firstly, due to the novelty and practical lack of applied technologies, the cost of implementing and maintaining the central bank digital currency system today cannot be established at all and can be unpredictably high [17, 19, 24].

Secondly, there is a significant risk of losing large amounts of "cash" central bank digital currency, because, unlike cash, unlimited large amounts of central bank digital currency can be in "cold" storage and will be irretrievably lost if the corresponding storage is damaged or stolen [24].

Thirdly, the use of central bank digital currency may become impossible during the time when the Internet is disconnected: during natural disasters, man-made disasters, and government locks during public unrest [24]. However, it is fair to note that cash also does not have absolute protection against loss in extreme conditions, and legal settlements with them may be difficult, for example, when cash registers stop working.

Fourth, the imperfection of the central bank digital currency technology (distributed registry or its equivalent) will inevitably cause errors in the operation of the new payment system, which, in addition, will be vulnerable to hacker attacks [17].

Further, anonymity of transactions on the central bank digital currency blockchain can be overcome by private individuals, with the result that these individuals will receive all the information about financial transactions of citizens [16]. Disclosure of expenses and personal income can lead to various negative consequences. In particular, commercial profiling can be a legal (but no less harmful) method of using private financial information for various purposes: from tertiary advertising to changing credit conditions for individuals (for example, due to their frequent purchase of alcohol). In addition, information on expenditures and income of a particular person may cause extortion or other criminal acts against him [12, 15, 18, 19]. Finally, the disclosure of this information will violate the fundamental human right to privacy (Article 12 of the Universal Declaration of Human Rights).

Finally, there is the possibility of losing state control over the movement of central bank digital currency across the state border, as a result of which capital withdrawal, offshore banking and other circumstances harmful to the state's economy will be possible [3]. In this regard, it is characteristic that among all the central banks that publicly studied the introduction of central bank digital currency, only two central banks (Canada - in connection with the tourism industry, and China - in connection with cross-border capital management) mentioned the topic of cross-border payments in central bank digital currency. And in both cases, this issue was considered as a problem, not an advantage [19].

Thus, the introduction of central bank digital currency both from the point of view of consumers and from the position of central banks is a controversial and far from ideal solution [19]. However, central bank digital currencies can get their niche in developed countries, gradually taking the place of aging cash, and in developing countries, where the banking sector and other methods of reliable value preservation are not developed.

5 Conclusion

Summarizing the study, we can conclude that central bank digital currency can be considered the next step in the development of money. Nevertheless, central bank digital currency is not the best alternative to other forms of money, which will constitute a significant competition for central bank digital currency if it is in a certain state. Both the prospects and risks of central bank digital currency are poorly understood today, primarily due to the fact that the advantages and disadvantages of specific central bank digital currency will be determined by its technological, legal and economic characteristics. But now it is already possible to identify a number of promising areas for the development of the regulatory environment for the introduction of central bank digital currency. In particular, it is important to determine its status at the international level, delineate the boundaries of authority of the issuer and users, and create reliable guarantees for their protection by optimizing the existing financial control system under the nascent institution. Finally, the issue of transboundary use of central bank digital currency is the least studied and contains opportunities and threats that researchers will have to identify in the future.

Acknowledgements. The research was conducted by the authors with the financial support of the Russian Foundation for Basic Research, Project no. 19-011-20091.

References

1. Adderley, R.: Bahamas information services. Press Release: Digital currency to be introduced, says DPM (2018). https://www.bahamas.gov.bs/wps/portal/public/gov/government/news/digital%20currency%20to%20be%20introduced%2C%20says%20dpm. Accessed 3 June 2019
2. Bakshi, I.: Cash deals cost 1.7% of GDP: Visa. Business Standard (2016). https://www.business-standard.com/article/economy-policy/cash-deals-cost-1-7-of-gdp-visa-116100500617_1.html. Accessed 3 June 2019
3. Coeuré, B., Loh, J.: Central Bank Digital Currencies. Bank for International Settlements (2018). https://www.bis.org/cpmi/publ/d174.pdf. Accessed 19 June 2019
4. Bank of Israel Report of the team to examine the issue of Central Bank Digital Currencies (2019). https://www.boi.org.il/en/NewsAndPublications/PressReleases/Documents/Digital%20currency.pdf. Accessed 3 June 2019
5. Bech, M., Garratt, R.: Central bank crypto currencies. Bank for International Settlements. BIS Q. Rev. (2017). https://www.bis.org/publ/qtrpdf/r_qt1709f.pdf. Accessed 3 June 2019
6. Berentsen, A., Schar, F.: The case for central bank electronic money and the non-case for central bank cryptocurrencies. Rev. Fed. Reserve Bank St. Louis **100**(2), 97–106 (2018). https://doi.org/10.20955/r.2018.97-106
7. Bjerg, O.: Designing new money - the policy trilemma of central bank digital currency. CBS Working Paper, June 2017. https://ssrn.com/abstract=2985381or. http://dx.doi.org/10.2139/ssrn.2985381. Accessed 19 June 2019
8. Bordo, M., Levin, A.: Central Bank Digital Currency and the Future of Monetary Policy (2017). https://www.hoover.org/sites/default/files/bordo-levin_bullets_for_hoover_may2017.pdf. Accessed 3 June 2019

9. Caginalp, C., Caginalp, G.: Opinion: valuation, liquidity price, and stability of cryptocurrencies. Proc. Natl. Acad. Sci. U.S.A. **115**(6), 1131–1134 (2018). https://doi.org/10.1073/pnas.1722031115
10. Garratt, R.: CAD-coin versus Fedcoin (2016). https://static1.squarespace.com/static/55f73743e4b051cfcc0b02cf/t/5908cf271b10e3feb2b593d2/1493749544157/cad-coin-final.pdf. Accessed 3 June 2019
11. de Venezuela, G.B.: Petro, hacia la revolución digital económica (n.d.). https://www.petro.gob.ve/files/petro-whitepaper.pdf. Accessed 3 June 2019
12. Golubitsky, S.: Bolivar cannot bear bitcoin: how Venezuela fell in love with cryptocurrencies and ignored Petro. New Newspaper (2019). https://www.novayagazeta.ru/articles/2019/03/25/79988-bolivar-ne-vyneset-bitkoyn. Accessed 3 June 2019. (in Russian)
13. Hossein, N.: Central bank digital currencies: preliminary legal observations. J. Bank. Regul. (2019). https://ssrn.com/abstract=3329993 or http://dx.doi.org/10.2139/ssrn.3329993. Accessed 19 June 2019
14. Iwamura, M., Kitamura, Y., Matsumoto, T., Saito, K.: Can we stabilize the price of a cryptocurrency? Understanding the design of bitcoin and its potential to compete with central bank money (2014). https://ssrn.com/abstract=2519367 or http://dx.doi.org/10.2139/ssrn.2519367. Accessed 19 June 2019
15. Kahn, C.M., McAndrews, J., Roberds, W.: Money is privacy. Federal Reserve Bank of Atlanta: Working Paper 2004–18 (2004). https://pdfs.semanticscholar.org/0878/9f519b6237124e04d219e1dd82de5ca8a962.pdf. Accessed 3 June 2019
16. Kashin, D.A.: Research on implementation experience of national cryptocurrency on example of El Petro in Venezuela. Master's thesis, Peter the Great St. Petersburg Polytechnic University, St. Petersburg (2015). http://elib.spbstu.ru/dl/2/v18-4981.pdf/info. Accessed 19 June 2019. (in Russian)
17. Kiselev, A.: Does the digital currency of central banks have a future? (2019). https://www.cbr.ru/Content/Document/File/71328/analytic_note_190418_dip.pdf. Accessed 3 June 2019. (in Russian)
18. Koning, J.P.: Why the fed is more likely to adopt bitcoin technology than kill it off (2013). http://jpkoning.blogspot.com/2013/04/why-fed-is-more-likely-to-adopt-bitcoin.html. Accessed 19 June 2019
19. Mancini-Griffoli, T., Peria, M.S.M., Agur, I., Ari, A., Kiff, J., Popescu, A., Rochon, C.: Casting light on central bank digital currency. International Monetary Fund (IMF) (2018). https://www.imf.org/~/media/Files/Publications/SDN/2018/SDN1808.ashx. Accessed 3 June 2019
20. O'Keeffe, D.: ECB shuts down estonian national cryptocurrency. Crypto Disrupt (2018). https://cryptodisrupt.com/ecb-shuts-down-estonian-national-cryptocurrency/. Accessed 3 June 2019
21. Pacces, A.M., Hossein, N.: The law and economics of shadow banking. In: Chiu, I.H., MacNeil, I. (eds.) Research Handbook on Shadow Banking: Legal and Regulatory Aspects, Edward Elgar (2017); European Corporate Governance Institute (ECGI) - Law Working Paper No. 339/2017 (2017). https://orbilu.uni.lu/bitstream/10993/29464/1/The%20law%20%26%20economics%20of%20shadow%20banking.pdf. Accessed 3 June 2019
22. Riksbank: The Riksbank's e-krona project, Report 2 (2018). https://www.riksbank.se/globalassets/media/rapporter/e-krona/2018/the-riksbanks-e-krona-project-report-2.pdf. Accessed 3 June 2019
23. Top 100 Cryptocurrencies by Market Capitalization (2019). https://coinmarketcap.com/. Accessed 3 June 2019

24. Wadsworth, A.: The pros and cons of issuing a central bank digital currency. Reserve Bank of New Zealand. Bulletin **81**(7), 3–21 (2018). https://www.rbnz.govt.nz/-/media/ReserveBank/Files/Publications/Bulletins/2018/2018jun81-07.pdf. Accessed 3 June 2019
25. Yakovlev, A.: Fedcoin: what is "state digital money"? Invest Forsyth (2019). https://www.if24.ru/fedcoin-chto-takoe. Accessed 3 June 2019. (in Russian)

Approaches Determining the Applicable Law Using Internet Technologies in the Digital Economy

K. K. Taran(✉)

Moscow State Institute of International Relations (University) of the Ministry of Foreign Affairs, Moscow, Russia
taran.kira@yndex.ru

Abstract. In the era of intensive development and growth of information technologies, new legal norms controlling legal relations when using new Internet technologies are formed. Moreover, digitization of economic relations accelerates the civilian turnover, and transactions are made much faster and more convenient for the parties. The forms of interaction between foreign counterparties are changing. The study considers current approaches that are used in Russia, the United States and the EU countries to determine the applicable law to legal relations when using the Internet. The author highlights restrictions on the choice of the applicable law in the states under consideration. On the basis of the studied approaches, their advantages and disadvantages, we can assume further trends in the norms of private international law.

Keywords: Conflict of laws · Legal relations on the Internet · International private law and the Internet · Internet technologies · User rights

1 Introduction

Relationships that develop on the Internet are often cross-border in nature. This is explained, firstly, by the fact that the parties to the contract may belong to different states, and, secondly, the server that serves the Internet resource may be located in the third state or in neutral space, and in this connection there will be difficulties in determining the nationality of the server (for example, if the server is located in space on a private satellite or on an abandoned ship in neutral waters). Also, taking into account the change in information transfer, it is difficult to determine its initial and final source. The implementation of the state's ability to subordinate websites to its rule of law, which can be accessed from the territory of this state, is not always effective, and can cause both legal uncertainty and lead to unfair legal consequences.

Digitization of the economic turnover and cross-border transactions need to revise and improve the rules of private international law. Classical conflict rules adapt to new legal relationships, but we need to know how effective they are. The study considers approaches of Russia, the United States and the EU countries, which have a number of similarities and differences related both to the development of the legal system of states and the introduction of Internet technologies.

S. I. Ashmarina et al. (Eds.): ISCDTE 2019, LNNS 84, pp. 622–629, 2020.
https://doi.org/10.1007/978-3-030-27015-5_74

Collision bindings that determine the applicable law to the contract concluded on the Internet, if such the right is not determined by the parties themselves, should help determine the rule of law that is most associated with relevant legal relations and suggests a suitable and most fair regulatory mechanism.

2 Methodology

In the course of the study, the author used the following general scientific methods: analysis - the study of key components of legal relations to determine the applicable law to legal relations when using Internet technologies; inductive and deductive methods; modeling methods and others. Special methods were also used: a comparative legal method for comparing approaches to the definition of the applicable law to legal relations when using Internet technologies in Russia, the United States and the EU, which made it possible to identify regulatory features of these states; logical and legal methods; methods of systematic interpretation of legal norms; synergistic method; systematization; and etc.

3 Results

The study will consider only the issues determining the applicable law to legal relations, despite the fact that the definition of judicial jurisdiction is very closely related to this issue.

Relations on the Internet can be divided into two groups: (1) Relations that are made through the Internet, i.e. which can be implemented without the Internet, for example, to conclude a contract of sale. The structure and characteristics of such relationships undergo changes (the type, time of receipt of the offer and acceptance, time of conclusion of the contract, and other issues), but the essence and purpose of legal relationship remains unchanged. An example would be legal relations in the field of e-commerce. (2) Relations that cannot be realized without the Internet. This group can include relations when using cloud storage of information, downloading music and other files from the Internet, and so on. Often legal relations on the Internet imply the transfer of information to the server/s, where it is not excluded that it is transferred (sorted) to another server.

Due to the fact that legal norms regulating relations, which can be implemented using the Internet, are only being formed, new legal relations are emerging, and the possibility of applying legal norms by analogy cannot be excluded.

The Civil Code of the Russian Federation and the EU Regulation Rome 1 allow the parties to independently choose the applicable law [1, 4]. It is assumed that this right will be applicable to the entire contract, and not to its specific part, including relations when using Internet technologies. In particular, this approach is peculiar to the United States, where legal relations can often be brought under the jurisdiction of a certain state and under the law of that state. In the US, attempts were made to summarize legal relations arising and implemented on the Internet under the principle of territoriality: the administration of the Attorney General of Minnesota in the mid-1990s was asked to subordinate disputes arising in cases of violations of criminal and civil law when people know that information will be distributed including through the state of Minnesota [6].

Russia and the EU use a similar approach: if all elements of legal relations are located in another country than the one whose right is chosen, the choice of the parties should not prejudice the provisions of the law of this other country, which are not allowed to retreat by agreement.

However, this free approach to the choice of the law has its own limitations: in Russia and the EU countries legal mechanisms are used to counter the circumvention of the law, requirements for compliance with the norms of direct application and non-contradiction to the rules of public law and order. In the United States, the chosen applicable law to contracts in which the consumer is a party must have a substantial connection with the contract. In Russia and the EU countries there are no requirements that the choice of the applicable law by the parties should be based on the substantial connection of the contract with the chosen law.

In all the legal procedures under consideration, the chosen applicable law cannot conflict with the rules of judicial jurisdiction.

In determining the applicable law in most countries, the status of the parties is taken into account. That is, if legal relations are formed between a merchant and a consumer, then legal mechanisms will be used that provide the greatest protection to the weaker consumer.

Article 6 of the EU Regulation Rome 1 [4] enshrines the application of the law of the country of the consumer's residence. Moreover, it is prohibited to reduce the level of consumer protection in the absence of the applicable law. This implies that, for example, in the case of a contract for downloading music or software, the relationship took place in the country of the consumer's residence, that is, the site offered to enter into an agreement, and the site should be active (i.e., to function in the territory given state and have feedback with the consumer).

Russia also has a special regulation of legal relations to which the consumer is a party, namely Art. 1212 of the Civil Code of the Russian Federation [1] guarantees compliance with peremptory norms of law aimed at protecting the consumer's rights. Russian law permits the choice of the applicable law to a contract to which the consumer is a party, however, if such a right is not chosen and if circumstances are established by the Civil Code of the Russian Federation, the law of the country of the consumer's residence is applied.

In the US, the situation is different: the official documents establish that the law chosen by the parties cannot be applied if its application contradicts the fundamental order of the state, which has substantially more interest than the right of the chosen state in relation to a particular issue; this right would also be applied if there is no choice of the parties. The US approach is interesting because it determines in advance that a particular state has an interest in a particular legal relationship in order to determine whether the law of that state will be applied. Judicial practice in cases involving consumers in the United States is very diverse. The law of the EU countries, on the contrary, chooses the applicable law to the contract, with the exception of provisions that harm the consumer.

In the US, legal relations on the creation/development of computer information, transactions related to computer programs, Internet contracts, data processing transactions, etc. are governed by the Uniform Computer Information Transactions Act 2000 (English UCITA, adopted in only two states: Maryland and Virginia). This law also allows the parties to choose the applicable law, with the exception of the provisions that relate to the consumer [7].

Despite the fact that conflict rights to consumer relations are fairly developed, difficulties may arise when it is not possible to determine the place of residence/location of the consumer, as he can make a transaction in one state, transfer funds from the accounts of the bank that is located in another state, and he himself moved to the third state - the place of the transaction.

In the EU Regulation Rome 1, as well as in Russian legislation, the applicable law is generally determined according to the law of residence of the party performing decisive execution under an agreement; in other cases the principle of the closest connection is used.

The EU Regulation Rome 2 [3] establishes legal approaches to the definition of the applicable law, to non-contractual obligations. The provisions of the EU Regulation Rome 2 also have many similarities with the regulation in Russia.

The Russian Federation is characterized by the approaches lex loci damni (the law of the place of harm) and lex loci delicti commissi (the law of the place of the unlawful action). Such approaches are used in the European Union: the first is common in many states; the second is rarely used. These approaches are enshrined in the Civil Code of the Russian Federation in Art. 1219: obligations arising from caused harm, the law of the country where the action or other circumstance took place that served as the basis for the claim for damages; if harm occurred in another country, the law of that country could be applied, if the injurer foresaw or should have foreseen the occurrence of harm in that country [1].

Taking into account the peculiarities of information transmission via the Internet, the place of committing unlawful actions that may cause adverse consequences will be considered the place where information is entered into the network. If we take as an example cloud storage of information, then information can be uploaded to the cloud storage, which will not be malicious or will not contain a virus, but entering it can reduce the value of the stored information (this can be fraudulent actions and actions aimed at causing harm to the party), for example, the substitution of data regarding concluded contracts and the order of obligations, the substitution of data associated with the invention, leveling its novelty. And the location of the server through which information is transmitted will not be considered the place of unlawful actions that may cause adverse consequences for the party (then, for example, if the virus is downloaded to the cloud storage, the law of the state from which the virus was downloaded will be applied). However, there are significant difficulties in determining where to download information, as the offender may use various programs that hide the place of loading, or download information from the territory of the state where the issue is not settled or insufficiently resolved. To solve this problem, it is possible to shift the downloading information/virus to the affected party while justifying their requirements, then this side has the advantage of choosing the most convenient material right. In this case, there will be a dilemma before the violator: continue to hide information about the place of downloading information, which may be beneficial for the affected party, or record and submit to the court the place of downloading the malicious program.

Such an approach is interesting and quite convenient for the parties and the court, but taking into account the scientific and technological progress, it may not be possible to establish the loading site. It may be practical in case of violation of the user's rights by a cloud company or another person to resort to the presumption used in Swiss law. It indicates the coincidence of the usual location of the offender's administrative center (or residence, if it is an individual) with the place of entry data to the network. But this presumption also does not exclude the deliberate choice of the place of the tort. And the choice is very difficult.

As for binding at the place of occurrence of harm, in this case there are difficulties concerning the parties: (1) A company operating on the Internet must foresee the possibility of such situations, the result of its actions and the possible effect in countries; (2) A user may face the consequences not in the country of citizenship/domicile, but in another state (since many technologies are cross-border). Such an approach implies a high level of responsibility of companies that operate on the Internet, for ensuring a high level of security for the user, but it is difficult for practical application, despite the fact that binding at the place of occurrence of harm is quite widely used.

When a virus is infected with data, the applicable law will usually be determined by the location of damaged databases and software.

In determining the applicable law, it is still necessary to consider what consequences the offender wanted. If the goal was data destruction and hacking, then the right of the place where the data owner is located should be applied. If the purpose of the offender was to obtain data for maintaining unfair competition and obtaining benefits, then in accordance with Art. 1222 of the Civil Code of the Russian Federation to obligations arising from unfair competition, the law of the country whose market is affected or may be affected by such competition is applied. However, Art. 1223.1 of the Civil Code of the Russian Federation, after committing an act or other circumstance that entailed harm or unjust enrichment, allows choosing the applicable law by the parties, but if all the circumstances relating to relations of the parties are connected with one country, the choice of the applicable law should not affect the imperative norms of this country. In the EU as a whole, a similar approach is used [1].

When determining the applicable law in transferring information, this is unlikely to be crucial in the event of a legal dispute, since this process is instantaneous.

It is necessary to pay attention to the fact that Art. 1219 of the Civil Code of the Russian Federation also indicates the possibility of applying the law, to which the parties have subordinated the contract, in case it was concluded by the parties in carrying out business activities [1].

Russian companies that offer a cloud storage service usually specify in their agreements with users the applicable law is the law of the Russian Federation, and the courts authorized to consider the dispute are the courts of the Russian Federation.

In the US, there is a slightly different tendency associated with the determination of the applicable law: the law largely depends on which court will be recognized as authorized to consider the dispute. In the USA, a number of approaches and presumptions are used, which allow, first of all, subordinating the proceedings in the case to US courts, which later may also have an impact on the definition of the applicable law. In the USA, the following approaches are used both individually and in combination, if the parties have not chosen the applicable law and, if the party is not a consumer.

- The place of conclusion of the contract. For example, in the case of Bodreau v. Scitex the court determined that the law of the State of Massachusetts is applicable, as e-mails about the contract were received in this state.
- The "best" right for a legal relationship.
- The public interest approach.
- The law of the place of execution.
- Minimum contact test (long arm rule).
- Purposeful submission (or the goal criterion). The courts in Europe are much less likely to apply this criterion compared to US courts, and they mainly use criteria related to the harmfulness of the act and the place of its occurrence. The goal criterion may soon be included in the legal system of the Russian Federation, which may be very promising for the development of the judicial system.
- Theory of loading and unloading information.
- Location of data processing and storage center. In the United States, the definition of the applicable law will also vary depending on the location of the data processing center and the location of the user. For example, if the data center of an American company of the relevant specialization is located in Singapore, then for a user from Australia, the applicable law will be Singapore's law, and Singapore's courts will have exclusive authority to deal with disputes between a user from Australia and a US company. If the user lives in the United States, then the applicable law will be the law of the United States, and the relevant courts of the given state will consider the dispute [4, 8].
- The place of operation. This criterion is preferred by companies that have a presence in different countries.

The European Union Regulation (EC) 2016/679 [5] has expanded its rights in the field of personal data processing, i.e. it is applied to the processing of personal data at the place of business, regardless of whether processing is carried out in the union or not. The conflict rules discussed indicate a closer approximation of conflict rules of the Russian Federation and the EU countries in determining the applicable law to legal relations when using Internet technologies. The US approaches are of great interest both for Russia and for the EU, since they began developing legal regulation of Internet technologies earlier than in other states, therefore these approaches are more diverse. But despite this there is no established practice that should be guided in most cases.

4 Discussion

The Russian and EU countries' approaches have much in common, the structure of conflict-of-law regulation largely coincides, new conflict-of-laws rules to resolve situations when the parties have not chosen the applicable law to the contract are underway.

In Europe, according to Czigler [2], greater attention is paid to commercial situations where the consumer is a party to a contract. And since most of transactions on the Internet are carried out by consumers, the rules of law on consumer protection will be guiding principles for determining the jurisdiction of the court and the applicable law. Moreover, when concluding a contract, the consumer should take into account both the activity/passivity of the site in the state (the passivity of the site does not prevent the consumer from contacting the company if he wants to place an order), as well as all circumstances related to the contract: advertising the company's offer with worldwide delivery, place of conclusion of the contract, location of the parties, place of the computer from which the transaction was made; place of payment, place of breach of contract, etc.

In the United States, an integrated approach is used, in which various collision bindings can be used and all the circumstances of a particular legal relationship are taken into account.

Svantesson (USA) [6] believes that in order to develop new legal norms in the field of private international law, it is necessary first of all to rethink the classical and established norms, since the current changes are a slight deviation from the well-established approaches. In fact, the regulation of relations when using Internet technologies is proceeding very slowly.

In the Russian Federation, adapting legal norms to new legal relations is under way. The Digital Code is being developed, which will probably lead to a whole new level of regulation of legal relations when using Internet technologies.

5 Conclusion

Legislation of different states regulates differently legal relations when using Internet technologies. In the event of a dispute, the parties have the advantage of choosing the state with the most advantageous material and procedural rules for court proceedings taking into account the future place of recognition and enforcement of the decision court. This possibility of choosing the law suggests the need to create a model law, which would include the procedural, material and conflict-of-law regulation of relations when using Internet technologies. Perhaps the best option for the most effective regulation of legal relations on the Internet, the definition of the applicable law would be to create a multilateral model convention that would reinforce dominant approaches of states. Most likely, such a document will be created, and the states will be able to come to a common denominator, as the relations arising when using Internet technologies are the same in most states.

Acknowledgements. The research was conducted by the authors with the financial support of the Russian Foundation for Basic Research, Project no. 19-011-20091.

References

1. Civil Code of the Russian Federation. Part 3rd no. 146, adopted on 26 November 2001. http://www.consultant.ru/document/cons_doc_LAW_34154/. Accessed 3 June 2019. (in Russian)
2. Czigler, T.D.: Choice-of-law in the internet age-US and European rules. Acta Juridica Hungarica **53**(3), 193–203 (2012). https://doi.org/10.1556/AJur.53.2012.3.2
3. Regulation (EC) No. 864/2007 of the European Parliament and of the Council of 11 July 2007 on the law applicable to non-contractual obligations (Rome II). https://eur-lex.europa.eu/legal-content/EN/TXT/PDF/?uri=CELEX:32007R0864&from=enU.C.C. Accessed 3 June 2019
4. Regulation (EC) No. 593/2008 of the European Parliament and of the Council of 17 June 2008 on the law applicable to contractual obligations (Rome I). https://eur-lex.europa.eu/legal-content/EN/TXT/PDF/?uri=CELEX:32008R0593&from=EN. Accessed 3 June 2019
5. Regulation (EU) 2016/679 of the European Parliament and of the Council of 27 April 2016 on the protection of natural persons with regard to the processing of personal data and on the free movement of such data, and repealing Directive 95/46/EC (General Data Protection Regulation) (Text with EEA relevance). https://eur-lex.europa.eu/legal-content/en/TXT/?uri=CELEX%3A32016R0679. Accessed 19 June 2019
6. Svantesson, D.J.B.: Jurisdictional issues and the internet – a brief overview 2.0. Comput. Law Secur. Rev. **34**(4), 715–722 (2018). https://doi.org/10.1016/j.clsr.2018.05.004
7. Uniform computer information transactions act. https://www.steptoe.com/images/content/1/4/v1/1468/2359.pdf. Accessed 3 June 2019
8. Vincent, M., Hart, N., Morton, K.: Cloud computing contracts white paper a survey of terms and conditions. Truman Hoyle (2011). https://ficpi.org.au/articles/White_Paper_June2011.pdf. Accessed 19 June 2019

Stablecoin as a New Financial Instrument

E. L. Sidorenko(✉)

Moscow State Institute of International Relations (University) of the Ministry of Foreign Affairs, Moscow, Russia
12011979@list.ru

Abstract. The author considers the legal and economic nature of stablecoin, the development stages of low-volatility assets, describes the existing models of their implementation in the public and private sectors. Particular attention is paid to the assessment of legal risks of mono-secured and multi-secured cryptocurrency on the example of projects in Venezuela and Russia, as well as the possibility of using stablecoin for the development of the financial system and evasion from economic sanctions.

Keywords: Cryptocurrency · Sanctions · Assets · Stablecoin · Finance · State policy · Legal risks

1 Introduction

The rapid development of digital technologies puts the world economy in front of the need to build a qualitatively new model of financial relations on the principles of universal trust and transparency. It was expected that the role of the driver of digitalization of the financial system will be taken by cryptocurrency created on the basis of distributed ledger technology (blockchain). The very architecture of the blockchain provides the ability to verify cryptocurrency transactions without any help of third parties all the time (24/7), which makes the cryptocurrency as convenient as possible for transnational exchange operations. But despite its undeniable technological advantages, it remains unattractive for large investors and business owners because of high volatility and vulnerability for the impact of information and political factors [3].

It is the high volatility of cryptocurrency that has brought to life the stablecoins as a tool for hedging risks of decentralized coins. Stablecoin was originally conceived as a tool to minimize risks of investors in the turnover of digital assets without withdrawing funds to a bank fiat account. As the volatility of cryptocurrency increased, so did the market's need for a "safe haven". According to the research conducted by the Center for Digital Economy and Financial Innovation of MGIMO (Moscow State University of International Relations), there is a stable inverse connection between the turnover of digital money and stablecoin, which indicates a trend of crypto-currency market strengthening by transferring funds into low-volatile digital assets.

The stability of the stablecoin exchange rate excludes its investment attractiveness, but, at the same time, it allows us to consider it as a potential means of payment, saving and exchange of assets, as well as a tool for evasion from economic restrictions and sanctions. Due to its last mentioned capacity, it is becoming increasingly attractive to

S. I. Ashmarina et al. (Eds.): ISCDTE 2019, LNNS 84, pp. 630–638, 2020.
https://doi.org/10.1007/978-3-030-27015-5_75

countries under economic sanctions of the United States and Europe. Many of these countries, including the Russian Federation, are starting to develop projects for the introduction of stablecoin, backed by the national currency, oil, gas, ore or rare earth metals (projects El Petro (Venezuela), Norilsk Nickel (Russia), etc.). In addition, in countries with strict sanctions regimes, ideas are being developed to attract financing for mining operations. But so far, no major mining company has started to produce stablecoin, backed by the warehouse and (or) the supply of a real asset, as well as not a single working platform for the production and exchange of stablecoins has been appeared.

Market demand for turnover of low-volatility digital assets has already led to a significant jump in the stablecoin economy. By June 2019, their total capitalization was about 3.5 billion U.S. dollars [11], and their share in the total cryptocurrency market was 2.7% [12].

The dynamics of the main (index) stablecoin also attracts attention. Tether increased its capitalization to $ 2.5 billion (since 2016), USD Coin – up to $ 260 million (since 2018), Dai – to $ 84 million (from 2018). To date, 3 432 130 225 USDT (tether coins), 83 579 195 DAI (Dai coins), 343 652 881 USDC (USD Coin), 175 852 162 PAX (Paxos Standard Token), 20 631 088 GUSD (Gemini Dollar), 31 979 207 EURS (STASIS EURS), etc. have been issued.

It is impossible to ignore the issue of Venezuela's own cryptocurrency (stablecoin) El Petro. Its price corresponds to the price of a barrel of oil set by OPEC for the previous day of the end of trading. The price is calculated from the average price of the OPEC basket, namely 15 different types of crude oil produced in OPEC member countries. The Venezuelan government has described its plans to use El Petro at the official website elpetro.gov.vz [8]. And despite the presence of some contradictions in the assessment of the legal nature of this tool, El Petro is already considered by the authorities as an effective means for avoiding sanctions, exchange and saving funds.

It is important to pay attention to the position of the Central Bank of the Russian Federation, which officially announced that it is considering the possibility of launching a digital currency on the basis of mature technologies and under the condition of the reliability and continuity of payments [1]. However, despite the market demand in the development of stablecoin, neither at the practical nor at the theoretical level there is no developed unified approach to the definition of the concept and features of stablecoin, assessment of its advantages and risks.

In modern economic and legal literature, these issues are partially solved, but only in the applied way in relation to clearly defined research tasks. Among these works are reports of the Bank for International Settlements "Processing with caution – a survey on central bank digital currency" [2], "Stable coins: From electronic money on blockchain to a cryptocurrency basket" [5], "The State of Stablecoins 2019. Hype vs. Reality in the Race for Stable, Global, Digital Money" [14], "2019 State of Stablecoin" [3] and others, as well as works by Sat, Krylov, Bezverbnyi, Kasatkin, and Kornev [15], Böhme, Christin, Edelman, and Moore [4], Coeuré and Loh [6], Ivantsov, Sidorenko, Spasennikov, Berezkin, and Sukhodolov [9], Dyson and Hodgson [7], Pernice, Henningsen, Proskalovich, Florian, Elendner, and Scheuermann [13], etc. However, these works do not fully address the most important issues for practice, and that is why this study is aimed at analyzing the nature of the stablecoin and "weaknesses" of its implementation practices.

2 Methodology

While carrying out this study, general scientific methods of assessing complex economic and legal phenomena were widely used: analysis, synthesis, abstraction, logical and systemic methods. Special legal methods (formal-legal, comparative-legal, historical-legal methods and method of legal modeling) and sociological methods (content analysis, generalization of practice, statistical method, etc.) were applied. This paper is the first to publish the results of the work of the Center for Digital Economy and Financial Innovation of MGIMO University to study the phenomenon of stablecoin and search for the most effective model of its implementation.

3 Results

The main distinctive features of stablecoin are its low volatility and security of assets with stable pricing. However, this, at first glance, this statement is questioned when referring to the practice of its implementation. There are two main models for ensuring low volatility of stablecoin:

(1) natural mechanism: direct reference of the price to the asset unit having a stable value. In this case, the volatility of stablecoin is directly determined by the stability of the price of the underlying asset (national currency, gold, oil, etc.). According to the research of MGIMO University, more than 95% of all developed stablecoins have a natural mechanism of implementation. 35% of them are based on the provision of coins with money (fiat stablecoins), 55% – on the provision with goods (commodity stablecoins) and 10% – on the provision with the commodity-money basket. Each of the proposed models has its advantages and disadvantages and can be implemented in specific industries and for specific purposes;

(2) artificial mechanism: management of pricing in the market using complex mechanisms to stimulate demand and supply, increase or decrease in currency issue, repayment and redemption of assets. Developers use the method of stock seigniorage. Through the introduction of special smart contracts, they can manage the currency issue, stimulate demand and supply depending on the value of the coin in a specific period. Currently, this method of coins stabilizing is rarely used (only in pilot projects) because of the instability of the crypto-currency market and the lack of algorithms for its control (projects Saga, Basis, Basecoin, etc.).

But independent from the model of price stability, one thing is obvious: the price and volume of coins in circulation is directly determined by the creator – issuer. On the one hand, it gives him unlimited control over the financial asset, on the other, it imposes on him the entire burden of responsibility.

Speaking about the economic advantages and legal risks of stablecoin, it is important to emphasize the significant differences between the following two types of assets: (1) mono-secured stablecoins – assets backed by one low-volatility asset. Depending on the nature of the asset, they are fiat-backed stablecoins (backed by a state currency) and tradeable stablecoins (backed by commodities: ore, oil, gas, etc.) [8]; (2) multi-secured stablecoins are backed by a basket of low-volatility assets.

Mono-secured stablecoins:
These assets are primarily of a centralized nature and generally linked to national currencies or raw materials.

There are two main models mono-secured stablecoins:

(1) model of settlement digital units (Bank Settlement Strategy);
(2) commodity option model (Anti-sanction Strategy).

The bank settlement strategy is based on the circulation of coins backed by the national currency and is considered by the central banks as a pilot model of digitalization of payments in the conditions of cash turnover reduction. Previously all non-cash payments existed mainly within the framework of the transactions of legal entities (B2B transactions), now they are beginning to penetrate into the sphere of relations of individuals (P2P transactions). In this regard, the regulators of the countries are interested in creating a digital model that would allow, firstly, to meet the needs of the market in non-cash and cryptocurrency P2P payments, secondly, to strengthen transmission processes in the monetary policy, and thirdly, to ensure the safety of this process.

Nowadays, two concepts of settlement digital units are considered [4, 6, 8]:

- retail stablecoins – assets available to a wide range of users;
- wholesale stablecoins – assets available to a limited circle of users, mainly participants of the organized auctions.

If the first type of assets is characterized by the presence of a bank account, the second – by the availability of reserves in the central bank. If in the first case the token is free in circulation and is considered as a low-volatility cryptocurrency, in the second case it is characterized as a requirement for the central bank expressed in the digital form.

The place of these tokens in the modern two-tier financial system is not the same: the first token covers the lower user layer of relations, while the second one does not go beyond the transactions s between financial institutions (correspondent accounts and deposits). In the case of retail tokens, the central bank liabilities can become a liquid asset, and the prevalence of calculations will lead to a reduction in transaction costs [9] and an increase in the transmission mechanism of the monetary policy [10]. The introduction of the second token can give a certain margin of safety to interbank interaction, ensuring transparency and reliability of transactions. However, independent from the design chosen as a basis for development of stablecoins, one thing is clear: each of the scenarios will face a number of legal risks and limitations. If the retail scenario is implemented, the following legal tasks will arise:

(1) differentiation of regulatory regimes of stablecoin and electronic money;
(2) bringing legislation in the field of financial, tax and currency control, banking and civil legislation in correspondence with changes in the architecture of calculations;
(3) validation of settlement transactions, changes in the accounting system;
(4) improvement of the anti-money laundering legislation and legislation on combating terrorist financing in terms of user identification, licensing (certification) of payment operators and other market participants.

Implementation of the wholesale scenario will cause no less legal problems. In this case, there will inevitably be difficulties associated with reformatting the legislation on securities and organized trading, the approach to payment operators will be radically changed, the regime of regulation of banking and exchange activities will require significant adjustments. Thus, taking into account the identified risks, it is hardly possible to talk about the operational implementation of one of the above mentioned designs of the stablecoin. The second model of mono-secured stablecoins (commodity option model/anti-sanction strategy) can hardly be considered as indisputable too.

The first country which has launched such a project was Venezuela. The government of the country repeatedly pointed out in its statements that the state coin El Petro is not a currency, but an asset regulated under the civil law. Thus, it tried to solve the dilemma between the functionality of the new currency and the interdiction of the Central Bank of Venezuela on the issue of cash surrogates.

Calling El Petro "a digital contract for the purchase and sale of oil" ("contrato digital decompra venta"), the state simultaneously creates an infrastructure for its widespread use and circumvention of sanctions. But this does not bring the expected results largely because of the controversial nature of this asset.

On the one hand, the state binds it to the low-volatile currency of OPEC, but on the other hand, it does not actually provide it with real oil reserves (undeveloped mining is not taken into account). On the one hand, the government sees in stablecoin the possibility of overcoming the economic crisis and the growth of investment attractiveness, on the other hand, it offers investors a low-volatility asset, burdened by the economic and political crisis in the country.

Finally, the legal nature of El Petro is not clear. On the one hand, it is close to government bonds in terms of its support by the state, on the other hand, it is an option contract for the development of oil fields of Ayacucho in the Orinoco Petroleum Belt.

The only indisputable advantage of this currency is that due to the lack of international regulation of cryptocurrencies, it is legally difficult to impose sanctions on El Petro, because such a currency is not associated with traditional views of financial transactions and commodity exchange.

Without denying the possibility of developing this project in the future, it is important to note that a positive result can be possible only if a single strategy will be developed on a number of issues: the specification of the nature of El Petro as a legal category, the study of the economic advantages of stablecoin for potential investors, the preservation of advantages of the asset as a settlement tool and the expansion of international cooperation with other oil exporting countries to push the El Petro into the system of international exchange. Only in this case this asset can become attractive for investors who want to hedge their risks in the fuel market.

The similar in its focus project (for evasion from of sanctions) was proposed by the largest Russian company – JSC Mining and Metallurgical Company "Norilsk Nickel". It offers to release stablecoin, backed by palladium. It is assumed that this asset will be sold on a special exchange platform. Users can be both individuals and legal entities. They can acquire, dispose of and change stablecoins for fiat money and cryptocurrency, and also exchange them for a real asset.

While supporting the company's desire to develop the infrastructure of the digital market and circumvent sanctions, it is important to note a number of legal and economic risks of this project implementation:

1. Like El Petro, stablecoin of MMC "Norilsk Nickel" (hereinafter – MMC NN) is aimed at attracting investors, which in itself does fundamentally differ from the idea of stablecoin as a low-volatility asset.
2. The exchange platform organized within the framework of this project is controlled by the issuer (MMC NN), which means that, firstly, the rate of stablecoin can be controlled, and secondly, the risks of the project are directly tied to the risks of the issuer.
3. The marketing component of the project is not clear. In particular, it is unclear what motivation should guide an investor participating in MMC NN project except for risk hedging. But in this case, it is not clear on the basis of what funds the token exchange platform will operate. Obviously, the commission received from the conversion will be distributed between the platform and the issuer. But how ordinary participants of the process will be stimulated? Is also relates to capacity producers for transactions (miners).
4. The trick of this project is that other tokens – coins for operations on the platform – will be included in the created ecosystem. In the end, they will overtake the stablecoin of MMC NN due to their turnover capacity. Stablecoins will become a kind of asset of the platform, while operational tokens will replace them in the transactions. Such a scenario is most likely, but then it is unclear whether the MMC NN project can be considered as a stablecoin project.
5. The question of how the exchange of cryptocurrency for palladium will be carried out also needs further scientific development. The fact is that in many countries, the circulation of rare earth metals is limited, and individuals may be criminally liable for its turnover.
6. The issue of user identification also needs further elaboration too. The company is interested in the wide use of these coins by the population of different countries, but the similarity of their status with derivatives requires the organizers to reflect the data on buyers in special registers as fully as possible. This, in turn, significantly limits the turnover of MMC NN stablecoins.

In general, without denying the importance of the development of the stablecoin sanctions scenario, it is important to take into account all possible legal risks as fully as possible. These are the reliability of the guarantor and the availability of the necessary security assets, the conflict of regulatory regimes of stablecoins, derivatives, cryptocurrencies and securities, the close relationship between the fate of the project and the bankruptcy of the issuer, external audit of the guarantor and conflict with the internal compliance procedures of the issuer, etc.

Multi-secured stablecoins:
Such projects are just beginning to develop, but it is already clear that they cannot be fully implemented on the basis of a market with a weak infrastructure. The economic advantages of this scenario are the expansion of the number of active participants in the crypto economy, the search for new sustainable forms of financial relations, the development of transnational contacts with further access to the creation of interstate settlement units, backed by the assets of ecosystem participants. But at the same time, the legal risks of implementing such projects are also obvious:

(1) high level of volatility of the assets providing stablecoin;
(2) the legal complexity of the design and production of coins;
(3) technical difficulties in creating and maintaining smart contracts;
(4) vulnerability to hacker attacks;
(5) the complexity of the procedure for assessing the reliability of stablecoin: verifiability of reserves, transparency of the code, the stability of relations between the platform, the custodian bank and issuers, the speed of transactions and the ability of the coin to restore prices after the collapse.

4 Discussion

Currently, there are various estimates of the further development of stablecoins. As A. Kiselev notes, "the introduction of (these) assets can have wide consequences for all participants of the monetary system… Their useful properties and possible risks depend on the system design (who and in what conditions has access to the currency) and the fact whether the currency has security and what is the exchange method for other types of money and so on" [10].

Among the key criteria that allow to predict the development of the project, the following ones should be mentioned: the goal (the launch of the payment system, the increase in exports bypassing sanctions, etc.), the degree of centralization (decentralization) of stablecoin, the expected turnover and jurisdiction of the country, the infrastructure availability of stablecoin by exchangers, etc.

At the same time, it is important to take into account such factors as the possible saturation of the market with stable assets. Some indirect facts already show the limitations of the market. In particular, the decrease in the market share of Tether in 2018 led to the growth of Paxos Standard and USD Coin, which means that the cryptocurrency environment already fixes the competition of assets close in nature.

A significant impetus to the development of the market of stablecoins can be the fixation of their legal status. If the Cases of the United States against Faiella [17] and Murgio [16], bitcoins traders were brought to criminal liability, regarding stablecoins this question remains open, as well as the question of whether the issuers of stablecoins have responsibility for their illegal manipulation.

According to the latest analytical reports [1], in the long term, stablecoins can become the basis of the ecosystem of digital assets, if the necessary infrastructure and legal conditions are created for them. It is expected that one of the most popular application areas will be the use of these assets in smart contracts, in the system of taxation and interbank cooperation.

5 Conclusion

The market demand for low-volatility digital instruments of risk hedging leads to the creation of a fundamentally new philosophy of relations between the parties of financial relations – the concept of stablecoins. A distinctive feature of these assets is their low volatility. However, depending on the participants of the ecosystem, the strategic plans of developers and the nature of the assets lying in the provision, it is possible to identify several possible programs for the development of stablecoins both within one country and the world as a whole. However, whatever the proposed concept is, it is a priori fraught with a number of serious legal obstacles because of the weak adaptability of the law to new digital trends.

Acknowledgements. The research was conducted by the authors with the financial support of the Russian Foundation for Basic Research, Project no. 19-011-20091.

References

1. Bank of Russia can create its own cryptocurrency – Nabiullina (2019). https://www.pravda.ru/news/economics/1421461-crypto/. Accessed 19 June 2019. (in Russian)
2. Barontini, Ch., Holden, H.: BIS Papers No. 101 Proceeding with caution – a survey on central bank digital currency. Bank for International Settlements (2019). https://www.bis.org/publ/bppdf/bispap101.htm. Accessed 19 June 2019
3. Blockchain 2019 State of Stablecoin (2019). https://www.blockchain.com/ru/research. Accessed 19 June 2019. (in Russian)
4. Böhme, R., Christin, N., Edelman, B., Moore, T.: Bitcoin: Economics, technology, and governance. J. Econ. Perspect. **29**(2), 213–238 (2015). https://doi.org/10.1257/jep.29.2.213
5. Bondar, D.: Stable coins: from electronic money on blockchain to a cryptocurrency basket. ForkLog Consulting (2018). https://forklog.com/pdf/FLC_Stablecoins_report_eng.pdf. Accessed 19 June 2019
6. Coeuré, B., Loh, J.: Central bank digital currencies. Bank for International Settlements (2018). https://www.bis.org/cpmi/publ/d174.pdf. Accessed 19 June 2019
7. Dyson, B., Hodgson, G.: Digital cash: why central banks should start issuing electronic money. Positive Money (2016). https://positivemoney.org/wp-content/uploads/2016/01/Digital_Cash_WebPrintReady_20160113.pdf. Accessed 19 June 2019
8. Hossein, N.: Central bank digital currencies: preliminary legal observations. J. Bank. Regul. (2019). http://dx.doi.org/10.2139/ssrn.3329993. https://ssrn.com/abstract=3329993. Accessed 19 June 2019
9. Ivantsov, S.V., Sidorenko, E.L., Spasennikov, B.A., Berezkin, Y.M., Sukhodolov, Y.A.: Cryptocurrency-related crimes: key criminological trends. Russ. J. Criminol. **13**(1), 85–93 (2019). https://doi.org/10.17150/2500-4255.2019.13(1).85-93. (in Russian)
10. Kiselev, A.: Does the digital currency of central banks have a future? Analytic note. Bank of Russia (2019). https://www.cbr.ru/content/document/file/71328/analytic_note_190418_dip.pdf. Accessed 19 June 2019. (in Russian)

11. LH-CRYPTO: The basics of crypto trading (2014). https://www.lh-crypto.com/reviews/currencies/tether.html. Accessed 3 June 2019
12. Micky News: Why new stablecoins are being developed at a rapid rate (2019). https://micky.com.au/why-new-stablecoins-are-being-developed-at-a-rapid-rate/. Accessed 3 June 2019
13. Pernice, I.G.A., Henningsen, S., Proskalovich, R., Florian, M., Elendner, H., Scheuermann, B.: Monetary stabilization in cryptocurrencies - design approaches and open questions. In: 2019 IEEE Crypto Valley Conference on Blockchain Technology (CVCBT) (2019). https://arxiv.org/pdf/1905.11905.pdf. Accessed 19 June 2019
14. Samman, G., Masanto, A.: The State of Stablecoins 2019. Hype vs. Reality in the Race for Stable, Global, Digital Money (2019). https://static1.squarespace.com/static/564100e0e4b08c9445a5fc5d/t/5c71e43ef9619ae6c83c30af/1550967911994/The+State+of+Stablecoins+2019_Report+2_20_19.pdf. Accessed 19 June 2019
15. Sat, D.M., Krylov, G.O., Bezverbnyi, K.E., Kasatkin, A.B., Kornev, I.A.: Investigation of money laundering methods through cryptocurrency. J. Theor. Appl. Inf. Technol. **83**(2), 244–254 (2016)
16. United States of America v. Anthony R. Murgio, et al.: Defendants, Case No. 15-cr-769 (AJN), Memorandum and Order. Southern District of New York, United States District Court (2016). https://www.plainsite.org/dockets/2x5hvau0h/new-york-southern-district-court/usa-v-murgio-et-al/. Accessed 19 June 2019
17. United States of America v. Robert M. Faiella, a/k/a "BTCKing", and Charlie Shrem, Defendants, case No. 14-cr-243 (JSR). S.D. New York, United States District Court (2014). http://www.leagle.com/decision/InAdvFDCO150521-000337/U.S.v.Faiella. Accessed 3 June 2019

Author Index

A
Afanaseva, E. P., 172
Aleshkova, D. V., 80, 109
Ashmarina, S. I., 62, 189, 514
Astafeva, O. V., 373, 382
Astafyev, E. V., 373
Averina, L. V., 261

B
Bakanach, O. V., 441, 458
Bakanova, I. G., 494
Barinova, E. P., 327, 366
Belanova, N. N., 39
Belozerova, O. A., 548
Bikmetov, R. Sh., 109
Blium, M. A., 342
Bogatyreva, I. V., 137, 390
Bortnikov, S. P., 564
Bulavko, O. A., 39

C
Chernousova, K. S., 144
Chernova, D. V., 252, 487
Chirkunova, E. K., 205
Chistik, O. F., 458
Chudaeva, A. A., 351
Chulova, E. S., 12
Czegledy, T., 298

D
Deltsova, N. V., 557
Dorofeeva, J. A., 574

E
Efimova, T. B., 54
Elkina, L. G., 116
Evtodieva, T. E., 398

F
Fedorenko, R. V., 298
Fomin, V. P., 233

G
Gerasimov, B. I., 342
Gerasimova, E. B., 342
Greshnova, M. V., 80
Guryanova, A. V., 47
Gusakova, E. P., 172

H
Horák, J., 416

I
Ilyukhina, L. A., 137
Izmailov, A. M., 514

K
Kamardin, I. N., 506
Karpova, N. P., 398
Karyshev, M. Y., 458
Kazakova, O. B., 281
Khalikova, E. A., 373
Khansevyarov, R. I., 71
Khisamova, Z. I., 308
Khmeleva, G. A., 205

S. I. Ashmarina et al. (Eds.): ISCDTE 2019, LNNS 84, pp. 639–641, 2020.
https://doi.org/10.1007/978-3-030-27015-5

Khvostenko, O. A., 335
Kitaeva, M. V., 359, 448
Kmecová, I., 526
Konovalova, M. E., 12, 180
Konovalova, V. G., 473
Kopylova, A. A., 27
Korneeva, T. A., 159
Korobejnikova, E. V., 382
Koroleva, E. N., 205
Kot, M. K., 548
Kozhuhova, N. V., 128
Krivtsov, A. V., 85
Krulický, T., 416
Kulikova, N. V., 214
Kurnikova, M. V., 205
Kuzaeva, E. Yu., 281
Kuzmina, L. I., 252
Kuzmina, O. Y., 12, 180
Kuzminykh, N. A., 281

L
Lazareva, N. V., 165, 433
Lebedeva, L. G., 514
Leybert, T. B., 373
Lobachev, V. V., 480
Lovcheva, M. V., 473
Lykov, A. A., 613

M
Machová, V., 198
Makeeva, M. N., 342
Mamai, I. N., 359
Mamai, O. V., 359
Mantulenko, V. V., 62, 317
Martiskova, P., 92
Mikhailov, A. M., 27
Mikhaleva, O. L., 144
Milova, I. E., 597
Mitrofanova, E. A., 390
Mitropolskaya-Rodionova, N. V., 366
Molotkova, N. V., 342

N
Nasretdinov, I. T., 487
Naumova, O. A., 159
Nazarov, M. A., 144
Nikitina, B. A., 19
Nikitina, N. V., 243, 506
Noskov, S. V., 494
Novikova, E. N., 487
Nurtdinov, I. I., 252

O
Opekunov, A. N., 506
Opekunova, L. A., 506
Osipova, I. A., 373

P
Panova, G. S., 604
Pecherskaya, E. P., 261, 269, 382
Pershin, M. A., 335
Persteneva, N. P., 214
Pertulisov, Yu. A., 150
Pimenova, E. M., 466
Pogorelova, E. V., 54, 225
Popok, L. E., 80
Potokina, E. S., 233
Proskurina, N. V., 441, 458

R
Rakhmatullina, A. R., 269
Razuvaeva, E. B., 116
Razzhivina, N. I., 597
Repina, E. G., 441
Revina, S. N., 589
Rosenberg, G. S., 433
Rowland, Z., 422
Ruslanova, T. V., 214
Ryazanova, O. E., 101

S
Sapova, O. A., 433
Schekoldin, V. A., 137
Serikbek, A., 290
Serper, E. A., 335
Sharafutdinova, N. S., 252, 487
Shchutskaya, A. V., 172
Shepel, V. N., 448
Sheremetyeva, E. N., 327, 366
Sidorenko, E. L., 308, 613, 630
Sidorov, A. A., 165
Sidorova, A. V., 541
Simonova, M. V., 128, 473, 480
Sivaks, A. N., 269
Smolina, E. S., 80, 150, 281
Smotrova, I. V., 47
Sosunova, L. A., 494
Spanagel, F. F., 548
Speshilova, N. V., 448
Starun, N. V., 109, 116, 165
Streltsov, A. V., 243, 466
Suieubayeva, S., 408
Šuleř, P., 198

Sumburova, E. I., 3
Svec, R., 92
Svetkina, I. A., 159, 351
Svistunov, V. M., 480
Syrova, K. P., 494

T
Tagirova, N. F., 3
Taran, K. K., 622
Tarasenko, V. V., 390
Tarasova, T. M., 261, 382
Tokarev, Yu. A., 514
Tokmakov, M. A., 581
Toymentseva, I. A., 398
Troshina, E. P., 317
Tuktarova, L. R., 39

U
Utegenova, A., 408

V
Valeeva, Y. S., 252, 487
Valinurova, L. S., 85
Vaulin, A. V., 225
Velinov, E., 189
Vochozka, M., 416
Vodopyanova, L. A., 150
Volkodavova, E. V., 71
Vrbka, J., 422

X
Xametova, N. G., 487

Y
Yakovlev, G. I., 71, 243, 466
Yashina, E. Z., 62
Yavorskiy, M. A., 597

Z
Zaichikova, N. A., 298
Zhabin, A. P., 71
Zherdeva, Yu. A., 3
Zhironkin, S. A., 180
Ziyadin, S., 290, 408
Zolotareva, V. P., 101
Zotova, A. S., 189, 327, 351

Printed in the United States
By Bookmasters